Anti-Evolution

Anti-Evolution

A Reader's Guide to Writings before and after Darwin

Tom McIver

The Johns Hopkins University Press

Printed in the United States of America on acid-free paper

Published by special arrangement with McFarland & Company, Inc., Publishers, Jefferson, North Carolina

Johns Hopkins Paperbacks edition, 1992

The Johns Hopkins University Press
701 West 40th Street
Baltimore, Maryland 21211-2190
The Johns Hopkins Press Ltd., London

Library of Congress Cataloging-in-Publication Data

McIver, Tom.
Anti-evolution: a reader's guide to writings before and after Darwin/Tom McIver.
p. cm.
Originally published: Jefferson, N.C. : McFarland, c1988.
Includes bibliographical references and index.
ISBN 0-8018-4520-3
1. Evolution—Controversial literature—Bibliography.
2. Creationism—Bibliography. I. Title.
Z5322.E9M37 1992
016.575—dc20 92-16755

To my wife, Beth

Table of Contents

Preface to the Johns Hopkins Edition

Many new creationist books have been published since the original 1988 edition of *Anti-Evolution,* and I have since learned of other, older books. I have not, however, attempted either to correct or update the text, as the most important omissions can be found in other sources now appearing, and existing errors of which I am aware are mostly minor.

Creationists are now concentrating more on lobbying at the local level, and many, though not all, of the more visible creationist groups are deliberately downplaying the religious basis of their anti-evolutionism; hence, creationism is not attracting as much public and media attention as it has in previous years. Opposition to evolution, however, remains vigorous, widespread, and deeply entrenched.

Rather than updating the text, I will mention here a few of the more significant books that have appeared since 1988. Based on preliminary excerpts and notices, I had listed Wendell Bird's *The Origin of Species Revisited: The Theories of Evolution and of Abrupt Appearance,* which was published in 1989. This two-volume book is an outgrowth of Bird's weighty Supreme Court brief in defense of the Louisiana creation-science bill. It attempts to make anti-evolutionist teaching less vulnerable to the legal challenges of unconstitutional violation of the religious establishment clause by arguing not for *creationism* but for presentation of the theory of "abrupt appearance." Bird maintains that discontinuist "abrupt appearance" of organisms in complex form is not the same as creationism, as it need not invoke a Creator or the supernatural, and it is an equally scientific alternative to evolution. His heavily footnoted book is an exhaustive, often repetitious, compendium of thousands of quotes by scientists, philosophers, and others who are critical of Darwinian theory or who object to the dogmatic and uncritical presentation of evolution. Bird claims that the religious basis of creationism is "incidental" to its scientific validity and that evolution is at least as religious as creationism. Bird argues that, in the interest of academic and religious freedom, the theory of "abrupt appearance" should be permitted in the schools alongside evolution.

Phillip E. Johnson's *Darwin on Trial* (Washington, D.C.: Regnery Gateway, 1991) is one of the most influential recent anti-evolution books. Johnson, a University of California–Berkeley law professor, attacks the logical and scientific inadequacies of evolutionist arguments but does not defend young-earth creation-science. He is a Christian theist who believes that God could have created directly without recourse to evolution but allows that He might have used some form of evolution. His chief objection is to the presentation of evolution as fact rather than theory; *macro*-evolution has never been demonstrated, he claims, and the "theory" of evolution is neither scientific nor strongly supported by available evidence. Macro-evolution is simply assumed by extrapolation from *micro*-evolution and by the unwarranted assumption that hierarchical *classification* of organisms proves relationships of *descent*. Johnson argues that Darwinian evolution is defended by scientists because it is the only purely "naturalistic" explanation available. Evolution, he says, has become a dogmatically held religious tenet, promoted in order to discredit belief in any supernatural influence. He considers creationism a Kuhnian paradigm equal in status to scientific naturalism. Like Macbeth (bibliography entry *1014*), also a lawyer, Johnson emphasizes the "tautology of natural selection," arguing that Darwinism is not subject to falsification and is thus not truly scientific. Though most scientists would disagree emphatically with his conclusions, Johnson presents his case cogently and persuasively.

Of Pandas and People: The Central Question of Biological Origins, by Percival

Davis, Dean Kenyon, and academic editor Charles B. Thaxton (Dallas: Haughton, 1989), has also been influential. All three authors are listed in this bibliography for previous works. The book's copyright is held by Thaxton's old-earth creationist organization, the Foundation for Thought and Ethics (FTE). The text was reviewed by several dozen, mostly creationist, authorities prior to publication. Originally designed as a supplementary school textbook for which state approval would be needed, the book was developed in part in a public school district field test, but evolutionist opposition has forced FTE to shift its efforts to local lobbying. *Pandas* does not openly advocate a Creator but rather emphasizes that the scientific evidence can be interpreted either in terms of evolution or as a product of "intelligent design" (and a designer)—the explanation it clearly favors. It focuses on problems of homology, origin of life theories, gaps in the fossil record, supposed limits of natural selection, and coded genetic information and biochemical similarities as evidence for intelligent design (relying heavily on Denton, bibliography entry *470*).

Henry Morris has a major new book, *The Long War against God: The History and Impact of the Creation/Evolution Conflict* (Grand Rapids, Mich.: Baker, 1989). Morris's Institute for Creation Research (ICR) continues to produce new books and to reissue and update existing publications, including many for children and many that are explicitly biblical. ICR sells large numbers of books at its popular "Back to Genesis" meeting series and at other creationist presentations and seminars held around the country. Recently, ICR successfully resisted an attempt by the State of California to deny renewal of its license to grant graduate degrees in science, and it intends to increase and expand its teaching, publishing, and promotional efforts.

The Evidence for Creation by Glen McLean, Roger Oakland, and Larry McLean (Springdale, Pa.: Whitaker House, 1989), presents standard creation-science arguments, but it is noteworthy for reportedly having been published in Russia with an initial press run of 60,000 copies. Oakland has lectured widely in the former Soviet Union, now fertile ground for creationist proselytizing. He also co-authored, with Caryl Matrisciana, *The Evolution Conspiracy* (Eugene, Ore.: Harvest House, 1991), which presents occultism and New Age beliefs as corollaries of evolution, Satan's chief weapon in his plot to destroy Christianity.

There are a few new non-creationist books that contain additional bibliographic information. Historian of science Ronald Numbers has completed a major work, *The Creationists* (New York: Alfred A. Knopf, 1992), which discusses a number of older creationist books not listed here, some of which are important sources. I am writing a book on various theories of creationism which will mention some texts not listed here and will also update and revise some of the information in this bibliography. Kevin Wirth, who developed the computerized bibliographic database for Students for Origins Research (bibliography entry *1557*), now has available an unpublished *Creation-Evolution Bibliography,* an annotated listing of books and articles on both sides of the issue. The National Center for Science Education plans to issue a volume consisting of reviews of some of the most influential recent creationist publications, an expansion of its 1984 booklet (ed. Stan Weinberg) in which thirty-one creationist books were reviewed.

Preface

This bibliography is intended as a guide for the understanding and evaluation of the plethora of anti-evolution books, pamphlets and other literature in circulation today.

Much of the research for this study was conducted at the Institute for Creation Research, which has perhaps the most comprehensive collection of anti-evolutionist and creationist literature in the world. Much material came from various creationist meetings and conferences I have attended, and from perusal of creationist journals and periodical literature. I also used the libraries of Ambassador College and Fuller Theological Seminary.

Robert Schadewald, a veteran observer of the creationist and Bible-science scene, very generously provided me with a number of the entries listed here, and with other valuable information.

Arnold Ehlert, librarian of the Institute for Creation Research until his retirement in 1988, graciously allowed me access to his files of anti-evolution sources. His article "The Literature of Scientific Creationism and Anti-Evolution Polemic" (1978; *Proc. of the Amer. Theol. Library Assoc.*) also contains useful bibliographic information.

Several secondary sources provided references for works I could not obtain first-hand. These sources include: Peter Bowler's *Evolution: The History of an Idea* (1984; Berkeley: Univ. of California Press); Michael A. Cavanaugh's *A Sociological Account of Scientific Creationism* (unpub. 1983 Univ. of Pittsburgh Ph.D. diss.); Daniel Cohen's *Waiting for the Apocalypse* (1983 [1973]; Buffalo, N.Y.: Prometheus Books); Martin Gardner's *Fads and Fallacies in the Name of Science* (1957 [1952]; New York: Dover); Willard Gatewood's *Controversy in the Twenties: Fundamentalism, Modernism, and Evolution* (1969; Nashville: Vanderbilt Univ. Press); Neal Gillespie's *Charles Darwin and the Problem of Creation* (1979; Univ. of Chicago Press); Charles C. Gillispie's *Genesis and Geology* (1959; Harvard Univ. Press); John C. Greene's *The Death of Adam: Evolution and Its Impact on Western Thought* (1959; Iowa State Univ. Press); David Hull's *Darwin and His Critics* (1973; Univ. of Chicago Press); George M. Marsden's *Fundamentalism and American Culture* (1980; Oxford Univ. Press); Milton Millhauser's "The Scriptural Geologists" (1954; *Osiris* II: 65–86) and *Just Before Darwin* (1959; Middletown, Conn.: Wesleyan Univ. Press); Ronald Numbers' "Creationism in 20th-Century America" (1982; *Science* 218: 538–544); Andrew D. White's *A History of the Warfare of Science with Theology* (1960 [1896]; New York: Dover); and the A.C.L.S. *Dictionary of Scientific Biography* (C. Gillispie, ed.; 1980; N.Y.: Charles Scribners Sons).

The National Center for Science Education, which publishes the *Creation/Evolution Newsletter* and monitors creationist activity in conjunction with its efforts to promote and defend the integrity of science education, provided a travel grant enabling me to attend the 1987 National Creation Conference.

I would like to thank my twin sister, Peggy Gregerson, who assisted me at several desperate moments in obtaining books I could not otherwise have gotten.

Finally, I thank my wife Beth, to whom this book is dedicated. She has worked tirelessly to support our family while I have spent much of my time and a large portion of our income (including a large portion of *her* income), pursuing this research.

Each time I came home with a stack of anti-evolution books, or a batch showed up in the mail, I promised her that it would definitely be my last purchase. But there was always more. Beth has patiently borne the burden of my expensive and unprofitable hobby.

Introduction

The approach in this annotated bibliography is descriptive rather than polemical. I make no attempt to refute or oppose the arguments and conclusions presented in these works, nor even, in most cases, to note any personal disagreement at all with the author's contentions. This does not mean I agree with them; on the contrary. When, for instance, I state that an author "refutes" certain evolutionist arguments, I am merely describing that author's own claims, not my own opinion. I state this at the outset in order to avoid any misinterpretation of my descriptions as endorsments of the anti-evolutionist arguments or conclusions.

Since this is a descriptive approach, I have attempted to be objective, examining each individual work, as much as possible, on its own merits, in order to find out what the author's actual position is. The creationist assault on evolution is serious and has far-reaching consequences. Though I disagree with the anti-evolution arguments and conclusions, I sympathize with their concerns, which we ignore at our peril. We need to pay attention to what it is they are saying, in order both to reply effectively to their scientific arguments, and to understand their real motivations—religious, moral, political and educational.

One of the most important aims of this collection is to differentiate between the diverse types and forms of anti-evolutionist and creationist theories and doctrines. I evaluate and describe (where possible) what type is being advocated in each work. My intention is to illustrate the enormous range and depth of the opposition to evolution. This opposition is broad—in fact fantastically diverse—and very deeply rooted. It is not monolithic. There is a great variety of creationist interpretations and theories, many of which are mutually contradictory. Rival creationists often oppose each other fiercely and dogmatically. (It is ironic that one of their chief criticisms of evolutionary theory is that evolutionists do not always agree amongst themselves.)

I have chosen to interpret "anti-evolution" very broadly rather than limit it to creationism. Protestant fundamentalist creationism constitutes by far the largest, most vocal, and certainly the most important opposition to evolution, but there are also many anti-evolutionist works by adherents of other religions (Catholic, Jewish, Islamic), as well as various non-religious challenges to Darwinian theory.

Although Darwinism is not synonymous with evolution, I have included some anti-Darwinian evolutionist works for two reasons. Some of the attacks on Darwinian theory seem to be motivated by extra-scientific concerns: for instance, desire for a non-materialistic theory of development or evolution. Another, more important reason is that many critics of Darwinism are quoted and appealed to by creationists as if they were hostile to evolution itself—as providing evidence that evolution is scientifically untenable. In many cases the creationists cite these anti-Darwinian critics misleadingly and unfairly, but since they continue to be used thus by creationists, it seems justifiable to include them in this "anti-evolution" collection— if only to present a fairer account of their actual views and criticisms.

I have also included many works advocating occult and psychic evolution, and evolution guided or manipulated by extraterrestrial beings, since these generally deny the adequacy or even the occurrence of naturalistic evolution.

"Creationist" works by Protestant fundamentalists fall into several general categories, though there are many subdivisions of each. "Strict" or "young-earth" creationism comprises the best-known group (indeed, the only type of creationism most of the public is aware of). Young-earth creationists believe that God created the world by divine fiat, *ex nihilo* (out of nothing), in six literal, 24-hour days. Some strict creationists insist this occurred at or around 4004 B.C.; others allow a few thousand years more.

Most strict creationists advocate Flood Geology, claiming that all or most of the geological record is the result of the worldwide Flood, which destroyed all terrestrial life except that taken aboard Noah's Ark, and catastrophically altered the earth. Flood Geology was originally a scientific attempt to explain earth history. As the science of geology developed, and scientists realized that the earth was far older than a literal interpretation of the Genesis chronology indicated, Flood Geology was abandoned. George McCready Price reinvented it in the beginning of this century as a means of returning to a literal interpretation of Genesis. Whitcomb and Morris's *Genesis Flood* (1961) marked the beginning of the modern popular resurgence of the young-earth Flood Geology. Since young-earth Flood Geology arguments are relatively well known (most of the recent creationist lobbying and publicity efforts have been spearheaded by "strict" creationists), I shall not attempt to describe them here in further detail.

"Old-earth" creationists, who are quite numerous, deny recent creation. They generally accept the standard geological and paleontological chronology, thus avoiding one of the most obvious conflicts with modern scientific theory, but at the price of seeming to deny the plainest, most literal interpretation of Genesis. One popular old-earth approach is the "Day-Age Theory," which interprets the six creation 'days' of Genesis as long ages. Another is the "Gap Theory," which maintains that the six days of creation were literal and recent, but that the geological ages preceded them: the original creation was subverted by Satan; God then destroyed the world, and re-created it in six days. "Progressive" creationism is less clearly defined, and may verge on either Day-Age creationism or theistic evolution. According to progressive creationists, God intervened directly at various times over the ages, creating new species or modifying existing ones. According to the "Framework" or "Literary" theory of creationism, the Genesis account employs poetic or literary devices, and the events or processes of the six 'days,' which are not necessarily chronological, may overlap. The "Revelatory" theory of creationism interprets the six days as a series of visions (or written accounts) revealed to Moses or some other seer. There are also numerous combinations and variations of all of these theories. Old-earth creationists may advocate a regional rather than a global Noachic Flood, or a "tranquil" Flood.

Though I began this study as a collection of modern creation-science, I ended up including many older works. These historical works illustrate the development, and the continuity, of a number of themes. Certain aspects of "modern" creation-science are not so new as many suppose. Conversely, there have been some interesting and significant shifts of emphasis, despite some of the dogmatic assertions of modern creationists; the fundamentalism of the 1920s, the creationism of previous decades and centuries, was not identical to today's.

Inclusion of these historical works introduces a potentially serious risk, requiring an explicit warning. The older works must be interpreted in their historical contexts, not solely in the context of modern creationism. One reason for including them is indeed because some of the arguments and themes have survived or reemerged in modern creationism. But another reason is to demonstrate the many differences in approach and concerns between the older and the modern works. And, of course, it is unfair to judge past scientists and writers in terms of present knowledge. The most obvious instance: Pre-Darwinian biological works must certainly be judged by entirely different scientific criteria than modern works.

Prior to Darwin's publication, in 1859, of the *Origin of Species,* there was no plausible scientific theory of evolution. Therefore, most scientists were, in one sense of another (and there are many senses of the term), "creationists." But many of these "creationist" scientists advanced theories which were later incorporated in the evolutionist framework; some became crucial elements of evolutionary theory. Thus, some of these historical figures may legitimately be claimed as ancestral to both creationists and evolutionists. Again, this illustrates the risk of judging works—especially historical works—by a single criterion: in this case, of classifying works as either "anti-evolutionist" or not. I have attempted to view each case, and describe each author's position, in proper perspective in order to minimize this risk.

I have included a few post-Darwinian works advocating inheritance of acquired characteristics (so-called "Lamarckian" evolution), since this type of evolution opposes (or is interpreted as opposing) the foundations of Darwinian evolution. Pre-Darwinian hypotheses of species transmutation relying on inheritance of acquired characters, on the other hand, would be considered "anti-creationist"—"evolutionist"—and are not included here.

I have not included purely theological works, but I do include some theological works in which the author also presents scientific arguments against evolution or in favor of "scientific" creationism.

Most of the entries in this bibliography are books, but I have also included many booklets, pamphlets and tracts, plus a few items in other media. I have not included journal or periodical articles (except for cases of separately distributed reprints in pamphlet form), nor do I include films, videos, and other visual media.

The bibliographic entries are arranged alphabetically by author. Works by the same author are arranged by date of publication (original edition if possible), except when publication date is unknown.

Names which appear in **boldface** are of authors with entries listed under that name in the bibliography.

Most of the entries described in this bibliography are works I have seen first-hand. For a number of the older books especially I have had to rely on secondary sources. I have indicated books described from secondary sources with the phrases "discussed in," "quoted in," or "reviewed in" those other sources. Secondary sources included as main entries in this bibliography are indicated by the author's name in **boldface.** Other (non-anti-evolutionist) sources mentioned are indicated by the author's name in regular roman type; these sources are listed in the Preface.

The Bibliography

1 Ackerman, Paul D. *It's a Young World After All: Exciting Evidences for Recent Creation.* 1986. Grand Rapids, Mich.: Baker Book House. *Ackerman:* Univ. Kansas Ph.D.; teaches psychology at Wichita State Univ.; president of Creation Social Science and Humanities Society and editor of Creation Social Science and Humanities Quarterly. This book is a review of young-earth (strict) creationist arguments. Includes moon dust argument (notes anti-creationist replies and gives Slusher's counter-argument), comet lifespans, **Gentry's** radiohalos, Poynting-Robertson effect, shrinking sun, flaws of radiometric dating, rock viscosity of moon craters, polystrate fossils, vertical whale fossil (with absurd drawing). Also includes **Setterfield**'s decaying speed of light theory and Dodwell's theory that astronomical calculations of ancient monuments prove asteroid impact tilted earth's axis in 2345 B.C., causing Flood. Photos, drawings; bibliog., index.

2 Acworth, Bernard. *This Bondage.* 1929. London: J. Murray. "A study of the 'migration' of birds, insects and aircraft, with some reflections on 'evolution' and relativity." *Captain Acworth:* British naval officer; respected amateur ornithologist; founder and later president of Evolution Protest Movement (first meeting 1932). **H. Morris** (1984) says this book is based on anti-evolution articles published in various journals. Acworth also wrote books such as *The Navy and the Next War* and *What We Are Fighting For.*

3 ________. *This Progress: The Tragedy of Evolution.* 1934. London: Rich & Cowan. Eloquent plea for creationism. Creation and evolution are "flatly contradictory." If creationism is accepted as true, then Christianity will regain the ground it lost in England due to the advance of evolution. Evolution results in bad morals and bad behavior. (**Morris** says orig. 1932.)

4 Acworth, [O.] Richard. *Creation, Evolution and the Christian Faith.* 1969. London: Evangelical Press. *Acworth:* nephew of **B. Acworth;** first secretary of Evolution Protest Movement. "If the Bible teaches creationism, and if Christianity depends on the truth of the Bible, then in principle, it is capable of being disproved. If evolution were to be proved false, that is to say, the Bible would be proved false." (Quoted in **Haigh** 1975.)

5 Adams, L. Mabry. *Ringing Bells of Harmony.* 1975. Priv. pub. Oversize book with many reproductions of old charts, time-lines, and illustrations. Maps of the Pre-Flood world; shows the Garden of Eden is now submerged in Indian Ocean. (In ICR Library.)

6 Adler, Mortimer J. *What Man Has Made of Man.* 1937. New York: Longmans, Green. Also pub. 1937 by Frederick Unger (New York). *Adler:* Univ. of Chicago neo-Thomist philosopher of Great Books fame; Catholic. Calls evolution a "popular myth"; not fact but "at best a probably theory, a history for which the evidence is insufficient and conflicting." The facts show only that some animal types no longer exist; they "do not establish the elaborate story which is the myth of evolution.... I say 'myth' in order to refer to the elaborate conjectural history, which vastly exceeds the scientific evidence." This history is a "concoction" of evolutionary philosophers and popularizers. (Quoted and discussed in Gardner; **Jauncey** 1961.)

7 ________. *Problems for Thomists.* 1940. New York: Sheed and Ward. Discusses how many "species" exist—that is, how many creative acts of God are required to explain the evolutionary "jumps" (gaps). Adler suggests four; certainly less than ten. Each "species" is fixed: consists of immutable essence, not derived from others but directly from God. Admits scientific evidence for this is "indecisive" but feels there is strong theological argument for it in Old Testament *Wisdom of Solomon* (considered apocryphal by Protestants). (Discussed and quoted in Gardner.)

8 ________. *The Difference of Man and the Difference it Makes.* 1967. New York: Holt, Rinehart and Winston.

9 Agassiz, Louis. *Methods of Study in Natural History.* 1873. Boston: Ticknor and Fields. Also pub. by J.R. Osgood (Boston) and Houghton, Mifflin (Boston). 21 eds. up to 1893. *Agassiz:* the great Swiss naturalist who proposed the glacial (Ice Age) theory in 1837 to explain geological features then believed caused by Flood. Later, he moved to Harvard Univ., and was the most famous scientist in America. Agassiz, who studied under **Cuvier,** never accepted Darwin's theory and remained a staunch creationist to the end, considering Darwin no more persuasive than Lamarck or Oken. Moore says Agassiz "states in Preface that this is a more popular presentation of his views in his *Essay on Classification* [1859], and an opportunity to protest against transmutation (evolution) theory."

10 ________. *Evolution and Permanence of Type: A Refutation of the Darwinian Theory of Evolution.* Undated pamphlet [1894]. Hawthorne, Calif.: Christian Book Club of America. (The Christian Book Club of America is now listed as an imprint of Noontide Press, the parent group of the Institute for Historical Review—see **Hoggan.**) Orig. published 1894 (posthumously) in *Atlantic Monthly.* Agassiz concluded that each species came into existence when God thought of them, and disappeared when He ceased to think of them. (Reprinted and discussed in Hull.) Agassiz is also listed as "contributor" to **J. Nott**'s *Types of Mankind* (Nott was a pro-slavery propagandist who took advantage of Agassiz's belief in separate creations of races, reprinting them to promote his own racist ideas).

11 ________. *Gists from Agassiz: Passages on the Intelligence Working in Nature; Essay on Evolution and Permanence of Type.* 1973 (c1953). Hawthorne, Calif.: OMNI Publications (Christian Book Club of America).

12 Akridge, Russell. *The Sun Is Shrinking* (pamphlet). 1980. El Cajon, Calif.: Institute for Creation Research. ICR "Impact" series #82. *Akridge:* Georgia Tech physics Ph.D.; Th.M. from New Orleans Baptist Theol. Sem.; ass't prof. of physics at Oral Roberts Univ.; *CRSQ* contributor; now lives in Georgia. Measurements show that the sun is getting smaller. If this decrease is extrapolated, it proves recent creation. (This argument has been popularized by **T. Barnes,** but Akridge does not mention him here.)

13 ________. *Radiometric Dating Using Isochrons* (pamphlet). 1982. El Cajon, Calif.: Institute for Creation Research. ICR "Impact" series #113. Demonstrates, by means of a parable, the allegedly faulty assumptions underlying radiometric dating.

14 Alexander, Charles D. *Creation No Accident* (pamphlet). 1965. Y.M.C.A. (John G. Eccles): Inverness, Scotland. 1965 Y.M.C.A. address: Inverness, Scotland. *Alexander:* Fellowship of Independent Evangelical Churches minister; journalist, editor, and pamphlet writer. Includes section proving "The Bible Ahead of Science." The coelacanth proves non-evolution. Population increase rate proves mankind only 6,000 years old. Evolution is the "Highroad to Atheism."

15 Allen, Eula. *A Trilogy of Creation.* 1963–1966. Virginia Beach, Vir.: A.R.E. Press. 3 vols. The trilogy consists of the three titles listed below. Allen is a follower of famed psychic Edgar Cayce of Virginia Beach; his books are occultic interpretations of the Bible and of prophecy.

16 ________. *Before the Beginning: A Study of Spiritual Creation.* 1963. Virginia Beach, Vir.: A.R.E. Press. "Based on twenty years' study of the Bible and the Edgar Cayce readings."

17 ________. *The River of Time: A Study of the Physical Creation.* 1965. Virginia Beach, Vir.: A.R.E. Press.

18 ________. *You Are Forever: A Study of Man's Struggle [. . .].* c1966. Virginia Beach, Vir.: A.R.E. Press. "Interpretation from the psychic readings of Edgar Cayce."

19 Allen, Frank E. *Evolution in the Balances.* c1926. New York: Fleming H. Revell. *Allen:* Ph.D., F.R.S.C.; physics prof. and dept. head at Univ. Manitoba, Canada.

20 Allford, Dorothy. *Instant Creation—Not Evolution.* 1978. New York: Stein and Day. Recommended by **J. Bergman** (*Bible-Science Newsletter* 1986): "An excellent general defense of creation by a secular publisher. The author is an M.D., and a medical researcher."

21 Allis, Oswald T. *God Spake by Moses.* 1979 [1951]. Phillipsburg, N.J.: Presbyterian and Reformed. *Allis:* a strict Calvinist; has degrees from Princeton, Princeton Theol. Sem., Univ. Berlin (Ph.D., D.D.); prof. at Princeton Theol. Sem. and Westminster Theol. Sem. Defends authorship of the Pentateuch (including Genesis) by Moses; refutes "higher criticism" of Bible. Supports *ex nihilo* six-day fiat creation as plainest, most direct interpretation. Rejects evolution as "unproved hypothesis." Flood was of "incomparable destructiveness," probably worldwide, since this would better account for geological phenomena than would vast ages. A theological argument, but includes allusions to scientific support for strict creationism.

22 Ambassador College. *The Origin of Life* (pamphlet). c1972 [1956]. Pasadena, Calif.: Worldwide Church of God. No author listed. Ambassador College: founded and led by **H.W. Armstrong** (Worldwide Church of God; *Plain Truth* magazine). Origin-of-life studies show evolution is impossible. Photos, diagrams.

23 Ambrose, Edward J. *The Nature and Origin of the Biological World.* 1982. New York: John Wiley & Sons. *Jack Ambrose:* London Univ. cell biologist. Anti-Darwinist. "A creative view of the origin of life and species no longer needs to be defended against evolutionary arguments. It is the reductionist evolutionist who is now in retreat." For evolution to proceed, "There must have been a steady input of new information which we have ascribed to the activity of Creative Intelligence..." "Although I believe that Divine acts of creation, as described so beautifully in the beginning of Genesis, gave rise to a physical and biological world that was entirely good, I also accept that evil powers have been at work." Book ends with an appeal to Christ Savior. (Reviewed and quoted in *CRSQ* 1985; also quoted in **Hayward** 1985.)

24 American Scientific Affiliation. *Modern Science and Christian Faith.* c1948. Wheaton, ILL.: Van Kampen Press. 2nd rev. ed. pub. 1950. **F.A. Everest,** ed. "Eleven Essays on the Relationship of the Bible to Modern Science." Symposium of the American Scientific Affiliation. The A.S.A., founded in 1941, is: "A group of Christian scientific men devoting themselves to the task of reviewing, preparing, and distributing information on the authenticity, historicity, and scientific aspects of the Holy Scriptures in order that the faith of many in the Lord Jesus Christ may be firmly established." ASA members have held various positions regarding evolution, from strict creation to theistic evolution. As ASA grew less favorable to strict creationism, some members (**Morris, Gish, Rusch, Klotz, Marsh**) formed their own strict creationist organization: the Creation Research Soc. Of the 11 essays in this volume, only the one by **Tinkle** and **Lammerts** (also founding members of CRS) is strict creationist, though most of the others are decidedly anti-evolution. **P. Stoner** has a chapter on the "perfect harmony" between modern astronomy and Genesis (origin of the universe proves Bible account correct). Edwin Gedney (Sc.M.; prof. at Gordon Coll. Theol., Boston) writes about geology and the Bible; discusses sudden appearance of fossils in the Cambrian, lack of transitional forms, apparent "directionalism" of forms (contradicting chance selection or mutation). Considers various creationist theories but argues for Day-Age creationism as harmonizing with the facts of geology. Tinkle and Lammerts discuss "biology and creation"—mostly genetics, claiming that no aspect of genetics allows for evolution of greater complexity. They also reject notion of pre-Adamic men, and insist that God directly created all the Genesis 'kinds.' **R.L. Harris** discusses Bible-science chemistry. **F. Allen** writes about physical science and the Bible, declaring the Genesis creation account non-mythical and fully in accord with science, appealing to entropy, relativity theory and the Big Bang, and equating prediction in physics with biblical prophecy. Also decries mechanism, especially evolution, stating that evolution cannot produce any new characters and that natural selection has been rejected. "The inevitableness of the iron laws of evolution, as they were often termed, held humanity in the bonds of a determinism so rigid that all conscious efforts to thwart or even to mould the resistless processes of development must be futile in the extreme." (Allen then notes that the uncertainty principle has rendered mechanism paradoxical.) George R. Horner (anthropology and archaeology dept. of Wheaton Coll., Ill.) discusses physical anthropology (chapter is adapted from his forthcoming textbook *Man in Space and Time*). Emphasizes difference between man and apes, and argues that all fossil humans down to *Pithecanthropus [Homo] erectus* are really human. "We, as Christians, can say that man from the earliest times has always been man—a view which is scientific and which follows logically from the data." Changes and racial variations do not require evolutionary explanation. Suggests that Noah's sons were not direct ancestors of human races, but only of Mediterranean families. In 1950 ed., the anthropology chapter is by Smalley and Fetzer (similar arguments). Other chapters on psychology, Bible prophecy and math, medicine, and archeology. Biog. notes. Photos, drawings, charts.

25 ________. *Teaching Science in a Climate of Controversy: A View from the American Scientific Affiliation.* 1986. Ipswich, Mass.: American Scientific Affiliation. Written by ASA's Committee for Integrity in Science Education: David Price (chairman; former high-school teacher), **John Wiester,** and Walter Hearn (adj. science prof. at New College Berkeley; editor of *JASA*). Committee consultants include **J.O. Buswell** (anthropology), **R. Fischer** (chemistry), **R. Newman** and H. Ross (astronomy), E. Olson (geochemistry), D. Wilcox (biology), and **D. Wonderly** (biology, geology). This large handsome booklet is a response to the Nat'l Acad. of Science's 1984 booklet denouncing creationism, and is likewise scheduled to be distributed to all U.S. high schools. It urges a mild approach, criticizing both young-earth creationism and non-theistic evolution (such as defended in the NAS booklet) as extreme and dogmatic positions. The tone is conciliatory and calm, but it clearly favors old-earth creationism over either "extreme." Appendix lists 14 recommended books—virtually all old-earth creationist. Criticizing "chance" origin-of-life theories, they argue that all experiments introduce "purpose," so natural chance element can never be demonstrated. The Piltdown hoax is compared with the Paluxy "man-tracks" to show fallibility and dogmatism of both extremes. Argues that dearth of pre-Cambrian fossils precludes definitive statements on natural origin of life, and that the common descent of man and other animals is not proven. Concludes that "too many problems remain unresolved and too many pieces of evidence are missing to say with certainty that we share a common ancestry with apes." Also suggests that non-theistic explanations relying on "chance" do not dispose people to good behavior, whereas theistic belief does. Photos, charts, drawings.

26 Anderson, Bernard W. *Creation versus Chaos.* 1967. New York: Association Press. Cited in **Bixler.** Cosmogonic warfare between God and evil Chaos during Creation.

27 Anderson, J. *The Course of Creation.* 1850. Cited in Millhauser as advocating Day-Age creationism.

28 Anderson, J. Kerby, and Harold G. Coffin. *Fossils in Focus.* 1977. Grand Rapids, Mich.: Zondervan / Richardson, Texas: Probe Ministries (co-pub.). Christian Free Univ. Curriculum series. Book politely advocates special creation. Usual creation-science paleontology arguments: Cambrian "explosion," lack of transitional fossils between major groups. Attractive format and good drawings; looks quite scientific and well-reasoned. Includes chapter abstracts. Many scientific references; biblical creation is urged in last pages. Like other volumes in this series, a "Reponse" is included, by **Russell Mixter** (brief and fairly un-critical).

29 Anderson, Norris. *Man—Evolved or Created* (pamphlet). c1975. Manila, Philippines: O.M.F. Publishers. Kayumanggi Study Series—apologetics and evangelism. *Anderson:* M.S.; teacher. Explicitly religious. Creationism is as scientifically valid as evolution. Strict young-earth creation, Flood Geology. Standard creation-science arguments. Includes study questions.

30 Andrews, Edgar H. *Is Evolution Scientific?* (booklet). 1981 [1977]. Welwyn, Hertfordshire, England: Evangelical Press. 3rd ed.; orig. 1977. *Andrews:* Ph.D. in fracture mechanics; D.Sc.; prof. of materials at Univ. London; macromolecules expert; president of Biblical Creation Society. Emphasizes philosophy of science. Says evolution is accepted because it "relieves us of the need to believe in God" and "provides comfort." Strict creationism.

31 ________. *From Nothing to Nature.* 1978. Welwyn, Hertfordshire, England: Evangelical Press. "A Young People's Guide to Evolution"—aimed at teenagers. Purpose or Chance? God and evolution (that is, chance) can't both be true. Evolution contradicts the Bible. Presents standard creation-science arguments; argues for young earth. Explicitly religious; includes many Bible references. Last chapter on the six days of creation. Drawings, charts, good photos.

32 ________. *God, Science & Evolution.* 1981 [1980]. Welwyn, Hertfordshire, England: Evangelical Press (distr. in U.S. by Creation-Life Publishers: San Diego, Calif.). 2nd ed. Orig. 1980. Standard young-earth creation-science; explicitly religious, though it includes many scientific references. Based on lectures given to both scientific and religious groups. Largely philosophical arguments: for the existence of the supernatural and miracles, the inadequacy of evolution, the conflict between evolution and scripture. Urges a re-unification of science and theology, and presents a "theology of science."

33 Andrews, E.H., W. Gitt, and W.J. Ouweneel (eds.). *Concepts in Creationism.* 1986. Welwyn, Hertfordshire, England: Evangelical Press. Lectures from the first European Creationist Congress (Heverlee, Belgium: 1984). Foreword by **W. Lang** (BSA). Contributors: **W. Gitt** (Germany), Andrews (U.K.), **Ouweneel** (Netherlands), Gerald Duffet (U.K.; head of biology—City of Ely Coll.), Eberhard Bretsch (Germany; math/computer), Dennis Cheek (U.S.; science chair—Bitburg Amer. H.S.), Chris Darnbrough (U.K.; biochem.—Glasgow Univ.), **David Rosevear** (U.K.), Hendrick Murris (Netherlands; biologist—Pieter Nieuwland Coll.), and **S. Scherer** (Germany). Biog. notes included. Topics presented as scientific, religious (biblical creation), and mixed. Emphasis on unity of knowledge: science and religion speak of same Truth. Most or all contributors strict creationists. Charts, photos.

34 Anonymous. *A Brief and Complete Refutation of the Anti-Scriptural Theory of Geologists.* 1853. London: Wertheim & Macintosh. Argues that "all the organisms found in the depths of the earth were made on the first of the six creative days, as models for the plants and animals to be created on the third, fifth, and sixth days" (quoted in A. White). **H. Miller** 1857 says the anonymous author describes himself as "a Clergyman of the Church of England."

35 Anonymous. *Genesis of the Earth and of Man.* 1854. Revelatory creationism: the six creation days were a series of trances during which Moses saw a recapitulation of history. (Quoted in **H. Miller** 1857.) "Christian philosophers have been compelled to acknowledge that the Mosaic account of creation is only reconcileable with demonstrated facts, by its being regarded as a record of *appearances;* and if so, to vindicate the truth of God, we must consider it, so far as the *acts* are concerned, as the relation of a revelation to the *sight,* which was sufficient for all its purposes, rather than as one in words; though the words are perfectly true as far as describing the revelation itself, and the revelation is equally true as showing man the principal phenomena which he would have seen had it been possible for him to be a witness of the events."

36 Arachim. *Pathways to the Torah.* 1985. Aish HaTorah Publications. Adapted from ed. published by Arachim (Beny Bracq, Israel), by Yeshiva Aish HaTorah staff under direction of Rabbi Yehuda Silver. *Arachim:* organization for the furtherance of Jewish awareness.

Book is in nine sections, including "Evolution," "Science vs. Scientism," "Archeology and the Torah" and "Prophecy." Strongly creationist, it presents a large number of standard anti-evolution quotes, citing **Moorhead** and Kaplan, **G.M. Price, Macbeth,** Einstein, Wald, Popper, **Trop,** Lipson, Prigogine, **Hoyle** and Wickramasinghe, **Klotz,** Simpson, Romer, **Nilsson, J.N. Moore,** A. Huxley, Urey, **Hitler,** etc. Aish HaTorah edition adds quotes from **Hitching** and Hanoka. Other sections deal with intricate mathematical studies of word patterns and distributions in text of Torah, proving supernatural origin. Loosely organized, largely bits of reprinted excerpts, with comments; many parts with Hebrew text also. Charts, diagrams, maps.

37 Archer, Gleason L. *Encyclopedia of Bible Difficulties.* 1982. Grand Rapids, Mich.: Zondervan. Archer is prominent member of the Int'l Council on Biblical Inerrancy; he acknowledges ICBI encouragement. An apologetics compendium of Bible passages alleged to present problems for inerrantist interpretation, with their solutions. Among "Recommended Procedures": be convinced that an adequate explanation exists, even if it is not yet found. Archer, like many ICBI members, is an old-earth creationist. Presents non-literal (though strictly inerrantist) interpretation of six-day creation. Cites **Dewar** 1948 on lack of Precambrian animals. Cites C-14 date of 12,800 BP for "Cretaceous" wood from Paluxy by R. Berger of UCLA. Pre-Adamic hominids may have existed—but not humans. Goodall and Patterson ape communication studies indicate these pre-Adamic races may have used languages and tools.

38 Armstrong, Garner Ted. *The Amazing Archer Fish Disproves Evolution!* (pamphlet). c1967 [1966]. Pasadena, Calif.: Ambassador College Press. *G. Armstrong:* son of Worldwide Church of god and **Ambassador College** founder **H.W. Armstrong;** now heads Texas WCG offshoot after final schism with father following sex scandal. Garner Ted used to do the WCG T.V. broadcasts and wrote many effective and hard-hitting anti-evolution pamphlets. Archer fish's behavior and design couldn't have evolved. "The BIGGEST false doctrine ever is EVOLUTION."

39 ———. *The Fable of the First Fatal Flight* (booklet). c1966. Pasadena, Calif.: Ambassador College. Bird examples which refute evolution. Impossibility of evolutionist origin of flight: can't fly if half-way evolved. Includes *Archaeopteryx;* design of woodpecker. Evolution is a "fowl" hypothesis. Comic strip of half-reptile, half-bird creature trying to fly. Scientific references, photos, pictures.

40 ———. *Some Fishy Stories About an Unproved Theory* (booklet). c1971 [1966]. Pasadena, Calif.: Ambassador College Press. Includes angler fish, archer fish, lungfish, coelacanth. All refute evolution. Good photos; cartoon of angler fish.

41 ———**, and Paul W. Kroll.** *A Theory for the Birds* (booklet). c1971 [1967]. Pasadena, Calif.: Worldwide Church of God. *Kroll:* theology M.A.; mass communications M.A.; writer for *The Plain Truth,* **H.W. Armstrong**'s widely-distributed free magazine. Birds refute evolution. Includes bird migration; bill types. Photos, pictures, charts.

42 ———**, and** ———. *A Whale of a Tale, or—the Dilemma of Dolphins and Duckbills* (booklet). c1968. Pasadena, Calif.: Ambassador College. Whales, dolphins, and duckbill platypuses refute evolution. Many good photos, mostly from L.A.'s Marineland. Nice platypus cartoon. Platypus is combination of different animal types—couldn't have evolved.

43 ———**, and** ———. *Our Awesome Universe* (booklet). c1973. Pasadena, Calif.: Ambassador College. Design of the universe proves creation. Many good photos.

44 Armstrong, Herbert W. *Did God Create a Devil?* (booklet). c1978 [1959]. Pasadena, Calif.: Worldwide Church of God. *Armstrong:* founder and leader of Worldwide Church of God, **Ambassador College,** and *Plain Truth* magazine; T.V. evangelist; father of **Garner Ted Armstrong.** Orig. 1959; still widely distributed. The origin and nature of the Devil, explained by reference to the Gap Theory of creationism. Ages ago, God created perfect world—not the Chaos of Gen. 1:2. Lucifer rebelled with one-third of the angels. God then made the earth Chaos; later re-made it. Adam was the first man, but these angels lived long before Adam. Includes many Bible references to support Lucifer's (Satan's) reign, pre-Adamic life, and the Gap. God didn't create a devil: He created a perfect angel with free will who rebelled.

45 _______. *The Missing Dimension in Sex*. 1981 [1964]. New York: Everest House. 3rd ed.; orig. 1964. Armstrong's sex education book, developed from his **Ambassador College** course on family life begun in 1949. Includes frequent attacks on evolution for being unable to explain sex. Tells how he presented a science librarian with his refutation of evolution; she confessed that he has utterly destroyed it with his arguments. Sex is good, but only between man and wife. Satan made sex shameful in Eden.

46 _______. *Mystery of the Ages*. c1985. New York: Dodd, Mead. Also pub. by Worldwide Church of God (Pasadena, Calif.). Also serialized in *Plain Truth* magazine. Unlike strict fundamentalists (who consider Armstrong's WCG a cult), Armstrong claims that the Bible is a great "mystery" never before understood until revealed to him. He explains its hidden messages. Presents Gap Theory creationism: the original creation may have been billions of years ago; much later, following Satan's rebellion, occurred the six-day re-creation (including creation of man). Largely concerned with Bible prophecy; other anti-evolution passages scattered throughout. Armstrong teaches that U.S. and Britain are the true descendants of the Lost Tribes of Israel and are fulfilling God's prophecies regarding Israel, His chosen people.

47 Arndts, Russell T., and William Overn. *Isochron Dating and the Mixing Model* (pamphlet). 1983. Minneapolis: Bible-Science Association. *Arndts:* inorganic chem. and physics Ph.D.—Indiana Univ.; chemistry prof. at St. Cloud State Univ.; president of Bible-Science Assoc. Radiometric dating is unreliable, and isochron method merely deludes evolutionists into thinking they have accurate geochronological technique.

48 Arthur, Kay. *Establishing Your Foundations*. c1985. Chattanooga, Tenn.: Precept Ministries of Reach Out. A volume in the Precept Upon Precept Bible Study Series. Kay and Jack Arthur founded Precept Ministries of Reach Out in 1970. This book is intended for home study course; with questions, exercises, diagrams. Heavily creationist. Includes sections reprinted from **Whitcomb** and **Morris;** endorses canopy theory; other creation-science sources quoted also. This Bible Study series also available in video. College credit available from Columbia Bible College; local training workshops and Precept Bible Study groups also available.

49 Austin, Steven A. *Depositional Environment of the Kentucky No. 12 Coal Bed (Middle Pennsylvanian) of Western Kentucky, with Special Reference to the Origin of Coal Lithotypes*. 1979. Unpub. Ph.D. diss.: Pennsylvania State Univ. Geology Ph.D. dissertation (Penn State). Argues for rapid formation of coal, and presents "floating mat" hypothesis. Masses of trees and vegetation are swept into great floating mats by flooding; waterlogged detritus (mostly bark) sinks to bottom and turns into coal. Creationist implications are not explicitly stated because of secular university situation, but Austin intended this as refutation of uniformitarianism and as evidence for formation of coal by Noah's Flood. Austin, now geology prof. at ICR, has admitted he was lucky in getting this thesis through Penn State.

50 _______. *Did the Early Earth Have a Reducing Atmosphere?* (pamphlet). 1982. El Cajon, Calif.: Institute for Creation Research. ICR "Impact" series #109. Cites evidence of oxidized Archaean minerals, which suggest that earth did not have strongly reducing atmosphere. Thus, he argues, origin-of-life scenarios are impossible, if early atmosphere was oxidizing. Austin also wrote other "Impact" articles for ICR and other creationist articles under the pseudonym "**S. Nevins**" before getting his Penn State Ph.D.

51 _______. *Ten Misconceptions about the Geologic Column* (pamphlet). 1984. El Cajon, Calif.: Institute for Creation Research. ICR "Impact" series #137. Mostly standard Flood Geology objections. Notes that geologic column was devised by creationist geologists.

52 ________. *Catastrophes and Earth History.* 1984. El Cajon, Calif.: Institute for Creation Research. ICR Technical Monograph No. 13. A collection of articles and documentations of catastrophic geophysical events. The purpose is to refute uniformitarianism—considered the basis of evolution. The material collected by Austin is not directly creationist, though he assumes that proof of catastrophism is evidence for creationism. (Not to be confused with 1984 book by non-creationists Berggren and Couvering with same title.)

53 Aw, S.E. *Chemical Evolution: An Examination of Current Ideas.* 1982 [1976]. San Diego, Calif.: Master Books. Orig. 1976: Univ. Education Press (Singapore). *Aw:* nuclear medicine dept. head at Singapore Gen. Hosp.; former biochem. prof. at Nat'l Univ. of Singapore. Critique of origin-of-life theories. Fairly technical biochemistry discussion, except for last chapter (on origin-of-life philosophy). Includes critiques of Fitch, Jukes, Ponnamperuma, and most well-known molecular evolution theorists. "Update" section of 1982 ed. includes critiques of Dickerson, J.W. Schopf, Popper. Index; diagrams.

54 Ayres, Clarence E. *Science: The False Messiah.* 1927. Indianapolis, Ind.: Bobbs-Merrill. *Ayres:* philosophy prof.—Amherst College (Mass.); assoc. ed. of *New Republic.* Criticizes faith in science as Truth. Compares "folklore" of science to religious folklore (authority of Bible). Science and folklore both based on axioms accepted by believers; facts "fit" because axioms are assumed true. Both "folklores" are proved by circular reasoning; both "verified by repetition and sanctified by faith." Emphasizes evolution especially: does not claim evolution is necessarily false, but insists it is "proved" and believed in the same way as Bible creation folklore. (Excerpted in Gatewood.)

55 Baerg, Harry J. *Creation and Catastrophe: The Story of Our Father's World.* 1972. Washington, D.C.: Review and Herald. *Baerg:* professional illustrator for Review and Herald, a Seventh-day Adventist publisher. His book has nice drawings on every page. Standard creation-science arguments; strict creationism, discusses the biblical Flood and Noah. Creation: ca. 3950 B.C. Flood: ca. 2650 B.C. Location of original Eden cannot be known because the Flood completely altered earth's features. Describes Adam and Eve and antediluvian world. Adam was 15 feet tall. Network of shallow canal-like seas may have caused the worldwide tropical climate. Satan caused species to degenerate with mutations and amalgamation (crossing). Flood precipitated either by extinction of burning moon, or tilting of earth's axis. Ice Age followed Flood. Peleg's name may refer to division of continents then. Concedes that many animals must have changed considerably after Flood. Presents Flood Geology explanations of fossil formation. Describes resettlement of earth by Noah's descendants, including cave-men. Muses about what earth will be like after the Millenium, when Edenic conditions are restored. "Further reading" references following each chapter are all Seventh-day Adventist. Chapter notes cite **E.G. White** frequently.

56 Baillie, Mrs. E.C.C. *The Protoplast: A Series of Papers.* 1853. London: Wertheim & MacIntosh. Two vols. Scriptural Geology; man coeval with oldest fossils. (Cited in Millhauser.)

57 Baker, Alonzo, and Francis D. Nichol. *Creation—Not Evolution.* 1926. Omaha, Neb. / Portland, Or.: Pacific Press. *Baker* and **Nichol:** assoc. editors of *Signs of the Times* (Seventh-day Adventist), and authors of *The San Francisco Debates on Evolution* (see under **Shipley**). Foreword by **G.M. Price.** Standard creation-science arguments. Chapters include "Evolution's Unsavory History," "The Flood," "The Crusade for the Missing Link," "Questions for Evolutionists to Answer," "Evolution a Philosophy and a Religion," "The Bible, the Crux of the Controversy," "Back to Creationism," and others denouncing evolution and affirming strict biblical creationism and Flood Geology. Many scientific references. Includes Arizona Indian drawings of dinosaurs. Drawings; bibliog.

58 Baker, Sylvia. *Bones of Contention: Is Evolution True?* (booklet). 1986.

Phillipsburg, N.J.: Evangelical Press. Orign. in *Evangelical Times* (Surrey, England): 1976. Pub. as booklet in Australia 1980 by CSF. 2nd ed. 1986; titled *Evolution: Bone of Contention. Baker:* studied biology at Univ. Sussex. Standard strict creation-science. Includes moon dust argument, decay of earth's magnetic field, ocean sedimentation rates, population increase extrapolation, former warmth of polar regions, Second Law. Many statements, however, contradict other creationist writings. States that for years, evolution was not popular, due to rival and "equally godless" theory of spontaneous generation. Wilberforce was ignorant, and botched his great debate with Huxley. (But Barker defends him in 2nd ed.) Peking Man was true *Homo sapiens.* Photos, pictures.

59 Baldwin, James Lauer. *A New Answer to Darwinism.* 1957. Priv. pub. (Chicago). "Recent scientific discoveries conclusively refute the Darwinian theory of evolution." Genesis special creation is proved true, but Flood Geology is wrong. "Yom" ('day') in Genesis refers to alternating periods of illumination from catastrophic eruptions, radioactive gases and cosmic electrical discharges (modified Day-Age view). Evolution exists, but it is like ontogeny: a progressive development of species in size and complexity. Appeals to Mendel, St. Augustine, **D'Arcy Thompson.** God created germinal species in electromagnetic matrix under the South Pole; the genes then gradually uncoiled, allowing for species development. Original human organism was a single polyploidal individual (pre-Adamic) in which both sexes were combined. Adam was first real male. Adam's 'rib' was the Y chromosome; genetic material was transferred to the X, to produce a female.

60 Bales, James D. *Man on All Fours* (booklet?). 1973. Priv. pub. *Bales:* Ph.D.—Univ. Calif.; prof. of Christian doctrine at Harding College (Searcy, Ark.). Discussed and quoted in Bales 1975: ". . .if the atheist is right, matter in motion created man with his religious nature. Man is such a religious being that when he turns from God he seeks, sooner or later, some substitute for God." Bales also wrote books on Martin Luther King, the Phoenix Papers, and atheism, which are distributed by Billy James Hargis's Christian Crusade.

61 _______. *The Genesis Account and a Scientific Test: or The Predictive Value of Genesis One Through Three* (booklet). 1975. Priv. pub. (Searcy, Ark.). Scientific predictions from Genesis include: the world was uninhabited prior to life; plants came before animals; lack of transitional fossils; functions for "vestigial" organs; "primitive" tribes are similar in intellect to us; life is not now evolving from non-life; etc. All predictions confirmed by science. Evolution flunks these tests. Seems to allow old-earth creationism (Day-Age or Gap Theory). Includes many scientific references.

62 _______. *Evolution and the Scientific Method* (booklet?). 1976. Priv. pub. (Searcy, Ark.). (May have been orig. pub. before 1976.)

63 _______. *Evolution and the New Inquisition* (pamphlet). Undated. Lufkin, Texas: Tract a Month.

64 Balsam, David. *A Thin Line Between Theory and Fantasy.* 1983. Brooklyn, N.Y.: Systematic Science. "Balsam" is pseudonym of **Josh Greenberger.** 1986 ed. of this book uses real name.

65 Balsiger, Dave, and Charles E. Sellier, Jr. *In Search of Noah's Ark.* 1976. Los Angeles: Sun Classic Books. Over a million copies in print. Also a 1976 Sun Classic feature film. *Balsiger* was historian and technical advisor for film; VP of Anaheim advertising agency; also writes books on Satan. *Sellier* was film's producer and screenplay co-author; also makes Sun Classic family films. Bible is historically accurate. Written, cultural, and scientific evidence for global Flood found worldwide. Reviews early accounts of Ark sightings, and endorses all the standard Ark-eology tales of its discovery on Mt. Ararat. Had Ark model tested in "well-known" hydraulics lab [Scripps Ocean. Inst.] for movie to prove stability. Advocates pre-Flood water vapor canopy. Population increase argument. Emphasizes **Navarra**'s discoveries of wood from the Ark; claims the (non-C-14) lab tests proved it was genuine—UCLA and Penn doctoral students confirmed test results; criticizes C-14 dating. Interviews **Bliss** (now at ICR). Endorses tale of EROS satellite photo of Ark, but debunks earlier claim by Sen. Moss of ERTS NASA evidence. CIA and KGB have satellite photos. Denounces Bart LaRue's illegal Ararat trips (LaRue made rival movie). Relies on **Cummings, J. Morris** and **Montgomery** Ark books; also on **Whitcomb** and **Morris,** and many other creationist works. Photos.

66 Balyo, John G. *Creation and Evolution* (booklet). 1975. Des Plaines, Ill.: Regular Baptist Press. *Balyo:* Cleveland pastor. Mostly quotes from scientists; anti-evolution probability arguments.

67 Baran, Michael. *Twilight of the Gods.* 1984. Smithtown, N.Y.: Exposition Press. The Deluge, Atlantis, UFOs, etc. Common date of 11,000 B.C. reported for these events.

68 Barclay, Vera. *Challenge to the Darwinians.* 1951. Newport, Mon., England: R.H. Johns. Barclay also wrote Danny the Detective series and other children's books.

69 _______. *Darwin Is Not for Children.* 1950. London: Herbert Jenkins.

70 Barnes, Thomas G. *Origin and Destiny of the Earth's Magnetic Field.* 1983 [1973]. El Cajon, Calif.: Institute for Creation Research. ICR Technical Monograph No. 4. Rev. and expanded ed.; orig. 1973. *Barnes:* D.Sc.—Hardin-Simmons Univ.; physics prof. at Univ. Texas El Paso; former Dean of Grad. Study and Research at ICR. Measurements of earth's magnetic field since 1835 show it has weakened over the years. These data are extrapolated—and Barnes assumes the decrease to be exponential—to demonstrate upper limit for age of earth of about 10,000 years, when field would have been impossibly strong. Largely technical, mathematical discussion; many scientific references. Diagrams, charts.

71 _______. *Physics of the Future: A Classical Unification of Physics.* c1983. El Cajon, Calif.: Institute for Creation Research. Urges that physics return to classical models. Rejects relativity and quantum mechanics. "Our business, as **Newton** said, is with the sensible causes of the phenomena." Also praises Maxwell's physics. All forces are reduced to electromagnetic forces and explainable by classical electrodynamics. The nuclear strong force is really magnetic, caused by high spin rate of protons and electrons. Denies mass/energy equivalence, wave/particle duality of light, black holes, and that light has mass. Full of equations. Diagrams.

72 _______. *Space Medium: The Key to Unified Physics.* 1986. Foreword by **H. Slusher,** who compares it with **Newton**'s *Principia.* Rejects Einstein's mediumless relativity. Instead uses the physics of Newton, Faraday, and Maxwell, which provide better explanations than relativistic physics and quantum mechanics. "The abandonment of the concept of a [light-propagating] medium in space is perhaps the greatest mistake of physicists in this century." Describes this medium as reactive, propagative, massless and non-mechanical (unlike the "old" ether theories). Many equations. Diagrams; index. In 1987, **Overn** and five other creationists established the Thomas G. Barnes Inst. of Physics in Minneapolis to promote the return to classical physics (i.e. pre-1920) and to oppose relativity and quantum mechanics. Planned research includes a new test of the Michelson-Morley experiment.

73 Barredo, Josemaria Gonzalez. *The Subquantum Ultramathematical Definition of Distance, Space and Time.* Washington, D.C.: MIAS Press. Rejects Einstein's space-time, relativity, quantum uncertainty and complementarity principles, infinity concept of infinitesimal calculus, set theory, Descartes' concept of number. Proposes that so-called elementary particles are really manifestations of trion configuration interacting with rest of the universe. Claims these principles can be understood by high school students, "especially science fair winners." Book advertisement also adds "denial of the 'evolution' theory." Recommended at 1985 and 1986 Creation Conferences.

74 Barrett, Eric C., and David Fisher (eds.). *Scientists Who Believe.* 1984. Chicago: Moody Press. *Barrett and Fisher:* leaders of RADAS (Radio Acad. of Science), a "growing group of integrated projects using science as a vehicle for evangelistic and pre-evangelistic ministries." RADAS broadcasts to USSR; also in other languages. Scientists who believe include astronauts **James Irwin** and Jack Lousma, Surgeon General C. Everett Koop.

75 Barth, Howard. *Who Made the Earth?* (tract). Undated. Minneapolis: Bible-Science Association. *Barth:* former editor of BSA publications. Strict creation. Barth also wrote BSA tracts *Who Made the Universe?* and *Who Made the Solar System?*

76 Bartoli, Giorgio. *The Biblical Story of Creation: In the Light of the Recently Discovered Babylonian Documents.* c1926. New York: Harper & Brothers. Copyright 1926 by Sunday School Times Co. *Bartoli:* Ph.D.; D.Sc.; D.D.; Italian; a nondenominational Christian; prof. of geology and chemistry; director of a mine in Sardinia. Primarily theological, but includes scientific and archeological arguments. Defends supernaturalism and Genesis against the bitter attacks of "infidel science." "Biblical Concordism, or the effort to reconcile the Bible and science, has utterly failed." "With Genesis stands or falls the whole fabric of Christianity." "The evolutionists, of course, reject the first chapters of Genesis because these flatly contradict their beloved theory..." If man evolved, then God is a liar. The Bible declares that "man is nothing." Most of book is presentation of Gap Theory creationism. Claims that writers of Augustine's time endorsed Gap Theory. God did not create world as Chaos; this Chaos was the result of previous sin of Satan. "Between the first creation, indicated by the first verse, and the description of chaos of the second verse, there occurred a cosmic catastrophe, an appalling cataclysm of worlds, whereby not only our earth was broken up into fragments, but even the solar system was displaced..." God then recreated the world in six days. Bartoli discusses in detail the creation of the angels, the rebellion of Satan and his fallen angels, and their destruction of the world. II Peter refers to this pre-Adamic destruction, not the Flood. The pre-Adamic earth was far superior, as attested to by fossils of giant creatures. Appeals to apocryphal Book of Enoch as supporting Gap Theory, also to Babylonian documents. Flatly rejects evolution. "The missing link between man and beast is still missing, and it will never be found. Fossil man does not exist, and it is useless to look for him. Infidel anthropology is no science at all.... Man was created by God, where and when the Bible tells us."

77 Barton, Jon, and John Whitehead. *Schools on Fire.* 1980. Wheaton, Ill.: Tyndale House. *Barton:* former English teacher; taught Bible as literature at Santa Monica (Calif.) H.S. Includes a section listing creationist organizations (ICR, CRSC, CRS, BSA) with descriptions; also Christian legal organizations.

78 Bartz, Paul A. (ed.). *Our Beautiful World.* 1985(?). Dubuque, Iowa: Kendall/Hunt. Our Science Readers Books series, from Bible-Science Association. *Bartz:* managing editor (also editor of *BSN*). **Nancy Pearcey:** research editor. Norman Hafley, Wes Chase: contributing editors. Bonnie Bartz: design ed. Science Readers series began as monthly leaflets in 1973, then was pub. in 1974 and 1976 in two, then four vols. Warren (Pat) Taylor was editor first two years; **Howard Barth** next two. **Jim** and **Darlene Robinson,** and Ken English were also editors before Bartz became chief editor. This volume for kindergarten. Orig. pub. 1974–76 (?), for kindergarten and grade 1. Title previously used for kindergarten leaflet series.

79 _______. *Our Wonderful World.* 1985(?). Dubuque, Iowa: Kendall/Hunt. Our Science Readers Books series, from Bible-Science Association. This volume for grade 1; orig. (1973) for grades 1–3.

80 _______. *Our Designed World.* 1985(?) Dubuque, Iowa: Kendall/Hunt. Our Science Readers Books series, from Bible-Science Association. This volume for grade 2.

81 _______. *Our Created World.* 1985(?). Dubuque, Iowa: Kendall/Hunt. Our Science Readers Books series, from Bible-Science Association. This volume for grades 3–4. Orig. pub. 1976(?), for grades 2–3.

82 _______. *Our Orderly World.* 1986. Dubuque, Iowa: Kendall/Hunt. Our Science Readers Book series, from Bible-Science Association. This volume for grades 5–6. Orig. pub. 1974–75 for grades 4–6, and/or 4–7.

83 _______. *Our Scientific World.* 1986. Dubuque, Iowa: Kendall/Hunt. Our Science Readers Books series, from Bible-Science Association. This volume for grade 7. Orig. pub. 1974–76(?) for grades 7–9, then for grades 8–12. Alphabetical arrangement of topics, from "Anthropology" (man created in God's image, not evolved from animals; Adam and Eve) to "Young Earth" ("the scientific belief in great earth ages does not rest on solid evidence, but developed historically as a rection against belief in the Bible"). "Where else can you send a student to do research on coal, the Galapagos Islands, or petroleum in the confidence that he can do his research without having evolution subtly thrust into his thinking? And more importantly, *Our Scientific World* supplies the student with important facts which help him see that the Biblical view is an intelligent and better alternative to evolution." Photos, drawings.

84 _______. *Our Amazing World.* 1985(?). Dubuque, Iowa: Kendall/Hunt. Our Science Readers Books series, from Bible-Science Association. This volume for grade 8. The same material as previous volume, with some additions. Includes sections on Francis Bacon, Boyle, G.W. Carver, **Cuvier,** Darwin, Faraday, Kepler, Mendel, **Newton,** Pascal, Pasteur, **Steno;** "Religion and Government," bombardier beetle. Standard creation-science arguments and examples, Flood Geology and recent creation. Photos, drawings.

85 _______. *Our Miraculous World.* 1987. Dubuque, Iowa: Kendall/Hunt. Our Science Books Readers series, from Bible-Science Association. This volume for grades 9–10. Standard creation-science arguments. Alphabetical arrangement of topics; includes classical creationist scientists, dinosaurs, coal formation, fossils, *Archaeopteryx,* biblical "kinds," biblical creation, design in nature, young earth, etc. Strict creationism. Many drawings.

86 _______. *Our Structured World.* c1985. Dubuque, Iowa: Kendall/Hunt. Our Science Readers Books series, from Bible-Science Association. This volume for grades 11–12. Bible-science: emphasizes Flood and creationism; full of anti-evolution material. Topics and material mostly identical with series volume for grades 9–10.

87 Barzun, Jacques. *Darwin, Marx, Wagner: Critique of a Heritage.* 1958 [1941]. Garden City, N.Y.: Doubleday Anchor. 2nd ed.; orig. 1941. *Barzun:* history prof. and dean of Grad. Faculties at Columbia Univ. Darwin, Marx, and Wagner epitomized the scientist, social theorist, and artist of their time, and they dominate 20th century attitudes. Their scientism and mechanical materialism has "seemingly made final the separation between man and his soul." Barzun, a romantic, strives to emphasize what Darwin did *not* prove, and criticizes his "obscure" writing style. Stresses the great indebtedness of all three to tradition and to previous theorizers—which Barzun claims none of them ever properly acknowledged. This unsympathetic analysis is often cited by creationists.

88 Baugh, Carl E. *Dinosaur: Scientific Evidence That Dinosaurs and Men Walked Together.* c1987. Orange, Calif.: Promise Publishing. Written "with" **Clifford Wilson.** Foreword by **R.L. Whitelaw.** "Special edition for The Southwest Radio Church" (Oklahoma City). *Baugh:* minister; head of International Baptist College (Missouri); archeology M.A.—Pacific Coll.; anthropology Ph.D.—Coll. of Advanced Education. Baugh is the most notorious Paluxy investigator and promoter. Baugh and many other creationists claim that the celebrated dinosaur tracks in the Paluxy River (Texas) also show contemporaneous human footprints. To avoid charges that all the mantracks were carved, Baugh excavated new tracks and invited the media and various experts to confirm their authenticity. Excavations were assisted by artist Robert Summers (best known for statue of John Wayne) and Baptist pastor Charles Hiltibidal. Baugh notes endorsements of his mantracks by **Hinderliter,** geologist John DeVilbiss (head of Office for Research on Origins), local geologist Billy Caldwell, James Hall of Liberty Univ., chemist Hugh Miller, Russell Humphreys, aerospace engineer R. Helfinstine, **D. Patten, Joe Crews, A.E. Wilder-Smith,**

and other creationists. Baugh christened the makers of the mantracks "Humanus Bauanthropus" (named, he says, after the royal Bau tribe of Fiji). Discusses biblical evidence of pre-Flood giants (many of the mantracks are oversize) and of dinosaurs. Noah took dinosaurs aboard the Ark; they survived for a while after the Flood, as attested to by the evidence from myth and folklore. Describes reports that brontosaurs still survive in the Congo—the creatures called "Mokele-Mbembe" by the natives. Also discusses a "Russian Paluxy": Soviet discoveries in Turkemenia of mantracks among dinosaur prints. Expresses suspicion of G. Kuban's refutations of the mantracks—points out that the newly-discovered stains which prove dinosaurian origin might have been artificially produced. Concludes with dramatic account of discovery of human tooth (dismissed by non-creationist geologists as tooth from a Cretaceous fish). "The authors of this book are not surprised at this strong evidence for creation. After all, they stand on the authority of Jesus Christ Himself Who declared of *all* Bible truth, including creation and the flood, 'Thy Word is Truth'. . ." Photos. Baugh and Wilson also co-authored 1987 book *Tracks Step on Evolution* (Oklahoma City: Hearthstone), which may be another version of this.

89 Baugh, Steven M. *A Comparative Study of The Genesis Creation Account—Myth or Marvel?* Undated. Priv. pub.(?). Apologetics; a response to "higher criticism" of the Bible, especially concerning the two creation stories in Genesis, and comparison with Mesopotamian myths.

90 Baurain, Thomas S. *The Bible and Science: Pace II 500001* (booklet). c1980. DFW Airport, Texas: Reform Publications. 34-page workbook from Accelerated Christian Education. "Self-Pac" of Basic Education, for supplementary course from ACE. Fundamentalist college science curriculum. Reading list includes 59 creation-science and no evolutionist books. **H. Morris, Gish, Frair** and Davis, **Rice** are required reading. **Whitcomb** and Morris, **Shute, Klotz, Howe** are required references. Includes test questions: "atheism, humanism and evolution" equals "false philosophies"; the greenhouse effect is produced by the (pre-Flood) water-vapor canopy; and on biblical creation. Baurain says he "rejects evolution in any form whatsoever." (Apparently there is separate booklet for PACE I.)

91 Baylee, Joseph. *Genesis and Geology*. 1857. Cited by Millhauser as advocating Gap Theory creationism.

92 Beardslee, Kenneth R. *A Challenge to Michigan Public Schools for the Teaching of Theories of Beginnings* (report). 1976. Priv. pub. Distributed in Jackson County, Mich.: churches and public places. Includes a questionnaire on creation/evolution two-model teaching, which the great majority of respondents favored.

93 ________. *Creation-Evolution and the Christian* (report). 1978. Unpub.(?) Report about teaching of creation-science. Advocates two-model approach; includes questionnaire page for responses. Quotes lots of creation-science. From Jackson County, Mich.

94 Beasley, Walter J. *Creation's Amazing Architect*. 1958 [1955]. London: Marshall, Morgan & Scott. Orig. 1955 (Melbourne: Australian Institute of Archeology). Also pub. 1955 by Gospel Literature Service (Bombay, India). Modern Science and the Bible Series No. 1. "How the Modern Science of Geology Was Anticipated by Nearly 3,500 Years." *Beasley:* F.R.G.S.; president of Australian Inst. Archaeol. Day-Age creationism: correlates and reconciles the geological ages and science with the Bible. Includes charts comparing the creation 'days' with the geological record. Lack of tree rings prior to Mesozoic proves that atmosphere was not cleared to reveal sun until then (day 4). Day 5: Mesozoic. Day 6: Cenozoic. Genesis rules out chance, purposeless evolution and proclaims a predetermined creation. The Great Architect planned earth for man's habitation from the very earliest ages. He designed natural resources for men to exploit. Scientific and biblical references. Drawings.

95 Beck, Horst W. *Biologie und Weltanschauung: Gott der Schöpfer und Vollender, und die Evolutionskonzepte des Menschen*. 1979. Neuhausen-Stutgart, Germany: Hänssler-Verlag. Wort und Wissen series, No. 1. (Wort und Wissen is also a German creation-science organization.) *Beck:* theologian at Univ. Basel. Other volumes planned for series by J. Scheven, D. Bierlein, P.C. Haegele, **W. Gitt, H. Slusher,** and H. Schneider.

96 ________ **(ed.).** *Die Debatte um Bibel und Wissenschaft in Amerika.* 1980. Neuhausen, Germany: Hänssler-Verlag. Wort und Wissen series, No. 8. (Mentioned in **Schirrmacher,** in connection with the emergence of the German creationist movement and its contacts with U.S. groups.)

97 Beierle, Fredrick P. *Giant Man Tracks* (booklet). 1974. Priv. pub. (Prosser, W.V.: Perfect Printing). *Beierle:* worked in nuclear industry. Giant human footprints amidst the Paluxy River dinosaur tracks. Beierle is an active Paluxy investigator.

98 ________. *Man, Dinosaur and History.* c1980. Priv. pub. (Prosser, W.V.: Perfect Printing). Foreword by **T. Barnes.** Man and dinosaur prints in same strata at Paluxy River near Glen Rose, Texas, proving that strict creationism is true and evolution false. Vindicates Bible account of Noah's Flood; shows that earth is 7–10,000 years old. Detailed account of many of the creationist investigations and digs; much background information. Also presents several other anomalous fossil finds from other parts of the country which similarly confound evolutionary chronology. Quotes **Corliss** at length. Many photos (including some of prints that are obviously carved). Beierle claims that "very little has been published" about these strange human tracks, even though they have been known for sixty years—attributes this to evolutionary conditioning. Discusses and reproduces R. Berger's UCLA C-14 analysis dating Paluxy wood at 12,800 yrs. old. Lots of photos.

99 Bell, Sir Charles. *The Hand, Its Mechanism and Vital Endowments, as Evincing Design.* 1837 [1833]. London/Philadelphia: Pickering. One of the Bridgewater Treatises, the famous series of eight volumes on natural theology. Leading British clerics and scientists (and many were both) were commissioned to produce books "On the Power, Wisdom, and Goodness of God, as manifested in the Creation," in the tradition of **Paley,** from a fund bequeathed by the Earl of Bridgewater. The Bridgewater Treatises are classic statements of the Argument from Design, which is still considered conclusive by creationists. None of the treatises advocate strict young-earth creationism, though. *Bell:* surgery and anatomy prof. at Univ. Edinburgh; surgeon at Waterloo; made important discoveries about nerve function. In this book, Bell stresses that the fossil record shows there have been separate sets of organisms in each successive geological age. "But the animals of the Antediluvian world were no monsters": each type was perfectly adapted to the particular conditions of its age. Man was created last of all.

100 Bellamy, Hans Schindler. *Moons, Myths, and Man.* 1937 [1936]. London: Faber & Faber. Rev. ed. 1937; also 1949. Also pub. 1936 by Harper (New York). Bellamy, a mythologist, was an English follower of the Austrian Hans Hörbiger, whose "cosmic ice theory" became a Nazi pseudo-science cult. Asserts that planetary and lunar orbits all slow down. Various small planets have been captured by the earth, becoming moons; these eventually fall into earth. Our present moon was captured 13,500 years ago, as proved by myths and racial memories. Collapse of earth's Tertiary moon caused Noah's Flood: a catastrophic ice rain and release of the immense "girdle tide" which had been raised up by the moon's low, rapid orbit. Paradise myths date from period of geological calm between moons. Bellamy claims that myths accurately reflect historical events. (Discussed and quoted in Gardner.)

101 ________. *A Life History of Our Earth.* 1951. London: Faber. Bellamy's presentation of Hörbiger's Cosmogonic (Cosmic Ice) Theory. The geological ages, and evolution, are caused by the various moons the earth has captured. A new moon is captured each age. Its orbit shifts; it becomes stationary over the earth; finally it disintegrates. Tectonic activity results only from moon effects. Magma and air tides pile up; anchorage masses are raised below the stationary orbit; rift systems form from the escape of moons, ore deposits from their collapse. Drastic rotation and orbital change cause ice ages and fossils. These catastrophes cause evolution: "life is what the Moons have made it." Mesozoic man escaped the tide hills of a former moon; Atlantis was destroyed by such a catastrophe. (Bellamy also wrote separate book on Atlantis.)

102 ________. *In the Beginning God.* 1945. London: Faber & Faber. "A new scientific vindication of the cosmogonic myths in the Book of Genesis." Earth has had series of moons, which slow and plunge into earth. Genesis is account of re-creation (not original creation) after last such catastrophe. Interprets Eve coming from Adam's rib as reference to Caesarean birth (the myth mixed up the sexes). In another book, *The Book of Revelation Is History,* Bellamy says the Apocalypse vision of St. John is a factual account of the Tertiary catastrophe. (Discussed in Gardner.)

103 Belloc, Hilaire. *Companion to Mr. Wells's Outline of History.* 1927 [1926]. San Francisco: ESA. Orig. 1926 (London: Sheed and Ward). Belloc, a Catholic, here refutes H.G. Wells' strongly evolutionist *Outline of History* (1920). According to Gardner, most of Belloc's arguments "are so ancient and flimsy that not even **Price** had the courage to exhume them." Wells demolished this attack in *Mr. Belloc Objects* (1926). (Also discussed in **Bergman** 1984.)

104 ________. *Mr. Belloc Still Objects to Mr. Wells's Outline of History* (pamphlet). 1927 [1926]. San Francisco: ESA. Orig. 1926 (London: Sheed and Ward). Belloc's counter-response to Wells' refutation. (Discussed in Gardner; **Bergman** 1984.)

105 Bennet, James E. *The Bible Defeats Atheism.* c1941. Grand Rapids, Mich.: Zondervan. *Bennet:* presented Bible School Lessons on radio in the 1920s, by mail in the '30s, and later in the *Christian Beacon.* This book is: "The story of the famous **Harry Rimmer** trial as told by the attorney for the defendant, James E. Bennet." Plaintiff brought breach of contract suit against Rimmer, who had offered $1,000 reward for proof of any scientific error in the Bible. (Also described in Rimmer's *That Lawsuit Against the Bible.*)

106 ________. *Evolution and Religion.* Western Voice. Also wrote *Atom and Evolution.*

107 ________. *In the Beginning God....* Undated. Priv. pub. (New York, N.Y.) Orig. pub. as Bible School Lessons in *Christian Beacon* (starting with 1958 issue). Exegesis of Genesis. This, "being a record revealed by God Himself through holy men of old, constitutes the only strictly scientific answer to the question of origins." "The only way that man can know anything at all about his creation is to have it revealed to him by his Creator." Advocates Gap Theory creationism. "Something terrific must have occurred in the period that must have elapsed between the records of the first and second verses. Personally, I believe that all the 'geological ages' did actually occur during those ages..." Plant seeds survived the destruction of the original creation, but animal life was re-created in the reconstruction. Literal interpretation of Genesis. Strongly denounces theory of evolution as materialist, atheist, and entirely devoid of proof. "There is no possibility of a spiritless beast evolving into a spirit-directed man." Stresses role of Satan, coming rapture and destruction of the world. Also rejects Freud's "evil theory of 'id'." Worldwide Flood, about 1550 years after six-day creation, destroyed all mankind except Noah. "If there were no Noah and no flood, then Jesus was completely wrong." Noah was 10–15 feet tall; therefore cubit (and hence Ark) was correspondingly larger.

108 Bennion, Howard S. *The Creation and Age of the Earth and the Origin of Life Thereon As Told by Its Creator* (booklet). 1970. Priv. pub.(?). *Bennion:* World War One veteran; engineer and research for Edison Electric (New York); later a Mormon patriarch in Utah. Largely a reprint of a 1923 article. Refers to *Book of Mormon, Pearl of Great Price,* and *Doctrine and Covenants;* Mormon teachings of God's descent from Planet Kolob, etc.

109 Benson, Clarence H. *The Earth – The Theatre of the Universe: And a Scientific and Scriptural Study of the Earth's Place and Purpose in the Divine*

Program. 1938 [c1929]. Chicago: Moody Press. Orig. c1929 by Bible Institute Colportage Assoc. (Moody). Some sections orig. in *Moody Monthly*. Bible and science "must harmonize." Book is mostly a strong presentation of Gap Theory creationism. Passing star caused break-up of a planet of which asteroids are remnants. This catastrophe associated with Satan's Fall. Also endorses Flood Geology of **G.M. Price,** however. "No fact in the world's history is better substantiated than the Deluge, and nowhere do men show their ignorance and their folly more conspicuously than when they attempt to forget it, for they cannot consistently deny it. In the first place there is the biblical account which we must believe even if it were not substantiated elsewhere." "If the Deluge goes, the Bible goes." Includes Flood legends from around the world. Suggests Deluge resulted from change in earth's axis. Also includes emphasis on astronomy: the Gospel in the stars. Diagrams.

110 _______. *Immensity: God's Greatness Seen in Creation*. c1937. Chicago: Scripture Press. Introduction by **L.A. Higley.** Companion volume to *Earth—The Theatre of the Universe*. Some chapters orig. in *Moody Monthly*. The vastness of the universe, which bespeaks of the Creator's might. Scientific and biblical references. Photos, diagrams.

111 _______. *The Greatness and Grace of God*. 1953. Chicago: Scripture Press. "Conclusive Evidence that Refutes Evolution: arranged to be used as a textbook in Christian Evidences."

112 Berg, Leo S. *Nomogenesis, or Evolution Determined by Law*. 1969 [1922]. Cambridge, Mass.: MIT Press. Orig. 1922. First English ed. 1926 (London: Constable). Introduction by **D'Arcy Thompson.** Foreword to 1969 ed. by T. Dobzhansky. *Berg:* prof. at Univ. Leningrad; "outspoken critic of Darwinism." Stresses importance of Law as opposed to Chance in biology. Proposes that organisms have developed polyphetically: from tens of thousands of primary forms. Subsequent evolution was not primarily divergent, as in Darwinism, but chiefly convergent. It is based on law (nomogenesis), not on chance variations. It proceeds by "leaps, paroxysms, mutations" rather than slow gradual variation. Presents many examples of convergence. Octopus has the same kind of eye as vertebrates; different grains produce the same set of varieties; etc. (Discussed in **Rendle-Short** 1942; unpub. **Frair** bibliog.)

113 Bergman, Jerry. *Teaching About the Creation/Evolution Controversy* (booklet). 1979. Bloomington, Ind.: Phi Delta Kappa Educational Foundation. Phi Delta Kappa Fastback series #134. *Bergman:* sociology and biology B.A., educ. psych. M.A. and evaluation and research Ph.D. from Wayne State Univ.; ass't. prof. of educational foundations and inquiry at Bowling Green State Univ. (Ohio). (Bergman told me Phi Delta Kappa didn't realize he was a creationist when they asked him to write this.) Discusses creation and evolution as different "belief structures" impossible to prove by science. Also discusses assumptions underlying evolution; dogmatic evolutionism in textbooks; limitations of empiricism. Favorable references to creation-science books. Advocates two-model approach in schools to insure fairness and objectivity.

114 _______. *Mankind—The Pinnacle of God's Creation* (pamphlet). 1984. El Cajon, Calif.: Institute for Creation Research. ICR "Impact" series #133. Wonder and design of the human body—couldn't have just happened by natural selection and random mutations. Bergman also wrote "Impact" #144: *The Earth: Unique in all the Universe,* and hundreds of other articles on diverse topics in diverse journals. Many pop up in several places. He has published in humanist and atheist journals as well as many creationist journals, plus scores of obscure periodicals.

115 _______. *The Criterion: Religious Discrimination in America*. 1984. Richfield, Minn.: Onesimus Publishing. Foreword by **Eidsmoe;** preface by **W. Bird.** Bergman interviewed over a hundred creationists with advanced degrees; all reported academic discrimination. Documents and/or alleges cases of discrimination: denials of admission, promotion, or tenure; firings, ridicule, death threats, etc. Names not given in many cases due to fear of further discrimination; many cases anecdotal. Protests against discrimination on basis of religious belief, and asserts superior performances of victims. Appendix, by **L. Sunderland** (1983 reprint), mostly concerns Bergman's own case. He was denied tenure at Bowling Green State Univ., allegedly for creationist beliefs, and was defended by the Creation Science Legal Defense Fund. (However, in a letter to a white supremacist journal, Bergman claimed he was denied tenure for racial as well as religious reasons—reverse discrimination for being a white male.)

116 Bergson, Henri. *Creative Evolution.* 1944 [1911]. New York: Modern Library. Orig. 1911; French. There is no transcendent Designer controlling the development of life, but evolution is a striving upwards of the creative life-force, the *élan vital.* Bergson's is the best-known of the vitalist theories, which deny the adequacy of materialist evolution.

117 Berkhof, Louis. *Systematic Theology.* 1955. Grand Rapids, Mich.: William B. Eerdmans. Cited in **Whitcomb** and DeYoung as a theological work which has "contributed significantly to the current renaissance of Biblical creationism," advocating the full historicity of the Genesis account and recent creation.

118 Berlitz, Charles. *Doomsday 1999 A.D.* 1981. Garden City, N.Y.: Doubleday. With collaboration, maps, and drawings by J. Manson Valentine. *Berlitz:* grandson of Berlitz language school founder; author of *The Bermuda Triangle* (1975). Ancient prophecies and modern physics show that world will be destroyed in 1999. Includes a chapter on worldwide Flood legends. Also chapter on Noah's Ark on Mt. Ararat; relies on **LaHaye** and **Morris, Balsiger** and Sellier, **Navarra,** and other creationist Ark-eologists. Includes new unpublished report of G. Schwinghammer, a U.S. fighter pilot stationed in Turkey, and repeats many other Ark-eology tales. Noah's Flood, and Atlantis and other myths, prove that the world has been destroyed before.

119 ________. *The Lost Ship of Noah: In Search of the Ark at Ararat.* 1987. New York: G.P. Putnam's Sons. Discusses ancient and modern claims of sightings of the Ark, with information about most of the recent expeditions. Includes many interviews of principal Ark-eologists from Berlitz's 37-year pursuit of Ark and Flood stories. Much of book's material provided by Ahmet Ali Arslan, a Turk who has interviewed and served as guide for many Ark searchers. Worldwide legends of the Flood are racial memories of great catastrophe which a few people survived in boats. Noah's Ark may be one such vessel. The Flood may have been caused by tilting of earth's axis (perhaps from close approach of Venus or a comet), resulting in immense tidal waves, disloding of polar ice, and violent upheavals. Equates Flood with catastrophe which occurred 11–12,000 years ago, causing sudden overflow of oceans and mass extinctions. Surviving species underwent a "sort of instant evolution." Discovery of a ship on Ararat would confirm all this. Gives a favorable review of the story of the three atheist scientists who set out to disprove the Ark rumors and thus the Bible, and were furious when they found the Ark. Implies that Genesis account is scientifically valid. Discusses the ship-like formation near Ararat believed by some Ark-eologists to be the Ark (but contemptuously dismissed by others). Investigator Fasold, quoted at length, says it is the remains of the Ark, which was a reed boat covered with cement made of asphalt with pumice and pitch. Cites account of atomic war in the *Mahabharata,* the Flood story in the Koran, **Velikovsky, Sitchin,** Hapgood's ancient maps, Hal Lindsey's Bible prophecy, Nostradamus and Cayce. Urges readers to take Noah's warning seriously. Photos, drawings; bibliog., index.

120 Berry, R.J. *Adam and the Ape: A Christian Approach to the Theory of Evolution.* 1975. London: Falcon Books. Criticizes both "Fundamentalists" (e.g. **H. Morris**) and "liberals" (especially **Teilhard**). Cites **D. MacKay** approvingly for reconciling biblical and scientific truth. Defends *ex nihilo* creation (though not a literalist). Man has long biological history, but Man (Adam) was created theologically also—perhaps in the Neolithic. God endowed Adam with spirituality. Accepts evolution, however:

"Widely-quoted criticisms such as **Kerkut** or **Moorhead** and Kaplan are largely about details." Evolutionist research will modify theories, but these changes will not deny evolution. Conservative theistic evolution, but similar to old-earth creationism.

121 Besant, Annie. *Riddle of Life: And How Theosophy Answers It.* 1911. London: Theosophical Publishing House. "An occult overview of creation and evolution." *Besant,* active in Indian politics, became the leader of the Theosophical Society after **Blavatsky.** Seven "Root Races." The first "Root Race" was invisible and ethereal "fire-mist" people. The second had astral bodies. The third were ape-like giants living in Lemuria. A sub-race escaped to Atlantis when Lemuria sank. Aryans, the fifth Root Race, came from the fifth sub-race of Atlanteans. The sixth Root Race is emerging from the sixth sub-race of Aryans in Southern California.

122 ________. *The Secret of Evolution* (pamphlet). 1930s. Harrowgate, England: Theosophical Publishing.

123 Best, Stan. *The Battle for the Schools.* c1982. Paso Robles, Calif.: Final Press. *Best:* missionary in Brazil; writes Christian children's adventure books set in the Amazon. In this book, Best describes the decline of strict Christian schools and Bible institutes. "Unitarianism, rationalism, evolution and the like had worked devastation on formerly dependable schools." Humanism and Darwinism have entered theological thought. Discusses Christian Heritage College situation: CHC (home base of ICR) foregoes WASC accreditation in order to retain biblical integrity and creationist emphasis. Quotes **H. Morris** that CHC was seriously hurt by unfavorable WASC evaluation in 1978.

124 Bettex, Frederick. *The First Page of the Bible* (booklet). 1926 [1903]. Burlington, Iowa: German Literary Board. Orig. German. *Bettex:* Lutheran. Day-Age creationism.

125 ________. *Modern Science and Christianity.* 1903. London: Marshall. Orig. German. Also pub. by Jennings and Pye (Cincinnati; 1901). Strongly anti-evolution. Large Bible-science treatise. Condemns evolution as atheist, demoralizing theory. Criticizes science and materialism. Chapters include "Progress," "Evolution and Modern Science," "Christians and Science," "Science," and "Materialism." Index.

126 ________. *The Six Days of Creation in the Light of Modern Science* (booklet). 1924 [1916]. Burlington, Iowa: Lutheran Literary Board. A "compact yet thorough expression of belief in creation" (**J.N. Moore** in 1964 *CRSQ*). Day-Age creationism.

127 Bhaktivedanta Institute. *Origins —Higher Dimensions in Science* (booklet). 1984. Los Angeles: Bhaktivedanta Institute. Handsome, nicely illustrated magazine format, from the International Society for Krishna Consciousness (ISKCON): the "Hare Krishnas." Solidly anti-evolution, but, as reincarnationists, ISKCON emphasizes rather than rejects mankind's kinship with the animals. They deny that life could arise by itself, and reject evolution as a materialist denial of pre-existing spirit. Includes "Big Questions about the Big Bang" (design of the universe); "The Mystery of Consciousness" (non-mechanistic Supersoul); "Life from Chemicals: Fact or Fantasy?" (couldn't arise by chance); "A New Look at Evolution" (standard creation-science arguments); "The Record of the Rocks" (modern man can be found all the way back; surviving Neandertals); "Higher Dimensional Science" (the Vedas were right all along; Vedic "inverse evolution" or unfolding from original superior ancestor). Color photos and illustrations on every page.

128 Bhaktivedanta Swami, A.C. *Life Comes from Life.* 1979. Los Angeles: Bhaktivedanta Book Trust. From taped discussions with Bhaktivedanta Swami Prabhupada in Los Angeles, with T.D. Singh, the director of the Bhaktivedanta Inst. Singh: organic chem. Ph.D. Castigates science for foolish materialist bias and for not recognizing Spirit. If evolution is true, why can't science create life? Scientists are tiny-minded idiots. Asserts mind/body dichotomy. Harangues against belief that life could originate without prior Mind, Spirit. Many Vedic references.

129 Bible-Science Association. *The*

Creation Alternative. 1970. Caldwell, Idaho: Bible-Science Association. Reprints of ten articles from the *Bible-Science Newsletter.* Co-edited by **Walter Lang** and **V. Raaflaub.** The Bible-Science Association was founded in 1963 by Rev. Walter Lang, who also began its *Newsletter* that year. The BSA is the second-oldest active U.S. creationist organization (after CRS), and its newsletter the oldest active creation-science publication. The BSA also claims to be the largest creationist organization, based on *BSN* circulation. This volume contains articles by **Chittick, Davidheiser,** Willard Ramsey, **McCone, Slusher, Burdick, Mulfinger, J. Read,** and Lang, on standard creation-science topics.

130 _______. *A Challenge to Education.* 1972(?). Bible-Science Association. Essays from 1972 BSA Creation Conference (Milwaukee). (**Lang** apparently edited most or all of the volumes of papers from the various creation conferences, though references vary. Some are listed under **Lang.**) Lang says publication was assisted by Gerald Mallman, a Kenosha, Wisc. science teacher.

131 _______. *Are You a Creation Evangelist?* (tract). 1973. Caldwell, Idaho: Bible-Science Association. The BSA has published numerous creationist tracts. 1973 and 1974 tracts listed here give no author, but were all "edited" by Walter Lang. This one argues a standard BSA theme: that many "who cannot be reached in conventional ways can be reached through an approach of Bible-Science relationships. In evangelism the goal is to reach people. Because the creationist approach provides an opportunity to reach people, it is an effective tool for evangelism."

132 _______. *Can Life Be Created Through Science?* (tract). 1973. Caldwell, Idaho: Bible-Science Association. No, it can't—but you can have eternal life.

133 _______. *Dinosaurs and Sin* (tract). 1973. Caldwell, Idaho: Bible-Science Association. Dinosaurs became extinct because of human sin. Evolutionists have trouble explaining why big animals came first, then smaller ones, because that "is not good evolution." Impossible to find absolutes in science because our minds and our nature are no longer perfect.

134 _______. *Does Time Have Creative Powers?* (tract). 1973. Caldwell, Idaho: Bible-Science Association.

135 _______. *Is Noah's Ark on Mt. Ararat?* (tract). 1973. Caldwell, Idaho: Bible-Science Association.

136 _______. *Is the World Really 4.5 Billion Years Old?* (tract). 1973. Caldwell, Idaho: Bible-Science Association.

137 _______. *Is This Your Ancestor?* (tract). 1973. Caldwell, Idaho: Bible-Science Association. Debunks proposed human ancestors: all frauds or modern humans. Usual cast of characters. Msgr. **O'Connell** claims that Peking Man is a hoax also—its "lost" bones are not really lost.

138 _______. *The Stones Cry Out* (tract). 1973. Caldwell, Idaho: Bible-Science Association. Creationism is a simpler explanation than overthrusting for out-of-order strata. Focuses on Chief Mountain in Glacier National Park.

139 _______. *Violence in the Bottom of the Grand Canyon* (tract). 1973. Caldwell, Idaho: Bible-Science Association. Pine and oak pollen found in Hakatai Shale of Grand Canyon, supposedly 600 million years old. This refutes evolutionary sequence. Noah's Flood explains the tremendous violence.

140 _______. *What You Should Know About Entropy* (tract). 1973. Caldwell, Idaho: Bible-Science Association. What you should know is that things run down, contrary to what evolution claims.

141 _______. *Dinosaurs and Catastrophe* (tract). 1974. Caldwell, Idaho: Bible-Science Association. Examples of dinosaur graveyard sites suggesting catastrophes.

142 _______. *Grand Canyon and the Bible* (booklet). 1974. Caldwell, Idaho: Bible-Science Association. Issue of *Five Minutes with the Bible & Science* series, a "daily reading magazine." The series was later incorporated into the *Bible-Science Newsletter.* Photos.

143 _______. *How Old Is the Grand Canyon?* (tract). 1974. Caldwell, Idaho: Bible-Science Association. Geological unconformities—hundreds of millions of years missing in Grand Canyon strata, according to evolutionists. Real age is

4–5,000 years old; formed by biblical Flood.

144 _______. *Human Footprints and Dinosaur Tracks* (tract). 1974. Caldwell, Idaho: Bible-Science Association. Both found together at Paluxy River (Glen Rose, Texas), thus refuting evolutionary sequence.

145 _______. *Is a Mountain Able to Witness for Christ?* (tract). 1974. Caldwell, Idaho: Bible-Science Association. Mt. Ararat is, because Noah's Ark is on it.

146 _______. *Lessons Taught by Minifossils in the Grand Canyon* (tract). 1974. Caldwell, Idaho: Bible-Science Association. Clifford Burdick found Precambrian pollen, thus refuting evolutionary sequence. Photos.

147 _______. *What Is the Age of this Lava Flow?* (tract). 1974. Caldwell, Idaho: Bible-Science Association. Radiometric dating is false: it gives old date for recent lava flow.

148 _______. *Would You Become a Bible-Believer if You Saw Noah's Ark on Mount Ararat?* (tract). 1974. Caldwell, Idaho: Bible-Science Association. Some of us would; some wouldn't (too biased). Recounts several tales of discoveries of Noah's Ark.

149 _______. *Fossil Men*. 1975. Caldwell, Idaho: Bible-Science Association. *Five Minutes with the Bible & Science* series. Cites **F. Cousins** 1966 and other creationists. Many biblical references.

150 _______. *Creationists: The Better Scientists* (pamphlet). 1978. Caldwell, Idaho: Bible-Science Association. Based on 1974 BSA pamphlet with same title. Subject is **T. Barnes**'s keynote address (same title) to 1974 Creation Conference. Maxwell united electricity, magnetism and optics, but left out gravity because it had negative energy. Einstein, however, tried to include it—but his relativity is of doubtful validity. Kelvin's "irrefutable" argument of earth's rate of heat loss proves earth less than two million years old. Nobel prizes go to older scientists now, because younger ones are evolutionists and not as productive. Gabor, e.g., won in 1971 for work done in 1940s; Gabor "does not accept" the theory of evolution. Creationist **Lammerts** developed better plant breeds more quickly than a rival UC Davis team, because they worked on evolutionist premise.

151 _______. *Repossess the Land*. c1979. Minneapolis, Minn.: Bible-Science Association. Essays and Technical Papers from 15th Anniversary Convention of Bible-Science Assoc. (Anaheim, Calif.). Sponsored by CSRC and San Fernando Bible-Science Assoc. (**Lang** 1984 says edited by Lang.) Papers by **Ackerman**, A.S. Anderson, H.L. Armstrong, J. Baumgardner, **Beierle,** V. Bigelow, **Bliss,** R.H. Brown, **Burdick,** D. Caster, **Chittick, Coffin,** D. Coppedge, G. Croissant, **Davidheiser,** D. Dean, **Deen, Faulstich,** G. Graf, **Hanson,** T.R. Ingram, D. Kaufmann, **Kofahl,** R. Koontz, **Lammerts, Lang,** P. Leithart, **Lubenow, McCone, Mulfinger, Nafziger, E. Nelson, Northrup, H. Otten, Overn, Read, Rusch,** J. Scheven, M. Tippets, **Whitehead, Whitelaw, E. Williams, P. Zimmerman.** Biog. notes included. Variety of standard creation-science topics.

152 _______. *Evolution: Fact or Fiction?* (pamphlet). 1981. Minneapolis: Bible-Science Association. "Evolutionists must believe that life exists on other planets for, if life evolved on earth purely through a series of accidents, opportunities for evolution abound in the vast reaches of space." But now we know there is no life on Mars. Other standard creation-science arguments also.

153 _______. *Science at the Crossroads*. 1984. Minneapolis, Minn.: Bible-Science Association. Essays from 1983 BSA Creation Conference.

154 _______. *Proceedings of the 11th Bible-Science Association National Conference*. c1985. Minneapolis, Minn.: Bible-Science Assoc. Collection of papers from conference held in Cleveland, 1985. Includes **Arndts, Bartz, Bergman** (censorship and discrimination against creationists), **Davidheiser** (evolution and ethics), **Deen** (criminal justice), **Fish** (creationism as a mission), **Frair** (turtle biochem.), **Hedtke** (roots of evolutionism), D. Kaufmann (animal adipose phylogeny), **Ouweneel** (evolution and humanities, with diagrams), **Kofahl** (new definition of science), **Overn** and Arndts (radiometric dating critique), and **Lang.**

Elmendorf's geocentric presentations are not included; nor is the formal debate on geocentricity, though it was the main event of the Conference.

155 Biblical Creation Society. *Life on Earth: Evolved or Created?* 1979. Glasgow, Scotland: Biblical Creation Society. Special issue of *Biblical Creation* (Journal of the Biblical Creation Society) devoted to 11-10-79 London conference. Includes **E. Andrews, S. Baker, D. Watts,** Keith Stokes, **D.B. Gower, N. Cameron.** Standard creation-science topics. The BCS, founded in 1978, is one of the major British creationist groups.

156 Bird, Wendell R. *Freedom of Religion and Science Instruction in Public Schools* (booklet). 1978. *Yale Law Journal* reprint. Reprint from 1978 *Yale Law Journal:* widely distributed and cited by creationists. *Bird:* summa grad. of Vanderbilt Univ. (first student exempted freshman year), majored in history, econ. and politics; editor of *Yale Law J.,* won Egger Prize for this article, which was written under supervision of Yale Law prof. R. Bork; later staff attorney at ICR; now with Atlanta law firm and Rutherford Inst.; chief counsel for Louisiana creation-science case, argued it before the Supreme Court. Argues here that exclusive teaching of evolution advances "other" religions (those founded on or which affirm evolution), and violates neutrality by advancing humanism while denigrating fundamentalist beliefs. No compelling state interest in teaching evolution to override these violations of freedom of religion. Only fair and workable remedy is to teach scientific creationism whenever evolution is taught to restore neutrality. Comprehensive legal argument; massively footnoted. Scores of creationist references and legal citations.

157 ________. *Evolution in Public Schools and Creation in Students' Homes: What Creationists Can Do* (pamphlet). 1979. El Cajon, Calif.: Institute for Creation Research. ICR "Impact" series #69 and #70. Creationists can read ICR books, read Bird's law articles, urge adoption of ICR and CRS texts in schools, get involved with curriculum and textbook adoption processes, and petition boards of education to pass "balanced treatment" resolutions. But be careful, warns Bird: push for scientific creationism, not biblical creationism.

158 ________. *Freedom from Establishment and Unneutrality in Public School Instruction and Religious School Regulation* (booklet). 1979. *Harvard Journal of Law and Public Policy* reprint. Widely distributed and cited by creationists. Argues that creation-science is religiously neutral, and that exclusive teaching of evolution violates neutrality because evolution is the foundation of many religions (humanism, liberal denominations, etc.). Criticizes tripartite test used to determine if laws violate establishment of religion clause, but also argues that creation-science is not unconstitutional by that test. "Scientific creationism is as scientific and as nonreligious as evolution. Scientific creationism must be distinguished from biblical creationism." Massively footnoted with legal citations.

159 ________. *Resolution for Balanced Presentation of Evolution and Scientific Creationism* (pamphlet). 1979. El Cajon, Calif.: Institute for Creation Research. ICR "Impact" series #71. Sample resolution to be adopted by boards of education. Calls for teaching of creation-science if and when evolution is taught. Teaching scientific evidences for creationism does not violate Constitution's prohibition against establishment of religion, but teaching only evolution does. Introductory note stresses that this is intended as a "resolution" and not legislation. ICR favors "education and persuasion" rather than legal coercion, which it sees as short-sighted and apt to backfire. (However, Bird's sample resolution was used as the model for the Arkansas creation-science law, and for similar bills.)

160 ________. *The Case for Creation Science* (audiocassette). Shreveport, La.: Creation-Science Legal Defense Fund. Audiocassette tape CMC-2 offered by *Creation* magazine (J. of Creation-Science Legal Defense Fund). CSLDF is chief legal and financial support for Louisiana creation-science bill and other cases. Bird, lead attorney for Louisiana case, here gives "thorough briefing on how the creation law will be defended in court." Mentions defense lawyers and witnesses, and outlines basic creation-science legal arguments. No legal technicalities or discussion of actual courtroom tactics though.

161 ________. *The Origin of Species Revisited: The Theories of Evolution and Abrupt Appearance.* To be publ. 1988. New York: Philosophical Library. 2 vols. Bird's 600-page Supreme Court brief in support of the Louisiana creation-science bill, here published as a book. ICR "Impact" series #173 and 176, *The Anti-Darwinian Scientists* and *More on the Anti-Darwinian Scientists* (1987, 1988), is an excerpt. Bird quotes various anti- and non-Darwinian scientists, and others critical of orthodox evolutionary theory. Creationists hail the law brief as a masterpiece of argumentation which conclusively and massively demonstrates that evolution is just as religious as creationism, and that creation-science deserves "balanced treatment" in schools if evolution is taught.

162 Birks, Thomas Rawson. *The Bible and Modern Thought.* 1862 [1861]. London: Religious Tract Society. Orig. pub. by Hitchcock (Cincinnati). *Rev. Birks:* rector of Kelshall (Hertfordshire, England). Includes chapter "The Bible and Modern Science." Bible uses "optical" (phenomenological) language: describes things how they appear—not in modern scientific idiom. Most geologic ages occurred after creation of Gen. 1:1 and before six-day re-creation (Gap Theory creationism). Suggests nearly 30 creations appearing during different earth stages. Harmony of science and Bible. (Discussed and quoted in **Wonderly** 1977, who describes Birks as similar to **Hitchcock** but with less scientific knowledge.)

163 ________. *Modern Physical Fatalism and the Doctrine of Evolution.* 1876. London: Macmillan. "Including an examination of Mr. H. Spencer's First Principles."

164 ________. *The Scripture Doctrine of Creation, with Reference to Religious Nihilism and Modern Theories of Development.* 1887 [1882]. London: Christian Evidence Committee of the Society for Promoting Christian Knowledge. Also pub. by Young (New York).

165 Bishop, T.B. *Evolution Criticized.* 1918. Edinburgh: Oliphants. Also pub. by Scripture Union Office (London).

166 Bixler, R. Russell. *Earth, Fire and Sea: The Untold Drama of Creation.* 1986. Pittsburgh, Penn.: Baldwin Manor Press. Foreword by **Patten;** 2nd Foreword by J. Rea (Bible prof. at **Robertson**'s CBN Univ.). *Bixler:* heads Christian T.V. station WPCB (Pittsburgh); organizer of 1986 Int'l Creation Conference (Pittsburgh). Bixler asserts he is a creationist who accepts a "quite literalistic view of Gen. 1," including creation in six literal days. But he rejects *ex nihilo* creation as un-biblical; says it is a later heretical interpretation—a gnostic combination of Zoroastrian, Greek, and Egypto-Christian scientism. Study of Hebrew texts compared with Greek and Roman Bible translations, Jewish traditional sources, ancient commentaries and careful exegesis shows that God created from pre-existing Chaos. Asserts dualism: Chaos (the Abyss) was evil, and resisted God mightily. Creation was a titanic battle. God forcibly restrained the Abyss and the Darkness. Admits this notion is similar to pagan cosmogonies, yet true. "Satan consistently displays the counterfeit, the half-truth." Creation was recent, but Chaos ancient; earth is thus "both old and young": strict creationist and evolutionist dating results are both right. Chaos may have been created by God originally, but Bible account begins later. Bixler favors a translation making Gen. 1:1 a dependent clause: "In the beginning of God's creating the heavens and the earth—the earth being a formless waste..." Denies Gap Theory, since Chaos predates Gen. 1:1, and rejects other "concordist" theories as ad hoc attempts. *Mabbul* ("waters above") is the pre-Flood vapor canopy. Inspired by **Patten** and **Velikovsky;** suggests Creation event was an ice-comet or planet hitting the fiery proto-earth. First four creation days involved cosmic catastrophes. Suggests that lesser light appointed to rule the night was Saturn, as in pagan cosmologies. Argues that gnosticism, Aristotelianism, and dispensationalism were all "*unbiblical* intrusions into the Church by the world," each caused by the triumph of scientism. States that God operates now just as he did during Creation, creating wine out of water (e.g.), contrary to the dispensationalist view. Interesting references and analyses of old sources and commentaries. Praises **Weston Fields**'s exegesis (except for his acceptance of *creatio ex nihilo*). Discusses many ancient writers claimed to be Gap Theorists, giving a more plausible rendering of their ideas as referring to pre-existent Chaos. Drawings. Bibliog., index.

167 Blake, H.D. *The Tyranny of Evolution.* Priv. pub. (Listed in SOR CREVO/IMS.)

168 Blavatsky, Helena Petrovna. *The Secret Doctrine.* 1888–1936 [1876]. Two vols. *Blavatsky:* Russian; married a Czarist general; interested in psychic and occult, spiritualism, Hinduism; founded Theosophical Society in New York in 1875. Aim of book is "to show that Nature is not 'a fortuitous concurrence of atoms,' and to assign man his rightful place in the scheme of the Universe; to rescue from degradation the archaic truths which are the basis of all religions; and to uncover, to some extent, the fundamental unity from which they all spring; finally, to show that the occult side of Nature has never been approached by the Science of modern civilization." "Root Races" of Theosophical doctrine. Third "Root Race"—ape-like, egg-laying creatures, some with third eye or four arms—came from Lost Continent of Lemuria. Fourth Root Race were human-like inhabitants of Atlantis. We are Fifth. Also describes Sixth and Seventh races.

169 Blick, Edward F. *Correlation of the Bible and Science* (booklet). c1976. Oklahoma City: Southwest Radio Bible Church. *Blick:* aerospace, nuclear and mechanical engineering prof. at Univ. Oklahoma; on ICR Technical Advisory Board. Strict creation; standard creation-science, Ark-eology, the Flood. "Evolution: Satan's Fairy Tale for Adults." Includes biog. note.

170 Bliss, Richard B. *A Comparison of Two Approaches to the Teaching of Origins of Living Things to High School Students in Racine, Wisconsin.* 1978. Unpub. Ed.D. thesis—Univ. Sarasota. *Bliss:* public school science teacher for 23 years; later Racine, Wisc. district director of science educ.; now ICR director of curriculum development and science education prof. Bliss developed "two-model" curriculum as teacher, then got Univ. Sarasota Ed.D. for his two-model research after joining ICR.

171 _______. *Origins: Two Models—Evolution, Creation.* 1978. San Diego: Creation-Life. Illustrated classroom "module." Standard creation-science. Largely the same arguments, examples, and illustrations as 1984 video/book of same name. Photos, pictures, charts on every page.

172 _______. *Evolutionary Indoctrination and Decision-Making in Schools* (pamphlet). 1983. El Cajon, Calif.: Institute for Creation Research. ICR "Impact" series #120. Insists that teaching evolution only is bad education; that "two-model" approach produces better students more interested in science. (These are the conclusions of his Racine teaching experience and of his Ed.D. research.) Bliss has also written other "Impact" articles.

173 _______. *A Video Guide to Origins: Two Models—Evolution, Creation.* 1984. El Cajon, Calif.: Institute for Creation Research. Accompanying teacher's guide to Bliss's ICR video *Origins: Two Models—Evolution, Creation.* Video is intended for public school use; this book comes in two versions: Christian School and Public School Editions. Standard creation-science arguments and examples. Emphasizes "process skills of science" and stresses openness, fairness, and freedom of inquiry. Presents assumptions and evidence seeming to support each "model." Includes transcript of video narration. (Video transcript opens with re-enactment of Scopes Trial, misquoting Darrow—having him call for both creation and evolution to be taught in schools.) Test questions, answers. Drawings, charts and photos, plus section of transparency masters. Bibliog.

174 _______**, and Cindy Carr.** *Dinosaur ABC's Activity Book* (booklet). 1986. El Cajon, Calif.: Master Books. *Bliss:* project director. *Carr:* writer. Doug Schmitt: illustrator. "A Creation Knowledge Book" for young children. Coloring-book format. Descriptions of various dinosaurs plus strict creationism teaching, Noah's Flood, Bible references. Very simple.

175 ________, Duane Gish, and Gary Parker. *Fossils: Key to the Present.* El Cajon, Calif.: Institute for Creation Research. ICR "two-model" study, "designed for use in public schools."

176 Blocher, Henri. *In the Beginning: The Opening Chapters of Genesis.* 1984 [1979]. Downers Grove, ILL.: InterVarsity Press. Orig. 1979; French *(Révélation des Origines). Blocher:* systematic theology prof. at Faculté Libre de Théol. Evangélique (Vaux-sur-Seine, France). Scholarly; discusses various interpretations of Genesis, including "literal" (strict creationism), "reconstruction" (Gap Theory), "concordist" (Day-Age), and "literary" (framework). Prefers literary: says Genesis account is non-chronological, thematic, overlapping. Long appendix on scientific hypotheses and creation: discusses various kinds of creationism, evolutionism and anti-evolutionism; wide range of sources. Accepts old earth, and evolution at least up to order level, but also truth of Genesis. Rejects naturalistic evolution; considers mathematical criticisms of evolution "unanswerable." Index.

177 Blumenfeld, Samuel L. *NEA: Trojan Horse in American Education.* 1984. Boise, Idaho: Paradigm Co. *Blumenfeld:* CCNY graduate; worked in N.Y. book pub. industry 10 years; former teacher; now supports the Christian Reconstruction movement. Book dedicated to **Pat Robertson** and Rev. Sileven. Includes chapters "The Impact of Evolution" and "Turning Children Into Animals." "Evolution is at the very basis of modern public education where the child is taught that he is an animal linked by evolution to the monkeys." Protests 1984 ruling of Texas Att'y Gen. that mandating creation-science was unconstitutional: Blumenfeld declares that "it clearly takes more faith to believe that the world arose spontaneously out of nothing than it does to believe in a Creator of superhuman intelligence and powers." Traces history of development of U.S. educational system from German evolutionist models. Sees progressive education as humanist, socialist, anti-Christian conspiracy. Asserts NEA (Natl. Educ. Assoc.) is deliberately promoting illiteracy in order to aid a socialist takeover.

178 Bluth, Christoph. *Der Ursprung des Menschen* (booklet). 1972. Augsburg, Germany: Verlag Lebendiges Wort. **Schirrmacher** 1985 says this is a booklet "rephrasing published creationist material" by a German scientist.

179 Boardman, William J., Robert Koontz, and Henry Morris. *Science and Creation.* 1973 [1971]. San Diego: Creation-Science Research Center. Orig. 1971. **Morris** and others formed CSRC in early 1970s. It is now a separate organization; the **Segraves** left Morris's group and kept the CSRC name; Morris re-named his group ICR. *Boardman:* chemist prof. at Biola College. *Koontz:* entomology Ph.D.—Oregon State Univ.; biology prof. at Biola Univ. Reference book for CSRC Science and Creation Series of booklets for public school use. Book contains standard creation-science arguments. Strict young-earth creationism. Many scientific, some biblical references. Photos, appendix of creation-science sources, index.

180 Boddis, George. *Evolution Versus Facts* (booklet). 1928 [1923]. Philadelphia: School of the Bible. Address to Baptist Minister's Conference (Philadelphia, 1923).

181 Boice, James Montgomery. *Genesis: An Expositional Commentary; Volume 1: Gen. 1:1–11:32.* 1982. Grand Rapids, Mich.: Zondervan. *Boice:* Harvard, Princeton Theol. Sem. degrees; D.Th.—Univ. Basel (Switz.); former ass't. ed. for *Christianity Today;* pastor of Tenth Presbyterian Church, Philadelphia. Includes chapters on the weakness of evolution, theistic evolution, Gap Theory creationism, and six-day (recent) creationism. Considers earth billions of years old; prefers "progressive creation"—a Day-Age interpretation. Cites many creationists favorably **(Filby, Thurman)**; also **Jastrow.** Critical of ICR and other strict creationists, but endorses worldwide Flood and cites many Arkeology tales **(Roskovitsky,** etc.) approvingly. Index.

182 Bole, Simeon James. *The Modern Triangle: Evolution, Philosophy, and Criticism.* 1926. Los Angeles: Biola Book Room. Introduction by **L.S. Keyser.**

183 Bonnet, Charles. *Considerations sur les Corps Organismes.* 1762. Amsterdam. *Bonnet:* Swiss naturalist and philosopher. Bonnet was chief spokesman for the "preformationist" school, which he defends in this book against the rival "epigenetic" theory of human embryology. The preformationists advocated *emboîtement* ('encasement'): the notion that the germ cell (egg or sperm) contains, in some miniaturized form, the complete individual. The adult form "unrolls" ('evolution' in its original usage—Bonnet was apparently the first to use the term this way) from this tiny homunculus. Each germ cell must carry within it the "preformed" homunculi of all its ancestors (through either the male or female line), back to the original human created by a direct act of God. **Marsh** 1950 points out that Bonnet's argument was the "first really scientific theory of creationism," and that it was "consistent and ingenious" —the "most extreme form of creationism imaginable." Preformationism was plausible before the discovery that germ cells carried a genetic information 'code'; further, Bonnet lived before geology had extended the age of the earth (and mankind) past the limited number of biblical generations. Marsh argues that Bonnet anticipated the modern theory of heredity; Bonnet did not insist that the homunculi resembled adult forms.

184 Boodin, John Elof. *Cosmic Evolution.* 1925. New York: Macmillan. *Boodin:* UCLA philosophy prof.; 1937 Faculty Res. Lecturer. "Materialism has substituted magic for sober thought. The whole process of evolution becomes a succession of miracles. . . . That any age should take seriously such an incoherent mixture of mysticism and science is evidence of nothing so much as a want of logical thinking. . . . By chance variation the structure of protoplasm is supposed to be built up from inorganic matter. . . . Chance is God. . . . Materialism offers the most astounding instance of credulity in history."

185 _______. *God and Creation: Three Interpretations of the Universe.* 1934. New York: Macmillan. Three conceptions of history: Preformation, Emergence, and Creation. Proposes a theory of creation which "does not fall back on magic" but which agrees with the traditional theory of Genesis that creative genius is necessary. A cosmological hypothesis of spiritual control showing significance of God in the universe. Cosmic religion.

186 Book Fellowship. *Darwin's Fatal Bee Sting* (tract). Undated. North Syracuse, N.Y.: Book Fellowship. Design of the bee and the hive, including popular "bee's knees" example. Fatal to Darwinism; proves Creator.

187 _______. *Science and the Bible: No Conflict!* (tract). Undated. North Syracuse, N.Y.: Book Fellowship. Cites astronomer Maurice Brackbill, Dr. Alter (Griffith Planetarium director), **G.F. Wright, Bell Dawson** and **W. Dawson** as all supporting Bible-science. No dates given for their anti-evolution arguments (which are mostly from the 1920s and earlier). Recommends **Stoner** and **Morris.**

188 _______. *Who Made All This?* (tract). Undated. North Syracuse, N.Y.: Book Fellowship. "Take your typewriter apart." It couldn't reassemble itself by chance. Design argument.

189 _______. *The Woodpecker* (tract). Undated. North Syracuse, N.Y.: Book Fellowship. "God's Flying Power Drill." Woodpecker carefully designed for special function. Evolution cannot explain such specialized organs—only adaptive when fully developed. Adapted from **Meldau.**

190 Booth, Ernest S. *Biology: The Story of Life.* 1954 [1950]. Mountain View, Calif.: Pacific Press. Rev. ed.; orig. 1950. *Booth:* head of Biology dept. at Walla Walla Coll. (Wash.); Seventh-day Adventist; editor of *The Naturalist,* which published creationist articles, and organizer of the Associated Nature Clubs of America; also wrote bird and animal field guides. Illustrations; bibliog.

191 Bosanquet, Samuel R. *Vestiges, etc., Its Arguments Examined and Exposed* (booklet). 1845. London: Hatchard. 2nd ed. *Bosanquet:* theologian. Refutation of Chambers' proto-evolutionary *Vestiges of the Natural History of Creation.* Attempts detailed examination of its scientific arguments, but also dismisses its logic: "Reason is the root of unbelief and heresies." (Discussed in Millhauser.)

192 Boschke, F.L. *Creation Still Goes*

On. 1964 [1962]. New York: McGraw-Hill. Orig. 1962; German (*Die Schöpfung ist noch nicht zu Ende;* Düsseldorf/Vienna: Econ-Verlag). *Boschke:* editor of *Angewandte Chemie* (Heidelberg). Contradiction between Genesis and modern science is only apparent. Earth and life are millions of years old. "The seven days [of Creation] may simply represent the seven stages of development of an uninhabited planet. The Book of Genesis expresses in the simplest and most comprehensible way the extremely complicated process of evolution."

193 Bosizio, Athanasius. *Geology and the Flood.* 1877. Mainz, Germany: F. Kirchheim. Orig. German. *Bosizio:* German Jesuit. Argued that geological strata were formed by Flood. Argued against fossil indices, unformitarianism, evolution. Flood waters were 16,000 feet deep; the subsiding waters continued geological work for hundreds of years after Noah landed. (Described and quoted in **B. Nelson** 1931.)

194 Bouw, Gerardus D. *With Every Wind of Doctrine: Biblical, Historical and Scientific Perspectives on Geocentricity.* 1984. Cleveland: Tychonian Society. *Bouw:* astronomy Ph.D.—Case Inst. Tech. (Case Western); teaches computer sci. at Baldwin-Wallace College; editor of *Bulletin of the Tychonian Society* (geocentrist journal). Insists on absolute authority and inerrancy of Bible. Rejects Copernican revolution—heliocentricity—and calls for return to Tycho Brahe's system: planets orbit the sun, but sun orbits earth. Thorough exegesis; demonstrates convincingly that the Bible teaches geocentrism. Refutes alleged Bible-science claims of heliocentric passages. Sun really did stand still for Joshua; cites legends of Long Day (and correspondingly legends of Long Night in opposite hemisphere). Heliocentrism based on astrological and philosophical biases. Doubts relativity (especially its moral consequences), but relies on relativistic claim that science cannot demonstrate any center or fixed reference point in universe; thus, cannot prove sun goes around the earth. But Bible *does* say where center is: the earth. Rejects evolution also. Many references. Diagrams, pictures.

195 Bowden, Malcolm. *Ape-Men: Fact or Fallacy?* 1981 [1977]. Kent, England: Sovereign Publications. 2nd ed.; enlarged. Orig. 1977. *Bowden:* British engineer. Refutes all proposed hominid ancestors. Standard strict creation-science arguments. True humans found in oldest layers as well. **Teilhard** was probably culprit in Piltdown hoax—but hoax was known at very high level: a pro-evolution propaganda attempt. Peking Man bones probably removed deliberately when found to be (modern) human. Dubois found gibbon skull and human leg bone; claimed this was Java Man—but hid bones of modern man found at same site. Nebraska Man. Australopithecines are all just apes (cites Zuckerman and Oxnard). Emphasizes that ape-men claims are based on artists' wishful reconstructions.

196 ________. *The Rise of the Evolution Fraud.* 1982. San Diego: Creation-Life. Co-pub. by Creation-Life and M. Bowden (Bromley, Kent, England). Foreword by **H. Morris.** Lurid conspiracy-theory view of evolution and Darwinism. Relies heavily on **Himmelfarb** 1959; also **R.E.D. Clark** 1967. Reveals true purpose of Darwin and his henchmen—Huxley, Hooker, Lyell—as "an attack upon Christianity." Exposes dark purpose of Huxley's secret X Club: evolutionist propaganda. Lyell's devious strategy was to undermine Mosaic account by luring Darwin into attacking it. Darwin was a devious sophist, not a real scientist. Alleges deliberate suppression of Mendel's genetics. Standard creation-science arguments; emphasizes evil influences of evolution. Appendices includes list of creationist organizations; quotes for creationist use; British Museum evolutionist propaganda; synopses of scientific evidence against evolution and for creation; ape-men fallacies (summary of Bowden 1981); Bible quotes; and endorsement of Lady Hope tale of Darwin's deathbed conversion and repudiation of evolution. Drawings, bibliog., index.

197 Bowman, Edward F. *"Malfunctions" of Nature?* (booklet). c1977 [c1976]. Priv. pub. (Kansas City, Mo.). "Citing certain events wherein Nature has apparently 'Violated some of its own laws.'" *Bowman:* amateur astronomer, lecturer and telescope maker; Seventh-day Adventist. "The earth is a very sick planet." This is due to man's sin, as told in Genesis. Describes wonderful conditions of Eden, the pre-Flood Water Canopy. Noah's Flood and Joshua's Long Day were major "malfunctions" of the once-perfect earth. Proposes that these disturbances were caused by cosmic cloud coming between earth and sun, resulting in shifts of earth's axis. Cosmic cloud perhaps remains of planet exploded by God—or Satan. Young-earth creationism. "I am pleased to detect a healthy trend toward the acceptance, by many educators, of a direct Creation as opposed to Evolutionary Theories." Diagrams; index.

198 Boyden, A. *Perspectives in Zoology.* 1973. Elmsford, N.Y.: Pergamon Press. **W. Frair** says this book was intended as a continuation of **Kerkut**'s protest against compacency and unwarranted assumptions by evolutionists. (Boyden was Frair's advisor.) "Well known as a scientist in the field of systematic serology, the author, who is an evolutionist, strongly objects to unjustified use of evolutionary philosophy, especially in the field of taxonomy" (Frair: unpub. anti-evolution bibliog.).

199 Brackman, Arnold C. *A Delicate Arrangement: The Strange Case of Charles Darwin and Alfred Russel Wallace.* 1980. New York: Times Books. Conspiracy theory: Darwin and his party conspired to claim undeserved priority over Wallace for discovery of evolution through natural selection. Cited often by creationists, notably **Ian Taylor** 1984.

200 Bradley, Abraham. *A New Theory of the Earth* (booklet?). 1801. Wilkesbarre, Penn.: Joseph Wright (printed for). Subtitled "or, The Present World Created on the Ruins of an Old World: Wherein It Is Shown from Various Phenomena, that the Earth Was Created at a Period of the Highest Antiquity; that it was Afterwards Destroyed by a Deluge; and, that After the Deluge, it Was Repeopled by a New Creation of Men, and Other Animals." (Listed in Schadewald files.)

201 Brady, James C. *Not Time Enough for Evolution* (tract). North Syracuse, N.Y.: Book Fellowship. Tract No. 1222. Proofs that earth is only a few thousand years old. Includes creation-science arguments of H. Armstrong, **Slusher, Barnes, M. Cook, Whitelaw, Hallonquist.** Impressed by authority of CRS scientists.

202 Brasington, Virginia F. *Flying Saucers in the Bible.* 1963. Clarksburg, W.V.: Saucerian Books.

203 Brazo, Mark William. *Theories of Origins: Do They Persist Despite Contrary Evidence?* 1983. Unpub. thesis: Institute for Creation Research. ICR biology M.A. thesis. Committee: **Cumming** (chair), **Gish, Parker.** "This propensity of ignoring scientific evidence thereby deferring to political and religious biases, has caused the continuance of origin theories that are inappropriately and unfortunately designated 'scientific.'" "Philosophical and religious preoccupations should be the anathema of science." Evolutionist theories of origins do persist even though evidence proves them false. Standard creation-science argument that evolution is believed in because of philosophical and other non-scientific biases.

204 Bree, Charles Robert. *Species Not Transmutable Nor the Result of Secondary Causes: Being a Critical Examination of Mr. Darwin's Work Entitled "Origin and Variation of Species."* 1860. London: Groombridge and Sons.

205 ________. *An Exposition of the Fallacies in the Hypothesis of Mr. Darwin.* 1872. London: Longmans, Green.

206 Brock, Fred R., III, and Michael J. Bardon (eds.). *The Biblical Perspective on Science.* c1972. New York: MSS Information Corp. "Custom-made book of readings"—mostly *CRSQ* reprints; also *Bible-Science Newsletter* reprints, and lectures. Includes **Kofahl, Barnes, Whitcomb, Morris, Slusher,** H. Armstrong, **H. Clark, E. Williams,** David Penny, **Northrup, Burdick, C. Clough** (on the reception of *Genesis Flood*), W. Springstead, **Patten** (Ice Age), **Cousins** (horse series), **Hedtke,** W. Meister (trilobite fossil with human print), Robert Wood, **R. Teachout** (Bible chronology), **M. Cook** (C-14), **Whitelaw, G. Howe, Klotz,** Julio Garrido, *U.S. News & World Report* (on medical ethics). Usual creation-science topics in cosmology, dating, geology, biology.

207 Bronn, H.G. *Untersuchungen*

über die Entwickelungsgesetze der organischen Welt während der Bildungs Zeit unserer Oberfläche. 1858. Stuttgart, Germany: Heinrich Georg. ("Investigations into the developmental laws of the organic world during the formation of the crust of the earth.") *Bronn:* German zoologist, geology and paleontology prof. at Freiburg, then Heidelberg. First German translator of Darwin's *Origin*. Seems to prefer vitalist theory; objects to spontaneous generation (in origin of life). Last chapter translated into English in 1859. Discussed and quoted by Hull, who calls this book a defense of special creationism. Bronn later admitted species variation.

208 Broocks, Rice. *Change the Campus, Change the World!* c1985. Gainesville, Fla.: Maranatha Publications. "A Battle Plan for Reaching This Generation." Foreword by Rosey Grier. *Broocks:* director of evangelism and second in command at Maranatha Campus Ministries (Maranatha Christian Churches); co-host of Maranatha's *Forerunner* T.V. show; a charismatic. Maranatha focuses its evangelistic efforts on college campuses (*Forerunner* magazine). Broocks aggressively advocates Dominion Theology, militantly urging Christians to "take over" every aspect of society and transform the world according to biblical principles. In 1984 he formed the Society for Creation Science, which provides materials and training for campus chapters. Includes information about starting chapters in this book. Broocks plans to have creationism taught at every major college, at first through campus religious clubs, later as regular courses. "Evolution teaches [students] that they are nothing more than animals with no real purpose to fulfill." Evolution "has done much to destroy the faith of our youth in a loving God who created them. It is critical that people know that they are more than biological chance: they are a special creation of God's with a destiny." Includes "Creation Scriptures" and reprint of ICR "Impact" articles #95–96. Photos.

209 Brooks, Jim. *Origins of Life*. 1985. Tring, Herts, England/Belleville, Mich.: Lion Publishing. *Brooks:* Ph.D., F.R.S.C., F.G.S.,; with Exploration Div. of Britoil (Glasgow, Scot.). Handsomely illustrated book; scientifically knowledgeable. Accepts geological ages and standard chronology of fossil record. Addresses origin of universe; emphasizes early earth and Precambrian fossils. Big Bang is true—but don't rely on it for proof of God: "This 'God of the Gaps' is a wrong and pathetic substitute for the infinite, all-powerful God of the Bible. . ." who Brooks endorses openly. In "Chance or Purpose?" chapter, says that 'chance' origin of life impossible. "Science and Creation" chapter: Bible "gives a true account of God's creation." Seems to prefer revelatory or framework creationism. Bibliog. recommends several creationist books. Full of beautiful photos, charts, diagrams, paintings.

210 Brooks, Keith L. ("edited by") *Overwhelming Mathematical Evidence of the Divine Inspiration of the Scriptures* (tract). North Syracuse, N.Y.: Book Fellowship. Tract No. 1219. "From the works of Dr. **Ivan Panin,** Harvard Scholar and Mathematician." Panin's Bible numerology, based on numerical value of each Hebrew and Greek letter (gematria), proves Bible was designed by supernatural Maker. Tract includes endorsements by creationists (and fellow Canadians) **W.B. Dawson** and **A. Brown.**

211 Broom, Robert. *Evolution: Design or Accident?* (pamphlet). 1930. Johannesburg, South Africa: Univ. Witwatersrand Press. *Broom:* Scottish M.D.; became famous paleoanthropologist—discovered *Plesianthropus* (now *Australopithecus*) at Sterkfontein, South Africa. This pamphlet and his 1933 book pre-date that discovery.

212 ________. *The Coming of Man: Was It Accident or Design?* 1933. London: H.,F.&G. Witherby. Relation of evolution to the spiritual. Attributes aspects of mentality to all life. Outside spirit agencies guide life to increasing mentality. Unconscious soul causes inheritance of acquired characteristics; evolutionary change too complex for natural selection and mutation only. Humans have conscious soul too: the purpose of evolution is to produce souls. Soul survives death in spirit world. Broom saw deliberate preparation for evolutionary changes, e.g. change in composition of atmosphere prior to mammals.

213 Brown, Arthur Edwin. *Canopy of Ice.* 1985. Farmington, Miss.: EDM Digest Co. *Brown:* electr. discharge machining (EDM) trade magazine publisher. Inspired by **Morris, Whitcomb, Velikovsky,** Ivan Sanderson, **Bellamy,** science fiction; but includes no creation-science references. Told in first person; uses fictional characters. When earth cooled, water stayed above as ice canopy. Brown never denies evolution, but affirms truth of Bible. Collapse of canopy caused Flood, and mass extinctions (especially mammoths). Canopied earth had greenhouse climate: Eden. Lifespans declined after canopy fall from unhealthy exposure to sun. The less sun the better: has chapter on "Methuselah children" raised in attic (slower maturation); also case from India. Sun and moon were not seen until canopy collapse. Prehistoric art shows sun and moon never depicted until 11,600 BP (Flood). Biblical Peleg ("earth divided") so called because moon first seen then. But cave art does show depictions of cracks in the canopy as it begins to deteriorate. Earth, flipping on axis within stable magnetic field, scrapes off ice from canopy. Photos, charts, drawings.

214 Brown, Arthur Isaac. *Evolution and the Bible* (booklet?). 1920s. Los Angeles: Research Science Bureau. Res. Science Bureau: **Rimmer**'s organization. *Brown:* M.D.; Vancouver, B.C. physician; then full-time preacher (Baptist). **H. Morris** 1984 says Brown was a Gap Theory creationist, but did not discuss Gap Theory in his books.

215 ________. *Evolution and the Blood Precipitation Test* (pamphlet). Undated [1926]. Glendale, Calif.: Glendale Commercial News Printing. Also pub. by Florida-Earth Printing (Orlando). Refuted evolutionist claims that G. Nuttall's 1904 anti-serum reaction tests, which indicated degree of biochemical similarity between species, proved common descent.

216 ________. *Science Speaks to Osborn.* 1927. Fort Wayne, Ind.: Glad Tidings Pub. A reply to H.F. Osborn's evolutionist book *The Earth Speaks to [W.J.] Bryan.*

217 ________. *Was Darwin Right?* (booklet). 1927(?). Glendale, Calif.: Glendale News Commercial. "A Reply to Sir Arthur Keith's Leeds, Eng., Address on 'Darwin.'"

218 ________. *God's Creative Forethought* (pamphlet). Undated [1927?]. Glendale, Calif.: Glendale Printers. Also pub. in 1930s by Research Science Bureau (Los Angeles).

219 ________. *Footprints of God.* 1943. Findlay, Ohio: Dunham. Also pub. by Van Kampen Press (Wheaton, Ill.). Advertisement: "presents astounding, scientific facts which clearly prove the absolute necessity of an all-wise Designer and Creator, and disprove the claims of evolution." Foreword by Charles M.A. Stine (VP of Du Pont).

220 ________. *Miracles of Science.* c1945. Findlay, Ohio: Fundamental Truth Publishing. 39 lectures on design in nature (mostly biological). Orig. radio talks, from Moody Bible Inst. "Miracles of Science" radio program and others. Aim of every talk: to show "indisputable, scientific fact of a personal, omnipotent Creator-God" and exalt Christ. Includes "bees' knees," Venus fly-trap, whales, beavers, water and atmospheric circulation, birds, etc. Each contains explicit Christian evangelizing. Some scientific quotes. Cites Leuba on loss of belief in God by college students. Similar to anthropic principle argument.

221 ________. *God's Masterpiece—Man's Body.* c1946. Findley, Ohio: Fundamental Truth Publishing Co. (Discussed in **Morris** 1984.) Created design of the human body.

222 ________. *God and You: Wonders of the Human Body.* 1940s? Findlay, OH: Fundamental Truth. *Brown:* "scientist, surgeon and Bible teacher." "Few books of this nature are free from the evolutionary hoax." This one is. Shows how human body proves all-wise Designer. "No speculative evolutionary hypothesis will suffice as an explanation of these wonders." Condensed edition (undated) pub. by Good News (Westchester, Ill.), titled *Wonderfully Made.*

223 ________. *Men, Monkeys, and Missing Links* (booklet). Undated. Findlay, Ohio: Fundamental Truth Publishers.

224 ________. *Must Young People Believe in Evolution?* (pamphlet?). Undated. Oak Park, Ill.: Designed Products.

225 Brown, Henry. *The Geology of Scripture, Illustrating the Operation of the Deluge, and the Effects of Which it Was Productive: with a Consideration of Scripture History.* 1832. Frome: W.P. Penny (printer). Illustrations.

226 Brown, J. Mellor. *Reflections on Geology, Suggested by Perusal of Dr. Buckland's Bridgewater Treatise.* 1838. London: James Nisbet. Also pub. in Edinburgh. Last of the four great Scriptural Geologists (Millhauser). Denounced geologists and their evil work. God can reduce long eras to a moment, so it is foolish to apply inductive chronology against revelation. "It is perfectly reasonable to suppose," says Millhauser in describing Brown's argument, "that fossils, with all their appearance of an extended pre-history, might have been created by divine fiat, in their present form, at the same instant as the hills wherein they lie." (Thus Brown anticipated **Gosse**'s theory.)

227 Brown, Walter T. *In the Beginning---* (booklet). 1986. 4th ed. Naperville, Ill.: ICR Midwest Center. *Brown:* West Point B.S.; mechan. engin. Ph.D. —MIT; former chief of science and technol. studies at Air War College; former assoc. prof. at USAF Acad.; USAF col. (ret.); director of ICR Midwest Center (renamed Center for Scientific Creation; no longer affil. with ICR, and since moved to Ariz.). Standard creation-science, Flood Geology, Arkeology. Comprehensive list of creationist arguments in concise form. "What You Can Do" section: how to spread creationism in your community. Includes scientific references, charts, tables, bibliog.

228 Bruce, Les. *On the Origin of Language* (pamphlet). 1977. El Cajon, Calif.: Institute for Creation Research. ICR "Impact" series #44. *Bruce:* Ph.D. cand. in linguistics; member of Summer Inst. of Linguistics and Wycliffe Bible Translators; visiting linguistics prof. at CHC (ICR affil.). Language doesn't evolve—it just changes.

229 Bryan, William Jennings. *The Menace of Darwinism* (booklet). 1922. New York: Fleming H. Revell. *Bryan:* 3-time presidential candidate; Sec'y of State under Wilson; progressive, anti-imperialist politician; later champion of fundamentalism and anti-evolutionism. "Menace of Darwinism" was one of his popular lectures; this booklet is reprint of a chapter ("The Origin of Man") from his 1922 book plus additional comments.

230 ________. *In His Image.* 1922. New York: Fleming H. Revell. Based on 1921 lectures at Union Theol. Sem. (Richmond, Vir.). A defense of the Bible and fundamentalist Christianity. Chapter "The Origin of Man" is a strong denunciation of evolution. Evolution contradicts the Bible. Darwinism "leads logically to war" and "to a denial of God." Teaching evolution erodes religious faith. Since religion is the "only basis of morality," evolution is its greatest threat. Impressed by Leuba's 1916 study that students lose religious faith in college. Explains how he learned that the German militarism responsible for the World War was based on evolutionist materialism and on Nietzsche's Darwinist philosophy (cites evolutionist V. Kellogg's *Headquarters Nights*). Implores taxpayers to deny support to teaching of evolution—atheists and evolutionists should build their own private schools. Evolution destroys belief in immortality of soul—the only stimulus to righteous living. Ridicules evolutionist notion that freckles turned into eyes and warts into legs, accidentally yet symmetrically.

231 ________. *The Bible or Evolution?* (booklet). Undated (1925?). Murfreesboro, Tenn.: Sword of the Lord. Bryan, a Presbyterian, defended the Tennessee anti-evolution law in the 1925 Scopes Trial. Major concern was evil effect of evolution on morals. If evil, it must be false. This booklet contains many of his standard arguments, including many used in the Scopes Trial. Ridicules Darwinism. Evolution is merely a hypothesis, which means a guess. Scientists—who comprise only one in 10,000 of the U.S. population—attempt to dictate education and religion. Quotes Fosdick that eyes supposedly evolved from freckles, legs from warts. Students lose faith in college. Teaching non-existence of miracles subverts Christianity. Atheists should build their own schools instead of subverting public schools. (During the trial, Bryan conceded that Day-Age creationism was possible.)

232 _______. *Bryan's Last Speech: The Most Powerful Argument Against Evolution Ever Made* (booklet). c1925. Oklahoma City, Okla.: Sunlight. Bryan's intended closing address for the Scopes Trial (not delivered at trial because of procedural shift by the defense which ended the trial early, but widely circulated afterwards). Bryan died a few days after the trial. (Also included in **National Book Co.** trial manuscript.)

233 _______, **and Lowell Harris Coate.** *The Dawn of Humanity; The Menace of Darwinism; and, The Bible and Its Enemies; Evolution Disproved.* c1925. Chicago/Pasadena: The Altruist Foundation. *Evolution Disproved* is by Coate; the other sections are by Bryan.

234 Bucaille, Maurice. *The Bible, the Qur'an, and Science.* 1983 [1976]. Indianapolis, Ind.: American Trust Publications. 10th ed. Orig. 1976; French. *Bucaille:* surgeon. The Bible is a divine revelation from God, but has been transmitted and recorded by fallible humans. The Qur'an is a later, perfect revelation: God's words to Mohammed were written down directly. The Qur'an (unlike the Bible) is absolutely accurate scientifically, beyond Mohammed's human ability. There is "absolutely no opposition between the data in the Qur'an on Creation" and modern knowledge of cosmogony. The Creation description in Genesis, on the other hand, is a "masterpiece of inaccuracy from a scientific point of view." Bucaille presents a detailed case for "higher criticism" of the Bible, explaining how various parts of the Bible were written by different authors at different times. The separate (and conflicting) Yahvist and Sacerdotal sources for Genesis are unarguable, he argues. Says that when Qur'an was written, science had not progressed since Jesus's time—yet the Qur'an contains no scientific inaccuracies. Islamic science flourished because of the Qur'an. Using his own translation, Bucaille presents examples showing how modern science is confirming Qur'anic passages. Includes prediction of life on other planets. Flood was not worldwide: Qur'an instead describes "several punishments inflicted on certain specifically defined communities," including Noah's people. Six 'days' of creation described in Qur'an are long periods; advocates Day-Age creationism of overlapping periods. Asserts that Qur'anic creation account is "quite different" from the biblical. Rejects recent creation of man—Bible is wrong. This book is very popular in France and Moslem countries, and has been translated into many languages.

235 _______. *What Is the Origin of Man?: The Answers of Science and the Holy Scriptures.* 1982. Paris: Seghers. Credits Darwin with some insight, but claims evolutionist followers have extrapolated recklessly from his theory. Relies heavily on **Grassé**'s arguments against Darwinism. Animals have evolved, but prior intelligent programming was necessary. Man was created separately but similar to the apes; scientists "do not possess one iota of evidence" that they are related, though there has been evolution within the hominid lineage. Darwin was motivated by sociological factors, and materialist evolution flourishes because of ideological, not scientific, reasons. Darwin never claimed that man descended from apes, though later evolutionists assert this in order to deny God and promote materialism. Notes that many Europeans are skeptical of Darwinism, but that it is accepted uncritically in America. Chance mutations are entirely incapable of accounting for evolution, which is "quite obviously" directed. Strongly denounces J. Monod's materialist "chance and necessity" evolutionism. Reviews biblical higher criticism and asserts scientific infallibility of the Qur'an. "In contrast to the Bible," the Qur'anic text is "none other than the transcript of the Revelation itself; the only way it can be received and interpreted is literally." Discusses various 'scientific' passages in the Qur'an, especially dealing with human reproduction and embryology.

236 Buckland, William. *Reliquiae Diluvianae.* 1823. London: J. Murray. "Observations on the Organic Remains Contained in Caves, Fissures, and Diluvial Gravel, and on Other Geological Phenomena, Attesting the Action of an Universal Deluge." Also pub. by Arno Press (New York; 1977 reprint ed.). *Rev. Buckland:* first geology prof. at Oxford Univ. From study of a Yorkshire cave,

Buckland argued that fossils found in caves were "relics of the Flood." Violence of the Flood destroyed fossils outside of caves. Buckland argued that only the top geological layers (loams and gravels) resulted from Flood; all lower layers predated it.

237 ________. *Geology and Mineralogy, Considered with Reference to Natural Theology.* 1836. London: Pickering. 2 vols. One of the Bridgewater Treatises. Buckland here retreats from his earlier position that the upper strata were produced by the worldwide biblical Flood, and admits there were several catastrophes in different ages. The only Bridgewater Treatise on geology, Buckland treats it "soundly, in the spirit of reconciliation" (Millhauser). "Geology has shared the fate of other infant sciences, in being for a while considered hostile to revealed religion; so like them, when fully understood, it will be found a potent and consistent auxiliary to it, exalting our conviction of the Power, and Wisdom, and Goodness of the Creator." (Quoted in **D. Young** 1982.) Scriptural Geologists and other literalists lamented Buckland's defection from a conservative (though not strictly literal) interpretation of the Bible and his admission of the inadequacies of Flood Geology. "Some have attempted to ascribe the formation of all the stratified rocks to the effects of the Mosaic Deluge; an opinion which is irreconcilable with the enormous thickness and almost infinite subdivisions of these strata, and with the numerous and regular successions which they contain of the remains of animals and vegetables, differing more and more widely from existing species, as the strata in which we find them are placed at greater depths." He presents here a Gap Theory interpretation (though he rejects the fallen angel scenario). Genesis alludes to "an undefined period of time, which was antecedent to the last great change that affected the surface of the earth, and to the creation of its present animal and vegetable inhabitants; during which period a long series of operations and revolutions may have been going on; which, as they are wholly unconnected with the history of the human race, are passed over in silence by the sacred historian. . ." The long pre-Adamic period alluded to in the first verse ends with the six-day re-creation. The "chaos" of the second verse "may be geologically considered as designating the wreck and ruins of a former world." He suggests that armor of early fish was created to protect them from great heat of early earth. Buckland's modified catastrophist views were important in the development of modern scientific geology.

238 Buckner, Louis, and Betty Buckner. *The Case for Creation Science: From a Layman's Point of View* (booklet). 1982. Priv. pub. (Talullah, La.). Earnest little booklet which strives to prove that creation-science is indeed scientific. Urges support for Louisiana's creation-science law. (B. Buckner was a plaintiff in preliminary motions of that bill.)

239 Bugg, George. *Scriptural Geology; or, Geological Phenomena Consistent Only with Literal Interpretation of Sacred Scriptures [etc.].* 1826, 1827. London: Hatchard & Sons. One of the four great Scriptural Geologists (Millhauser). Accepts Bible as final authority in science. Bugg dismisses **Cuvier, Buckland,** and every scheme proposed to reconcile geology and the Bible. Insists on the old Diluvial geology instead. Denies that fossils represent extinct species, and insists that all animals were herbivorous before the Fall. Bugg was outraged at the theories of geology: "Was ever the 'word of God' laid so deplorably prostrate at the feet of an infant and precocious science!" Millhauser characterizes Bugg's tone as "angry fear." If the six-day Creation were questioned, Bugg realized, people might then consider it merely an invention of priests and kings to keep their subjects in awe. "Geology is the last subject to which the adversaries of *Revelation* have resorted."

240 Bullinger, Ethelbert W. *Witness of the Stars.* 1967 [1893]. Grand Rapids, Mich.: Kregel. Orig. 1893 (London: Lamp Press). *Bullinger:* Anglican clergyman; descendant of Swiss reformer J.H. Bullinger. The Gospel writ in the stars: the constellations depict Bible themes. This work has influenced other Bible-science interpretations of the stars: **Rand, H. Morris.** Mankind was 2,500 years without written revelation from God—but we had His revelation in the stars.

Messages of constellations include The Coming Redeemer, The Redeemed, and Second Coming themes. Charts, star maps. Bullinger later served on the Committee of the Universal Zetetic (Flat Earth) Society.

241 Bunker, Laurence H. *Natural Law and Science in the Bible* (booklet). Undated. Stanmore, Middlcsex, England: Chosen Books. "A self-teaching book." *Bunker:* tutor—Inner London Educ. Authority (City of Westminster College). Program tested by 5th Form students of Independent Grammar School. Listing of Bible-science questions and responses, with references. Responses included some creationist affirmations.

242 Burdick, Clifford. *Canyon of Canyons* (booklet). c1974. Minneapolis: Bible-Science Association. Introduction by **W. Lang.** *Burdick:* Ph.D.—Univ. of Physical Science (Arizona); graduate study at Univ. Ariz. (Tucson); consulting geologist; CRS Board of Directors. The Grand Canyon: formed by the Flood. Inner Gorge of Canyon shows tremendous violence of Flood—Precambrian strata tilted vertically and intruded with granite and lava. Winds sent by God to dry up Flood waters brought in sediments composing horizontal formations above gorge. Canyon carved out when Flood waters trapped upstream rushed down through newly-deposited sediments. Includes history and description of Canyon; also Burdick's claim of Cambrian and Precambrian pollen samples from Canyon. Many photos.

243 ________. *Footprints in the Sands of Time.* Forthcoming. Burdick, after reading R. Bird's 1939 report on dinosaur tracks and mystery prints at Paluxy River (Glen Rose, Texas), searched for the "manprints." Found [carved] manprints in Ariz. shop (now at Columbia Union Coll.). Describes several other manprints found at Paluxy. Rejoices in collapse of geologic timetable and of evolution if manprints are genuine. (Excerpted in 1986 *Ark Today.*)

244 Burkholder, James. *Humanism: The Religion That Claims No God* (tract). Undated [1980s]. Crockett, Ky.: Rod & Staff. "Prepared and submitted by the publication board of the Eastern Pennsylvania Mennonite Church." Humanism is the attempt to deny God. "The evolution theory has thoroughly saturated the textbooks of most schools for the last few decades. As a result, children have grown to adulthood and feel no responsibility to God, the Creator." How else could the universe have come to be? "Man is the crowning work of God's six-day creation."

245 Burnet, Thomas. *The Sacred Theory of the Earth: Containing an Account of Its Original Creation, and of All the General Changes Which It Hath Undergone, Or Is to Undergo, Until the Consummation of All Things.* 1759 [1681]. London. 7th ed. Orig. 1681; Latin. Creation, Flood, and final destruction by fire. Burnet attempts to confirm scriptural account by natural law explanation, arguing that all events of earth history can be accounted for by natural processes. He refused to invoke miraculous intervention (and disagreed with **Newton,** who advocated an interventionist God). Burnet, however, relied primarily on reasoning ("philosophy") rather than observation for his theory. Scripture tells us there was a worldwide Flood. It must have involved eight times the volume of water of our present oceans. Burnet postulated that the earth was originally created as a perfect sphere, with a smooth crust of land forming over the ocean layer—the Abyss. It was a perfect and uniform Eden, but it then began to dry up and become wrinkled. Pressure built up in the Abyss from this dessication and shrinkage, and from vaporization caused by absorption of heat. The waters of the Abyss burst out when the crust finally fragmented and collapsed. The fragmented chunks of crust form the present rough topography. We live on degenerate wreck of a planet. Burnet did not deny that the Flood was God's punishment for sin, but he tried to argue that it was also the result of a natural process. Burnet supposed that natural processes cause great cumulative change over time (like modern earth science, and unlike the prevailing orthodox religious view), but he assumed these processes were degenerative only. In a later work, *Archaelogiae Philosophicae,* Burnet defended himself against charges of heterodoxy which arose from his insistence upon natural causation for the Flood, and attempted to reconcile his theory with the Genesis account.

246 Burton, Charles. *The Deluge and the World After the Flood.* 1845. London: Hamilton, Adams. Rev. Burton discusses the Flood, the Ark, etc., in great detail.

247 Bushey, Clinton J. *In the Beginning: An Interpretation of the Scriptures* (booklet). 1935. Olivet, Ill.: Alumni Print Shop. *Bushey:* science prof. at Olivet College (Ill.). "True science has never conflicted with our Bible, and never will." Knows that evolution cannot be true because it conflicts with Bible. Largely quotes from **G.M. Price**. Strict creationism.

248 Buswell, J. Oliver, Jr. *A Systematic Theology of the Christian Religion.* 1962. Grand Rapids, Mich.: Zondervan. *Buswell:* Presbyterian; former president of Wheaton College. **Morris** 1984 calls him "ardent and capable defender" of Day-Age creationism.

249 Butler, Samuel. *Evolution Old and New: Or the Theories of Buffon, Dr. Erasmus Darwin and Lamarck, as Compared with that of Charles Darwin.* 1911 [1879]. New York: E.P. Dutton. Butler, the novelist, once admired Darwin, but became convinced (after reading **Mivart**) that Darwin stole idea of evolution from Buffon and Lamarck—and that Lamarck's version was superior anyway. By conscious choice animals learned new adaptations which became habits, then instincts. Also wrote other anti-Darwin books such as *Luck or Cunning?*

250 _______. *Unconscious Memory.* 1911 [1880]. New York: E.P. Dutton. Lamarckian argument inspired by famous Viennese physiologist Ewald Hering's neo-Lamarckian lecture. Butler turned against Darwin's theory as too random and materialistic. Urged for return to Erasmus Darwin and Lamarck: evolution by will of organism, not chance processes.

251 Caleb Curriculum. *Ape-Men: Fact or Fantasy?* (booklet). c1984. Herrin, Ill.: SAC [Student Action for Christ] Publishers. "PLUS, Scientists Who Are Creationists." Caleb Curriculum series. Consists of articles orig. pub. in *Issues & Answers,* a free evangelical newspaper distributed by students in many public schools. Each article designed as a 5-minute classroom lesson. Strict creationism; standard creation-science arguments. Alleged ape-men all either pure ape, true humans, or hoaxes. Repeats claim that evolutionists "berated" **Bryan** with Nebraska Man during the Scopes Trial. Also design of the eye, lack of all transitional fossils. Authors credited in some articles: Lane Anderson, Christopher Barnes, Ginny Hastings, and Dan I. Rodden (founder and president of The Caleb Campaign). Many drawings.

252 _______. *Dinosaurs: And the People Who Knew Them* (booklet). c1985. Herrin, Ill.: Caleb Publishers. Caleb Curriculum series. Coloring book. Shows that dinosaurs co-existed with man. Noah took young dinosaurs aboard the Ark.

253 _______. *Evolution & the 'New Age'* (booklet). 1985 (?). Herrin, Ill.: Caleb Publishers. Caleb Curriculum series. "Shows the theory of evolution to be the basis for the occult, Eastern Religions, UFO-ology, and the New Age Movement. Also answers the common question: 'Couldn't God *use* evolution to create the world?'"

254 _______. *Origins: What Does Science Say?* (booklet). c1985. Herrin, Ill.: Caleb Publishers. Caleb Curriculum series. Collection of articles orig. in Caleb Campaign's *Issues & Answers.* Attempts to be cute and humorous. "We *encourage* you to make photocopies of pages from this curriculum to distribute to students in school, Sunday School, churches, etc." But requests royalty payment (as with all booklets in this series) in order to develop the curriculum and promote Christian faith. Authors listed include Tim Hastings, Ginny Hastings, Ginny Gray, Randy Rodden. "Evolution: Good Fairy Tale, Bad Science." Standard creation-science arguments; quotes many creationists. Young-earth creationism. Many drawings, photos.

255 Callaway, Eloy E. *In the Beginning.* c1971. Dayton, Ohio: ITB Association. Listed in Schadewald files.

256 Callaway, Eugene Charles. *The Harmony of the Eons.* 1934. Priv. pub. (Atlanta, Ga.). Includes section defending Gap Theory creationism and pre-Adamic races; also defense of pre-millennialism. Mostly theological.

257 Cameron, Nigel M. de S. *Evolution and the Authority of the Bible.* 1983. Exeter, England: Paternoster Press. *Cameron:* chairman of Biblical Creation Society (Glasgow); lives in Edinburgh. Cameron "develops the thesis that the theory of evolution in any form is incompatible with an evangelical approach to Scripture, but he is by no means tied to a literalistic reading of Genesis" (**Blocher** 1984).

258 ________. *Thinking Again About ... Evolution* (pamphlet). Undated. Glasgow, Scotland: Biblical Creation Society. Genesis says there was no evil, suffering, or death prior to man's Fall in Eden; thus evolution and the Bible cannot both be true, and cannot be harmonized.

259 ________, **(ed.).** *In the Beginning ...: A Symposium on the Bible and Creation.* 1980. Glasgow, Scotland: Biblical Creation Society. Volume includes J.G. McConville, Ranald Macaulay, D.A. Carson and **E.H. Andrews.**

260 Camp, Robert S. (ed.). *A Critical Look at Evolution.* 1972. Atlanta, Ga.: Religion, Science and Communication Research and Development. *Camp:* president of Religion, Science and Communication Research and Development Corp. Contributors: Thomas Warren, Moody Coffman (physics Ph.D.—Texas A&M), **D. England,** H. Douglas Dean (biology Ph.D.—Univ. Alabama; USC lecturer [now at Pepperdine]), **J.W. Sears,** Russell Artist (biology Ph.D.—Univ. Minn.), **J. Bales.**

261 Campbell, George Douglas [Duke of Argyll]. *The Reign of Law.* 1867. London: Alexander Strahan. 5th ed. *Campbell:* the Duke of Argyll (Scotland). Strong critic of Darwin. Nature shows purpose; purpose means Mind. We naturally use language implying purpose, because we know it intuitively. Perception of will in nature: anthropopsychism. "Nothing is more certain than that the whole Order of Nature is one vast system of Contrivance." Anti-materialist, but doesn't advocate miracles; Creative power is lawful. Admitted secondary causes of development through which Creator works, but remained skeptical of biological evolution. Creative (saltational) births possible. Illustrations.

262 ________. *The Unity of Nature.* 1885. New York: G.P. Putnam's Sons. (Discussed in Gillespie.)

263 ________. *Organic Evolution Cross-Examined; or, Some Suggestions on the Great Secret of Biology.* 1898. London: John Murray. Discussed and quoted at length in Gillespie, who describes Argyll's creationism as "nomothetic creation": creation not by miracle or divine interventions, but through secondary causes—general laws. Nomothetic creation was a position of many scientists who rejected special creationism, but could not entirely abandon the creationist paradigm to embrace positivistic evolution: evolution entirely by natural means.

264 Camping, Harold. *Adam When?* 1974. Alameda, Calif.: Frontiers for Christ. Camping has written for *CRSQ.*

265 Carazzi, D. *Il Dogma dell' Evoluzione.* 1920. Italy. **Field** 1941 says Carazzi was an "Italian biologist quoted by **Vialleton** [who] rejected evolution" in this book.

266 Carlile, Warrand. *Geological Confirmation of the Truth of Scripture.* 1850. Glasgow, Scotland: W.G. Blackie (printer). Attack on Chambers' *Vestiges.* Carlile said the author of that work, "in order to gratify an insensate and almost insane propensity for theorizing ... would destroy his own soul, and those of his reader." (Quoted in Millhauser.)

267 Carroll, Charles. *The Negro a Beast; or, In the Image of God.* 1900. St. Louis, Mo.: American Book and Bible House. *Carroll:* said he spent fifteen years and $20,000 writing this book. "The Bible as it is!" "The Negro created a beast, but created with articulate speech, and hands, that he may be of service to his master—the White man." Rejects notion that Negro is descended from Ham. Negro was created before Adam as one of the Genesis "beasts." Presents biblical and scientific proofs. "All scientific investigation of the subject proves the Negro to be an ape, and that he simply stands at the head of the ape family..." Has mind, but no soul. Quotes **Guyot** a lot, and **Winchell** 1890 (on inferiority of Negro); also de Gobineau. Long section on differences between whites and Negroes. The "serpent" in Eden was a Negro (Carroll also wrote *The Tempter of Eve;* 1902). Demands a return to "primitive Christianity" and belief in literal truth of the Bible. If Genesis is rejected, then Redemption message will be too. There are only two origin theories, and they are in "absolute conflict": biblical creation and atheistic evolution. Has chapter "The Theory of Evolution Exploded; Man was Created a Man, and Did Not Develop from an Ape." What good is biblical teaching on Sunday if atheistic evolutionist "filth" is forced upon children the rest of the week? Genesis prohibits mating with "beasts" (Negroes): they were created a different 'kind'—not of the same flesh. Cain's offspring were cursed by the sin of "amalgamation." God destroyed mankind in the worldwide Flood because man had corrupted his kind by amalgamation with Negro beasts. Negroes were taken aboard the Ark along with the other animals. God had not intended Christ to be crucified; this too resulted from the sin of amalgamation. Drawings. **Destiny Pub.** 1967 is based on this book.

268 Carron, T.W. *Evolution: The Unproven Hypothesis.* c1957. Worthing, England: Lindisfarne Press. Standard creation-science arguments. Cites **G.M. Price**. Chapter on "Evil Consequences of Evolution." Evolution part of anti-God conspiracy. Creation possibly millions of years ago—no hint of age in Bible. Java Man: Thigh bone found by DuBois probably human; said skull was from giant gibbon. He admitted 30 years later that he had found two other skulls (Wadjak skulls) there also.

269 Carter, Russell Kelso. *Alpha and Omega.* 1894. San Francisco: O.H. Little. *Captain Carter:* early advocate of pre-Flood Water Canopy theory; friend of **Vail.** Carter believed that the Bible prophesies the Canopy will be restored at the Millenium. Mentioned in **LaHaye** and **Morris** 1976 as discounting Nouri's Arkeology claim.

270 Carver, Wayne. *The Panorama of the Ages* (booklet). Undated. San Antonio, Texas: Christian Jew Foundation. Orig. radio sermons by Carver. Strong affirmation of strict creationism. Blasts the scoffers of the Last Days described in II Peter as "evolutionary uniformitarians" who reject supernaturalism and believe the earth is old. "Creation is a Scientific Fact." Describes and endorses the pre-Flood water vapor canopy as explanation for Edenic conditions. Its collapse, and the waters of the deep, caused the worldwide Flood. II Peter also foretells second destruction of the world, by nuclear devastation.

271 ________. *The Science of Creation: Radio Sermons by Wayne Carver.* Undated. San Antonio, Texas: Christian Jew Foundation.

272 Caullery, Maurice J.G.C. *Le Probleme de L'Evolution.* 1931. Paris: Payot. *Caullery:* Sorbonne prof.; parasitologist; gave 1916 Harvard lecture "Problem of Evolution" (same?); wrote other pamphlets titled "The Present Problem of Evolution." "Without doubt today we feel farther from representing how evolution had been effected than 40 years ago when the writer began to study zoology." "The greatest difficulty at the present time is to reconcile this [genetic] stability with the mutability that the very notion of evolution supposes." Impossible to imagine development of symbiosis, or of complex structures, by gradual stages. Discusses examples of insect parasitism. Apparently rejects the "teleological" explanation (creationism) after noting its attraction, but emphasizes difficulties of Darwinism. He suggests instead either "preadaptation," or changes induced in the egg by the environment. (Quoted by **Lammerts** in 1974 *J. Christian Reconstruction;* **Shute** 1961.)

273 Causton, F.G. *Tooth Replacement in Fish & Evolutionary Theory* (pamphlet). Undated. England: Evolution Protest Movement. EPM pamphlet #29.

274 Chaffer, Richard. *The Creation, or The Agreement of Scriptural and Geological Science.* 1856. Gap Theory creationism. (Cited in Millhauser.)

275 Chalmers, Thomas. *Evidences and Authority of the Christian Revelation* (booklet). 1815. Orig. pub. 1814 as *Encycl. Brit.* article. *Chalmers:* evangelical divinity prof. at Univ. Edinburgh; founded Free Church of Scotland; respected for his eloquence and his work with the poor. In this booklet Chalmers proposed the Gap Theory of creationism.

The new science of geology showed the earth to be far older than strict creation allowed. Catastrophists had harmonized geology with the Bible, but had become less biblically literalist, and were being challenged by Huttonians. Chalmers suggested that the first verse of Genesis described an initial creation, followed by the ages of geology. Gen. 1:2 alludes to the destruction of this first creation. Gen. 1:3 describes God's re-creation of the world in six literal days.

276 ________. *On the Power, Wisdom, and Goodness of God as Manifested in the Adaptation of External Nature to the Moral and Intellectual Constitution of Man.* 1833. London: Pickering. 2 vols. One of the Bridgewater Treatises. Gap Theory creationism. Chalmers' Gap Theory became extremely popular, especially in England, as a means of reconciling Genesis with geology. Also wrote *Natural Theology* (1857).

277 Chambers, Claire. *The SIECUS Circle: A Humanist Revolution.* 1977. Belmont, Mass.: Western Islands. SIECUS (Sex Information and Education Council of the U.S.) is humanist conspiracy with "vast spheres of influence . . . skillful technique of deception." "Massive compendium" of information on the "organized humanist network" which is "dedicated to the proposition that all men are not created, but evolved from inanimate matter, and therefore owe no allegiance to God." Has blurbs by Max Rafferty (former Calif. superintendent of schools), and C. Rice (Notre Dame Law Sch.). Names all known SIECUS members, organizations, and their communist affiliations. Goals of UN are "population control, scientific breeding, and Darwinism." Theosophy is Satanism. Humanists (ACLU) stage-managed the Scopes Trial. (Publisher is arm of the John Birch Society, though JBS is nowhere mentioned in book.) Index.

278 Chapman, James Blaine, and Basis William Miller. *Evolution Has Failed.* c1922. Kansas City, Mo.: Nazarene Publishing House. (Listed in Ehlert files.)

279 Charles, J.D. *Fallacies of Evolution.* Quoted and frequently cited in **C. Lehman** 1933.

280 Charroux, Robert. *Legacy of the Gods.* 1974 [1964]. New York: Berkley. Orig. 1964; French. Dedicated to Cocteau, who supposedly admired author's previous book. Spacemen from Venus; atomic war between Atlantis and Mu; the Bible, Noah and the Flood, Ezekiel; **Velikovsky,** the Pyramids; Holy Grail; etc., etc.

281 Chase, Hiram. *A Treatise on Cosmogony and Geology.* 1849. New York: Rev. Hiram Chase. Subtitle: *Together with the Changes Produced Upon the Surface of the Earth Since Creation, and the Causes of Those Changes; Viewed in Accordance with the Literal Sense of Divine Revelation. Hiram:* minister. "The object of this work is not only to vindicate the literal account of the Mosaic history of the creation, against the figurative and far-fetched theories which assign to the world vast ages of existence before the days of Adam; but also to *try to show,* that *all the changes* the earth has undergone since its creation, (in the rising of mountains and the sinking of valleys, in the formation of the different series of earths, from the lower stratum of the *transition rocks* to the *upper tertiary,*) could as easily have been accomplished within six thousand years, as within any other given time."

282 Chatelain, Maurice. *Our Ancestors Came from Outer Space.* 1979. New York: Dell. Orig. French (Doubleday held rights). Chatelain moved to Calif. from Morocco (he deplored the social chaos after France pulled out). In aerospace work: U.S. Navy, Air Force, NASA, Convair, North American Aviation (telecommunications and radar; worked on Apollo program communication system). ETs came to earth 65,000 years ago; started sudden evolution of man by improving mankind's intelligence through "insemination and mutation," then initiated man into secrets of civilization and science.

283 Chesterton, Gilbert K. *The Everlasting Man.* 1925. New York: Dodd, Mead. *Chesterton:* journalist, novelist, and lay expositor of Christianity. This book is argument against comparative religion claims that all religions are basically the same. Includes anti-evolution arguments, especially stressing the unbridgeable gap between man and animals, which is too great for transitional forms. (Discussed and quoted in Gardner; section on Satanism excerpted in 1985 *SCP Newsletter*.)

284 Chestnut, D. Lee. *The Monkey's on the Run: ... Yes, It's About Evolution!* (booklet). 1969. Priv. pub. (Phoenix, Ariz.). *Chestnut:* electr. engineer (43 years with GE) and business analyst. Retired from GE; now lectures for Christian Business Men's Committee Int'l on nuclear physics and the Bible. Orig. Presbyterian; now Conservative Baptist. Discusses Scopes Trial and 1968 Arkansas case. Standard creation-science arguments. Scientific references; relies on **Moorhead** and Kaplan a lot.

285 _______. *The Atom Speaks.* 1973 [1951]. San Diego: Creation-Science Research Center. Orig. 1951 (Grand Rapids, Mich.: Eerdmans). Also booklet series with same title. The atom speaks and "Echoes the Word of God." Nuclear physics proves Designer. Biblical End of World prophesies nuclear destruction. Includes chapter "Nuclear Science Supports Creationism." Photos, diagrams.

286 Chi-Lambda Fellowship. *The Christian Faith and Evolution.* 1982. Lincoln, Ill.: Journal for Christian Studies. Special issue of *A Journal for Christian Studies* (Chi-Lambda Fellowship): vol. II, no. 1. John Loftus, ed. Based at Lincoln Christian Sem. (Lincoln, Ill.).

287 Chick, Jack T. *Big Daddy?* (booklet). c1972. Chino, Calif.: Chick Publications. One of Chick's many comic-book format tract booklets. Perhaps the most popular and widely-distributed piece of creationist literature ever. Chick is also notorious for savage anti–Catholic material. College student stands up to obnoxious evolutionist professor who ridicules his Christianity. Student hero calmly and boldly reduces prof. to stammering idiocy with creation-science arguments, then leads class to Christ. Includes drawings of alleged ape-men series which is copied (unacknowledged) from Time-Life evolution book, but deliberately mislabeled by Chick (says all are fakes or true humans). Nasty tone. Acknowledges assistance of **Davidheiser** for arguments.

288 _______. *The Ark* (booklet). c1976. Chino, Calif.: Chick Publications. Comic-book format. Chick's drawings compare favorably with secular comicbooks. Story of Russian attempts to sabotage efforts by the Christian heroes to locate Noah's Ark atop Mt. Ararat (the commies know it really is there). Includes descriptions of Ark and Flood. Repeats Ark-eology tales: **Navarra** and Czarist expedition (evidence destroyed by Communists). Cites creation-science books.

289 _______. *Primal Man?* (booklet). c1976. Chino, Calif.: Chick Publ. Comicbook format. Acknowledges help from **Segraves, Kofahl** and **Davidheiser**. Christian heroes convince exploitative makers of evolutionist movies of truth of creationism, with standard creation-science arguments (Paluxy, ape-men hoaxes, etc.). But film-makers continue to make evolutionist movies for the money. Chick's portrayal of Christians vs. atheist-evolutionists fits neatly into comic-book good-guy/bad-guy format. Simple, mean, and crude. Effective artwork.

290 Chittick, Donald E. *The Controversy: Roots of the Creation-Evolution Conflict.* 1984. Portland, Ore.: Multnomah Press. Multnomah Critical Concern series. *Chittick:* Ph.D.; former chem. prof. at George Fox College (Quaker); R&D director of Pyneuco (converts biological waste into fuel). Chittick is proving that creationist assumptions result in better science—better synthetic fuels. His own research is based on creationist assumption of rapid formation of coal. Standard creation-science arguments, attractively presented. Scientific, philosophical, and biblical proofs. Gödel's theorem: truth or falsity of logical system can't be proved within system (e.g. creation/ evolution); must go to absolute reference outside system—and Bible is absolute truth. "A miracle is not supernatural"— rather, it is simply a "noncustomary act" of God's will. Emphasizes necessity of creationism for morality. Strict creationism; Flood. Index.

291 Clark, Gordon H. *The Philosophy of Science and Belief in God.* 1964. Nutley, N.J.: Craig Press. *Rev. Clark:* Ph.D.—Univ. Penn.; also grad. study at Sorbonne (Paris); philosophy prof. at Butler Univ.; strongly Calvinist. Philosophy of science—logical rather than empirical arguments. Demonstrates limitations of science, and concludes with appeal to Christian revelation to go beyond science. Does not discuss evolution directly. Clark is championed by his disciple John W. Robbins of the Trinity Foundation in Maryland.

292 ________. *The Biblical Doctrine of Man.* c1984. Jefferson, Md.: Trinity Foundation. Opening chapter on "Creation and Evolution." Strongly and openly creationist: Bible "definitely asserts" special creation of Adam and Eve. The "most vigorous attack on the divine creation of man has been the theory of evolution. In the United States its exponents have obtained governmental compulsion of its teaching and governmental prohibition of the theistic view. Perhaps this is an indication of the intellectual weakness of the evolutionary theory.... Today the organized educators use legal force to ban the view they dislike. This method of legal repression may be subconsciously supported by the suspicion that scientific theories are tentative only.... One can say that in biology Darwinianism is not universally accepted.... The evolutionists must rely on political restraint." Says Piltdown hoax was based on one tooth, and used as proof of evolution until final disproof. Denounces behaviorism. Rest of book is theological and philosophical argument. Clark also wrote *A Christian Philosophy of Education* (1987).

293 ________. *In the Beginning* (pamphlet?). John Robbins' *Trinity Review* essay series #40.

294 Clark, Harold W. *Back to Creation.* 1929. Angwin, Calif.: Pacific Union College. "A Defense of the Scientific Accuracy of the Doctrine of Special Creation, and a Plea for a Return to Faith in a Literal Interpretation of the Genesis Record of Creation as Opposed to the Theory of Evolution." *Clark:* Seventh-day Adventist; studied under **G.M. Price** at Pacific Union College in 1920, then taught biology there himself for 35 years (now emer. prof.). Strict creationism.

295 ________. *Genes and Genesis.* c1940. Mountain View, Calif.: Pacific Press. Bibliog.

296 ________. *The New Diluvialism.* 1946. Priv. pub. (Angwin, Calif.: Science Publications). By the late 1930s, Clark began to realize **Price**'s Flood Geology had certain shortcomings. Though he remained a strict creationist and Flood Geologist, he updates, in this work, Price's *New Geology* by adding discussion of Post-Flood glaciation. Also introduces his Ecological Zonation Theory: proposing that the systematic order of fossil strata results from burial by the Flood of different life zones or ecological communities (a departure from Price's assertion that the strata could and did appear in any order). According to R. Numbers, when Price learned of Clark's "apostasy," he denounced him for years, and addressed a "vitriolic pamphlet," *Theories of Satanic Origin,* to him. Clark, however, insisted he was still a literal creationist, and that his disagreement with Price was "merely a matter of interpretation of details and never a question of fundamental concepts of creationism or diluvialism" (1966).

297 ________. *Creation Speaks.* c1947. Mountain View, Calif.: Pacific Press. "A Study of the Scientific Aspects of the Genesis Record of Creation and the Flood." Purpose of book: "to present the viewpoint of a creationist with respect to scientific problems involved in a literal interpretation of Genesis." Begins by affirming Bible is the inspired Word of God, and is historically and scientifically true. "Any true scientific theory regarding the origin and early history of earth and its life must agree with a plain, simple, obvious rendering" of Genesis. "Clear distinction must be made between the *facts* of science and theories by which scientific data are interpreted." Facts are absolute and unchanging; theories reflect assumptions. "Belief in evolution or creation ... is based upon certain assumptions. The evolutionist must assume that the earth is very ancient.... The creationist, on the other hand, if he accepts the Genesis record literally, is committed to a short chronology." Clark attempts a "positive" treatment of creationism, rather than simply a debunking of evolution. Corrects common accusation that creationists still believe in fixity of species, but affirms there are "ultimate units" which cannot mix. Original plan of Creation has been corrupted by Satan; this involved much mixing and variation within 'kinds.' Some animals became carnivores. Parasites are clear examples of degeneration. Describes the six Creation days. Mentions Water Canopy favorably. Presents Flood Geology and standard creation-science arguments. Mentions importance of **E.G. White**'s strict creationism in molding Seventh-day Adventist views.

298 _______. *Wonders of Creation.* 1964. Mountain View, Calif.: Pacific Press.

299 _______. *Evolution and the Bible.* 1965. Washington, D.C.: Review & Herald. Filmstrips with accompanying scripts.

300 _______. *Crusader for Creation: The Life and Writing of George McCready Price.* 1966. Mountain View, Calif.: Pacific Press. "Incorporating biographical material prepared by R. Lyle James." Biography of Clark's mentor and hero, creation-science pioneer **Price.** Traces development of Price's Flood Geology; presents and summarizes his major arguments. Discusses important influences on Price, his relationship with various fundamentalist leaders, his role at several Seventh-day Adventist institutions, and the reception of his theories. Mentions that Price attended W. Bateson's famous 1921 Toronto lecture criticizing Darwinian explanations. Downplays the significance of Price's accusation that Clark betrayed him when he criticized and modified certain aspects of his Flood Geology in the 1940s—says that Price heard a distorted account of his views, which were merely attempts to update Price's model in the light of new knowledge. Includes bibliog. of Price's books, and excerpts from most of them.

301 _______. *Genesis and Science.* c1967 [c1940]. Nashville, Tenn.: Southern Publishing Assoc. Orig. c1940; priv. pub. (Angwin, Calif.). Small book intended for young and lay readers. Strict creationism; standard creation-science arguments. Earth 6,000 years old. The Flood affected C-14 concentrations—old C-14 dates thereby explained as in error. Chronology of Flood. Worldwide Flood myths. Evolution is an old pagan idea. Includes scientific examples but not references. Charts, diagrams, index.

302 _______. *Fossils, Flood, and Fire.* 1968. Escondido, Calif.: Outdoor Pictures. Foreword by **E. Booth.** Standard creation-science Flood Geology and paleontology arguments. Lots of photos, mostly of geological formations (most taken by Booth). Topics include: a historical survey; the "delusion" of uniformity; the Flood Theory (classic Flood Geology and **Price**'s re-invention); formation of coal by Flood. Tertiary vulcanism caused sudden (single) Ice Age by clouding atmosphere with dust (inverse greenhouse effect). Mentions that he differs in some details from **Whitcomb** and **Morris.** Favors Price and other Seventh-day Adventist creationists, but endorses CRS. Photos, charts, diagrams, maps; index.

303 _______. *Questions on Evolution and Creation.* 1971. Angwin, Calif.: Life Origins Foundation. (Listed in Ehlert files.)

304 _______. *Why Creation Should Be Taught in the Public Schools ...* (pamphlet). 1973. Caldwell, Idaho: Bible-Science Association. "Controversy in California." Reprint from *BSN* (1973). Orig. presented at public meeting in Cresent City, Calif., 1973. Discusses BSCS curriculum and various attempts by creationists to counter the teaching of evolution in schools. Argues that the "only reasonable procedure" is to present both scientific "models," leaving students to choose the most satisfactory explanation.

305 _______. *The Battle Over Genesis.* c1977. Washington, D.C.: Review and Herald. Informative and fairly comprehensive discussion of the history of creationist and evolutionist ideas and theories. Includes the Greek philosophers; ancient creation myths; pagan, Islamic, and Jewish influences on early Christian philosophy and theology; medieval mysticism; the development of modern science; the "golden age of creationism" (the Reformation and Counter-Reformation); classical Diluvialism; geological and Darwinist challenges to creationism; creationist "detours" (old-earth creationist theories); the rise of the new creationism and Flood Geology pioneered by **G.M. Price;** fundamentalist advocacy of creationism; neo-Darwinism and other controversies in evolution. Useful section on the modern creationist movement; describes various interpretations and theories. Includes biog.

notes on many of the historical figures discussed in book. Bibliog., index.

306 ________. *New Creationism.* c1980. Nashville, Tenn.: Southern Publishing. Similar to Clark's other books; update of same arguments. Adds some discussion of modern biology. Neither creation nor evolution can be proved—data can be interpreted to fit either "model." "Diluvialists have generally considered Precambrian rocks as having been in position before the Flood, but recent studies have made some revision necessary." (Precambrian may be Flood-deposited also.) C-14 either not present before Flood, or destroyed by it, thus accounting for apparent C-14 dates of over a few thousand years. "Recommended Reading" list is all Seventh-day Adventist, except for **Morris.** Index.

307 Clark, Marlyn E. *Our Amazing Circulatory System: By Chance or Creation?* (booklet). c1976. El Cajon, Calif.: Institute for Creation Research. ICR Technical Monograph No. 5. Other printings titled *Optimal Design in Cardio-vascular Fluid Mechanics: Our Amazing Circulatory System. Clark:* theor. & applied mechanics MS—Univ. Illinois; theor. & applied mechanics and bioengineering prof. at Univ. Illinois Urbana-Champaign.

308 Clark, Robert E.D. *Creation.* 1958 [1946?]. London: Tyndale Press. *Clark:* organic chem. Ph.D—Cambridge Univ.; active in Inter-Varsity Christian Fellowship in England.

309 ________. *Scientific Rationalism and Christian Faith.* 1951 [1945]. London: Inter-Varsity Christian Fellowship. 3rd ed. Orig. 1945. "Dealing with the consequences of the thinking of Prof. J.B.S. Haldane and Dr. J.S. Huxley, the author relates evolution, dialectic materialism, and agnosticism while noting its impact on and implications for religious faith. This book is a study of once-born and twice-born scientific rationalists" (**J.N. Moore** in 1964 *CRSQ*).

310 ________. *Darwin: Before and After.* 1967 [1948]. Chicago: Moody Press. Orig. pub. 1948 (also 1966) by Paternoster (Exeter, Devon, England). "An Evangelical Assessment." Influential book; much cited by creationists. The rise of evolution; the scientific and personal factors which led Darwin to write the *Origin.* The harmful effects of Darwinism on society and political thought. Evolution can't "transform the fundamental structures"—superficial change only. Fundamentalist arguments (strict creation) are often "rubbish," but such opposition to evolution is still valuable. Quotes **McCabe:** fundamentalists say horse evolution is asserted from one fossil—*Eohippus*—which is really a large California rat trapped by the Deluge. Entropy ("morpholysis") argument.

311 ________. *The Christian Stake in Science.* 1967. Chicago: Moody Press. Orig. pub. by Paternoster (Exeter, England). **Hayward** 1978 says Clark includes a strong argument for the Revelatory Theory of Genesis (creation revealed in six days).

312 ________. *The Universe: Plan or Accident?* 1972 [1949]. Grand Rapids, Mich.: Zondervan. Orig. pub. 1949 by Paternoster (London); also 1961 rev. ed. Scientifically knowledgeable; relatively non-dogmatic. Old-earth approach. Creation occurred billions of years ago. Admires **A.R. Wallace**'s proto-"anthropic principle" argument. Doubts existence of extraterrestrial life, but not worried about the possibility. Clark also wrote *Science & Christianity: A Partnership* (c1972; Mountain View, Calif.: Pacific Press), which does not, however, deal specifically with evolution.

313 Clark, Robert T., and James D. Bales. *Why Scientists Accept Evolution.* 1966. Grand Rapids, Mich.: Baker Book House. Also pub. by Presbyterian and Reformed (Nutley, N.J.). *Clark:* space medicine. Analysis of Darwin and five contemporaries: Hutton, Lyell, Spencer, Huxley and Wallace. "Anti-supernatural bias" of many 19th-century scientists. These scientists rejected God first, and then accepted evolution as a substitute. Scientists now accept evolution largely because their teachers did—it is the new conformity. It is accepted uncritically; it is heresy to question it. Bibliog.

314 Clarke, Adam. *Commentary on the Scriptures.* 1810–26. England. Bible Commentary (Authorized version). Millhauser says Clarke claimed that thorns and disagreeable things were formed in a minor special creation after the Fall.

315 Clarke, G.C. Leopold. *Evolution and the Breakup of Christendom; or, World Conditions Traced to Modern "Science."* 1930. London/Edinburgh: Marshall, Morgan & Scott.

316 Clayton, Charles Lincoln. *Where Darwin Erred: or The Fallacy of Natural Selection.* 1933. Boston, Mass.: Meador Publishing. Supplants Darwin's theory: "An intelligent creative and sustaining force, or God, operating in the germ cell and other cells of organic beings, forms, sustains, and evolves them—when they evolve." (From Table of Contents:) "The truths of God and religion not dependent upon the Bible—the Bible a great book of moral and religious authority— The statement in Genesis that God formed all creatures an immortal truth— Many modern scientists have failed to attain it— Divine truth safe whether the Bible harmonizes with evolution or not..."

317 Clayton, John N. *The Source: Eternal Design or Infinite Accident?* c1976, c1983. Priv. pub. (South Bend, Ind.). *Clayton:* public school teacher (earth science, etc.); geology(?) M.S.; Church of Christ member; produces and distributes creation-science newsletter, tapes, filmstrips, etc. Clayton has given his "Does God Exist?" lecture series on many campuses, and made it into a film/video series. Book is aimed at high school students. Old-earth creationism, with old-earth proofs. Full of Bible references. Design of cosmos and of earth proves that God created. Science in the Bible proves it is God's Word. Genesis creation is same order as geological record. Clayton rejects standard Gap Theory. Suggests long ages before six-day creation, though. (Clayton advocates a modified Gap Theory, with elements of Day-Age creationism, and perhaps some theistic evolution). Standard anti-evolution arguments. Charts, pictures; bibliog., index.

318 ________. *Evidences of God.* 1977–1984. Superior Printing. 3 vols. Articles from Clayton's *Does God Exist?* newsletter: 1972–76, 1976–79, 1979–83.

319 ________. *Dandy Designs.* 1984. Priv. pub. Design articles from Clayton's bulletins 1974–82 "which show God's wisdom and planning in the creation.

320 Cleveland, Leroy Victor (ed.). *An Anti-Evolution Compendium* (booklets). 1954. Henniker, N.H.: High Way Press. Set of booklets: vols. 24(2), 25(1–2), 26(1–2) of *The High Way. Cleveland:* Sec'y of U.S. Div. of EPM; lives in Canterbury, Conn.; wrote Foreword to **Meldau** 1959.

321 Clough, Charles A. *A Calm Appraisal of The Genesis Flood.* 1968. Unpub. thesis: Dallas Theological Seminary. Dallas Theol. Sem. Th.D. thesis. *Clough:* math B.S.—MIT; MIT grad. study; meterologist and civil defense advisor—USAF Res.; pastor of Lubbock Bible Church (Texas); participated in Paluxy searches. Includes a "detailed study" of all major reviews of **Whitcomb** and **Morris**'s *Genesis Flood.* Conclusion: "complete failure on the part of TGF critics to refute the work at its basic foundation: exegesis of Genesis 6–8." Agrees with Whitcomb and Morris that the historical sciences should be reconstructed to fit with Genesis. (Quoted in author's article in 1974 *J. Christian Reconstruction.*)

322 ________. *Laying the Foundation.* 1977 [1973]. Lubbock Texas: Lubbock Bible Church. Cited in **Whitcomb** and DeYoung as a work which "contributed significantly to the current renaissance of Biblical creationism."

323 Clube, Victor, and B. Napier. *The Cosmic Serpent: A Catastrophic View of Earth History.* 1982. New York: Universe Books. *Clube and Napier:* Univ. Edinburgh astronomers. First half of book about astronomy; second half its relation to myth, history and religion. Authors propose that close approaches of a comet profoundly altered terrestrial and human history. Interstellar comets have subjected earth to "episodic bombardments controlled by the galaxy" (cites lunar soil analyses). Some become asteroids. Discusses effects of impacts on earth. The "mystery" of short-term comets: too many, according to standard theories of their origin. The authors examine myths and legends in a manner similar to **Velikovsky** (though they reject much of his particular astronomical theory); they also shorten Egyptian chronology by some 400 years. Clube and Napier propose a cometary origin for religion: the awesome comets were seen as gods, also dragons. Stress the polytheistic and astronomical nature of prehistoric religions as supporting cometary origin. A great comet produced spectacular displays; its breakup inspired tales of gods and cosmic battle. Discuss references to comets in Genesis; claim comet precipitated the Flood. The great comet approached earth in 2500 B.C. and 1369 B.C. **Whiston, Bellamy,** and **Velikovsky** were definitely on the right track. Authors stress the erratic nature of the fossil record, and discuss interpretation and patterns of mass extinctions. Comparing **Cuvier** with Darwin, they argue that the triumph of Darwinism unfortunately led to the abandonment of catastrophic theories and consideration of cosmic encounters. Implication is that cosmic catastrophes are the most significant factor in evolution.

324 Cockburn, Patrick. *An Enquiry Into the Truth and Certainty of the Mosaic Deluge.* 1750. London: C. Hitch and M. Bryson. *Cockburn:* vicar of Long-Horsely, Northumberland, England. An examination of the arguments of Isaac Vossius.

325 Cockburn, Sir William. *The Creation of the World* (booklet). 1840. London: J. Hatchard & Son. *Cockburn:* D.D.; Dean of York. According to A. White, Cockburn frequently denounced geology in press and from the pulpit of York Minster, though he had little knowledge of it. This booklet is addressed to R.J. Murchison; it refutes his book on the Silurian System. Argues for rapid formation of geological strata by Flood and by simultaneous submarine volcanic eruptions. "We may then, sir, easily believe from what we see, that the floodgates of heaven were opened, and that the clouds poured rain upon the earth in vast superabundance, and covered it with water. We may easily believe that this great catastrophe was attended with numerous violent volcanos both in sea and on land." "To me it appears, that if you allow a sufficient number of volcanos acting with immense power and at short intervals, every phenomenon or alteration now known may have been produced in a few months after the catastrophe began." Says "the doctrine of the earth's vast antiquity is but a bubble, which vanity has blown." Alternating series of eruptions entombed distinct sets of organisms and formed various strata, to which were added land animals and plants swept away by the Flood. Marine volcano pulverized crust and formed the Cambrian. Other volcanoes formed Silurian. Old Red Sandstone formed from torrential rain sediments. Active animals fled to higher levels; man to the highest peaks. Trilobites in lowest levels—they "had no feet." Fossils carried all over the world by Flood. Suggests that the Flood tilted the earth's axis. Cockburn challenged the BAAS to debate him on age of the earth.

326 ________. *The Bible Defended Against the British Association* (pamphlet). 1844. London: Whittaker. White says this pamphlet went through five editions in two years. Cockburn, he says, denounced the Brit. Assoc. for the Advancement of Science for conducting geological investigations.

327 ________. *A New System of Geology* (booklet). 1849. London: Henry Colburn. Addressed to **Sedgwick.** Argues against successive creations; claims there was just one creation and one great convulsion which left world in present state. Points out that his theory does not contradict the Bible, and accounts for all geological facts. "The account given by Moses *must,* then, be accurately true—I do not say it *may* be true, but it *must* be true; and its truth is most wonderfully established by the discoveries of geologists." The modern theories of geology are, however, all wrong. World was originally tranquil: equator equaled the ecliptic; no seasons. Marine volcanoes formed Silurian before the Flood. Old Red Sandstone formed from Flood rains. Mammals flee Flood waters; man flees furthest of all. Liquid coal ejected by volcanoes, then covered with vegetation uprooted by Flood. Different species found in separate strata, due to differing survival abilities or growth conditions. Fossils of sea animals found worldwide, mixed with land animals.

328 Coder, S. Maxwell, and George F. Howe. *The Bible, Science and Creation.* 1966 [1965]. Chicago: Moody Press. Rev. ed. Orig. 1965. Also pub. 1966 by Victory Press (London). *Coder:* dean of education and VP of Moody Bible Inst. (Chicago); former business executive. First section of book asserts divine origin of Bible. Authors present many examples of scientific knowledge contained and predicted in the Bible, and present Bible-science examples from many scientific disciplines. They lament that "natural man" is incapable of acknowledging the Bible's supernatural origin. Notes that God exterminated the Canaanites and their children as an "act of mercy" since their sins were so grievous. Much of book concerns creation-science; strongly rejects evolution as contrary to God's Word. Presents Flood Geology, Gap and Day-Age theories as three creationist alternatives; does not claim to know which one is true. Also includes sections on archeological confirmation of the Bible, and statistical proof of Bible prophecy. Index.

329 Coffin, Harold G. *Creation–Accident or Design?* c1969. Washington, D.C.: Review and Herald. *Coffin:* Ph.D.; paleontology prof. at Andrews Univ.; prof. at Geoscience Research Inst. 512-page presentation of Seventh-day Adventist creation-science. Some sections written by **E. Booth, H.W. Clark,** Robert H. Brown, Ariel Roth, and Edward E. White. Relies heavily on authority of **E.G. White;** cites her frequently. Also many other scientific, theological, biblical and creationist references. "Seventh-day Adventists reject the concept that any form of terrestrial life existed prior to creation week. Thus they have always held that the fossils must be interpreted largely in terms of the Deluge described in . . . Genesis." Describes the literal days of Creation. Adam was twelve feet tall (quotes White). Virility of organisms was greater at creation than now; Genesis 'kinds' may have crossed before the Flood. Some of the bizarre fossil forms, including ape-men, may be the result of crossing between 'kinds.' God declared these degenerate products of amalgamation "corrupt." Describes effects and evidences of the Flood, fossil formation, mountain-building, glaciation, radiometric dating, speciation, adaptation, genetics, "missing links and living fossils," and other standard creation-science topics. Arrangement of fossils is due largely to ecological zones, but some were transported long distances by the Flood. Concedes that radiometric dating contradicts "inspired testimony"–calls for further research. White stated that many new species have been produced by hybridization since the Flood. "It is Satan's desire to bring discredit upon the Creator, to cause discomfort to man, and to support his counterfeit of the creation story by working through the laws of genetics to bring about thorns on roses, stingers on nettles, parasites, predators, and the host of other ugly and degenerative changes." Affirms that only God can create life; man will "certainly never" be able to–not even Satan. Cites **Kerkut;** Austin H. Clark's proposal in *The New Evolution: Zoogenesis* (1930) of explosive development then stasis. Creation and evolution are both accepted on faith. Believers in biblical creation are "transformed morally," though, which attests to its truth. Stresses that Seventh-day Adventist belief in Sabbath day is based directly in Creation account. Photos, drawings (by **Baerg**), diagrams, charts; index.

330 ________. *Origin by Design.* c1983 [1982]. Washington, D.C.: Review and Herald. Also pub. 1982 (Minneapolis: Burgess). Written "with" Robert H. Brown; A. Roth apparently also assisted. Edited by **Gerald Wheeler.** Same topics and sections as 1969 book. Adds many recent scientific and creation-science references. Does not cite **E.G. White** (except one endnote). Many biblical references, but does not mention Seventh-day Adventism in main text. Coffin describes his flotation experiments (at GRI) supporting Flood origin of coal, and other evidences for worldwide Flood. Presents model to explain Flood origin of Yellowstone; compares with Mount St. Helens. Suggests that there was much volcanic activity during and immediately following Flood. "Creationists do not believe man has developed gradually from ape, but rather we believe a Creator fully formed him and that any change since then has been degenerative. Accordingly, we should expect to encounter the remains of modernlike or fully human beings in the oldest sediments that contain hominoid bones." Explains how C-14 data can be correlated with Flood and recent creation model. "The general theory of evolution is contrary to some of the basic laws of science." "Since both Creation theory and evolution theory lie outside the realm of science, we cannot make a decision on the basis of which one is science and which one is not." We must decide which fits the evidence best. Many photos, charts, drawings; index.

331 ________. *Creation: The Evidence from Science* (booklet). Undated. Anacortes, Wash.: Outdoor Pictures. Also pub. by Life Origins Foundation (Wash.).

332 Cohen, I.L. *Darwin Was Wrong—A Study in Probabilities.* c1984. Greenvale, N.Y.: New Research Publications. Cohen's conclusion, based on math and probability concepts: "there is no possibility that evolution was the mechanism that created" the approximately six million species. This study refutes Darwinism; shows instead the "evolution of thought of a design engineer." This Designer creates new, more mature blueprints, with new systems of sub-structures. Now we can do the same, with genetic engineering. Declares that "every single concept advanced by the theory of evolution (and amended thereafter)" has been proven false. No creation-science works cited. "There are a good number of scientists who now reject the concepts of evolution—not on religious grounds, but on strictly scientific grounds." But fear of ridicule and persecution keep them from speaking out. Scientists should "recognize the patently obvious impossibility of Darwin's pronouncements and predictions ... let's cut the umbilical cord that tied us down to Darwin for such a long time. It is choking us and holding us back."

333 Cole, D. *Evolution Not a True Science* (pamphlet). Undated. England: Evolution Protest Movement. EPM pamphlet #5.

334 Cole, Glenn Gates. *Creation and Science.* 1927. Standard Publishing. Quoted in **Marsh** 1950. "It must be granted ... that the word 'creation' belongs to a realm outside of that which is comprehended in science. But that does not authorize the scientist to deny that such a realm exists. This is the weakness of much of modern science, that it assumes there are no laws or processes outside of the scientific laws or processes." Urges scientists to acknowledge existence of the supernatural realm, since problems of religion, morals, law, and society cannot be solved by reference to natural law alone.

335 Cole, Henry. *Popular Geology Subversive of Divine Revelation!* 1834. London: Hatchard and Son. "A letter to the Rev. **Adam Sedgwick** ... being a Scriptural refutation of the geological positions and doctrines promulgated in his lately published commencement sermon [Cambridge Univ.—1832]..." Scriptural Geology; calls for return to the old diluvial catastrophism.

336 Colley, Evelyn C. *Evolution Nonsense: A Christian's View in Verse* (booklet). c1983. Priv. pub. "I shall present as nonsense/ This vague hypothesis/ In contrast to Creation/ As found in Genesis." Largely inspired by reading **H. Morris.** Strict creationism.

337 Comparet, Bertrand L. *Cain-Satanic Seed Line.* (Listed in Identity book list, in Christian Defense League Report.)

338 Conant, J.E. *The Church, the Schools, and Evolution.* Chicago: Bible Colportage Association. *Conant:* D.D. Mentioned in **W.B. Dawson** 1932 as containing a large number of quotations as to the inadequacy of evolutionist explanations.

339 Conservative Baptist Pastor's Fellowship of Oregon, and The Western Conservative Baptist Theological Seminary. *Creation, Evolution, and the Scripture.* 1966. Portland, Ore. "12 lectures by six prominent men of science and theology." Sponsored by Conserv. Bapt. Pastor's Fell. Oregon and West. Conserv. Bapt. Theol. Sem. (In Ehlert bibliog.)

340 Conybeare, W.D. *Outlines of the Geology of England and Wales.* 1822. London: William Phillips. Rev. Conybeare was convinced that geology could be reconciled with the Bible. He adopted the Gap ("interval") Theory of creation, but he considered other theories of reconciliation plausible as well.

341 Cook, Charles. *Exploding the Evolution Dogma Myth.* 1981. Priv. pub. (Grand Terrace, Calif.). *Cook:* toolmaker; deacon of Baptist Church. "William Jennings Bryan Vindicated." Intended as a point-by-point refutation of

S.J. Gould's 12-8-80 "Peoples Magazine" [sic] interview. Emphasizes 1980 Chicago Field Museum conf. and punctuated equilibrium as proof of evolution's trouble. Denounces ACLU at length as communist. Focuses on Scopes Trial to correct propaganda version. Darrow was the ignorant buffoon, **Bryan** the wise hero, of the trial. Stresses that creation/evolution debate is struggle to the death. Attempts standard creation-science arguments (several are headlined on cover under caption "TRUTH"). Urges Christians to resist evolution and secular humanism mightily. Cites **Chambers** 1977 often. Urges political action. Reprints much of Bryan's intended closing address.

342 ________. *If Evolution Is False, the Bible Must Be True!* (unbound booklet). 1986. Priv. pub. (Grand Terrace, Calif.: Center for Creation Studies). Cook has recently formed the Creation Studies Ministry (orig. called Center for Creation Studies), which distributes books and publishes Cook's essays, of which this is one. "How disheartening it is to see loved ones led astray by the intellectual duplicity of evolutionism as propounded from classroom lecterns when, actually, truth resides with creation world view!" Declares that "there has never been one shred of empirical scientific evidence validating the Darwinian theory." Evolution is a "non-theistic religious belief system ... posing as science. The Bible, evolution's intractable enemy, must be discredited for it stands in the way of evolution's complete dominance of our society." Insists there are only two views on origins; they are directly antithetical. Evolution leads to secular humanist view of man's autonomy, which is by definition amoral. Creation results in democracy; evolution in "one form or another of Totalitarianism." Bible is absolutely true; other claimed revelations are false. Urges parents and students to arm themselves with God's Truth—creationism—in order to resist Satan. Quotes many long passages from **M. Denton.**

343 ________. *The Scope's Trial: A Nation Deceived* (unbound booklet). 1986. Priv. pub. (Grand Terrace, Calif.: Center for Creation Studies). "Pick up any current college textbook on anthropology, geology, or biology and you will discover that the dogma of evolution is treated as an established fact, proven by science to be so. How can this be? We conclusively show that there are *no* scientific evidences at all proving evolution to be true." Accuses evolutionists of arrogance and "intellectual intimidation." "If men evolved from the beast, the sin nature is an inherited animal characteristic and cannot be due to the fall of man through disobedience. This denies the need of a Redeemer, and thus the atonement idea of Christ is foolishness." Quotes many evolutionist textbooks. Quotes heavily from trial transcript **(National Book);** argues that the trial was a tremendous ACLU propaganda event aimed at discrediting biblical supernaturalism. (Cook also publishes reprint of **Bryan**'s intended closing address under the title *Bryan vs. Darrow*.)

344 ________. *Bondage to Decay: Evolution's Nemesis* (booklet). Priv. pub. (Grand Terrace, Calif.: Center for Creation Studies). "A discussion of the amazing accuracy of the Bible" followed by discussion of second law of thermodynamics, which "unequivocally demonstrates the impossibility of evolution."

345 ________. *Evolution: An Intellectually Bankrupt System* (booklet). Priv. pub. (Grand Terrace, Calif.: Center for Creation Studies). "Evolution, as taught in our education system in America, is materialistic to the core, and it is without *any* genuine scientific evidence whatsoever.... Mr. Cook identifies and refutes every well known so-called evidence our educators repute to validate the general theory of evolution and thus justify its total dominance of the classroom.... It is believed that no thinking person can read this expose of the dogma of evolution without categorizing the surmise as the greatest delusion ever perpetrated on the human mind." (CSS brochure.)

346 ________. *God's Young Earth Signature: Uniformitarianism Falsified* (booklet). Priv. pub. (Grand Terrace, Calif.: Center for Creation Studies). Evidence for recent creation, especially **Gentry**'s radiohalos.

347 ________. *How God Did It!: The Amazing Genetic Code Message System* (booklet). Priv. pub. (Grand Terrace, Calif.: Center for Creation Studies). Darwin's natural selection replaced **Paley**'s design argument, but recent knowledge of DNA and computer technology have brought it back.

348 _______. *Law of Two Givens: A Hidden Subtlety* (booklet). Priv. pub. (Grand Terrace, Calif.: Center for Creation Studies). Evolution and creation as strict dichotomy. Evolutionists must insist on their theory; otherwise they would have to become creationists and believe in supernatural miracles. They seek to avoid this at any cost.

349 _______. *The New Physics Revelation: A Lesson Every Biologist, Geologist, Astronomer, et al., Should Learn* (booklet). Priv. pub. (Grand Terrace, Calif.: Center for Creation Studies). **Newton**'s physics led to mechanistic philosophy. Coupled with Darwinism, it resulted in belief that everything is predetermined. Quantum mechanics proves this mechanistic view is false.

350 _______. *On Defining the Term Secular Humanism* (booklet). Priv. pub. (Grand Terrace, Calif.: Center for Creation Studies). Secular Humanism is founded upon evolution. "This insidious belief system has a stranglehold on the education system in America, not to mention the print and TV media."

351 _______. *Science, Religion, and Evolution* (booklet). Priv. pub. (Grand Terrace, Calif.: Center for Creation Studies). Evolution is "a religious belief system posing as science." It is as religious as creationism. Creationism is as scientific as evolution. Discusses limits of science.

352 _______. *Spontaneous Generation: Evolution's Achilles Heel* (booklet). Priv. pub. (Grand Terrace, Calif.: Center for Creation Studies). Shows that spontaneous generation, which evolutionists must subscribe to, is "ludicrous."

353 Cook, Melvin A. *Prehistory and Earth Models.* 1966. London: Max Parrish. *Cook:* physical chem. Ph.D.—Yale; metallurgy prof. at Univ. Utah; president of IRECO Chemicals; explosives expert (won Nitro-Nobel award for development of safe slurry explosives, and is acknowledged as a leader in blasting technology); Mormon. Fairly technical. Difficult reading for laymen, but cited admiringly by creationists. Includes standard creation-science arguments, especially critiques of radiometric dating. Also discusses other age indicators (all either unreliable or suggesting young earth), catastrophic formation of coal and oil, paleomagnetism, fossil record. Cook is well known for claim that helium content of atmosphere proves recent formation. Last two chapters are on "Life, Evolution and the Fossil Record" and "Critique on Evolution." Argues for catastrophic "ice cap" model of continental drift—pressures from ice load caused breakup of Pangaea. Equations, charts, diagrams, photos. Bibliog. of scientific references (no creationist works listed).

354 Cook, Melvin A., and M. Garfield Cook. *Science and Mormonism.* 1967. Utah: Deseret Books.

355 Cooper, Thomas. *On the Connection Between Geology and the Pentateuch, in a Letter to Professor Silliman.* 1833. Boston: J. Hall. Also pub. by Times and Gazette (Columbia, S.C.). (Silliman: Yale Univ. geol. and chem. prof. In his 1829 *Geological Lectures* Silliman explained that the upper sedimentary deposits were remains of Noah's Flood, but that lower strata were deposited by earlier deluges. Silliman favored Day-Age creationism.)

356 Cooper, Thomas. *The Bridge of History Over the Gulf of Time: A Popular View of the Historical Evidence for the Truth of Christianity.* 1877 [1871]. London/New York/Cincinnati.

357 _______. *Evolution, the Stone Book, and the Mosaic Record of Creation.* 1893 [1878]. New York: Hunt and Eaton. Orig. 1878 (London: Hodder and Stoughton). Day-Age creationism.

358 Coppedge, James F. *Evolution: Possible or Impossible?* c1973. Grand Rapids, Mich.: Zondervan. *Coppedge:* Ph.D.; calls himself director of Probability Research in Biology (Northridge, Calif.). "Molecular Biology and the Laws of Chance in Nontechnical Language." Mostly extended demonstrations of standard creationist probability arguments: the fantastic odds against the simultaneous coming together by pure chance of all the units of a protein, cell, etc. Also reviews other standard creation-science arguments. Lots of very big numbers and scientific footnotes (very respectable references). Includes some personal correspondence with **H. Morowitz;** acknowledges assistance of other non-creationists (Ostrom, Pauling, S. Fox, etc.), plus many creationists. Strict creationism. Charts, diagrams, photo; index.

359 Corliss, William R. *Ancient Man.* 1978. Glen Arm, Maryland: Sourcebook Project. One of Corliss's several Handbooks of scientific anomalies. Corliss's approach is "Fortean" (after **C. Fort**) – he reports and describes anomalies and scientific mysteries, entertaining all unorthodox hypotheses without insisting on any particular interpretation. Among the books he distributes are many creation-science and anti-evolution works; he also includes (and praises) some anti-creationist and anti-paranormal books. This volume is a compendium of curious reports from various sources (old, new, scientific, anecdotal) of evidences of humans in early geological ages and strata. Includes stone circles and alignments, ancient astronomical observatories, mysterious ruins and inventions, ancient inscriptions, etc. Creationists use Corliss as a source of puzzles which contradict evolution and orthodox science.

360 _______. *Unknown Earth: A Handbook of Geological Enigmas.* 1980. Glen Arm, Maryland: Sourcebook Project. Compendium of geological anomalies. Includes discordant radiometric dates, catastrophic floods, fossil bone beds, frozen mammoths, doubts about paleomagnetism, strata anomalies, theories of non-biological origin of coal, gaps in the fossil record, continental drift contradictions, etc.

361 _______. *Incredible Life: A Handbook of Biological Mysteries.* 1981. Glen Arm, Maryland: Sourcebook Project. Compendium of biological anomalies. Includes tautology of evolution (natural selection), degeneracy of species, sea serpents, "latent" life, thermodynamics and life, recent mammoth survivals, flu from space, recapitulation theory doubts, "superorganisms," inheritance of acquired characteristics, Yeti and Sasquatch, etc.

362 Corvin, R.D. *Home Bible Study Course.* 1976. Charlotte, N.C.: PTL Club. PTL Club is televangelist Jim and Tammy Bakker's TV network. Bakker is Assembly of God. This series is mentioned in **Bixler** 1986 as espousing Gap Theory creationism in vols. 1 and 5.

363 Cottrell, Jack. *What the Bible Says About God the Creator.* 1983. Joplin, Mo.: College Press. Theology; not creation-science.

364 Cottrell, Ron. *The Remarkable Spaceship Earth.* c1982. Denver: Accent Books. *Cottrell:* aerospace engin. M.S. – Univ. Arizona; former teacher at Cal State Northridge; has worked with North American Aviation, Saturn V rocket, Hughes Aircraft, Northrop. "God, Creation and Man . . . A Spiritual Odyssey." Attractive loose-leaf color photo book with text. Mostly shots of and from space. Basically anthropic principle argument: earth is exquisitely designed for us to live on. Emphasis on Creator; includes Bible references. Affirms Genesis days – but does not mention age of earth or Flood. Bibliog. includes young-earth creation-science books. Admires space program and NASA. Dedicated to Christ Community Church (Tucson).

365 Courville, Cyril Brian. *The Recapitulation Theory.* 1941. Priv. pub. (Loma Linda, Calif.). *Courville:* M.D.; neurology prof. at Loma Linda Univ.; founder and director of Cajal Neuropathology Lab., Los Angeles; described as the "world's greatest neuropathologist" by colleagues; Seventh-day Adventist; brother of **D. Courville.** Article of same title by Courville in 1941 *Bull. Deluge Geology & Related Sciences.* Article "The Human Embryo and Theory of Evolution" in 1942 *Ministry* magazine. Refutes the theory, often claimed as evidence for evolution, that human embryos repeat the evolutionary sequence ("ontogeny recapitulates phylogeny"). Among his many works on medical topics, Courville wrote *Injuries of the Skull and Brain in Myth and Legend* (1967; New York: Vantage).

366 Courville, Donovan A. *The Exodus Problem and Its Ramifications* (2 vols.). 1971. Loma Linda, Calif.: Challenge Books. *Courville:* M.D., embryology specialist; Seventh-day Adventist; former editor (with **G.M. Price**) of

Bulletin of Deluge Geology (defunct); has contributed to *CRSQ*. Velikovskian history: shortens Egyptian chronology by claiming overlap of six hundred years. Tries to reconcile Velikovskian scheme with Palestinian archeology by extending Bronze Age and shortening Iron Age. Exodus dated to end of Egyptian Middle Kingdom (1450 B.C.), followed by Hyksos (asserted to be the biblical Amalekites). Dispersion of Babel also confirmed.

367 Cousins, Frank W. *The Antiquity and Descent of Man (The Evidence and How It Is Presented)* (pamphlet). Hayling Island, Hamps, England: Evolution Protest Movement.

368 ________. *Fossil Man: A Reappraisal of the Evidence: With a Note on Tertiary Man.* 1966. England: Evolution Protest Movement. Epigraph: "The greatest derangement of the human mind is to believe because one wishes it to be so" (Pasteur). Standard creation-science arguments against transitional hominids.

369 ________. *The Anatomy of Evolution.* c1980. Priv. pub. (England). 2 vol. bound MS copy in **Ambassador College** Library. Largely anti-evolution scientific quotes and citations. Includes tables, charts, pictures. (**Dankenbring**'s Triumph Pub. plans to publish a book with this title—the same?)

370 Cove, Gordon. *Who Pilots the Flying Saucers?* (booklet). Undated. Priv. pub. (England). In ICR Library. "Some Flying Saucers Friendly and Harmless but Others Are Definitely Hostile." Some are good angels ("God's Inspectors"); others demons. Maybe from Mars, Venus or Moon—Bible says many "dwelling places" in the universe. Satan may have taken over planet(s) for his headquarters. Discusses full moon "lunacy." Behavior of UFOs proves they are not piloted by humans—therefore supernatural beings. Sinister connection between UFOs and spiritism; Home's classic levitations. Ezekiel and other Bible descriptions are accurate. Antichrist will imitate the Second Coming to deceive humanity, arriving in a Flying Saucer.

371 Craig, William Lane. *The Existence of God and the Beginning of the Universe.* 1979. San Bernardino, Calif.: Here's Life Publishers. *Craig:* philosophy Ph.D.—Univ. Birmingham (England); theology Ph.D.—Univ. Munich; philosophy prof at Westmont College (Santa Barbara). Cosmogony. Scientific evidence of beginning of universe as consistent with theology. Not concerned with biological evolution. (Craig has article in 1986 *Brit. J. Philos. Science* on cosmogony.)

372 Creation Evidences Museum. *Creation Evidences in Color: Creation Evidences Museum* (booklet). 1986. Glen Rose, Texas: Creation Evidences Museum. **Carl Baugh:** museum director and excavation director (also heads International Baptist College in Missouri, and claims scientific degrees). Martha Baugh: editor. **Clifford Wilson:** archeologist. Robert Chambers: artist. Coloring-book format, illustrating Paluxy manprints and other finds. Baugh is the most credulous of the Paluxy diggers, finding "manprints" everywhere among the genuine dinosaur tracks. He has established his Creation Evidences Museum near the site. Booklet shows Paluxy manprint digs, dinosaur bones, **Dougherty,** Caldwell and **Burdick** manprints, Nebraska Man tooth, Glen Rose trilobite, Ordovician iron hammer, Moab Man fossil, etc.

373 Creation Science Association of Canada. *So You're an Evolutionist* (booklet). Vancouver, B.C., Canada: Creation Science Assoc. of Canada. What kind of evolutionist? Points out that there are many theories of evolution. They can't all be right. They may all be wrong, which leaves the alternative: Creation. (Reviewed in 1982 *CRSQ*.)

374 Creation Science Movement. *Particulars of the Creation Science Movement* (pamphlet). 1981. England: Creation Science Movement. Lists 227 pamphlets published by the Creation Science Movement (formerly called the Evolution Protest Movement), one of the oldest and most active British creationist groups. Also lists several CSM books.

375 Creation Science of Ontario. *Overwhelming Evidence for a Young Earth* (tract). Undated. Toronto, Ont., Canada. 33 reasons to believe in young earth, from creationist and non-creationist sources (misinterprets the latter).

376 Creation Society of Santa Barbara.

How to Start Your Own Creation Group. Undated [1976]. Santa Barbara, Calif.: Creation Society of Santa Barbara. Creation Soc. Santa Barbara was Univ. Calif. S.B. student organization; later became **Students for Origins Research.** Lengthy report containing many reprints of creation/evolution articles, and sample creation-science organization constitution. Apparently written by **D. Wagner.**

377 Creation-Science Research Center. *This Wonderful World: Order and Design* (booklet). c1971. San Diego: Creation-Science Research Center. Science and Creation Series: Book 1. Series consists of eight booklets, grades 1–8, in student and teacher editions, plus overall reference book (**Boardman,** Koontz and Morris 1973). "Suitable for use in either public or private schools." Series coeditors: Jimmy Phelps (leadership & human behav. Ph.D.—U.S. Int'l Univ.; ass't. super. of Santee, Calif. school dist.) and **H. Morris.** Managing editor: **K. Segraves.** Consulting editors: W.J. DeSaegher (Ph.D.—UCLA; English dept. chair at U.S. Int'l Univ. Calif. Western Campus) and Peter A. Stevenson (phys. science M.A.—Middle Tenn. State Univ.; principal of Christian H.S., San Diego). "The *Science and Creation Series* was introduced into 28 states through the adoption process during 1973 and 1974, thus forcing Science teachers to read the creation material as a viable alternative to the evolutionary interpretations of science" (1987 CSRC report). Morris complains that series was rushed by Segraves for 1971 Calif. textbook adoption deadline. Book 1 authors: R. Kofahl, K. Segraves, Carella DeVol (elem. teacher), Patricia Schiefer (reading spec.—primary grades).

378 _______. *Our Changing World: The Nature of Physical Processes* (booklet). c1971. San Diego: Creation-Science Research Center. Science and Creation Series: Book 2. Authors: William Beckman (chem. Ph.D.—Western Reserve); Edna Dudeck (2nd gr. teacher—Ramona, Calif.), Marita Danielson (2nd gr. teacher—Dehesa, Calif.). Matter and energy. Processes of nature are conservation and decay—not evolution.

379 _______. *The World of Long Ago: The Testimony of the Fossils* (booklet). c1971. San Diego: Creation-Science Research Center. Science and Creation Series: Book 3. Authors: R. Koontz, Lynn C. Mohr (principal of Escondido Lutheran School), Arlene E. Cook (writer; former teacher). Geological strata and fossils are evidence of different ecological communities, not different ages. Catastrophic burials by flood. Includes trilobite fossil inside human print. Paintings, photos.

380 _______. *The Living World: Structure of Living Systems* (booklet). c1971. San Diego: Creation-Science Research Center. Science and Creation Series: Book 4. Authors: H. Douglas Dean (biology Ph.D.—Univ. Alabama; biology prof. at Pepperdine Univ.), Marjorie Dyer (former supervisor—San Diego State Lab. School), Marie Oexner (social science teacher). Discusses ecosystems, classification, genetics, law of biogenesis and disproof of spontaneous generation. Also criticizes Miller and Fox origin-of-life experiments, and stresses that Kornberg and Khorana experiments did not create life.

381 _______. *Man and His World: Origin and Nature of Man* (booklet). c1971. San Diego: Creation-Science Center. Science and Creation Series: Book 5. Authors: **R.C. McCone,** Everett W. Purcell (aeron. engin. M.S.—USC; engineer with Philco Ford), Claire Liker (elem. teacher), Willard W. Threlkeld (psych. social worker—San Diego City Schools). Man's origin and nature; culture. Includes **Navarra**'s discovery of Noah's Ark, limitations of science (especially evolution), frozen mammoths and catastrophism, man and dinosaur co-existence (Paluxy), culture as degenerative. Photos, paintings.

382 _______. *World Without End: The Origin and Structure of the Universe* (booklet). c1971. San Diego: Creation-Science Research Center. Science and Creation Series: Book 6. Authors: **W. Boardman,** William Stallo (6th gr. teacher—San Diego), William Tomes. Astronomy, cosmology, cosmogony.

383 _______. *Beginning of the World: Creation, Evolution, and Modern Science* (booklet). c1971. San Diego: Creation-Science Research Center. Science and Creation Series: Book 7. Authors: **Davidheiser,** Patricia Ann Ennis (San Diego

State Univ. M.Ed.; 7th gr. teacher), Robert Olson (San Diego State Univ. M.Ed.; 8th gr. teacher). Strongly and overtly creationist. Simple presentation of standard creation-science arguments: thermodynamics invoked, transitional fossils disparaged. Many drawings, paintings. Teacher editions include background material and references, discussion questions, activities.

384 ________. *The World and Time: Age and History of the Earth* (booklet). c1971. San Diego: Creation-Science Research Center. Science and Creation Series: Book 8. "Age and the History of the Earth." Authors: **Chittick, Boardman,** John Blyth (8th gr. teacher—San Diego), R. Olson, Howard Nordeen (8th gr. teacher—San Diego). Points out radiometric dating problems; discusses other methods giving far younger dates (amount of given substances in ocean, atmosphere, or crust, divided by input rate). Includes study and test questions, experiments. Photos, drawings.

385 Crews, Joe. *How Evolution Flunked the Science Test* (booklet). 1980. Washington, D.C.: Review and Herald. *Crews:* pastor, missionary, evangelist; Seventh-day Adventist. Wrote books such as *Creeping Compromise* and *Reaping the Whirlwind* (Satan's battle against Seventh-day Adventist believers).

386 Criswell, W[allie] A. *Did Man Just Happen?* c1972 [1957]. Grand Rapids, Mich.: Zondervan. Rev. ed. Orig. 1957. *Criswell:* pastor of First Baptist Church, Dallas (largest Protestant church in U.S.). Man was created specially by God, he didn't just happen. Standard creation-science arguments presented in simple, derogatory fashion. Relies largely on ridicule. Ape-men hoaxes. Uses (**W.J. Bryan**'s) preposterous evolution examples: eyes came from pimples, legs from warts, etc. Elaborates on humiliation of Bryan at Scopes Trial by scientists who accost him with Nebraska Man evidence [actually never mentioned at trial].

387 ________. *Why I Preach That the Bible Is Literally True.* c1969. Nashville: Broadman Press. Includes two chapters on creation/evolution, with creation-science arguments. Earth is more than 6,000 years old—genealogical gaps in Bible. Genesis is foundation of whole Bible revelation. If evolution is true, then no Fall, no Salvation, no Savior. Flood tales are debased versions of Genesis—stories can't get better than Bible. If we believe evolution, we will act like animals and lose all pride. Evolution is the "grand myth . . . big lie" of our era. If evolution were true, scientists wouldn't disagree about it. People believe in evolution in order to avoid the Gospel and God, and because it is the basis for humanism.

388 Crofton, Denis. *Genesis and Geology; or, An Investigation into the Reconciliation of the Modern Doctrines of Geology with the Declarations of Scripture.* 1853. Boston, Mass.: Phillips, Sampson. Also pub. by William Collins (Glasgow). Introduction by **Edward Hitchcock.** *Crofton:* prof. at Amherst College (Mass.).

389 Crofut, William J. *Creationism Is for Catholics* (booklet). 1984. Skanateales, N.Y.: Catholic Creation Ministries. "A Rebuttal to The Diocesan Resource Paper 'Catholics and Creationism' issued by The Vicariate for Education: Roman Catholic Diocese of Syracuse, Nov. 1982." *Crofut:* founder of Catholic Creation Ministries (Skanateales, N.Y.). Rebuttal to 1982 paper written by Lemoyne Coll. profs. Advocates strict creationism (contrary to the official Catholic position and the opinion of the Papal Acad. of Sci., which accepts biological evolution). Cites **Morris, Whitcomb, Lammerts, Wysong, Gish, Parker, O'Connell.** Says 1950 *Humani Generis* of Pope Pius XII supports creation, not evolution. Standard creation-science. Quotes anti-evolution articles (Fr. Morrison) from *Homiletic and Pastoral Review.*

390 ________. *Does Chemical Evolution Explain the Origin of Life?* (booklet). 1985. Skanateales, N.Y.: Catholic Creation Ministries. Monograph No. 3 of Catholic Creation Ministries. Rebuttal to Richard Dickerson's 1978 *Sci. Amer.* article. Scientifically naive; relies on creation-science sources **(Gish, Austin, Barnes, Akridge, Levitt).** Quotes **Gentry** as an "evolutionist." Complains that Dickerson is recklessly speculative: points out all his assumptions, his "ifs," "may have beens" and similar phrases.

391 Crowley, Dale. *Scientists Against Evolution* (pamphlet). Evangel Missionary Fellowship. (Listed in SOR CREVO/IMS.) Presumably father of **D. Crowley, Jr.** Crowley Sr. wrote 1970 autobiography *My Life a Miracle* (Wash., D.C.: National Bible Knowledge Assn.) and is described by son as fighting against evolution in 1923. Crowley Sr. was editor of *Capitol Voice* and director of "Right Start for the Day" radio ministry. Has several books of radio sermons; also *Can America Survive???: Our God-given Freedoms Threatened by Deadly Enemies Within and Without* (c1975; National Bible Knowledge).

392 Crowley, Dale, Jr. *History's Wildest Guess: The Hard Facts About Evolutionism.* c1975. Washington, D.C.: National Bible Knowledge Association. Book inspired (and published by) Crowley's father. Based on *Capitol Voice* articles and "Right Start for the Day" radio broadcasts. Also inspired by, and rebuttal to, insidious evolutionist propaganda of Time-Life books: the "greatest single influence on world opinion" for evolutionism. Mentions great influence of **Rimmer** also. Foreword by Thomas H. Christie: science dept. head at Washington-Lee H.S. (Arlington, Vir.). Extremely derogatory tone. Evolution is "Satanic plot" against God; an evil fraud, deliberate lie, mere guess, pagan religion. Earth is probably young. Worldwide Flood. Attempts all standard creation-science proofs. Many biblical references; also scientific references. Includes section on anti-evolution quotes by scientists. Also section on questions to confront evolutionists with. (In 1978 the Nat'l Bible Knowledge Assoc.—Crowley—sued the Smithsonian for promoting evolution, demanding that it either cease its biased displays or devote equal resources to promoting creationism.)

393 Cuénot, Lucien. *Invention et Finalité en Biologie.* 1941. Paris: Flammarion. French. Directed evolution (anti-Darwinian). Inter-connections so complex in living organisms, and lead to such specific results, that "the idea of a finalist direction is born invincibly"—though this really only explains the "obscure by the more obscure." "Germinal invention": germ cell possesses mysterious power of inventing its own structure. (Quoted in **Jaki** 1978, who says he ended up endorsing pantheism; discussed in **Rostand** 1962.)

394 Culp, G. Richard. *Remember Thy Creator.* c1975. Grand Rapids, Mich.: Baker Book House. *Culp:* botany M.S.—Mich. State Univ.; D.O.—College Osteopathic Med. & Surgery; then practiced osteopathic med. in Harrisburg, Ore.; now lives in Ind.; co-editor of *Christian Family;* Mennonite. Culp challenged evolutionary indoctrination while Purdue biology student, then as Mich. and Mich. State grad student. Accuses evolutionists of dogmatism, even dishonesty. Includes many anecdotes of university intolerance and discrimination against creationist students (mostly from personal experience). Standard creation-science arguments. Describes the six creation days of Genesis; favors strict creationism. Attacks M. Gardner for bias. Crossing-over during meiosis shows power of the Cross. Photos, diagrams, charts; index.

395 Cumbey, Constance E. *The Hidden Dangers of the Rainbow: The New Age Movement and Our Coming Age of Barbarism.* c1983. Shreveport, La.: Huntington House. Rev. ed. *Cumbey:* Detroit lawyer. Best-selling warning of dangers of New Age Movement, with its occultic practices which fulfill biblical prophecies of Satanic perversion of belief and coming of Antichrist. Lists and describes New Age front organizations. Evolution mentioned as part of propaganda campaign; also corollary belief in perfectability of man. Condemns New Age belief in ability of man to become god, UFOs, reincarnation, TM, Eastern religions, various cults (Theosophy, etc.). In a later book, *A Planned Deception: The Staging of a New Age Messiah* (1985; E. Detroit: Pointe), Cumbey accuses **P. Robertson** of promoting New Age occultism.

396 Cumming, Kenneth B. *Design in Ecology* (pamphlet). 1984. El Cajon, Calif.: Institute for Creation Research. ICR "Impact" series #131. *Cumming:* ecology Ph.D.—Harvard; has worked on fish ecology and management; now biology prof. at ICR. Ecology as evidence of design in nature. (Cumming also advocates biblical "stewardship" rather than an exploitative "dominion over nature" concept, though the latter has been the more typical fundamentalist attitude.)

397 Cummings, Violet M. *Noah's Ark: Fact or Fable?* 1972. San Diego: Creation-Science Research Center. *Cummings:* wife of Ark-eologist and Ararat Explorer Eryl Cummings. Foreword by **T. LaHaye;** Preface by **H. Morris.** Comprehensive compilation of Ark-eology tales and sources, by an active believer who accepts almost all claims. Much background information on people and events; covers all significant Ararat Ark expeditions. Describes and endorses story of discovery by The Three Atheists and Haji Yearam (1856), James Bryce (1876), Turkish Commissioners (1883), Prince Nouri (1887), **Roskovitsky** the Russian Aviator, Alexander Koor and the Czarist Expeditions (1915–16), the Turkish soldiers (1915), **Carveth Wells** (1932), Hardwicke Knight (1936), World War II aerial sightings, Reshit the Kurd (1948), **F. Navarra** (1955, 1969), George Greene (1953), **Vandeman**'s Archeological Research Foundation and SEARCH teams. Details Eryl Cummings' own extensive searches and researches; cites many story references. Photos, drawings, maps.

398 _______. *Noah's Ark: Fable or Fact?* c1973. San Diego: Creation Science Research Center. Also pub. 1975 by Spire-Pyramid. "Adapted from" (shorter version of) Cummings 1972.

399 _______. *Has Anybody Really Seen Noah's Ark?* c1982. San Diego: Creation-Science Research Center. "An Affirmative Definite Report." "Ark-ivist": Phyllis E. Watson. Foreword by **D. Gish.** Detailed compendium of Ark-eology tales and reports; an update of Cummings 1972. Still believes most tales: includes vindication of refuted **Roskovitsky** sighting and Czarist expedition. Many newspaper reports and secondhand confirmations. Georgie Hagopian's 1905 discovery. **Vandeman**'s ARF/SEARCH teams. **Noorbergen** cited frequently. 1933 German April Fool's hoax, which the Cummings naively believed until massively refuted by their own laborious research. New material on Michigan Mounds as proof of memory of Flood in America (**Donnelly,** B. Fell cited). Rich source of references for alleged and published claims. Gives credence to tale of National Geographic and Smithsonian expeditions (1964–68), in which pieces of Ark and bodies were recovered and secretly taken to Smithsonian Museum. Photos, drawings, maps; index.

400 Curtis, George Ticknor. *Creation or Evolution?: A Philosophical Inquiry.* 1887. New York: D. Appleton. 2 eds. in 1887. *Curtis:* New York lawyer. "The whole doctrine of the development of distinct species out of other species makes demands upon our credulity which is irreconcilable with the principles of belief by which we regulate, or ought to regulate, our acceptance of any new matter of belief." Rules of evidence: "every fact in a collection of proofs from which we are to draw a certain inference must be *proved independently* by direct evidence, and must not be itself a deduction from some other fact." Each link must have its own logical justification and proof. Also, the several facts must be arranged in proper relationship to one another. Further, the whole collection must then be consistent with the inference to be drawn. Finally, "the collection of facts from which an inference is to be drawn must not only be consistent with the probable truth of that inference, but *they must exclude the probable truth of any other inference.*" Mere piling on of great quantities of indirect evidence does not constitute proper proof.

401 Custance, Arthur C. *Without Form and Void: A Study of the Meaning of Genesis 1:2.* 1970. Priv. pub. (Brockville, Ontario, Canada). *Custance:* oriental lang. (Greek & Hebrew) M.A.; anthropology (medical physiol.) Ph.D.—Ottawa Univ.; head of Human Engin. Lab of Defence Res. Board (studied combat heat stress); member Canad. Physiol. Soc.; Fellow—Royal Anthrop. Inst. This book is acknowledged as the most authoritative defense of Gap Theory creationism. The Gap Theory holds that there is a tremendous gap between first two verses of Genesis, in which all the geological ages occurred. God created perfect earth, then it became corrupt, and God destroyed it—made it "without form and void." Following verses (the six-day creation describe a later re-creation. Custance argues that the Gap Theory antedates the 19th-century conflict of geology with the traditional young-earth interpretation of Genesis: that it was a respected theological opinion rather than a late attempt at harmonization. Cites Masoretic, Septuagint, and Vulgate translations. Claims that many ancient authorities supported Gap Theory. Bibliog., index.

402 ________. *Genesis and Early Man.* c1975. Grand Rapids, Mich.: Zondervan. Vol. 2 of Doorway Papers series. The Doorway Papers are monographs by Custance (60 in all), which he orig. published separately. This vol. consists of seven papers: on fossils, primitive cultures, intelligence of early humans, supposed evolution of human skull, fallacy of anthropological fossil reconstructions, origin of language, and cultural anthropology and the Bible (kinship; cultural patterns). Articulate; filled with scientific and anthropologial references. Though he accepts ancient earth as Gap Theory believer, Custance insists he is a biblical literalist. "I am convinced that we do not yet need to surrender the position that Scripture seems to me to take rather clearly, namely, that the human race began with Adam's creation only a few thousand years ago." Without literal interpretation of Genesis "we undermine the logical basis" of salvation. So-called "primitive" societies have really only degenerated from the original advanced civilization. Ancient man was highly intelligent. Scholarly attack on human biological and cultural evolution. Drawings, photos.

403 ________. *Noah's Three Sons.* c1975. Grand Rapids, Mich.: Academie (Zondervan). Vol. 1 of Doorway Papers series. "Human History in Three Dimensions." Based on paper presented to ASA in 1950s; pub. by Canadian Gov't. as *Does Science Transcend Culture?* (1958), also Custance's 1959 Ph.D. thesis. "My basic thesis is that the tenth chapter of Genesis, the oldest Table of Nations in existence, is a completely authentic statement of how the present world population originated and spread after the Flood in the three families headed respectively by Shem, Ham, and Japheth." The Semites were appointed to care for man's spiritual needs (they devised both true and false religions). Japhethites are Indo-Europeans; their contributions are in the intellectual realm. Hamites, who comprise all other people (Negroid, Mongoloid, etc.), care for the physical needs. Hamites, who were masters of technology, spread first and fastest, pioneering and creating civilization in both Old and New World. The application of Japheth's philosophy to the technology of Ham produced science; Japheth's philosophy applied to Shem's religious insights produced theology. The great Jewish scientists are those who embraced Japheth's culture. At the Millenium, Shem will become the spiritual head of mankind (they defaulted to Japheth when Jesus was on earth). Fossil hominids are remains of post-Flood dispersion; the most degraded specimens of people driven to harshest environments. They are all Hamites. Custance stoutly denies that racial differences imply any notion of inferiority. Only standard for equality is recognition of equality under God. Men cannot be considered equal to each other (they are not), or of equal worth to society (a notion which leads to totalitarianism). Flood occurred about 2500 B.C. Ham was a dark-skinned mulatto; he also married a mulatto. His surprise at seeing Noah's exposed white skin led to his filial neglect. Sumerians were black. Gomer: Celts; Magog: Russians; Madai and Javan: Indians, Persians; Chittim: Macedon; etc. Quotes Whorf on language as affecting worldview: God confused tongues to make different groups dependent on each other, and attempts at enforcing unity will fail. Drawings.

404 ________. *Evolution or Creation?* c1976. Grand Rapids, Mich.: Zondervan. Vol. 4 of Doorway Papers series. Five papers (orig. pub. 1959–1972). "I believe that God acted creatively, in the most distinct and positive manner conceivable, throughout the whole of geological history, introducing new species as they became appropriate, and removing others when they ceased to be." Preparation of earth for man – "supernatural selection." Notes his view is similar to **Agassiz.** Advocates Gap Theory creationism: earth was created geological ages ago, but "something went wrong and a catastrophic judgment brought that older world to a disastrous end, leaving it ruined and desolate, as Genesis 1:2 describes it. Then followed a re-creation . . . of six literal days." This catastrophe may have tilted earth's axis 40 degrees. Biological convergence constitutes an impossible problem for evolution—presents many examples and quotes. Survival of the unfit: (planned) cooperation in nature, not ruthless struggle. Man is not an animal; man is unique and superior. Filled with biological and anthropological references; also biblical references. Knowledgeably presented; dignified and scholarly tone. Describes fitness of the environment, discusses teleology. Primitive monotheism is the pristine condition; polytheism is degenerate. Declares naturalism (scientific materialism) a false religion. Its creed is evolution, and it is extremely dogmatic. Custance emphasizes direct interventions of Creator God; man as creature made for God (not a superior animal), dependent on salvation through Christ because of Fall. Drawings by the author.

405 ________. *Time and Eternity: and Other Biblical Studies.* 1977. Grand Rapids, Mich.: Zondervan. Vol. 6 of Doorway Papers series. Includes "Time and Eternity and Instantaneous Creation in Scripture in Light of the Theory of Relativity." Other papers discuss linguistics, Cain's wife, etc.

406 ________. *The Flood: Local or Global?* 1979. Grand Rapids, Mich.: Zondervan. Vol. 9 of Doorway Papers series. Argues for regional Flood. "Strict adherence to the literal wording of Genesis 6–8 leaves little alternative than to view this flood as universal as far as man is concerned but local insofar as geography is concerned." Includes worldwide Flood traditions. Other papers discuss the problem of evil, art, and Christian scholarship.

407 ________. *The Seed of the Woman.* 1980. Priv. pub. (Brockville, Ontario, Canada: Doorway Publications). "Dr. Custance is not seeking to prove the fact of the creation of Adam, or the formation of Eve out of his body, or the literal fact of man's Fall by eating a forbidden fruit and his consequent loss of physical immortality, or the truth of the Virgin Birth. . . . All these things he takes as established and their truths he accepts without equivocation. The thesis of this volume is that each of the historical facts which lies behind the Creeds of our Faith is essential to the whole Plan of Redemption."

408 ________. *Two Men Called Adam: A Fresh Look at the Creation/Evolution Controversy from a Different Point of View—the Theological.* c1983. Priv. pub. (Brockville, Ontario, Canada: Doorway Publications). Christ is the Second Adam. (Vol. 3 of Doorway Papers, *Man in Adam and in Christ,* argued that there are two "species" of man: fallen man, and man who has accepted Christ.)

409 Custer, Stewart. *Problems of Evolution* (booklet). 1964. Greenville, S.C.,: Bob Jones Univ. Press.

410 Cuvier, Georges. *Essay on the Theory of the Earth.* 1827 [1813]. Edinburgh, Scotland: W. Blackwood. Orig. French. Also pub. in Arno Press reprint ed. (1977; New York). Cuvier, a brilliant comparative anatomist, was the major founder of the science of paleontology. Unlike his chief opponent, Lamarck, who believed in transformation (evolution) but not extinctions, Cuvier argued that fossils prove extinctions occurred but he adamantly refused to accept evolution. Cuvier's studies convinced him that animal anatomy and form was so harmoniously balanced (his idea of the "correlation of parts") that animal types could not change into new forms. A catastrophist, he argued that the many extinct species he described were wiped out by successive regional—not worldwide—floods caused by tremendous geological disruptions. Surviving species in other areas then repopulated the devastated regions. Cuvier considered evolution to be "contrary to moral law, to the Bible, and to the progress of natural science itself" *(Dict. Sci. Biog.).*

411 Dale, Lloyd. *The Value of Creation* (pamphlet). Undated [1981?]. Minneapolis: Bible-Science Association. From 1981 lecture. Chronicles Dale's 1979–80 struggle to reverse non-renewal of teacher contract in Lemmon, S.D. for pushing creationism in his classes. Dale is lobbying in South Dakota for creation-science bill. Proves that evolution makes communism possible. (Dale is also chair of South Dakota Moral Majority.)

412 Daly, Reginald M. *Earth's Most Challenging Mysteries.* 1972. Nutley,

N.J.: Craig Press. Also pub. by Baker. *Daly:* physics/math teacher at various colleges; nephew of R.A. Daly (Harvard geologist). Mysteries include the Ice Age, origin of coal and oil, submarine canyons, extinctions, marine fossils atop mountains. Defends Flood Geology as only reasonable theory to explain these. Ice Age followed Flood. Uprooted vegetation buried by Flood became coal; buried animals became oil. Cites **G.M. Price, B. Nelson, Velikovsky, Fairholme, Rehwinkel, Howorth, Morris,** and other creationists. Fairly technical geology; many scientific references. Includes Paluxy prints, world Flood myths, the pre-Flood vapor canopy. Repeats **Rimmer**'s claim that *Hesperopithecus* was flaunted at Scopes Trial. Defends Bible in conclusion; argues that evolution is taught as a religion—a deliberately anti-Christian faith. Bibliog.

413 ________. *It's Science Fiction—It's a Fraud* (booklet). 1984. Little Rock, Ark.: James J. Kelly ("co-publisher" with Daly). *Cover:* "Evolution is a quasi-religion camouflaged as 'science.' It's unconstitutional to use our taxes to brainwash students with irreligious, one-side-only" [sic]. Quotes many geologists on inadequacies of geological theories; also quotes **G.M. Price** frequently to refute evolutionist implications of geological column, and C. Hapgood. Cites film **Footprints in Stone** as demonstrating man co-existed with dinosaurs; also cites other fossil anomalies as evidence of man in Mesozoic and Paleozoic strata. Ridicules plate tectonics theory at length: "Seldom if ever has science reached a peak of absurdity equivalent to geology's theories that continents ride on 'plates' and crash into each other with force sufficient to underthrust and uplift the world's biggest mountains." Argues strongly for Gap Theory creationism. The first two verses of Genesis imply "an original creation long before the first day, in the undefined 'beginning,' followed by a catastrophic judgment that plunged the earth into a state of devastation and ruin from which it was restored during the six days." Praises and quotes **Custance**'s defense of Gap Theory, and sharply criticizes **Morris** and **Whitcomb** and other young-earth Flood Geology advocates. Agrees with **Ussher** and **Bullinger** that Creation was 4004 B.C.; the Flood 2348 B.C. The earth is old, but man is recent. "All so-called hominoid fossils (supposed to be older than 6000) are frauds..." Emphasizes circular reasoning in using fossils to date rocks. II Peter refers to pre-Adamic cataclysm, not to Noah's Flood. Does not want evolution taught at all—says appeals for balanced presentation "may sound plausible as an evolution-creation compromise, but it won't work."

414 Dametz, Maurice G. *Dead at the Top; or, The Present Status of Evolution.* 1945. Denver: Denver Bible College Press. Described by Schadewald as heavy-handed but light-weight evolution-bashing.

415 Dana, James Dwight. *Science and the Bible.* 1856. *Dana:* respected geologist and zoologist; studied under **Silliman** at Yale; married Silliman's daughter; became Silliman Prof. of natural history and geology at Yale; wrote many geology texts and manuals. In this work he endorses Day-Age creationism as a means of harmonizing the findings of geology with the Bible.

416 Dankenbring, William F. *Did God Create the Universe?: A New Look at the Creation/Evolution Controversy.* Undated [1970s]. Unpub. MS. Undated MS copy in **Ambassador College** Library. References up to 1970. *Dankenbring:* theological journalism M.A.; now heads Triumph Publishing. Thanks Ambassador College faculty for help with book, especially Stig Erlander (biochem. Ph.D. —Iowa State; chem. and biology prof.). Consulted with faculty at Texas as well as Pasadena branch. Dankenbring wrote 1973 article on creation/evolution for **H.W. Armstrong**'s *Plain Truth,* and is presumably a member of his Worldwide Church of God. This work largely incorporated into author's *First Genesis*—seems to be a preliminary version.

417 ________. *The First Genesis: The Saga of Creation vs Evolution.* c1979 [c1975]. Altadena, Calif.: Triumph Publishing. Expanded and rev. ed. Orig. c1975. 1979 ed. adds Preface by **Wernher von Braun.** Asserts that the Bible agrees with all established scientific facts in biology, geology, and physics. Argues strongly for Gap Theory creationism and against young-earth creationism. "Evolutionists often lump all Creationists in the same bag, not realizing there are broad and vast differences of thought among Creationists about Creation itself." Rejects Flood Geology—even multiple floods could not have created the geological column. But accepts worldwide Flood, and endorses Ark-eology tales. Presents standard creation-science arguments against chance origin of life and evolution, examples of design in nature, adaptation as evidence of purpose. "The geologic record indicates that God created new forms of life at various stages of His Divine plan." Proposes that the Cretaceous and maybe the Pleistocene extinctions were cosmic cataclysms. They resulted from the cosmic conflict when Satan rebelled with a third of the angels against God. Suggests that asteroids are remains of planet which collided with earth, causing these destructive upheavals. Cave-men were pre-Adamic non-human beings. Neanderthals may be remnants of pre-Adamic beings who survived into Adamic times: the biblical pre-Flood *Nephilim*. Quotes a variety of creationists and anti-Darwinists, inc. von Braun on creationism. Some drawings, photos; bibliog.

418 ________. *The Creation Book for Children.* c1976. Altadena, Calif.: Triumph Publishing. Foreword by **Wernher von Braun.** Lots of beautiful color photos—animals, stars, landscapes—plus other illustrations. Brief text; mostly on standard argument from Design in nature, with Bible quotes. Includes Gap Theory presentation. Many photos from **Ambassador College.**

419 ________. *Beyond Star Wars.* 1978. Altadena, Calif.: Triumph Publishing. "Star Wars really happened! Long ago great battles raged in the universe. A great war caused vast destruction throughout the cosmos and upon the earth. Super beings battled for control of the universe, space, and time." Gap Theory creationism: Satan and his angels rebelled against God; massive cosmic warfare resulted. God utterly destroyed the pre-Adamic world ruled by Satan. Biblical "Elohim" engaged in UFO Star Wars. Current UFOs are a diversion to scare people into expecting outer-space invasion, then to reject Christ at Second Coming. Includes chapter on Joshua's Long Day (suggests comet caused it). Other chapters on Hezekiah and the Sun Dial; Lost Continent of Atlantis (like **Velikovsky,** says Atlantis existed 900, not 9,000 years before Solon); Maps of the Ancient Sea Kings; Great Pyramid (follows Tompkins; Cheops = biblical Job; Pyramid was a memorial to the Flood); Tower of Babel Cataclysm; frozen mammoths; living Neanderthals (Soviet accounts of yeti, abominable snowman). "There can be no doubt that pre-Adamic races of mankind— from the so-called man apes of the Australopithecines and Homo erectus to Neanderthal man and Cro-Magnon man—walked the earth before Adam was ever created."

420 Darby, Joel. *Is It Really Noah's Ark?* (tract). North Syracuse, N.Y.: Book Fellowship. Tract No. 1212. Includes 1970 SEARCH Ararat expedition with Melvin Marcus (Univ. Michigan geography dept. chair) and William Farrand (Univ. Michigan geology prof.); **Navarra**'s wood samples. (Darby wrote other **Book Fellowship** tracts also such as *Did the Sun Stand Still?)*

421 Dargan, Sam. *The Creation Controversy* (report). c1983. Unpub. (Florence, S.C.). Unpub.(?) paper in ICR Library. Basic creation-science, Flood Geology, problems of evolution. Admits that creation-science use of 2nd Law of thermodynamics is faulty; says however there are theological reasons for supposing that the world is not running down since Christ entered it.

422 Davidheiser, Bolton. *Evolution and Christian Faith.* 1969. Phillipsburg, N.J.: Presbyterian and Reformed. Foreword by **C. Ryrie.** Includes undated Addendum (references up to 1974). *Davidheiser:* zoology Ph.D.—Johns Hopkins; biology prof. at Westmont College (Santa Barbara), and Biola College (La Mirada, Calif.); came from Mennonite background but did not accept Christ until after Ph.D. Knowledgeable about biology; avoids or corrects many common creation-science errors. Often repetitious; some sections overlap. Long section on history of evolutionary thought: emphasizes throughout that acceptance of evolution results in rejection of Bible and of Christianity. Presents generally accurate accounts of attitudes and statements of many evolutionists and theologians; also describes reaction of ASA to evolution, Scopes Trial. Says Wilberforce was foolish and ignorant. Criticizes tendency of some creationists to equate biblical 'kind' with taxonomic order as conceding too much to evolution. Critical section on the mechanisms and evidences of evolution; many scientific references and quotes. Argues that "exclusion principle" (natural selection operating on even small variations because of differential fitness values) prevents diversity, and hence inhibits evolution. Gives many examples of imperfect biological adaptations; argues that these wouldn't be produced by natural selection. (This is contrary to typical creationist argument of perfection of Design in nature.) Well-documented section on conflicting evolutionist statements as to whether or not man descended from monkeys or apes. Urges reconsideration of **Gosse.** Says that demons are real. Favors young earth. "If man evolved and there never was a first pair of human beings, then the Biblical account of the fall of man is not true but is a myth or an allegory." If man evolved, sin is just animal nature, and Christ was not our Savior. Index.

423 ________. *Science and the Bible.* 1971. Grand Rapids, Mich.: Baker Book House. Bible-science; largely standard creationist arguments. Cautions that Lady Hope story (Darwin's conversion) probably not true. "Genuine" spiritism—ESP—is activity of demons. The only "proper" contact with the supernatural is prayer to God via Christ. Also discusses Laban's flocks—God caused crosses from sheep carrying recessive genes.

424 ________. *To Be As God: The Goals of Modern Science* (booklet). 1977 [c1972]. Nutley, N.J.: Presbyterian and Reformed. Introduction by **R. Rushdoony.** Mostly about genetic engineering—playing God. Warns about rush by scientists to try to control birth and death; new techniques will not result in the Utopia they imagine. Includes much anti-evolutionism. Public is repeatedly told evolution is true; therefore, there must be something wrong with it or they wouldn't strive to persuade us with such "missionary zeal." Mentions contemporary human and dinosaur tracks, **Burdick**'s Precambrian pollen, and problems of C-14 dating as evidence for young earth.

425 Davidson, David. *The Great Pyramid: Its Divine Message.* 1941 [1924]. London: Williams & Norgate. 9th ed. *Davidson:* structural engineer from Leeds, England. Pyramidology, based on **Piazzi Smyth.** Structure of the Pyramid encodes the 6,000 year prophetic history of the world and the truths of the Bible. Adamic race began 4000 B.C. Discussed and quoted in **Noone** 1982, who says Davidson attempted to disprove R. Menzies' Pyramid theory, but converted to it after intensive research of the Pyramid's architecture and mathematics. Davidson, an Anglo-Israelite, said the "Final Tribulation" of the Anglo-Saxon-Celtic race would begin in 1928 and end in 1936. From 1936 to 1953 this race—the true descendants of Israel and God's favored race—would assemble together under God's protection to fight a coalition of world powers seeking to destroy them.

426 Davidson, Phil. *The Origin of Human Races: A Creation Perspective.* Forthcoming. Santee, Calif.: Master Books. Expected 1988. "Phil Davidson": pseudonym of Dave Phillips, who is active in the San Fernando Valley (Calif.) Bible-Science Association chapter (headed by **Coppedge**'s son). This book is a creationist interpretation of biological anthropology. It is largely about adaptation (micro-, not macro-evolution) to regional and environmental conditions. Davidson agrees with most anthropologists that it is impossible to define human "races" strictly; that different traits are distributed in different clines. Created human pair contained enough genetic variability to permit this adaptation.

427 Davies, L. Merson. *The Problem of Man's Origin.* 1930. Officers' Christian Union. (Listed in SOR CREVO/IMS). *Davies:* British lieutenant-colonel; paleontologist (foraminifera specialist); Fellow—Royal Soc. Edinburgh, F.R. Anthropol. Soc.; F. Geol. Soc.

428 ________. *The Bible and Modern Science.* 1953 [1943]. London/Edinburgh: T.&A. Constable. 4th ed.; orig.

1943. According to **Morris** 1984, Davies, a geology Ph.D., was a Gap Theory advocate: "the only geologist about whom I have ever heard who gave any credence to the gap theory."

429 Davies, L. Merson, and Douglas Dewar. *Is Evolution a Myth?* (booklet). England: Evolution Protest Movement. (Listed in SOR CREVO/IMS.) Cf. **Dewar**, Davies and Haldane 1949: same title; this may be shorter version.

430 ________. *Evolutionists Under Fire* (pamphlet). Undated. Edinburgh, Scotland: Darien Press.

431 Davies, Thomas Alfred. *Cosmogony; or, The Metaphysics of Creation.* 1857. New York: Rudd & Carlton. "Being an analysis of the natural facts stated in the Hebraic account of the creation, supported by the development of existing acts of God toward matter." *Davies:* major-general; 1872 presidential candidate.

432 ________. *Answer to Hugh Miller and Theoretic Geologists.* 1860. New York: Rudd & Carlton. Refutation of **Miller**'s concordism and the speculative theories of contemporary geologists.

433 ________. *Genesis Disclosed.* 1874. New York: G.W. Carlton. "Being the discovery of a stupendous error which changes the entire nature of the account of the creation of mankind. Also showing a divine law, plainly laid down, proving the error that all men have descended from Adam and Eve."

434 ________. *A Biblical Discovery: Am I Jew or Gentile?; or, The Created Origin of the Races.* 1893. New York: G.W. Dillingham. Presumably some form of British-Israelism.

435 Davis, John J. *Paradise to Prison: Studies in Genesis.* 1975. Grand Rapids, Mich.: Baker Book House. Cited in **Whitcomb** and DeYoung 1978 as having "contributed significantly to the current renaissance of Biblical creationism."

436 Davis, W.H. *Science and Christian Faith.* 1968. Abilene, Texas: Biblical Research Press. (Listed in **B. Thompson** bibliog.) Presumably anti-evolution.

437 Davis, Willard O. *Evolution and Revelation.* Undated [1967]. Austin, Texas: Firm Foundation. *Davis:* speech M.A.; USC Ph.D. candidate. Concerned about evolutionism in college—kids are losing their religious faith. Mostly standard creation-science; also discusses Higher Criticism of Bible, materialism, effects of evolutionism, proof of Bible.

438 Dawn Bible Students Association. *Creation* (booklet). 1952. East Rutherford, N.J.: Dawn Bible Students Association. 7th ed. Jehovah's Witnesses. Asserts fixity of species. Advocates Day-Age creationism; says "fundamentalists" (youth-earthers) are wrong. "Days" of creation may also overlap, e.g. days 4 and 5. The sun "rules" on the fourth day—but it was created earlier. "That the Bible does not attempt to give us further details is strong evidence of divine overruling in its writing.... Were the Genesis account of creation merely the guesses of an ambitious human, he could not have restrained himself from the urge to relate many details which would have no other foundation than his own imagination."

439 ________. *The Creator's Grand Design.* Undated. East Rutherford, N.J.: Dawn Bible Students Association. Undated, but not earlier than late 1960s. Advocates Day-Age creationism. Describes the six "days" (eras) of Creation as harmonizing with modern science. Sun and moon created earlier, but only began to "rule" on the fourth day. Asserts that "all species are fixed: there is no evolving from one to the other, even though there may be many varieties of each species." "Man was created to be king of earth; and when the grand design of the Creator concerning him is completed, the earth will be filled with perfect humans, exercising their original God-given dominion." "The fact that man was created in the image of God is a strong refutation of the theory of human evolution." At what stage could an ape or missing link have acquired thought or morality? Stresses that the "biblical viewpoint cannot be harmonized with the Darwin theory of human evolution." "If God did not create man and give him his law, then he has no divine law to guide him in his behavior." Stresses significance of 1914 as end of old world order—the church abandoned hope of converting world to Christianity. Denies existence of separate, immortal soul. Satan continually tries to deceive man into believing there is no death (reincarnation, the soul). God will resurrect the dead; the righteous will go to heaven, but hell does not exist. Christ's Second Coming "does not mean the destruction of the earth." However, the old, evil world will be totally transformed when Christ returns—"invisibly"—to establish his Kingdom, just as the Flood destroyed the old order and inaugurated a new world.

440 _______. *The New Creation*. 1941. East Rutherford, N.J.: Dawn Bible Students Association. "Author's Foreword" by Charles Taze Russell (Jehovah's Witnesses founder). Last vol. of 6-vol. "Studies in the Scriptures" series, "written before the turn of the century" which accurately forecast the calamitous events beginning in 1914. This vol. 737 pages. First chapter deals with seven creative days of Genesis; Vail's Canopy Theory; mentions earlier pre-Adamic creation. Coming millennium described in rest of book.

441 Dawson, Sir John William. *Archaia, or Studies of the Narrative of the Creation in Genesis*. 1857. Montreal. Also pub. 1860. *Dawson:* LL.D.; F.R.S.; F.G.S.; Canadian geologist and paleontologist; educated in Scotland; became principal of McGill Univ. (Canada); president of Geol. Soc. Amer., of Brit. Assoc. Advancement of Science, and of AAAS; Presbyterian; also trained in theology and Bible languages. Brought up in fundamentalist atmosphere, Dawson remained strongly opposed to evolution. In addition to his scientific works, he wrote a hundred or so religious articles and a dozen popular books on the relationship of science to religion *(Dict. Sci. Biog.)*. **Wonderly** cites biog. of Dawson by C.F. O'Brien (*Sir William Dawson: A Life in Science and Religion;* 1971) which lists 22 of his books, most dealing with the agreement between science and the Bible. Dawson advocates Day-Age creationism in this and other works.

442 _______. *The Story of the Earth and Man*. 1887 [1872]. London: Hodder and Stoughton. 9th ed. Orig. 1872. At least 11 eds. 1873 ed. pub. by Harper (New York). Describes the geological ages, their fossils and characteristic life-forms. Refutes evolution in final two chapters on "Primitive Man Considered with Reference to Modern Theories as to His Origin." "This evolutionist doctrine is itself one of the strangest phenomena of humanity.... [T]hat in our day a system destitute of any shadow of proof, and supported merely by vague analogies and figures of speech, and by the arbitrary and artificial coherence of its parts, should be accepted as a philosophy, and should find able adherents to string upon its thread of hypotheses our vast and weighty stores of knowledge, is surpassingly strange." Evolution "has no basis in experience or in scientific fact, and ... its imagined series of transmutations has breaks which cannot be filled." It is embraced by those seeking "deliverance from all scruples of conscience and fears of a hereafter." "It reduces the position of man, who becomes a descendant of inferior animals, and a mere term in a series whose end is unknown. It removes from the study of nature the ideas of final cause and purpose..." Evolution by survival of the fittest is the "basest and most horrible of superstitions. It makes man not merely carnal, but devilish." Evolutionists either suffer a "strange mental hallucination," or else "the higher spiritual nature has been wholly quenched within them." "We have, therefore, to choose between evolution and creation; bearing in mind, however, that there may be a place in nature for evolution, properly limited, as well as for other things, and that the idea of creation by no means excludes law and second causes." Evolutionists assume creationism relies on special miracles "contrary to or subversive of" ordinary natural law, "but this is an assumption utterly without proof, since creation may be as much according to law as evolution..." Dawson argues for lawful successive creations. "There is no necessity that the process should be instantaneous and without progression." Paleontology shows that "new species tend rapidly to vary to the utmost extent of their possible limits, and then to remain stationary for an indefinite time." Though life is ancient, man is recent, going back no further than the Post-glacial period. The Genesis Flood account describes the great Post-glacial deluge. Dawson says that Darwin's arguments, though wrong, are usually fair and accurate—unlike Spencer's. "Men must know God as the Creator even before they seek Him as a benefactor and redeemer. Thus religion must go hand in hand with all true and honest science." Drawings, charts; index.

443 _______. *The Bible and Science* (booklet). 1874. New York: N.Y. Tribune. N.Y. Tribune extra No. 26. Six lectures by Dawson; also contains a lecture by H. Crosby.

444 _______. *The Dawn of Life.* 1875. Montreal. "Being a history of the oldest known fossil members, and their relation to geological time and to the development of the animal kingdom." Also pub. 1875 in London (Hodder and Stoughton), titled *Life's Dawn on Earth: Being the History of the Oldest Known Fossil Remains.* Illustrations.

445 _______. *Nature and the Bible.* 1882 [1875]. New York: Robert Carter. Also pub. by Wilbur B. Ketchum. Orig. lectures delivered at Union Theol. Sem. (New York). Divine creation is asserted throughout the Bible. But the Bible deals with phenomenological descriptions of nature. Its authors were free from both mythological superstitions about nature and from scientific hypothesizing; thus its remarkable absence of error. "Perhaps there can be no surer test of a true revelation from God than to ask the question, Does it refuse to commit itself to scientific or philosophical hypotheses, and does it grasp firmly those problems most important to man as a spiritual being and insoluble by his unassisted reason? This non-committal attitude as to the method of nature and the secondary causes of phenomena is, as we shall see, eminently characteristic of the Bible." (Quoted in **Ramm** 1954.) Index.

446 _______. *The Origin of the World According to Revelation and Science.* 1893 [1877]. London: Hodder and Stoughton. 6th ed. Orig. 1877 (Montreal; also New York: Harper Bros.). Biblical and scientific references. "Evolution" may be okay "if understood to mean the development of the plans of the Creator in nature." Objects, however, to atheistic interpretation of evolution, which is anti-Christian. Many parallels between Genesis and geology: both testify to a beginning; both exhibit the progressive character of creation; both affirm unity of nature; both agree ocean preceded dry land, and hint of igneous conditions before; both show that man is culminating point of all creation. (Quoted in **Ramm** 1954.) Index.

447 _______. *The Antiquity of Man and the Origin of Species* (pamphlet?). 1880. Princeton. From *Princeton Review* (1880). Elsewhere (1896 interview), Dawson said, "I know nothing about the origin of man except what I am told in Scripture—that God created him."

448 _______. *The Chain of Life in Geological Time.* 1890 [1880]. London: Religious Tract Society. 3rd ed. 1888. "A sketch of the origin and succession of animals and plants." Illustrations.

449 _______. *Fossil Men and their Modern Representatives.* 1888 [1880]. London: Hodder and Stoughton. Orig. 1880 (Montreal: Dawson Bros.). "An attempt to illustrate the characters and condition of prehistoric men in Europe, by those of the American races." Illustrations.

450 _______. *Facts and Fancies in Modern Science.* 1882. Philadelphia: American Baptist Publication Society. Allows for possibility of theistic evolution, but says there is no proof of it in geological record. Still favors Day-Age creationism. Biblical and scientific references. Drawings.

451 _______. *Modern Science in Bible Lands.* 1892 [1888]. Montreal/London. In the "Continental Era" following the Ice Ages, in which the earliest traces of man have been found, the lands were raised much higher than now. A great subsidence then occurred, and land occupied by man was submerged by ocean. Dawson equates this with the Flood. Subsidence can be dated by measuring Nile delta deposition, since most of Mediterranean was dry land during Continental Era.

452 _______. *Modern Ideas of Evolution as Related to Revelation and Modern Science.* 1910 [1890]. New York: Fleming H. Revell. 6th ed. Orig. 1890 (London: Religious Tract Society). "Discusses the apparition of species in geological time, monistic evolution, agnostic evolution, and theistic evolution" (**J.N. Moore** in 1964 *CRSQ*). Doctrine of evolution has "stimulated to an intense degree that popular unrest so natural to an age discontented with its lot, ... which threatens to overthrow the whole fabric of society as at present constituted" (quoted in **G.M. Price** 1931).

453 ________. *The Meeting-Place of Geology and History*. c1894. New York: Fleming H. Revell. Also pub. in Montreal; London (Religious Tract Soc.) Discusses various theories of the Flood, and "sternly rejects" worldwide Flood. Denies that Flood destroyed all animal or human life. Rather, it appeared universal as far as the biblical narrator could tell. Likewise, the Genesis Table of Nations concerns only descendants of Noah's sons and does not say anything about others who may have survived Flood. It concerns one series of migrations centered in Mesopotamia, and not other areas or other times. (Quoted and discussed in **Ramm** 1954.)

454 ________. *The Historical Deluge in its Relation to Scientific Discovery and to Present Questions* (booklet). 1895. New York: Fleming H. Revell. "The Deluge thus becomes one of the most important events in human history; so that any attempt to discuss the history of primitive man, or his arts or his religion, without reference to this important factor, must necessarily be fallacious."

455 Dawson, William Bell. *Forethought in Creation* (booklet). c1925. Chicago: Bible Institute Colportage Association. The Evangel Booklets series: No. 29. *Dawson:* son of **J.W. Dawson;** geology prof. at McGill Univ. (Canada). The harmonious adaptation of species to their environments disproves chance evolution and can only be the result of intelligent planning. These adaptations could not have evolved gradually, because there would have been no use for them initially. Dawson also wrote books on tides, currents, and Bible prophecy.

456 ________. *Evolution Contrasted with Scripture Truth* (booklet). c1926. Chicago: Bible Institute Colportage Association.

457 ________. *The Bible Confirmed by Science*. Undated [1932]. London: Marshall, Morgan and Scott. By "accepting what the Bible states, we will invariably be pointed to the right road" and directed away from scientific error. Contrasts ancient myths, primitive religions, idolatry, and pantheism with biblical doctrine of creation, which frees mankind from fear of evil in nature. "A remarkable point in Biblical references to nature, is that we find no definite *explanation* anywhere of natural things. The writers of the Bible do not go beyond the description of what they actually see around them, and the correct way in which they describe what they do see is beyond praise." The writers of the Bible must have been divinely guided, because they scrupulously avoid all mythological notions. "Evolution should be classed as philosophy and not as science." Darwin made evolution acceptable, but the causes he suggested have proved to be "entirely inadequate." "The essential contrast between the doctrine of Evolution and the Bible, and the need to choose between them is thus clear"; the spread of unbelief is due to evolutionary teaching. Warns that evolutionary thinking, by assuming that man is responsible for his own progress, logically leads to infanticide, and to abandonment of prayer and religion. "Such views indeed, do away with morality itself; for if it is only an impersonal 'cosmic force' that we call God, we are liable to lose all distinction between right and wrong." Discusses "forethought in creation": exquisite adaptations—whole series of things that are "just right"—requiring supernatural planning. States that paleontological succession confirms Genesis order of creation. All types of creatures continue to survive in the present, but they show degeneration from their original state. It is the meek and humble representatives of the ancient creatures which survive, thus disproving the evolutionary struggle for existence. Biblical miracles are "reasonable": merely the overriding of laws of nature by God's higher laws.

458 ________. *Is Evolution True?: Error, and the Way of Truth* (booklet). 1932. London: Marshall, Morgan & Scott. (Listed in SOR CREVO/IMS.)

459 ________. *The Bible and the Antiquity of Man* (booklet). London: Bible League. Cave-men are not ancient, but merely wanderers and pioneers contemporary with the beginnings of civilizations in other areas.

460 ________. *Miracles and Laws of Nature* (booklet). London: Bible League.

461 Deen, Braswell D., Jr. *Evolution:*

Fact or Fiction?—Secular or Non-Secular? (pamphlet). 1981. Minneapolis: Bible-Science Association(?). *Deen:* judge—Georgia Court of Appeals. 1981 BSA conference paper, one of several creationist conferences Deen has lectured at (another is included in BSA 1985 conference volume). Deen flamboyantly and alliteratively denounces evolution and secular humanism as the root cause of all of society's evils—especially crime. Describes himself as an expert on "human origins from a law-science perspective," and advocates inclusion of young-earth creation-science in school curricula.

462 de Grazia, Alfred. *Chaos and Creation: An Introduction to Quantavolution.* 1981. Princeton, N.J.: Metron Publications. *De Grazia:* Ph.D.—Univ. Chicago; social theory prof. at NYU. A friend of **Velikovsky**'s. Neo-catastrophist, neo-Velikovskian. First book in ten-volume "Quantavolution" series.

463 DeHaan, M.R. *Genesis and Evolution.* c1962. Grand Rapids, Mich.: Zondervan. *DeHaan:* M.D.—Univ. Ill. Med. Sch.; Western Theol. Sem.; founder of Radio Bible Class. Hard-core anti-evolution. No conflict between *true* science and Bible properly interpreted. "Not a single statement in the entire Bible has ever been disproved by *true* science..." Repeatedly insists that if evolution is true, "it would automatically disprove the Bible, and reduce it to an antiquated compilation of superstitions, fables, and fancies unworthy of a place in human history." Evolution renders original sin, and Christ's redemption, utterly pointless. Sees it as deliberate attack on biblical Christianity. Has two chapters advocating Gap Theory creationism. Says Darwin and modern evolutionists explain giraffe got long neck from stretching. Condemns birth control as "race suicide." Many personal anecdotes and parables.

464 de la Peyrère, Isaac. *Praeadamitae.* 1655. Amsterdam/Basel. 5 Latin eds. in 1655. English ed. 1656 *(Men Before Adam). De la Peyrère:* French theologian; Huguenot. "The book was banned and burned everywhere for its heretical claims that Adam was not the first man, that the Bible is not the history of mankind, but only the history of the Jews, that the Flood was a local event, that Moses did not write the Pentateuch, and that no accurate copy of the Bible exists." The Pope ordered Peyrère to recant. He did so, saying his heresies resulted from his Calvinist upbringing, but he continued to collect evidence for his theory. Peyrère originally developed his pre-Adamite theory from a strange interpretation of Paul's Roman Epistle, then from historical information, and finally from anthropological evidence from many cultures. He was both a scientist and a "kabbalistic messianist." His Bible analysis was influential in the development of Higher Criticism. In the related volume *Du Rappel des Juifs* (1643) Peyrère argued "that the Jews were about to be recalled, that the Messiah is coming for them, that they should join the Christians, and with the king of France rebuild Zion": the Jews would then rule the world from Jerusalem. His pre-Adamite theory later became the basis for 19th century theories of polygenism and modern racism (Negroes and Indians not descended from Adam). (*Encycl. Judaica* 1972.)

465 Deluc, Jean Andre. *Treatise on Geology.* 1809. London. *Deluc:* Swiss naturalist; lived in England; Calvinist. Claimed Hutton's eternal earth theory denies creation and promotes atheism. Deluc attacked the Huttonian system on both religious and geological grounds (Bowler). He argued that the Neptunist theory corresponded with a literal interpretation of Genesis and was therefore geologically true. Though he refuted Hutton's vast ages, Deluc realized that the geological record required more time than six literal days. He "modified the Wernerian [Neptunian] theory to support the biblical story, with six periods of deposition corresponding to the days of creation and a comparatively recent deluge" (Bowler). Deluc said he showed "the conformity of geological monuments with the sublime account of that series of operations which took place during the *Six Days,* or periods of time, recorded by the inspired penman" (quoted in **D. Young** 1982). Flood: 2200 B.C. The ancient continents subsided in a catastrophic convulsion, and the modern continents were exposed. Also published letters to Hutton (1790-91) and book addressed to Blumenbach (1798).

466 DeMar, Gary. *God and Government: A Biblical and Historical Study.* c1982. Atlanta, Ga.: American Vision Press. Vol. 1. Foreword by **J.W. Whitehead.** Arranged in ten lessons for group study, with discussion questions. American Vision is dedicated to restoring America's biblical foundations to all aspects of life, including science, education, and (the focus of this study) civil government. ("Civil" government is contrasted with personal, family, church, and school governments.) "Who has jurisdiction over every aspect of American society, Jesus Christ or the State?" Asserts we must recognize sovereignty of Christ and give Him total allegiance, or submit to tyranny and slavery. Government must be "reconstructed according to His demands." Discusses Creation and the Fall, Satan's false promises, the Genesis Dominion mandate, Babel as the epitome of pagan humanist attempt to deny God and create world government. Relies on literalist interpretation of Genesis. Emphasizes that the United States was founded as a Christian nation on biblical principles. Quotes **Rushdoony, H. Morris, G. North, F. Schaeffer, F.N. Lee,** Calvin. Drawings, photos. Includes long list of Christian Reconstructionist works, plus bibliog. Also reply forms for various Christian reconstructionist organizations, inc. I.C.E. (North), Chalcedon (Rushdoony), Plymouth Rock Found. **(Walton),** Freedom Council **(P. Robertson),** Coral Ridge Ministries **(D.J. Kennedy).**

467 Dennert, Eberhard. *At the Deathbed of Darwinism.* 1904. Burlington, Iowa: German Literary Board. Orig. German *(Am Sterbelager des Darwinismus).* Dennert also wrote books on plants, and on Nostradamus.

468 ________. *Bibel und Naturwissenschaft.* 1904. Stuttgart, Germany: M. Kielmann.

469 ________. *Moses oder Darwin?* 1907. Stuttgart, Germany: M. Kielmann. From *Christentum und Zeitgeist* (journal?).

470 Denton, Michael. *Evolution: A Theory in Crisis.* 1986 [1985]. Bethesda, Md.: Adler & Adler. Orig. 1985 (London: Burnet). *Denton:* M.D.; Australian; molecular biology researcher at Prince of Wales Hosp. (N.S.W., Australia). Asserts that evolution is unproven and unconvincing; and that Darwinism is under heavy attack. Does not admit to any religious motivations. Full of scientific references. Presents many of the usual anti-evolution arguments in an impressive, carefully documented format. "The overriding supremacy of the myth [Darwinian theory] has created a widespread illusion that the theory of evolution was all but proved one hundred years ago and that all subsequent research—paleontological, zoological and in the newer branches of genetics and molecular biology—has provided ever-increasing evidence for Darwinian ideas. Nothing could be further from the truth." Darwinism—the only scientific theory of evolution—is still a "highly speculative hypothesis entirely without direct factual support"; it has become as dogmatic as the biblical account of creation it replaced. Modern science and philosophy is built on Darwinian paradigm that man is the product of mindless chance selection. It is a "cosmogenic [sic] myth" based on naturalism to replace the Genesis myth. Denton argues for a "typological" interpretation of organisms, with no transitions. Acknowledges influence of Plato and Aristotle (archetypes); also **Cuvier, Agassiz,** and **Owen.** Includes chapters on classification (hierarchical arrangements support typological rather than descent schemes) and molecular biology. Evolution occurs at species level, but not at higher taxonomic levels; higher taxa are unrelated. Uses Dayhoff's protein molecular biology data to demonstate that various organisms are equally distant from each other—no descent hierarchy (seems not to acknowledge that all living species may have undergone similar molecular evolution). Applauds cladistic approach for abandoning search for ancestral relationships. This book is hailed by creationists as a secular scientific work which demolishes evolution. Many good drawings, charts; index.

471 de Purucker, G. *Man in Evolution.* 1941. Point Loma, Calif.: Theosophical Univ. Press. *De Purucker:* with Int'l Theosophical Headquarters (Point Loma, Calif.). Revision of author's *Theosophy and Modern Science* (orig. 1927 lectures).

Anti-materialism. 'Evolution' as defined by Theosophy: inner urge unfolds destiny of spiritual entity. Theosophy endorses this 'evolution' but not "transformism" (random materialist change); rejects Darwin and ape-ancestry theory. Stresses missing links between phyla. Humans are a separate stock started by god-like extraterrestrials. This stock was fully human from its origin, but also the source from which all mammals (and all life, in prior Globe-Round ages) derived. Cites F. Wood Jones. Cosmic oneness.

472 Destiny Publishers. *In the Image of God.* c1967. Merrimac, Mass.: Destiny Publishers. Based on **C. Carroll** 1900; quotes Carroll at length. Genesis teaches that Negroes were created as "beasts" prior to Adam. "The Bible stands as an impregnable bulwark against Christendom's modern slogan that all men, regardless of color, are blood brothers.... The simple truth regarding the origin of races demonstrates conclusively that the Negroes and the white race do not have a common ancestry." God made Negroes to be servants of Adam and his race. Negroes helped Noah care for the other animals on the Ark. Anglo-Saxon-Celtic peoples are the modern descendants of Israel: the Covenant People. Cites many Bible passages proving the "beasts" are bipedal servants of "man" and distinct from other animals. Presents scientific evidence of physiological differences which demonstrates that Negroes were created to be beasts of burden. Genesis commands mandatory segregation and forbids miscegenation. Mixing of created 'kinds' is abhorrent to God. Cain's sin was mating with a Negro; his offspring were mixed blood. Mixing creates confusion. The "sons of God" (Gen. 6) were angels who committed the sin of mating with humans (cites Book of Enoch). This sexual perversion produced evil giants; earth was so defiled God sent the Flood to purify it. But the spirits of these diabolical creatures still exist as demons. Even Noah's line was tainted: Ham was of mixed blood (he mutilated Noah). Cites A. Hislop's *The Two Babylons* (1916) to prove that Nimrod—founder of all false religions—was black. The "atheistical theory of evolution" is a "spurious doctrine" which aligned itself with apostasy and led to "disbelief in the scientifically accurate but simply-worded statements of the Bible." This evil anti-Biblical error spread quickly through the schools. "The whole theory of evolution—postulating, as it puts it, man's descent from the ape—is not only irrational and unscientific; it is completely unscriptural." Describes Satan's desperate attempts to corrupt humanity in these Last Days: the clamor for integration is Satan's opening wedge. Suggests that atomic research has enabled demon spirits to fashion mechanical bodies and torment us as UFOs. Index.

473 Devine, Bob. *God in Creation* (booklet). c1982. Chicago: Moody Press. *Devine:* radio announcer/engineer for WCRF, Cleveland. Design and adaptation in nature as proof of creation. Includes porous egg membrane, the tundra, monarch butterfly, red and white blood cells, giraffe, skunk cabbage (produces heat in winter), salmon, 17-year cicada, snowshoe rabbit. Color photos for each example. Biblical references.

474 Dewar, Douglas. *Difficulties of the Evolution Theory.* 1931. London: Edward Arnold & Co. *Dewar:* studied science at Cambridge Univ., then law; went into Indian Civil Service; became authority on birds of India; Fellow—Zool. Soc.; was first secretary of Evolution Protest Movement, later EPM president. This book considered highly authoritative; much cited by creationists. Dewar was a Day-Age creationist, and wrote for the *Catholic Herald.*

475 ________. *Man: A Special Creation.* Undated [1936]. London: Thynne & Co. Also 3 eds. pub. by Victoria Inst. Evolution has captured the press, which won't publish attacks on evolution or even the slightest criticism. "Thus the public is permitted to hear only one side of the case ... the average man ... is led to believe that evolution is a law of nature as firmly established as is the law of gravity." (Quoted in **Field** 1941.)

476 ________. *More Difficulties of the Evolution Theory.* 1938. London: Thynne & Co. A reply to Morley Davies' book *Evolution and Its Modern Critics,* which was a reply to Dewar 1931. Dewar "goes into an extensive critique of the recapitulation hypothesis, so-called vestigial organs, dating of geological deposits, absence of fossils in the Precambrian rocks, sudden appearance of vast numbers in the Cambrian, origin of families, and finally so-called 'evolution' within the family" (**Lammerts,** in 1974 *J. Christian Reconstruction*). Early presentation of standard creation-science arguments. Modern man predates all hominid fossils proposed by evolutionists; Dewar presents Calaveras, Olduway, Abbeville, (etc.), finds as examples.

477 ________. *A Challenge to Evolutionists* (booklet). 1948 [1937]. Uplift Books. Orig. 1937 (London: Thynne). A report of Dewar's side of a debate against **J.J. McCabe** of Rationalist Press Assn. According to **Field** 1941, McCabe "threatened legal proceedings if his part of the debate were published."

478 ________. *The Transformist Illusion*. 1957. Murfreesboro, Tenn.: DeHoff Publications. Written in 1948; some material added 1951. (Dewar died 1957.) Introduction by **James Bales.** Standard creation-science arguments in authoritative presentation. Many quotes and references: all scientific (no biblical references). Cites many European anti-Darwinians. Quotes from his debates with Shelton, Morley Davies, **McCabe.** Dewar says the facts he presents "show that it is an illusion to believe that blind natural forces have caused life to emerge from inert matter and then gradually to assume to varied forms of living organisms." "It is high time that biologists and geologists came into line with astronomers, physicists and chemists and admitted that the world and the universe are utterly mysterious and all attempts made to explain them have been baffled." Lack of Precambrian fossils; lack of transitional fossils; critique of fossil hominids (including Calaveras, Clichy, Piltdown, and an account of Dubois' deceptive Java Man claims); conflicting evidence from blood precipitation tests; embryology; metamorphosis; parasitism; instincts; bird anatomy; violation of entropy law; mutual adaptations; the "astounding diversity of the minute details of organs." Fossils are "hostile witnesses" for evolution: "Not a single fossil of vital importance for the support of the theory has come to light." Declares that neither the Darwinian theory nor any materialistic theory of evolution can be defended in light of the scientific evidence. "The rocks cry out 'Creation!'" Charts; index.

479 ________, **L. Merson Davies, and J.B.S. Haldane.** *Is Evolution a Myth?* 1949. London: C.A. Watts/Paternoster Press (co-pub.). Written debate between Dewar and **Davies** of EPM and Haldane for the Rationalist Press Assoc. Arranged by New Zealand branch of EPM, which challenged Wellington Rationalist Assoc. Subject agreed upon: "Organic evolution, the theory that existing animals and plants, and also mankind, are descended from simple forms of life." (Haldane declined to include origin of life.) Six letters from each side. Polite but serious exchange; standard creation-science biology arguments. Haldane argues well, frequently citing **Dewar** and Shelton 1947.

480 ________, **and Frank Finn.** *The Making of Species*. 1909. London/New York: John Lane the Bodley Head. "We would emphasize that is it not Darwinism we are attacking, but that which is erroneously called Neo-Darwinism. Neo-Darwinism is a pathological outgrowth on Darwinism . . ." (Cf. Dewar's later, explicitly anti-evolutionist books.) **Field** says this book was specially commended by T. Roosevelt, and that Dewar was an evolutionist until his 1931 book.

481 ________, **and H.S. Shelton.** *Is Evolution Proved? A Debate Between Douglas Dewar and H.S. Shelton*. 1947. London: Hollis and Carter. Edited by **Arnold Lunn,** who wrote Introduction. Lunn quotes many European anti-Darwinist scientists (notably von Uexhull) [sic—von Uexküll?]; also Spengler's *Decline of the West* and A.N. Whitehead. He exposes the "foolish conspiracy" which presents evolution as demonstrated fact. Lunn says that "Darwin's everlasting title to fame is that he provided the atheist with a plausible if untenable answer to **Paley**'s argument from design." Index.

482 De Wit, J.J. Duyvene. *The Paleontological Record and the Origin of Man* (report). 1963. Mimeo version of 1963 address to Scientific Society of Univ. of Orange Free State (South Africa); also pub. in that society's *Proceedings*. *De Wit:* biology M.S.—Utrecht (Netherlands); Ph.D.; physiology prof.—Free Univ. Amsterdam; zoology prof.—Univ. Orange Free State. De Wit, a strict Calvinist, became a follower of **Dooyeweerd.** He similarly urged a reformation of biology (and all other sciences) by examining its philosophical and theoretical basis and reconstructing it according to Christian presuppositions. Strongly anti-evolutionist, he asserted evolution was hopelessly unscientific. "No fossil documentation whatsoever with respect to the assumed animal ancestors of man has been found." Argues that isolated populations cannot be source of new species, as they contain fewer genes than original pool. If man evolved from amoeba, he would have fewer genes than the amoeba. Evolutionist belief in spontaneous generation (which requires suspension of natural law) is pure animism; shows it is a non-Christian religious faith. De Wit's Christian biology recognizes that God created groups of laws specific for each species. De Wit argued against primacy of DNA; his research showed that DNA was responsible for intra-species variations only; the cytoplasm—outer cell layer—accounted for differences between species. De Wit urged recognition and merging of Dooyeweerd's "cosmonomic" anti-evolutionist biology with American fundamentalist creationism. **(Verbrugge** has 3-part series on De Wit in 1984–85 *CRSQ;* also includes correspondence with **G.F.Howe.)**

483 ________. *A New Critique of the Transformist Principle.* 1964. (South Africa?): J.H. Kok.

484 Dexter, David. *Evolution Refuted; or, Where Do We Stand and Why.* c1934. Grand Rapids, Mich.: Paul C. Clark.

485 DeYoung, Donald B. *Design in Nature: The Anthropic Principle* (pamphlet). 1985. El Cajon, Calif.: Institute for Creation Research. ICR "Impact" series #149. *DeYoung:* astrogeophysics Ph.D.—Iowa State Univ.; physics prof. at Grace College (Winona Lake, Ind.); visiting geophysics prof. at ICR. The anthropic principle as the return of **Paley**'s argument from design: the universe shows it was specially designed to be man's home.

486 Diamare, Vincenzo. [Title unknown]. 1912. Italy. According to **Field** 1941, Diamare was "Director of the Institute of Osteology and General Physiology in the University of Naples [who] rejected evolution in a book published in 1912."

487 Dick, Thomas. *The Christian Philosopher; or The Connection of Science and Philosophy.* 1888 [1826]. New York: G.&C.&H. Carvill. 4th U.S. ed. 1829. Harmony between nature and Revelation. The universe is designed to serve and delight man. Heavy on design argument, but does not deal directly with creationism. Great scientists worship God: **Newton,** Boyle, etc. Also wrote *The Philosophy of Religion; or, an Illustration of the Moral Laws of the Universe* (1829; Brookfield, Mass.: Merriam).

488 Dickey, C.R. *The Bible and Segregation.* Merrimac, Mass.: Destiny Publishers. The Bible commands the races not to overstep their bounds. God created the Negroes separately from Adam's race. Quoted in **Destiny** 1967.

489 Dickson, Robert E. *The Fall of Unbelief.* 1982. Winona, Miss./Singapore/New Delhi: J.C. Choate. *Dickson:* former missionary in Brazil; now in Antigua, West Indies directing a Bible and Leadership program. Introduction by **James Bales.** A comprehensive volume of Christian apologetics, with very strong emphasis on strict creationism (about 200 pages refuting evolution, plus Bible-science chapter). Presents all standard creation-science arguments, including young-earth dating proofs and Flood Geology, with lots of quotes and references. Acknowledges assistance of **B. Thompson** and **W. Jackson.** Also discusses biblical archeology, prophecy, miracles, divine inspiration and inerrancy of scripture, etc. Refutes old-earth creationist views. Index.

490 Dillow, Joseph C. *The Waters Above: Earth's Pre-Flood Vapor Canopy.* c1981. Chicago: Moody Press. *Dillow:* Th.D.—Dallas Theol. Sem.; scientist and theologian with Inherit a Blessing in Vienna. Foreword by **H. Morris.** Based on Th.D. thesis (under **C. Ryrie**); also inspired by **R. Rusk.** Handsome and impressive volume on the Water Canopy described in Genesis—the "water heaven" or Firmament which produced the Flood when it collapsed. "This book assumes that the Bible is the inerrant, authoritative Word of God; therefore, it provides a framework for scientific investigation of the ancient earth." Main purpose of book is to "demonstrate the truthfulness of the Bible in its statements on creation and the Flood." A "normal exegesis" of Genesis tells us God placed a literal ocean of water in the sky. There is no biblical indication it was sustained miraculously, so Dillow proposes a natural-law explanation of how it could have been supported. He presents a canopy model and tests its scientific predictions. It must have consisted of vapor, in order to remain aloft. The Canopy caused a strong greenhouse effect, preventing seasonal and weather changes and creating a worldwide warm climate. No rain (Gen. "mist"); no rainbow until after Flood. Demonstrates how Edenic pre-Flood conditions can be explained by canopy model: great longevity of Old Testament patriarchs from increased atmospheric pressure and shielding from cosmic radiation. Chapters on biblical evidence for canopy; canopy in world mythology and racial memory; critique of other creationist canopy models; analysis of atmospheric conditions under canopy. Canopy contained equivalent of 40 feet of water, which, along with Waters of the Deep, caused the Flood when it collapsed. This also caused sudden and permanent climate change; frozen Siberian mammoths. Stresses polemical intent of Genesis in discussion of hermeneutics and Bible-science. Dillow's model is endorsed by ICR Canopy specialist **L. Vardiman.** Many equations and tables. Comprehensive bibliog. contains many creationist and canopy sources. Photos, charts; index.

491 Dimbleby, Jabez B. *The Date of Creation: Its Immovable and Scientific Character*. 1902. London: E. Nister. *Dimbleby:* a founder of British chronological and astronomical associations. Uses eclipse cycles, planetary orbits, lunar and solar cycles, Chaldean and Hebrew calendars, and other astronomical data. Calculates that all cycles began together at Creation: 3996 B.C. All the planets were then in a straight line. Mentions that geologic ages occurred *before* creation. Inclination of earth's axis has remained constant since Creation Week —refutes claim it was tilted during the Flood. Affirms accuracy of **Ussher**'s dates. Diagrams, drawings. Also predicted End of World (Second Coming) for 1898 in several other books. **(Higley** 1940 calls him "John B. Dimbleby" and says he calculated Creation date as Sept. 20, 3999 B.C.)

492 Dixon, A.C. (ed.). *The Fundamentals: Vol. II.* Undated [1910]. Chicago: Testimony Publishing Company. *Dixon:* D.D.; pastor of Moody Church (Chicago). *The Fundamentals* is a series of twelve paperback books, each 125–130 pp., which sets forth the "fundamentals" of Christianity. The series was conceived in 1909 by Los Angeles millionaire Lyman Stewart (Union Oil), who, with the aid of his brother, distributed free copies to every pastor, missionary, evangelist, theology professor and student, and YMCA and YWCA secretary in the English-speaking world (three million vols., pub. 1910–1915). Stewart, who is not named in the vols., hired Dixon as editor; many of the greatest American and British conservative theologians and scholars contributed articles. Dixon edited the first five vols., Meyer the next five (see separately under **Meyer**), and R. Torrey the last two. This series helped define and publicize the emerging fundamentalist movement. Several authors attack evolution, but they do not advocate recent creation or Flood Geology. In general, political issues (prohibition, e.g.) are avoided, and controversial doctrines such as dispensationalism and premillennialism are downplayed. Personal testimony, prayer, and soul-saving are emphasized. Series contributors include **Pierson, Wright, Mauro, J.M. Gray, Bettex,** Dixon, **Scofield, Gaebelein,** J. Orr, Erdman, and Warfield. In this vol., Wright shows that archeology confirms the historicity of Genesis.

493 ________. *The Fundamentals: Vol. IV.* Undated [1910?]. Chicago: Testimony Publishing Company. This vol. inc. "Science and Christian Faith" by Rev. Prof. James Orr (D.D.; United Free Church Coll., Glasgow, Scotland). Orr denies that true science conflicts with the Bible, and refutes those such as A. White who claim that science and Christianity are opposed. Argues that miracles can be scientific. Cites **Wallace** on design of the universe. Declares "it is certain that the world is immensely older" than **Ussher**'s date of 6,000 years. Suggests a Day-Age view: Genesis does not give "a detailed description of the process of the formation of the earth in terms anticipative of modern science—terms which would have been unintelligible to the original readers—but a sublime picture, true to the order of nature, as it is to the broad facts even of geological succession.... There is no violence done to the narrative in substituting in thought 'aeonic' days—vast cosmic periods—for 'days' on our narrower, sun-measured scale. Then the last trace of apparent 'conflict' disappears." Acknowledges that considerable evolution has occurred, but argues that scientists now tend to admit there are definite limits to evolution. Warns against equating "Darwinism" with evolution; says modern evolution theory rejects Darwin's theory of natural selection and slow, gradual change in favor of sudden mutations. This new "evolution" is thus "but a new name for 'creation'"—the only difference being that it acts from within instead of externally. Asserts that origin of life, of consciousness, and of rationality and morality in man are steps in this process which require special acts of God. "It is hopeless to seek to account for life by purely mechanical and chemical agencies, and science has well-nigh given up the attempt." We must also presuppose that man—the "crown and explanation" of the whole creation—was made in God's image by a direct act of creation. Man's origin may be recent and "as sudden as Genesis represents."

494 ________. *The Fundamentals: Vol. V*. Undated [1910?]. Chicago: Testimony Publishing Company. This vol. inc. "Life in the Word" by **P. Mauro.** Mauro affirms the timelessness and inerrancy of the Bible, including science. Presents **Wallace**'s argument; says this shows that "if the universe is assumed to be the work of an intelligent Creator, it would follow that everything in this inconceivably vast and complex universe has been planned and arranged with special reference to making this little earth of ours a place suitable for the habitation of living beings, and especially of mankind." Darwin's theory of evolution is "directly contrary to the great and immutable law declared nine times over in the first chapter of the Bible in the brief but significant expression, 'after his kind.'" Not a single fact supports it. Science proves that life cannot arise from non-life. "If the Bible does not give us a truthful account of the events of the first six days recorded in its first chapter, it is not to be trusted as to *any* of its statements." "The theory of organic evolution, promulgated by Darwin and Wallace, has nothing to recommend it except that it offers an alternative to the acceptance of the account of the origin of species given in the Bible."

495 Dixon, Jeane. *My Life and Prophecies*. 1969. New York: William Morrow. Story of psychic seer Dixon, "as told to" **Rene Noorbergen.** Dixon claims to have predicted important events with stunning accuracy and no mistakes, and to have counseled Roosevelt and other world leaders. Noorbergen offers not the slightest hint of editorial criticism, and lavishes praise on Dixon's professed faith in the Bible, accepting her as a "biblical" prophet with God-given psychic power (but cf. Noorbergen 1976). Noorbergen, in his Prologue, says Dixon's fantastic ESP powers were the norm from creation until the Flood. He cites **G.M. Price** and **Rehwinkel,** and endorses Texas (Paluxy) giant manprints.

496 Dixon, Malcolm. *Science and Irreligion* (pamphlet). 1952. (England?): InterVarsity Fellowship. Quoted in **J.G. Williams** 1970, who says Dixon is a "Biochemist of the University of Cambridge." Quote concerns formation of enzymes by other proteins—how could this process have arisen initially? Concerned with other difficulties of evolution also.

497 Dolen, Walter R. *The Chronology Papers* (booklet). c1977, c1978. San Jose, Calif.: The Becoming-One Church. Calculation of biblical dates. Creation: 3971–3970 B.C., Sept.–Oct. Flood: 2316–2315 B.C. (1656 years after Creation). Christ died on Wednesday.

498 Dolphin, Lambert, Jr. *Lord of Time and Space*. 1974. Westchester, Ill.: Good News Publishers. *Dolphin:* born-again physicist; affiliated with Stanford Research Institute. Elasticity of time and space; relativity. Includes chapter on "Evolution and Entropy." "The general theory of evolution is not well established scientifically nor are its assumptions consistent with laws of physics such as the Second Law of Thermodynamics." Evolutionary theory shows itself to consist largely of emotionally-held "religious statements"—especially when creation-science is also presented. Also wrote *Astrology, Occultism and the Drug Culture.*

499 Donnelly, Ignatius. *Atlantis: The Antediluvian World.* 1971 [1882]. Blauvelt, N.Y.: Rudolf Steiner Publications. Also "modern revised ed." edited by Egerton Sykes (c1949; New York: Gramercy [Crown]), with Foreword by **Bellamy.** (Sykes: head of Hörbiger Inst. in England; ed. of *Atlantis* magazine; attempted expedition to Ararat to locate Noah's Ark.) Orig. 1882 (Harper & Bros.). *Donnelly:* two terms in U.S. Congress (Minn.); populist reformer; popularizer of Atlantis; also believed comet caused the Ice Age and the biblical catastrophes, and that Shakespeare's works contain coded messages. Plato's Atlantis was the biblical Eden where the first civilization arose, in the mid-Atlantic. Noah's Flood is the account of the sinking of Atlantis (the rest of the world was not flooded). Discusses worldwide Flood myths. Civilization was brought to other regions by Atlantean survivors. Endorsed by Prime Minister **Gladstone.**

500 Dooyeweerd, Herman. *A New Critique of Theoretical Thought.* 1957. Philadelphia: Presbyterian and Reformed. *Dooyeweerd:* strict conservative Calvinist philosopher; prof. of jurisprudence at Free Univ. Amsterdam. Inspired by Calvinist renewal in Holland led by turn-of-the-century prime minister Kuyper, Dooyeweerd called for a total reworking of philosophy and science on Dutch Reformed basis, rejecting non-Christian assumption of man's autonomy of thought and recognizing complete submission to God's will. This book is a massive (1948 pp.) treatise on his new "cosmonomic" philosophy. Entire cosmos is governed by God's law (nomos). Each species is governed by different set of laws. Nature is arranged in a hierarchy of increasing complexity: time, space, motion and energy, biotic realm, human sensorium, logic, Genesis cultural mandate and psychology, semantics, social intercourse, economics, esthetics, and law. God created laws for each hierarchical level. Claims to solve mechanist-vitalist debate by recognizing hierarchy of each sphere of God-given natural law. Scientific facts can be interpreted either from evolutionary naturalism frame of reference or from biblical creationist perspective. Christians accept Bible on faith; science is thus a religious activity. Dooyeweerd rejected evolution as unscientific as well as opposed to Christian presuppositions, and apparently believed the fossil record is best explained by the Flood. Reductionist philosophy of evolution is really animistic and requires miracles. We can't understand man by studying animals. With **De Wit,** he believed that cytoplasm contains higher level of information than DNA. (Son-in-law **M. Verbrugge**'s address is the Herman Dooyeweerd Foundation in La Jolla, Calif.)

501 d'Orbigny, Alcide C.V.D. *Prodrome de Paléontologie Stratigraphique Universelle des Animaux Mollusques et Rayonnés.* 1850–1852. Paris. 3 vols. *D'Orbigny:* French paleontologist; prof. at Paris Natural History Museum. D'Orbigny made important contributions in classifying invertebrate fossils and describing their stratigraphic distribution. He divided these strata into 27 stages according to characteristic fossils. D'Orbigny observed that most species were not found in the following stage, but were replaced by new ones. A follower of **Cuvier,** he elaborated on his theory of catastrophism. D'Orbigny argued that there were successive destructions in earth history followed by new creations of animal species: a distinct creation for each of the 27 stages.

502 Dougherty, Cecil N. *Valley of the Giants.* 1979. Priv. pub.? (Cleburne, Texas: Bennett Printing). *Dougherty:* chiropractor in Glen Rose, Texas. Affirms human footprints in Cretaceous limestone of Paluxy River in Glen Rose alongside dinosaur tracks.

503 Dov, Pincas. *Genesis, Mother of Sciences.* 1953. Chicago. "Pincas Dov" is pseudonym of Philip Warsaw. Mentioned in M. Cavanaugh 1983 diss. as a

Jewish creationist work.

504 Dow, T.W. *Truth of Creation.* c1967. Washington, D.C.: Celestial Press. Non-religious anti-evolutionism; crackpot science. Asserts existence of a spontaneous generation process: "continued creation of very simple organic beings may still occur in the present world." Strongly objects to nuclear fusion in interior of the sun—only the surface is burning. Also objects to expanding universe (Hubble's Law); law of angular momentum; anti-matter; continental drift; etc. Endorses extraterrestrial life. Human races developed—were created—"completely independently" from each other. Emphasis is on physics. Rambling.

505 Downing, Barry. *The Bible and Flying Saucers.* 1970. New York: Avon. *Downing:* B.D.—Princeton Theol. Sem.; Ph.D.—Univ. Edinburgh; pastor of Northminster Presbyterian Church (Endwell, N.Y.). Explains biblical miracles as UFO activity; attempts to go beyond **von Däniken.** Von Däniken's extraterrestrial explanation is correct, but Downing suggests that the Exodus and other biblical events were "deliberately caused by beings from another world *with religious motives.*" They successfully attempted to mold a select group of earthlings to commitment to a particular religious tradition. The Bible says that angels ("messengers") actually *caused* biblical religion, under God's direction; they are not supernatural, they are extraterrestrial missionaries. Jesus repeatedly stresses that he comes from another world. Cites **Newton**'s belief in reality of angels and heaven. "As the scientific revolution continued, however, and the age of Darwinian evolution emerged, it no longer seemed possible for a scientist to be committed to the Biblical world view." Evolution is based on "chance"; it says man is just a freak accident. But the space age allows us to consider life on other worlds once more. The pillar of cloud and pillar of smoke in Exodus was a UFO—its description matches modern reports of a mother ship class of UFOs. It parted the Red Sea by compressing the waters with the exhaust (reverse gravity) of its anti-gravity propulsion system; the Israelites crossed in the shadow of the anti-gravitational beam. Manna from heaven, the power display and presentation of Commandments on Mt. Sinai explained as UFO activity. (Ezekiel's "wheels" were probably UFO-induced psychic visions, however.) In the New Testament, the UFO activity is less blatant; more covert. Jesus was really an "undercover agent" from outer space who came to promote the new religion. The virgin birth "cover story" allowed him to claim he was both a naturally-born earthling and from another world. The Star of Bethlehem, the descent of the Holy Spirit at Christ's baptism, and Christ's Transfiguration, Resurrection, and Ascension were all caused by UFOs. Christ was carried away in a flying saucer. Examining the biblical descriptions in light of Einsteinian relativity, Downing suggests that Heaven is a parallel universe in our midst, "invisible due to its different spatial curvature." The Resurrection of the dead may involve some sort of human "energy precipitate" which can be reconstituted in a new body. "The UFO theory I have outlined is certainly consistent in explaining the Biblical data," claims Downing, but he insists it is only tentative. Relies on UFOlogists Keyhoe, F. Edwards, Vallee.

506 Drake, W. Raymond. *Gods and Spacemen in the Ancient East.* 1973 [c1968]. New York: Signet (New American Library). Spacemen taught primitive humans civilization, and are remembered as gods in lore of India, China, Egypt, Babylon, etc. Hebrew "God" is really a spaceman ruler—not the Cosmic Absolute God. Chapter on Exodus: the miracles and plagues were interventions by the supermen from the stars. Psycho-electrical science, Atlantis, hollow earth, cosmic forces, reincarnation—all considered plausible. War between Saturn, Jupiter and earth with death-rays and nuclear devastation; also a Velikovskian comet disaster. Filled with citations and examples from ancient myths and legends; also scientific references. Bibliog., index. Drake has written many other similar *Gods and Spacemen* books: *Gods and Spacemen in the Ancient West; . . . of the Ancient Past.*

507 Driesch, Hans. *The History and Theory of Vitalism.* 1982 [1905]. Philadelphia: Porcupine Press. Orig. 1905; German. From his research on sea-urchin embryos, Driesch concluded there must be a non-mechanistic (non-Darwinian) guiding force he named "entelechy." Driesch cut embryos in half, yet both halves continued to function and developed normally; he considered it impossible to explain this mechanistically. His Aristotelian notion of entelechy is a form of vitalism; he said he had become a convinced vitalist by 1895. Driesch delivered the Gifford lectures at Univ. Aberdeen; later became interested in parapsychology.

508 Duncan, Homer. *Evolution: The Incredible Hoax.* c1978. Lubbock, Texas: Missionary Crusader. *Duncan:* attended Moody Bible Inst., Dallas Theol. Sem.; director of Missionary Crusader. Hardcore creationist proselytizing; totally uncritical acceptance of all anti-evolution arguments. "The theory of evolution is one of Satan's master strokes." Consists mostly of quotes—by creationists, anti-evolutionist scientists and scientists dissatisfied with evolutionist theory—indiscriminately mixed together. Useful collection of such quotes and sources. 170 references cited; mostly scientific, some biblical. Recommends young-earth creation-science books and organizations. Mentions all standard creationist arguments. Men believe in evolution because they seek to avoid recognition of God and because it is Satan's strategy. "The conflict between creation and evolution is a major part of the gigantic battle between God and Satan. It is a battle for the minds and souls of men." "The Biblical account of creation and atheistic evolution are diametrically opposed to each other. Both positions cannot be true.... I Believe in Creation Because I Believe in God ... in Christ ... in the Holy Spirit ... in the Bible ... [and] in the Gospel of Christ." Appendices include more quotes, and excerpts from **H. Morris, Meldau, K. Taylor**, and **DeHaan.**

509 ________. *Secular Humanism: The Most Dangerous Religion in America.* 1984 [1979]. Lubbock Texas: Missionary Crusader. Rev. and enlarged ed. Orig. 1979. Orig. distr. by Christian Focus on Government. Introduction by Sen. Jesse Helms; blurbs by Archibald Roberts, **Criswell, Rice, LaHaye,** J. Robison, other well-known evangelists. Includes long quotes from Barbara Morris *(Change Agents in the Schools* and *The Religion of Humanism in Public Schools),* **Schaeffer, Whitehead** 1977; quotes many other creationists and anti-Darwinians. Reprints **Kennedy** 1983 anti-evolution sermon. Major section on "Humanism and the Theory of Evolution." The aim of the religion of secular humanism is to destroy biblical Christianity and make the government America's God. All our troubles began when Satan rebelled in Eden, and transferred his evil to Adam and Eve. "Satan's lie is that man is sufficient in himself and that he does not need God. The theory of evolution ... is based on this lie." Claims that Russians tried to start a new race in the 1920s by impregnating apes with human semen in order to prove evolution. Duncan says he has not met one student who learned any fact in school which contradicted evolution: "All such facts have been utterly repressed." Recommends ICR, CSRC, and several creation-science books. "Communism is humanism in political disguise." Lists humanist agencies, including the UN and "colleges and universities." Urges political action. Bibliog.

510 ________. *Evolution: Fact or Fantasy.* 1986. Lubbock, Texas: MC International. "A gigantic battle is being fought between Biblical Christianity and Secular Humanism..." Says he supports right to believe in other religions, but humanists and other atheists are flagrantly violating rights of Christians. Denounces Nat'l. Acad. Science anti-creationist booklet and 1985 Calif. Board of Educ. call for stronger evolutionist emphasis in textbooks. Refutes at length 13 evolutionist statements from NAS booklet, quoting many creation-science works. Refutes other claims of humanists and Satan. Evolution is believed because men are brainwashed, deceived by Satan, and because they wish to escape God's authority. Cites and recommends many creationist sources.

511 ________. *Evolution: True or False?* (booklet). 1987 [c1986]. Lubbock, Texas: MC International. A condensation of *Evolution: Fact or Fantasy?* Includes many scientific quotes; lists creation-science resources. Summary of standard arguments.

512 du Noüy, Pierre Lecomte. *Human Destiny.* c1947. New York: Mentor (New American Library). *Du Noüy:* French physicist. Purpose of book: to sustantiate faith in the high destiny of man by "giving it a scientific basis." States that natural evolution is one of the best demonstrated facts of science (though its mechanism is unknown), but opposes materialistic evolution, arguing instead for his "telefinalist hypothesis." Divine intelligence gives purpose and direction to evolution. Evolution is in "absolute contradiction to the science of inert matter." It runs counter to thermodynamics and probability. Therefore it must have been "willed" toward an end, a goal. Man is the goal of evolution. With the development of the human brain, our evolution is no longer physical, but psychological: spiritual and moral. Because of this evolutionary leap (the human brain), man must no longer obey the law of nature. Original sin is the tendency of all men to follow nature: "regression toward the animal." This regression is "contrary to the directing Will and represents absolute Evil. On the contrary, anything which tends to deepen the chasm between man and beast, which tends to make man evolve spiritually, is Good." Genesis is an almost Darwinian account of selection of the morally fit.

513 Dwight, Thomas. *Thoughts of a Catholic Anatomist.* 1927. London: Longmans, Green. "How very few of the leaders in the field of science dare to tell the truth as to the state of their own minds! How many feel themselves forced in public to do lip service to a cult that they do not believe in!" (Quoted in **J.G. Williams** 1970.) Dwight also wrote anatomy books.

514 Dye, David L. *Faith and the Physical World: A Comprehensive View.* 1966. Grand Rapids, Mich.: William B. Eerdmans. *Dye:* physics Ph.D.; radiation research at Boeing. *Ridenour* 1967 praises this work; says that "Dye points out the failures in scientism and the need for a world view of life philosophy that includes faith in God." Discusses "Christian presuppositions." **Whitcomb** and DeYoung 1978 say Dye is an advocate of the double-revelation theory, qualifying belief in the literality and historicity of Genesis by acceptance of some evolution or uniformitarianism.

515 Eads, Buryl R. *Let the Evidence Speak.* 1982. Atlanta, Ga.: Peachtree Press. "An excellent book in support of creation using largely a secular orientation (**J. Bergman,** in 1986 *Bible-Science Newsletter*).

516 Edershein, Alfred. *The World Before the Flood and the History of the Patriarchs.* 1875. London: Religious Tract Society. Also pub. by Revell (New York); some eds. undated. "The History of Israel and Judah Before the Flood, Revealing What the Bible Teaches, Defending Against the Insidious Attacks Arising from Misrepresentation and Misunderstanding of the Word of God."

517 Edgar, Morton. *The Great Pyramid and the Bible* (booklet). Chester Springs, Penn.: Laymen's Home Missionary Movement. Excerpted from 1000+ pp. 1910 book by John and Morton Edgar. Jehovah's Witnesses. Measurements of Great Pyramid passages contain biblical truths. Bible refers to Pyramid: parable of rejected cornerstone, etc. Pyramid reveals world chronology and God's Plan; the different historical Dispensations. Pyramid's base = Creation. Entrance = the Flood. First ascending passage = Old Law (Jewish) Covenant. Queen's Chamber passage = New Covenant (Millennial). Grand Gallery = Gospel Age. King's Chamber = 1914 A.D. Descending Passage = the evil world. Other symbolism explained also. Cites **Piazzi Smyth** and other Pyramidologists.

519 Effertz, Otto. *A Criticism of Darwinism.* 1894. Priv. pub. (New York). Translated and printed by Henry W. Cherouny. *Effertz:* M.D.

519 Eichman, Phillip. *Understanding Evolution: A Christian Perspective.* (Distr. by **J. Clayton;** presumably anti-evolution.)

520 Eidsmoe, John. *The Christian Legal Advisor.* 1984. Milford, Mich.: Mott Media. *Eidsmoe:* J.D. – Univ. Iowa; M.A. – Dallas Theol. Sem.; M.Div. – Lutheran Brethren Sem.; visiting prof. at O.W. Coburn Sch. Law (Oral Roberts Univ.). Foreword by **J. Whitehead.** Comprehensive manual on protecting and asserting fundamentalist religious rights. Chapters on creation/evolution. Objects to the "disestablishment" of creationism as a result of the Scopes Trial, and calls for the disestablishment of evolution. Discusses censorship of **Velikovsky** as example of religious persecution by evolutionists. Claims that evolution is a "religious tenet" as well as bad science, and that it relies on many faith assumptions. Praises CRS, ICR and other creation-scientists; recommends many creationist books and creation-science organizations. Urges adoption of two-model approach. Assails "pure fabrication" of dialogue and plot of "Inherit the Wind" (but Eidsmoe himself gets facts about the Scopes Trial wrong). Asserts his belief in creation of six literal (24-hour) days—in contrast to **W.J. Bryan,** who compromised on length of creation days. Bibliog., index.

521 Elam, Edwin Alexander (ed.). *The Bible Versus Theories of Evolution.* 1925. Nashville, Tenn.: Gospel Advocate.

522 Ellacott, Stephen W. *Mathematical Problems in the Evolutionary Model* (pamphlet). 1977. London: Newton Scientific Association. (Listed in Ehlert files.)

523 Ellsworth, Ward E., and Kay O. Ellsworth. *Identity Crisis; and "Science" Says: My Brother's Keeper? or My Keeper's Brother?* (booklet). 1975. Priv. pub. (Issaquah, Wash.). *W. Ellsworth:* M.A.—Western Wash. State Coll.; elementary school teacher. *K. Ellsworth:* B.A.—W. Wash. State Coll.; teacher. The Ellsworths (husband and wife), Church of Christ members, have been active in "attempting to destroy the evolutionists' monopoly in the public schools, replacing it with the 'Both Viewpoints or Neither' concept." There are only two basic viewpoints: creation and evolution. Ward presents both in his classes. Since creationism is at least as scientific as evolution (or, evolution is as religious and unscientific as creationism), intellectual honesty demands that both be taught. Includes letters by Ellsworth to Issaquah Curriculum Council, and presentations to state House and Senate. The Ellsworths were active supporters of state bills mandating balanced treatment for creation-science (reprinted here). Long section of quotes from various school library books, textbooks and other materials demonstrating the extent of dogmatic evolutionist indoctrination. (Inc. quotes from H. and N. Schneider, who wrote section of **D. Steele** 1973.) Advocates young-earth creationism; recommends creationist books.

524 Ellwanger, Paul. *Creation Science and the Local School District* (pamphlet). 1979. El Cajon, Calif.: Institute for Creation Research. ICR "Impact" series #67. *Ellwanger:* South Carolina respiratory therapist; heads Citizens for Fairness in Education (since moved to Texas). Consists of unsigned section and reprinted article from South Carolina about Ellwanger, plus section by Ellwanger himself. Ellwanger's efforts, using **W. Bird**'s sample resolution, eventually led to the Arkansas creation-science bill.

525 Elmendorf, R.G. *How to Scientifically Trap, Test and Falsify Evolution* (booklet). 1978. Bairdford, Penn.: Bible-Science Association of Western Pennsylvania. *Elmendorf:* Cornell grad; heads own engineering, design and fabrication company. "Backed Up By a $5000 Guarantee!" Second Law of Thermodynamics is unsurmountable barrier to evolution. $5000 for proof of evolution—but evolution is uphill process, and 2nd Law proves processes go downhill. Evolution is a religion, not science. Extremely simplified language and examples; lots of cute stick-figure drawings. "Creative trinity": apparent exceptions to the law—living organisms—require energy (open system), plus structure and intelligence (energy conversion mechanism), and coded genetic instructions. Elmendorf, a whimsical eccentric, refers to himself as a "windmill-tilter." (He is also a confirmed geocentrist, and offers $1000 for proof the earth moves.)

526 England, Don. *A Christian View of Origins.* c1972. Grand Rapids, Mich.: Baker Book House. *England:* medicinal chem. Ph.D.—Univ. Mississippi; chem. prof. at Harding Univ. (Searcy, Ark.); Church of Christ deacon. Shows good knowledge of biology. Advocates old-earth creationism. Presents various creationist theories (including subsets and variants), lists objections to each. Doesn't commit himself to any particular view, but seems more sympathetic to "multiple gaps" interpretation (Gap Theory variant:

gaps between the six days of creation). Reasoned tone.

527 ________. *A Scientist Examines Faith and Evidence.* c1983. Delight, Ark.: Gospel Light Publishing. Introduction by **J.W. Sears.** Largely on how to approach and interpret the Bible. Nature's and Bible's truths: no conflict—but often conflict between theories. Faith is not dependent on science, which changes. Refutes most commonly-offered Bible-science "proofs"; shows that such Bible passages are mostly poetic, though he accepts a few as genuine. The Bible contains no bad science. Genesis is scientifically and historically accurate; England argues, however, that we shouldn't try to harmonize the Bible with science, or consider our fallible interpretations of Genesis as absolute truth. We can't prove theories based on silence (Gap Theory, inc. "multiple gap") or on loose and unwarranted exegesis (Day-Age creationism)—though we should take note of what Genesis does *not* say as well as of what it does. The Bible doesn't give us the age of the earth; there is no biblical reason to insist on Flood Geology or recent creation (though the earth *may* be young). The Bible does refute evolution: "There is no way, allegorically or otherwise, by which the Genesis account of the origin of the first man and the first woman can be brought into harmony with modern theories on the origin of man as expressed in general biological evolution." "Genesis One reads like history, and it is taken to be a factual, yet not exhaustive, account of creation events." Bible prophecy is true. Charts.

528 Enns, Abe. *Evolution: Science or Speculation?* 1979. Priv. pub. (Rosenort, Manitoba, Canada). *Enns:* B.Sc.; B.Th.; on the Council (along with **Patten** of W.B. Borrowes' N. Amer. Creation Movement (Victoria, B.C., Canada; EPM affil.); apparently a teacher in Manitoba.

529 Enoch, H. *Evolution or Creation?* 1976 [1966]. Welwyn, Herts, England: Evangelical Press. Rev. and enlarged ed. Orig. 1966 (India: Union of Evangelical Students); 1st British ed. 1967 (London: Evangelical Press). *Enoch:* zoology prof. (ret.) at Presidency College, Univ. Madras (India). Foreword by Sir **C. Wakeley.** Purpose of book is to "set before the public the truth of the Word of God as against the erroneous philosophy of evolution." Presentation of standard creation-science arguments, with explicit biblical references. Quotes many evolutionist and anti-evolutionist scientists. Morphology, paleontology, age of the earth, geology, embryology, mutations, origin of life, bad effects of evolution on civilization and religion, the need to embrace creationism. Advocates Flood Geology. Accepts Ussher's date of 4004 B.C.; says **Newton** did too. Concludes that Genesis account of man's creation 6,000 years ago is "literally, historically and scientifically true, for no demonstrable scientific fact so far known (without assumptions) can contradict that date." Argues that exposure to evolutionism leads to rejection of God, and loss of morality. Has section on "Darwin's Recantation" (story of his deathbed conversion and repudiation of evolution) in 1967 ed.; accepts story as true. 1976 ed. omits this section without explanation, but adds chapter on dating methods and other new data supporting creationism. Photos (inc. carved Paluxy "manprints"); bibliog.

530 Epp, Theodore H. *The God of Creation.* 1972. Lincoln, Neb.: Back to the Bible. Also pub. in 2-vol. ed. *Epp:* director of Back to the Bible Broadcast. Primarily theological, but includes much anti-evolution material, Flood Geology arguments, pre-Flood Water Canopy. "I repeat: evolutionists have not been able to find links between nothing and matter, between matter and animal, or between animal and man, because there are no links. God created each completely and distinctly apart from the others.... In our scientific age, it is unbelievable that even our educated people have accepted such an unscientific theory as evolution." It is accepted because it is an atheistic attempt to explain the world without God. Refutes Gap and Day-Age creationism; insists on strict young-earth creation. Chapter on "The Flood and Science" relies on **Whitcomb** and **Morris.** Also quotes other creation-scientists throughout the book. Extended discussion of II Peter: the uniformitarian evolutionist "scoffers" who, in the Last Days, ridicule

Bible-believing creationists. Includes photo section on Flood Geology.

531 _______. *Science and the Flood* (booklet). 1972. Lincoln, Neb.: Back to the Bible.

532 Epperson, A. Ralph. *The Unseen Hand: An Introduction to the Conspiratorial View of History.* c1985. Tucson, Ariz.: Publius Press. Exposes the great Conspiracy: its history and its modern forms. A massive compendium: includes sections on communism, Federal Reserve System, Trilateral Commission, population control, the Rothschilds, world government, and many other sinister manifestations of the Conspiracy. Also anti-laetrile campaign; states that Hunza has no cancer. "Science versus Reason" chapter is about creation-science, which the Conspiracy is trying to suppress. Evolution was a socialist, humanist, Masonic plot. Cites many creationist arguments, including Paluxy, Nebraska Man, moon dust, shrinking sun, population increase rate, probability, the Second Law. Quotes **Morris, F. Hoyle;** some scientific references. Lists ICR, the John Birch Society, etc., as good organizations. Index.

533 Erickson, Lonni. *The Teaching of Evolution in Public Schools: A Comparison of Evolution and Special Creation* (booklet). 1980. Lyons, Col.: Scandia Publishers. (In Ehlert files. Presumably favors creation-science.)

534 Ernst, Elizabeth. *Hummy and the Wax Castle* (booklet). c1984. El Cajon, Calif.: Institute for Creation Research. *Ernst:* author. **Richard Bliss:** project director. Steve Pitstick: illustrator. Young children's book about bees. Wonderful and complex design of bee proves creation, not evolution. Hummy the Bee explains all this to Johnny. Drawings on every page.

535 Estep, Howard C. *Evolution: True or False?* (booklet). 1969. World Prophetic Ministry. (Listed in SOR CREVO/IMS.) Estep also wrote *The Anabaptist Story.*

536 Evans, Jervice Gaylord. *Christianity and Science Versus Evolution and Infidelity* (booklet). 1895. Galesburg, Ill.: Corville & Barnes. Also wrote books on tobacco and Christianity, and politics and Christianity.

537 Everest, F. Alton. *Dust or Destiny.* c1949. Chicago: Moody Press. *Everest:* electrical engineer; assoc. director—Moody Institute of Science. Book parallels Moody Inst. of Science film (1959) of same title. (From Moody Inst. of Science "Sermons from Science" film series. First film 1945; series is still in production. Series is very widely shown in public schools, on T.V., and in 132 nations. The films affirm God as Creator and refute materialistic chance evolution.) Design of the human body; design of animals. Explicitly endorses **Paley**'s design argument. Bible references at conclusion. Drawings.

538 _______. *Hidden Treasures.* 1951. Chicago: Moody Press. "The story of the wonder microuniverse about us, hidden by the vagaries of size." Parallels Moody Inst. of Science film ("Sermons from Science" series) of same title. Drawings.

539 Faber, George Stanley. *Treatise on the Patriarchal, Levitical and Christian Dispensations.* 1823 [1816]. London: C. & J. Rivington. Millhauser says Faber imposed mystical interpretation on Day-Age scheme of geological and cosmic eras corresponding to Creation 'days.' Also wrote books about Bible prophecy.

540 Fabre, Jean Henri. *Social Life of the Insect World.* 1912. New York: Century. Also pub. by Unwin (London). *Fabre:* great French entomologist; especially well-known for his pioneering studies of instinct. Regarded evolutionary theory as "a solemn hoax"; "A theory exploited in big words but destitute of even little facts." "Although his works were admired by Darwin, Fabre was all his life opposed to evolution, remaining convinced of the fixity of species" *(Dict. Sci. Biog.).*

541 Fairholme, George. *General View of the Geology of Scripture.* 1833. Philadelphia: Key & Biddle. "In which the unerring truth of the Inspired Narrative of the early events of the world is exhibited, and distinctly proved, by the corroborative testimony of physical facts, on every part of the earth's surface." Also (orig.?) pub. in London. *Fairholme:* geographer; one of the four great Scriptural Geologists (Millhauser). Fairholme rejected all "speculative" modern geology as anti-biblical. Insists on the absolute reliability of the sacred Mosaic account. Denounces continental—especially French—theories as wild, absurd, atheistic philosophies. Ridicules Lamarck's development hypothesis. Quotes and praises **Linnaeus, Newton, G. Penn, S. Turner.** Asserts modern theorists err by addressing only "secondary causes"—even **Werner.** Discusses Lyell's new theory as prime example of this preoccupation with secondary causes. This reliance on "mere natural law" leads to theories of gradual "chemical" development of the earth and of life, instead of acknowledging fiat *ex nihilo* creation by God. Rocks, as well as plants and man, were "created in their mature and perfect forms": i.e. with appearance of age. Argues that the great interdependent chain of life must have been created simultaneously, as a whole. Earth's geology was shaped by original creation out of nothing, by the formation of the sea bed during the third day of creation (separation of water and land), and by the Flood. On the third day, the waters were drawn into the Abyss. The Flood was a worldwide catastrophe. Rejects theory that earth had worldwide warm climate; asserts that because all organisms "invariably float" when drowned, the great currents, tides and winds of the Flood carried their remains all over the earth prior to fossilization. Mentions rains of fish from volcanic eruptions; discusses Guadalupe skeletons as evidence of man's presence in lower strata; appeals to Book of Enoch; suggests that original language was Hebrew. "The great end of the study of geology ought to be, a *moral,* rather than a *scientific* one..."

542 ________. *New and Conclusive Physical Demonstration of the Fact and Period of the Mosaic Deluge.* 1837. London: T. Ridgway & Sons. Described and quoted in **B. Nelson** 1931; Nelson is impressed with Fairholme's strict creationism and Flood Geology. Fairholme argues that no early convulsions of the earth (as supposed by geologists) could have occurred, since there could be no Deluge prior to "moral guilt"—i.e. since creation of mankind (quoted in A. White 1895).

543 Fairhurst, Alfred. *Organic Evolution Considered.* c1897. Cincinnati: Standard Publishing. 3rd ed. has 1911 Preface. *Fairhurst:* A.M.; natural science prof. at Kentucky Univ. 474-page criticism of evolution. "The doctrine of evolution is not science." "My object ... is to promote the belief in Theism..." Attacks evolution for undermining belief in theism. Insists on mind/body duality; mind (spirit) as primary fact of our existence; necessity of teleological explanation of mind as created by Supreme Intelligence. "Mind exists, and God exists as its necessary Author." Many scientific references; few biblical references except in concluding chapter "Genesis and Geology," but refers often to Creator. Appends long critical review of E. Cope's *Primary Factors of Organic Evolution.* Also includes "Is the Scientific Doctrine of Evolution Compatible with Christian Faith?" (orig. presented 1901 at conf. of Disciples of Christ). Affirms divine fiat creation; miracles as acceptable scientific explanation. "If the Bible and Christ and Christianity were products of evolution by natural causes, then Christ was only an erring man who mistook his own nature and mission, who died in vain, did not rise from the dead, and our faith is in vain." Supports Day-Age creationism, but says detailed comparison of Genesis and geology is useless (purpose of Bible is moral). Bible employs phenomenological language. Accepts geological chronology, but species appear as if by substitution rather than transformation. Discusses embryology, vestigial organs, natural selection, sexual selection, instinct, fossil gaps and stasis, fossil humans, design in nature, evil and altruism, future life, etc. Moral and mental aptitude of savages contradicts natural selection. Evolution cannot account for development of flight, electric organs, the eye, etc. If spontaneous generation occurred, it should occur now. Argues that parent-offspring fertility prevents evolution of varieties. Index.

544 ________. *Atheism in Our Universities.* c1923. Cincinnati: Standard Publishing.

545 Fano, Giulio. *Brain and Heart.* 1926. Oxford, England: Oxford Univ. Press. Orig. Italian. Foreword by Prof. E.H. Starling. **Field** 1941 says Fano was "Director of the Institute of Osteology and General Physiology in the University of Rome," and that he attacked evolution in this book. Diagrams; bibliog.

546 Faulstich, Eugene W. *Master Designer—Master Historian—Master Builder* (report). Undated. Priv. pub. (Ruthven, Iowa). *Faulstich:* electrical engineer. Now heads Chronology-History Research Institute, intended as graduate level school devoted to Bible apologetics, emphasizing creation-science and study of biblical chronology; affiliated with **W. Lang**'s Genesis Institute (but not endorsed by BSA). Thesis: "Old Testament Hebrew Scripture Can Scientifically Be Proven to Be Historically Accurate Since Creation, and That Jesus Was Predestined in That History As the Messiah and Saviour of the World." Assumes no gaps in Genesis genealogies. Computer analysis shows Bible chronology is accurate; number patterns prove supernatural origin and plan of history chronicled and foretold in Bible. Numerology: significant time-spans occur as unusual sums of days, etc. Cycles of days, weeks, months and years only aligned every 2395 years, including 4001 B.C.: date of Creation. Unusual planet alignment on 4th day of creation verified by Harvard Univ.

547 ________. *Moses the Astronomer and Historian Par Excellence* (booklet). Undated. Rossie, Iowa: Chronology-History Research Institute. Moses' genealogy and history is perfect (not Luke's). Creation: 4001 B.C. (Sunday, March 17). Flood: 1656 years later. Confirmation from Harvard Center for Astrophysics of extraordinary alignment of Earth, moon, Venus, Mars and Mercury. Computer shows half-day discrepancy in planetary motions because of Joshua's Long Day—proves sun stood still then. Discovers numerical patterns in bible chronology, especially 7s and 3s. Includes astronomical computer program. Many chronology tables.

548 ________. *History, Harmony & the Hebrew Kings.* 1986. Spencer, Iowa: Chronology Books. Large, impressive-looking loose-leaf volume. Computer studies of biblical chronology correlated with astronomical data: eclipses, conjunctions, calendrical cycles; plus numerological evidence. Largely inspired by Edwin Thiele's *Mysterious Numbers of the Hebrew Kings.* Thanks Owen Gingerich of Harvard for providing key data—he confirmed 4001 B.C. planetary alignment, which Faulstich claims is unique. Describes chronology for reign of each king of Israel and Judah; also correlates with Assyrian chronology. Many chronological tables. Extensively referenced (mostly historical and chronological studies).

549 ________. *How Can You Know for Certain?* (pamphlet). Undated. Which religion is true? Faulstich's scientific confirmation of Genesis chronology and Creation proves Bible is true. Charts.

550 Fellowes, Francis. *Sacred History of the Deluge Illustrated, and Corroborated by Tradition, Mythology, and Geology.* 1836. Philadelphia: Key & Biddle.

551 Ferch, Arthur J. *In the Beginning.* c1985. Washington, D.C.: Review and Herald. Ed. by **G. Wheeler.** *Ferch:* Ph.D.—Andrews Univ.; head of religious studies and theology at Avondale Coll., Australia; Seventh-day Adventist. Commentary on Genesis. "The cataclysmic beginning of our world revealed in the book of Genesis *guarantees* the supernatural end of our planet when Eden lost will become Eden restored." Discusses literary structure, but insists on recent, six-day *ex nihilo* creation. Theological (not scientific) arguments. Denies immortality of soul. Emphasizes Flood as worldwide destruction of life.

552 Ferguson, Les. *It's About Time.* c1976. New York: Vantage Press. *Ferguson:* South Dakotan; amateur fossil hunger, geologist and archeologist. Old-earth creationism.

553 Ferris, A.J. *The Conflict of Science and Religion.* Undated. Vancouver, B.C., Canada: Association of the Covenant People. No dated references, but seems several decades old. Strongly anti-evolution. Science has destroyed the theory of evolution. Asserts that the two Genesis creation accounts describe different creations. Gen. 1: Creation of man (interpreted as Day-Age creationism). Gen. 2: Later, Adam and Adamic race were created. Cain bred with pre-Adamic races: Negroes, Mongols and coloreds; offspring were Latins and Teutons. Ham also interbreeds later. Pure Adamic race maintained through Shem (Israel) and Japheth (Slavs). Regional Flood destroys the Seth line only. Adamic race will be first to attain immortality; other races will be taught by them. (Assoc. of the Covenant People believe a type of British-Israelism; preach that Nordic-Celtic peoples are true descendants of Israel—the Chosen People.) Author presumably same as Alexander James Ferris who wrote *Britain-America Revealed as Israel* (8th ed. 1941; London); also other books on British-Israelism, and on the Great Pyramid.

554 Fides, Anthony. *Our Origin, Creation or Evolution.* Tape cassette. Distr. by Keep the Faith (Catholic). *Fides:* Catholic priest.

555 Field, A.N. *The Evolution Hoax Exposed.* 1971 [1941]. Rockford, Ill.: TAN Books and Publishers. Orig. 1941, then titled *Why Colleges Breed Communists.* TAN ed. reprint of 1971 ed. from Christian Book Club of America (Hawthorne, Calif.). Catholic. Shrilly denounces evolution as sinister anti-Christian propaganda. Evolution is not science; it is "a very low-grade religion, with its hymns played in jazz and syncopated cacaphony, and its sanctuaries adorned with cubist art." "Darwinian monkey-man materialism" has produced nothing but dirt and degradation. Accuses colleges of destroying faith by spreading evolution. Covers usual creation-science arguments: ancient pagan roots of evolution; lack of experimental proof; morphology as Design; no true vestigial organs; embryology; proposed ancestral hominids all fakes or apes. Emphasizes evil offspring of evolution: communism, determinism, psycho-analysis, League of Nations, etc. Evolution was invented in atheist revolutionary France; a Satanic conspiracy. Includes many quotes of anti-evolutionist scientists. Cites doubts about Piltdown Man. Rejects Malthus: no struggle for existence; no natural selection. Discusses Raglan's cultural diffusion: savages don't evolve; they are degenerates. Deplores evolutionist encouragement of race mixing: implies we should maintain racial purity.

556 Fields, Weston W. *Unformed and Unfilled: A Critique of the Gap Theory.* c1976 [1973]. Nutley, N.J.: Presbyterian and Reformed. Orig. pub. 1973 by Light and Life (Winona Lake, Ind.). Orig. M.Div. thesis (Grace Theol. Sem.). Grace Brethren. Acknowledged as definitive critique and refutation of Gap Theory creationism; a response to **Custance**'s 1970 Gap Theory defense. Strict creationism: includes strong young-earth arguments. Refutes claim that ancient commentators and church fathers advocated Gap theory; says they may have supposed an interval between first two verses of Genesis, but not long ages with ruin-reconstruction scenario. Cites Rosenmuller 1776 and Dathe 1791 as first true Gappers; debunks claims of earlier supporter. Well-argued; scholarly. Contains much bibliog. information.

557 Fields, Wilbur. *Retracing Paluxy River Tracks* (booklet). 1977. Joplin, Mo.: Ozark Bible College Bookstore. (Listed in Ehlert files.) Also wrote *Exploring Exodus* (c1976; Ozark Bible College).

558 ________. *The Paluxy River Explorations (1977–1979).* 1980. Priv. pub. (Joplin, Mo.). Rev. ed. Not copyrighted. The Paluxy "manprints"—humans contemporaneous with dinosaurs proved by human prints alongside dinosaur tracks.

559 Filby, Frederick A. *Creation Revealed: A Study of Genesis Chapter One in the Light of Modern Science.* 1964. Westwood, N.J.: Fleming H. Revell. Also pub. by Pickering & Inglis (London). *Filby:* Ph.D. (London); Fellow—Royal Inst. Chem.; inorganic chemistry prof. at S.E. Essex Tech. College. Progressive creation. Science "really proves" the "Wisdom and Glory of God." "I do not think the six days are primarily meant to be consecutive"—but almost so. Genesis deals with "six great topics which finally lead up to the coming of man and the completion of God's work." Time periods of creation 'days' may overlap; Day 4, e.g., refers back to the beginning of creation—sun and moon existed all along, but their fulfillment occurs on fourth 'day.' Age of man began 15,000 years ago. Considers existence of pre- and "co-Adamic" men likely. Biblical *nephilim* may have been offspring of humans and co-Adamic beings possessed by demons: "Hence the Flood." Suggests compromise between **Ussher**'s chronology and age shown by astronomy. Bibliog., index.

560 ______. *The Flood Reconsidered.* 1970. London: Pickering & Inglis. Genesis is authentic—a "sober, historical account." Christ knew world history. Argues for a regional Flood, 4000–3500 B.C.—not worldwide, but more than a local Mesopotamian one—which flooded Asia and Europe. Declares **Whitcomb** and **Morris**'s Flood Geology "absurd"; likes **Patten** better (planetary fly-by, resulting tidal Flood). Presents chart comparing 'short' (to 8000+ B.C.). vs 'long' chronology. Discusses worldwide Flood tales, frozen mammoths, Ark-eology tales (doubts some). Biblical Ark account is accurate; not all species were taken. Discusses the two rebellions: of the Flood and Babel. We are ignoring warning of the coming Fire Judgment by believing in uniformitarian geology. Sons of God (demons) mating with women: giant offspring.

561 Filmer, W.E. *Morals and Evolution* (pamphlet). Undated. England: Evolution Protest Movement. EPM pamphlet #33. Filmer also wrote other EPM pamphlets—on **Nilsson** (#44), the British Museum (#19), etc.

562 Finleyson, John. *God's Creation of the Universe As It Is, in Support of the Scriptures*... 1837 [1832]. London. Includes Excerpts from work by Richard Brothers. Geocentric. Planets are ice. Comets sent by God for particular purposes. British-Israelism. Prophesies immiment collapse of all non-British empires. Strict creation; worldwide Flood. Earth 5507 years old; Creation 3670 B.C. Fanatical tone.

563 Fischer, Robert B. *God Did It, But How?* 1981. Grand Rapids, Mich.: Academie Books (Zondervan). *Fischer*: analytical chem. and electrical engin. Ph.D. —Univ. Illinois; VP of acad. affairs at Biola Univ. "Creation, Science and Christian Faith." Favors Day-Age creationism, but not dogmatic. Accepts old earth. Concedes that evolution is "useful" and more than a mere "theory," but also has weaknesses. Mostly philosophical and theological; some scientific discussion, but no references or quotes. Attitude towards origins is primarily question of world-view or paradigm choice: either theistic, supernatural and miraculous, or purely naturalistic.

564 Fish, Robin D. *I Believe in God the Creator* (booklet). (1980s?). Minneapolis: Bible-Science Association. *Fish:* former BSA staff member; minister. This booklet "written especially to show Junior High students what is wrong with evolution and that the Bible does indeed offer an intelligent alternative..." Evolution is a religion, based upon desire to escape from God. Creationism is also religious—but it is better science than evolution. Presents summary of standard creation—science arguments. Drawings.

565 Fisher, James C[ogswell]. *The Mosaic Account of Creation* (pamphlet). 1858. Philadelphia: Merrihew & Thompson. Response (apparently favorable) to **H. Miller**'s Day-Age creationism.

566 Fix, William R. *The Bone Peddlers: Selling Evolution.* 1984. New York: Macmillan. *Fix:* behavioral science M.A. —Simon Fraser Univ. (Canada). Harsh critique of theories of human evolution. Persuasive presentation of bias and wishful thinking in construction of evolutionist phylogenies. Asserts that evolutionists are engaged in a coverup. Their dogmatic denials of creation-science are suspiciously defensive and emotional; the creation/evolution debate is a clash of religious ideologies. Declares that "human evolution is simply *not* proven." Presents standard creation-science arguments as refutations of evolution. Emphasizes that all proposed human ancestors have all been disqualified; many are hoaxes. Relies on various anti-Darwinist arguments: **Macbeth, Hitching, Broom,** etc. States that Paluxy Cretaceous human prints may be genuine. Stresses uniqueness of man. Plato's idealist approach was right; Darwinian materialism is wrong. "Whatever happened to such concepts as intelligence, design, and purpose?" Materialistic explanations of evolution as mere "accidents" are driving people to accept fundamentalist creationism, which Fix also ridicules. Fix proposes a "compromise": psychic evolution. Millions of spirits came down to earth, infused into early hominids, then further modified ontogeny to produce a whole generation of true humans. Argues that this solves all evolutionary puzzles, and retains the "dignity and singularity of man": man was never just an animal. Advocates spirit beings, astral bodies, Tibetan occultism; cites many paranormalists to support body/soul distinction, and evolution by spirits modifying physical bodies. Bibliog., index. Fix also wrote *Pyramid Odyssey* (1978; New York: Mayflower), which is quoted in **Noone.**

567 Fleischmann, Albert. *Die Descendenztheorie.* 1933 [1901]. Leipzig, Germany: A. Georgi. *Fleischmann:* comparative anatomy prof. at Erlangen Univ. (Germany). **Morris** 1984 calls this "a strong creationist work." The theory of evolution "suffers from grave defects which are becoming more and more apparent as time advances. It can no longer square with practical scientific knowledge, nor does it suffice for our theoretical grasp of the facts." "The Darwinian theory of descent has not a single fact to confirm it in the realm of nature. It is not the result of scientific research, but purely the product of imagination." (Quoted in **D. Zimmerman** 1976; Victoria Inst. vol. 65. **Field** 1941 also quotes letter from Fleischmann to **L.M. Davies** rejecting evolution.) (**Graebner** 1921 also cites Friedrich Pfaff, natural science prof. at Erlangen Univ., as refuting common ancestry of man and apes, and advocating recent and sudden appearance of man and the truth of the biblical account of creation.)

568 ________. *Die Darwinische Theorie.* 1903. Leipzig, Germany: G. Thieme.

569 Fleming, Sir John Ambrose. *Evolution or Creation?* Undated [1933]. London/Edinburgh: Marshall, Morgan & Scott. *Fleming:* physicist, prof. of electrical engineering at Univ. London; inventor of electron tube making radio broadcast possible; Fellow—Royal Soc.; former president of Evolution Protest Movement. Evolution is often accepted uncritically, but it is so blatantly opposed to the Bible we must examine it critically. If creation is false, so is the rest of the Bible.

570 ________. *Modern Anthropology versus Biblical Statements on Human Origin* (pamphlet?). 1935. Westminster, England: Victoria Institute. 2nd ed.; rev. Includes argument that rate of human population increase was much less in prehistoric times. Evolution requires low [sic] death rate; thus if evolution were true, population increase would have been high always. Also argues that *Pithecanthropus [Homo] erectus* fossils are misinterpreted.

571 ________. *The Origin of Mankind Viewed from the Standpoint of Revelation and Research.* c1935. London/Edinburgh: Marshall, Morgan & Scott.

572 Fleming, Kenneth C. *God's Voice in the Stars: Zodiac Signs and Bible Truth.* 1981. Neptune, N.J.: Loizeaux Bros. God created the constellations to illustrate Christian themes. "The grand truth is that God made the stars for a witness to mankind of a coming Redeemer . . ." Demonic forces later corrupted the meanings of the stars (astrology). Drawings.

573 Fleming, Partee. *Is God's Bible the Greatest Murder Mystery Ever Written?* 1980. Memphis, Tenn.: A-M Press. Bizarre, incoherent. Book contains many endorsements of Fleming by local politicians. Says dinosaurs and other extinct animals never existed; fossils are examples of Jesus's "wit."

574 Flindt, Max H. *On Tiptoe Beyond Darwin.* 1962. Priv. pub. Preliminary version of *Mankind—Child of the Stars;* cited by J.R. Greenwell in 1980 *Skeptical Inquirer.*

575 ________, **and Otto O. Binder.** *Mankind—Child of the Stars.* 1974. Greenwich, Conn.: Fawcett. Foreword by **von Däniken.** *Flindt:* worked as lab technician under Teller, Seaborg; at Lockheed and Lawrence Radiation Lab. *Binder:* wrote under NASA contract; hon. M.S. from NASA (1963); writes UFO books. UFOs; mankind is hybrid offspring of spacemen and sub-humans (Genesis "Sons of God, daughters of men"). Various anti-evolution sources cited: **Macbeth, Velikovsky,** etc. **Wallace** was right; Darwin was wrong.

576 Flori, Jean, and Henri Rasolofomasoandro. *Evolution ou Creation?* 1974 [1973]. Damarie les Lys, France: Editions SDT. 2nd ed. Authors are French Seventh-day Adventists. Strict creationism. Apparently contains many standard creationist arguments. (Discussed in **Blocher** 1984.)

577 Forssberg, C.F. William. *Dissertation on Noah's Ark.* 1943, 1945. Priv. pub.? Discussed and quoted in **Cumming** 1972. Describes in detail the appearance and construction of the Ark. Refers to Ark-eology tales and traditions from Mt. Ararat region. Forssberg is "certain" it is located on Ararat (though doubtful of claimed sightings.) Ark was lacquered, finished, and perched atop the mountain to survive intact for future discovery, when it will "expose false religion, evolution, and atheism..." Its discovery will climax all archeology and prove the Bible true.

578 Fort, Charles. *The Book of the Damned.* c1941 [1919]. New York: Ace. By "damned," Fort means anomalous data excluded by orthodox science. A collection of reported anomalies, largely of things falling from the sky—frogs, worked stone, carbonaceous material, etc. Witty style; blasts the scientific establishment. Fort attracted many literary supporters. It is unclear whether Fort's preposterous hypotheses were intended seriously. Planet Genesistrine is source of falling things; life may have begun there, and it influences evolution here. Fort's *New Lands* (1923) includes Prince Nouri's discovery of Noah's Ark. Fort alluded to the "tautology" of natural selection before creationists did.

579 Foster, David. *The Intelligent Universe: A Cybernetic Philosophy.* 1975. New York: G.P. Putnam's Sons. God as computer programmer. Appeals to DNA code, A. Eddington, Einstein, **J. Jeans,** Gödel, quantum mechanics. Universe is "one vast mind." "The total universe, inclusive of all aspects of matter and mind, shows a construction virtually indistinguishable from that of an electronic computer, and all of its workings are in the nature of intelligent data processing." Second Law of Thermodynamics is "anti-cybernetic," but a separate cybernetic process is imposed. Man can't evolve if he is "comfortable"—he must "wake up" and "aim" to evolve to greater intelligence.

580 Fothergill, Philip G. *Historical Aspects of Organic Evolution.* 1953. New York: Philosophical Library. *Fothergill:* botanist at King's College (Newcastle-upon-Tyne, England). An evolutionist, but sympathetic to criticisms of Darwinian evolution. Considers purely naturalistic evolution unsatisfactory; argues that Darwin's theory of natural selection is a reflection of the naturalistic, mechanistic philosophy of the times. Quotes and discusses many anti-Darwinians; approves of **McDougall, Lotsy,** others. Says he argued at length about evolution with **Dewar;** lists the leading anti-evolutionist biologists. Diagrams, tables; bibliog., index.

581 ________. *Evolution and Christians.* 1961. London: Longmans, Green. "It is right that any theory which seeks to prove that man is *merely* a product of evolution and nothing more should be combated and exposed for all its fallacy." **Jaki** 1978 says book shows that the history of modern biology might have avoided its inordinate craving to "deify natural selection" if Darwin had learned earlier than he did that mechanistic science is insufficient to explain everything. Presents evidence supporting evolution, but also mentions criticisms and difficulties. Considers evolution the "penultimate" expression of all life; but man alone the "ultimate." Last two chapters on evolution from Catholic viewpoint. Generally affirms Papal opinions regarding evolution; discusses ways of reconciling these with evolutionary biology. Adam may have been first true human, or first of race with spiritual capacity; Eve may have been Adam's daughter (also discusses various chromosomal theories of Eve's origin from Adam). Photos, drawings, charts; bibliog., index.

582 Fox, Norman. *Fossils: Hard Facts from the Earth.* c1981. San Diego: Creation-Life. ICR's Two-Model Children's Books series (**Richard Bliss:** Project Director). "Designed for use by children in the elementary grades in public schools, these two-model books contain creation/evolution discussions on an introductory scientific level" (ICR catalog). This book for grades 5-8. Emphasizes Flood as explanation for fossils. Includes Paluxy claims. Agate Springs fossils said to include animals of different ecological zones (thus the Flood). Says invertebrates appear "complete and complex" in Cambrian—no mention of any precursors.

583 Frair, Wayne, and Percival Davis.

A Case for Creation. 1983 [c1967]. Chicago: Moody Press. 3rd rev. ed. Orig. c1967. Foreword by **J. Klotz.** *Frair:* Ph.D.—Rutgers; biology prof. at King's College (Briarcliff Manor, N.Y.); research on biochemistry of turtle phylogenies. *Davis:* M.A.—Columbia Univ.; life science prof. at Hillsborough Community College (Tampa, Fla.). Succinct statement by trained biologists supporting strict creation. "This book is written to show that evolutionary doctrine is wrong." Warns Christians not to assume that evolution no longer poses a threat. Creationists must argue for creation carefully and intelligently, since evolution is advocated by men of genius. Avoids many common creationist errors and misinterpretations. Arguments mostly biological; cautiously accepts Flood Geology but admits much evidence is uncertain. Discusses homology, embryology, genetics, biogeography, transitional fossils, animal behavior, human evolution. Claims evolutionary assumption has hindered rather than aided taxonomy. Careful, moderate tone; plea for more and better creation-science research to counter evolutionist research. Some Bible references. Includes nice bibliog. essay, appendix of creationist organizations. Drawings; index.

584 Frangos, Apostolos. *From the Ape?* 1985. Athens, Greece. In Greek. *Frangos:* Greek Assoc. of Scientists for the Scientific Truth; speaker at 1986 Int'l Creation Conference; proposes International Federation to demonstrate that evolution is dogma not science. Book contains standard creation-science arguments (creationist references and footnotes in English); other scientific references. Photos, charts.

585 French, Joel, and Jane French. *War Beyond the Stars: Angelic Encounters.* c1979. Harrison, Ark.: New Leaf Press. *J. French:* with NASA chapter of Full Gospel Business Men Fellowship Int'l (Houston); staff engineer with NASA contractor; shared testimony with astronaut T. Stafford; runs Teen Challenge Center. Gap Theory creationism: heavenly war in pre-Adamic times—Lucifer and one-third of the angels rebel. Man later created on earth where dethroned Satan once ruled. UFOs are Ezekiel's "chariots of God"—supernatural space vehicles. **Hitler** was controlled by Satan, but there was godly intervention in WWII as well (e.g. Dunkirk); also a mysterious stranger (Archangel Michael) who persuaded Nazi High Command to make bad strategy decisions, providentially affecting the war's outcome. UFOs were also active in Israeli Six-Day war on both sides.

586 Frese, Elmer. *Lord Kelvin's & Dr. Barne's [sic] Age of the Earth* (booklet). Undated. Priv. pub. (Bellevue, Neb.). Comic-book style illustration of lecture by **T. Barnes** (ICR) to a college crowd. Comments of "audience" mostly in doggerel of strange syntax. An odd presentation.

587 ________. *The Peking Man Fraud* (booklet). Cartoon style format.

588 Friedrich, Orval. *Early Vikings and the Ice Age* (booklet). c1984. Priv. pub. (Elma, Iowa). Also c1987 version with 1986 Appendix. *Friedrich:* minister—Elma, Iowa. Viking evidences in Central North America (especially Iowa, Min., adjacent Canada, Missouri, Arkansas River, Dakotas, Okla.). Describes altars, weapons, inscriptions. Kensington Runestone is genuine. Bible is true. Pre-Flood Water Canopy, Flood (2347 B.C.) Velikovskian comet near-miss causes collapse of Canopy; ice blankets northern America. Adjacent land pushed up; ice melts but still trapped in inland sea. Finally bursts through near New Madrid, Mo. (site of 1811 quake). Then Vikings arrive, colonize, and enjoy a good Christian life; they later disappear. Photos, drawings.

589 Frysinger, W. Malsin. *The Weakness of Evolution.* 1925. Pentecostal Publishing Co. (Listed in SOR CREVO/IMS.) Apparently same Frysinger who wrote 1915 book on socialism.

590 Gabler, Mel, and Norma Gabler. *A Parent's Guide to Textbook Reform.* 1978. Washington, D.C.: Heritage Foundation. Heritage Foundation special supplement (Winter 1978). The Gablers are the well-known textbook monitors; they have been lobbying and scrutinizing schoolbooks submitted to Texas selection hearings since 1962 in order to protest anti-American, evolutionist, humanist, Marxist influences. They now head Educational Research Analysts (Longview, Texas).

591 _______, and _______. *Scientific Creationism* (Handbook #10). 1985. Longview, Texas: Educational Research Analysts. Introduction by **D. Gish:** neither creation nor evolution is scientific, but creation-science is superior. Compilation of various creation-science reprints, newspaper accounts of the Gablers' efforts, creation/ evolution polls, textbook analyses and ratings (especially of BSCS books). Includes inserts, ads for various creationist organizations. Has piece by **H. Morris:** "Evolution—The Established Religion of the State." Also reprints by **J.N. Moore, W. Brown, von Braun.** Presentations to State Board Educ. by **T. Barnes,** John Grebe (basic and nuclear res.—Dow Chem; 100 patents—synthetic rubber, styrofoam, Saran, "inventor" of petrochemical industry), Richard LeTourneau (engineer; president of LeTourneau College). Gablers claim that divergent and convergent evolution contradict each other logically. Science is the collection of facts. Fearful of Federal regulation of education, starting with biology. UNESCO writers—including Soviets—worked on BSCS curriculum team, though Texas law requires U.S. loyalty. **Norris Anderson** worked with BSCS; says BSCS writers admitted there was no evidence for human evolution.

592 _______, and _______. *General Assortment* (booklet). Undated [1980s]. Longview, Texas: Educational Research Analysts. Bound set of reprints on many textbook topics, including evolution and humanism, articles on creation/evolution polls, etc. "The Gablers ... DO NOT want anyone's creation story taught in public schools. Whether evolution, creation, or myth, these are in the area of the metaphysical. They do want students honestly and accurately taught scientific, empirically testable evidence for both the gradual appearance of life on earth and the sudden appearance of life on earth."

593 Gaebelein, Arno C. *The Conflict of the Ages.* 1983 [1933]. Neptune, N.J.: Loizeaux Bros. Rev. ed.; revised and edited by David A. Rausch. Orig. 1933. *Gaebelein:* editor of *Our Hope* (premillennial). Largely a warning against Bolshevism. Orig. ed. quoted Protocols of Elders of Zion, but omitted in rev. edition to avoid "misunderstanding." (Book was endorsed by right-wing anti-Semitic groups.) Rausch says Gaebelein was actually active in reporting Nazi persecution of Jews. First chapter, "The Great Enigma," about creation/evolution; refutes spontaneous generation. "To offer as a solution of the enigma of life, such theories as these—dead matter producing life, and life shaping itself into hundreds of thousands of different forms, through self-made laws, devoid of free will—is a gross insult to the intelligence of any human being." Includes many quotes by scientists rejecting Darwinism (relies heavily on **Graebner's** *God and the Cosmos*). Advocates Gap Theory creationism. Second chapter on Lucifer's pre-Adamic rebellion—the origin of all lawlessness. The earth "once sustained a gigantic animal creation and an equally gigantic vegetation." Geology confirms that it was later devastated by an enormous cataclysm. It remained submerged in ice and water for an incalculable period. Illuminati, Freemasonry, Sovietism, and other conspiracies all stem from Satan's rebellion.

594 Gale, Barry G. *Evolution Without Evidence.* c1982. Albuquerque, N.M.: Univ. of New Mexico Press. Legitimate history of science: examines Darwin's development of his theory. Cited by some creationists because of emphasis on Darwin's presentation of his theory as an argument which was not scientifically well-supported. Stresses the weaknesses of the theory as Darwin presented it; suggests he was forced to publish it "too soon." Bibliog., index.

595 Galusha, Walter T. *Fossils and the Word of God.* 1964. New York: Exposition Press. Written in form of a debate with atheist evolutionist co-worker. Bible always agrees with true science. Gap Theory creationism. Initial creation was followed by a catastrophe. Then, God created first man and woman in second creation (source of animal fossils, though pre-Adamites—the cave-men—survived). Second creation destroyed by second catastrophe. Adam and Eve created 6000 years ago in third creation; world later destroyed in Noah's Flood (2130 B.C.). On Fourth Creation Day of this creation, excess water was made into pre-Flood crystal (ice) Canopy shield. There were no carnivores in Eden: boa constrictors perhaps swallowed watermelons. Antediluvians had electricity but not internal combustion engines. Noah talked to the animals; they helped him build the Ark. The Flood broke the canopy sun shield. God shut off heat—"heat is some sort of a wave length"—freezing the Ark. Stone Age followed dispersal of Babel. God divided man into four colors and wants them to stay separate; the Devil, however, "will try to get them to unite and in this way defeat God's purpose."

596 Gange, Robert A. *Origin of the World* (booklet). c1983. Princeton, N.J.: Genesis Foundation. This booklet incorporated into 1986 book. *Gange:* physics Ph.D.; 25 years on staff at Sarnoff Research Center (Princeton, N.J.); electrical engineer (laser, missile, computer, holography, cryophysics); "honored seven times" by NASA; heads Genesis Foundation; has been featured on **P. Robertson**'s *700 Club.*

597 _______. *The Origin of Life* (booklet). c1984. Princeton, N.J.: Genesis Foundation. Incorporated into 1986 book.

598 _______. *Origin of Man* (booklet). 1985?. Princeton, N.J.: Genesis Foundation. Incorporated into 1986 book.

599 _______. *Origins & Destiny.* 1986. Waco, Texas: Word Books. This book incorporates Gange's three Origins booklets (1983–85). Back-cover blurb by Nobel physicist Wigner (applauds Gange's antimaterialism). Many scientific references; some creationist works cited. Old-earth creationism. Big Bang proves creation (beginning from nothing); so does Design of universe. Refutes reductionist explanations and "materialistic myth." Emphasizes entropy, information theory, statistics, "New Generalized Second Law of Thermodynamics," DNA code. "Accidents" couldn't form life; Intelligence (the God of the Bible) must have designed it. Debunks ape language claims. Includes bombardier beetle, Nebraska Man at Scopes Trial [false]. The Bible is ahead of science. Urges acceptance of Bible's eternal truth. Technical appendices. Bibliog., index.

600 Gardner, E.L. *Chains and Rounds* (booklet). 1966. London: Theosophical Society. From Appendix to *The Web of the Universe.* Theosophy: cycles of existence. Metaphysical; no scientific references. Seven Root-Races: the 3rd (black), 4th (yellow) and 5th (white) are all in "carnation" now simultaneously. Division of the sexes occurred in 3rd Root-Race; the reunion will occur in the next "round." We are in 5th sub-race of 5th Root-Race, in which Mind is dominant.

601 Gatewood, Otis. *There Is a God in Heaven.* 1970. Abilene, Texas: Contact. About evolution; presumably anti-evolution. Bibliog.

602 Geisler, Norman L. *The Creator in the Courtroom: "Scopes II".* c1982. Milford, Mich.: Mott Media. Written "in collaboration with" A.F. Brooke II and Mark J. Keough (both grad students at Dallas Theol. Sem.). Foreword by **D. Gish.** *Geisler:* Ph.D.—Loyola Univ.; systematic theology prof. at Dallas Theol. Sem.; defense witness at Arkansas creation-science trial. Account of the 1981 Arkansas trial. History and contents of Act 590; legal briefs; summary of testimony of all witnesses for and against; text of judge's decision; many samples of media coverage. Geisler argues strenuously against plaintiffs (evolutionists); analyzes and denounces legal errors and logical fallacies in decision and testimony. Geisler is not a young-earther; he says the law's wording requiring "recent" creation caused unnecessary trouble. Includes text of CSLDF **(Bird** and **Whitehead)** complaint against Att'y Gen. Clark's handling of case (Bird wanted to head defense; was rebuffed and persuaded some creationists not to testify). Geisler says, however, that Clark did good job under impossible circumstances. Motif of book is [false] claim that Darrow urged presentation of both creation and evolution in schools. In his testimony, Geisler admitted he believed UFOs were satanic deceptions. Useful though biased account.

603 _______. *Cosmos: Carl Sagan's Religion for the Scientific Mind.* 1983. Dallas, Texas: Quest Publications. Accuses Sagan of trying to impose secular "humanist religion, which denies God, on public." Geisler also wrote (with J.Y.

Amano) *Religion of the Force* (1983; Quest) about anti–Christian religion propagated in the film *Star Wars.*

604 ________. *Is Man the Measure?: An Evaluation of Contemporary Humanism.* 1983. Grand Rapids, Mich.: Baker Book House. Includes chapter "The Scientific Inadequacies of Secular Humanism." Lists creationist scientists of past – and notes that most "tended to be" creationist before 1860. Mentions **Thaxton**'s origin-of-life criticisms. Cites Blum, Yockey, Wald on impossibility of chance origin of life. Discusses C. Patterson's "Evolutionism and Creationism" talk at Amer. Mus. Nat. Hist. N.Y.; and Patterson's letter cited by W. Guste in 6-3-82 plaintiff's pre-trial brief on Louisiana creation-science case.

605 ________**, and J. Kerby Anderson.** *Origin Science: A Proposal for the Creation-Evolution Controversy.* 1987. Grand Rapids, Mich.: Baker Book House. Foreword by Walter L. Bradley. "It is the proposal of this book that a science which deals with origin events does not fall within the category of empirical science." Science may deal either with regularities ("operation science") or with singularities. Origins – both evolutionist and creationist views – belong properly to "singularity science." There are two types of scientific explanation: in terms of primary causes and secondary causes. Evolutionists rely exclusively on secondary cuases (the only proper explanation for operation science), but creationists legitimately appeal to primary cause for origins. Authors deny that religious basis of creationism makes it inherently nonscientific. They emphasize the "supernatural roots of modern science" – its Christian origins; they also argue that humanism is religious. Discusses views of various early scientists and philosophers. Affirms **Paley**'s design argument; denies it was refuted by Hume. Notes that **Newton** relied too much on supernatural intervention instead of operation science. Claims that science is now acknowledging origin science (praises **Thaxton,** Bradley and Olsen 1984). "The detailed analysis [of creation] is yet to be done by creatinists. However, it seems clear that if creationist views are to gain scientific credibility, then they must follow the principles of origin science and build a positive case for a primary cause, rather than relying on the ineffective means of pointing out flaws in various evolutionary hypotheses." Appends Geisler's update of Paley's 1802 argument (orig. pub. in 1984 *Creation/Evolution*); Judge Gee's dissenting opinion in Louisiana balanced-treatment law; summary of creation-science evidence from **W. Bird**'s Supreme Court brief. Bibliog., index.

606 Gentet, Robert E. *Dinosaurs Before Adam?* (booklet). c1974 [1963]. Pasadena, Calif.: Ambassador College. Reprinted from *Plain Truth* **(H.W. Armstrong).** Lessons from destruction of the dinosaurs. Gap Theory creationism: pre–Adamic creatures (including dinosaurs) were destroyed before the six-day re-creation. Pictures.

607 Gentry, Robert V. *Creation's Tiny Mystery.* 1986. Knoxville, Tenn.: Earth Science Associates. *Gentry:* Seventh-day Adventist; geophysicist; former prof. at Columbia Union College (Takoma Park, Md.); former guest scientist at Oak Ridge Nat'l Lab.; former member of **G. Vandeman**'s Noah's Ark team; Paluxy investigator and supporter; defense witness at Arkansas creation-science trial. Foreword by W. Scott Morrow (another Arkansas witness). Gentry is acknowledged expert on pleichroic halo research, and is considered among the most reputable creationist scientists, publishing in many legitimate journals. Gentry claims halos, from radioactive decay in rock crystals, prove Precambrian rock was initially solid – created suddenly, not gradually cooled – because halos from radioactive parent elements are not present. Claims related radiohalo evidence proves young earth and Flood also. Book title refers to disparaging reference to Gentry's halo anomalies during Arkansas trial. Includes reprints of all Gentry's major scientific articles; account of Gentry's struggle against evolutionist bias and persecution; extended critical discussion of Arkansas trial; reprints of correspondence to and from Gentry; explanation of halo evidence and its creationist implications. Gentry repeats his melodramatic challenges for evolutionists to falsify his theory. Color radiohalo photos, drawings; index.

608 Gibbs, Jessie Wiseman. *Evolution and Christianity.* 1931 [1930]. Priv. pub. (Memphis, Tenn.). Strongly creationist.

609 Gibson, William S. *The Certainties of Geology.* 1840. London: Smith, Elder. Mentioned in Millhauser as advocate of Gap Theory creationism.

610 Gilbert, Dan. *Evolution, the Root of All Isms.* 1942 [c1935]. San Diego: The Danielle Publishers. "The time has come to test the truth of evolution by a simpler and a surer standard": by comparing it with God's Word, and by judging its fruits. "Who can reasonably defend the evolutionary dogma, even though it be attested by every scientist on earth, if it acts as the tap root from which has sprung the upas tree of atheist-communism? Who can honorably defend as true on 'scientific' grounds a doctrine which proves itself false – on humanitarian and moral grounds – by poisoning human life and civilization with the lethal gases of communism and free-love?" Exposes reliance on Darwinism by socialists, Bolshevists, free-love advocates, psychoanalysis, determinists (anti-free-will), etc. "The 'mechanistic' theory of evolution is the cornerstone of atheism." Cites Leuba's study on loss of faith of students. Gilbert was later shot in a domestic scandal. He also wrote books about Hollywood devil-worship, what really happened at Pearl Harbor, **Hitler**'s disappearance, Stalin, *Crucifying Christ in Our Colleges, The Vanishing Virgin.*

611 Gillespie, William Honyman. *The Theology of Geologists.* 1849. Edinburgh: A.&C. Black. Cited in Millhauser as denouncing geology for contradicting the Bible.

612 Ginsburgh, Irwin. *First, Man, Then, Adam!: A Scientific Interpretation of the Book of Genesis.* 1975. New York: Simon and Schuster. *Ginsburgh:* physics Ph.D.; in research dept. of "major oil company." No real conflict between science and Bible. Includes chart summarizing Ginsburgh's modified Day-Age creationism chronology. Day 1: Universal Black Hole. Day 2: Big Bang. Origin of civilization: 5700 years ago. Adam and Eve crash-landed spaceship, from Pleiades, in Eden. They cross-bred with uncivilized Stone Agers (Genesis "Sons of God" and humans). Enoch repaired spaceship, then took off to summon rescue mission (origin of Messiah concept). Suggests all "miracles" result of spaceship activity. After Flood, no other hybrids around, so second cross-breeding with earthlings occurred; lifespan was then shortened. Psychic powers, UFOs are vestiges of this history. Argues his theory is superior because it has single assumption (Eden = spaceship), and solves all major puzzles. Declares some belief in God. Tree of Knowledge was the spacepeople's computer. Circumcision marked descendants of extraterrestrials. Relies on L. Ginzberg's 1972 *Legends of the Bible.*

613 Gisborne, Thomas. *Considerations on the Modern Theory of Geology, and Their Consistency or Inconsistency with the Scriptures* (booklet). 1827 [1836]. London: T. Cadell. Millhauser says Gisborne expresses resentment against modern theories of geology for being anti-biblical, and relies instead on old-fashioned diluvial catastrophism.

614 Gish, Duane T. *Evidence Against Evolution* (booklet). 1972. Wheaton, Ill.: Tyndale House. *Gish:* chem. B.A. – UCLA; biochem. Ph.D. – U.C. Berkeley; formerly with Upjohn Co.; now biochem. prof. and VP of ICR; acknowledged as most active and effective creationist debater. "Handy pocket sized compilation of concise arguments against evolution written for nonspecialists" (**J.N. Moore** 1983). Bibliog.

615 _______. *Evolution? The Fossils Say NO!* 1979 [1972]. San Diego: Creation-Life. 3rd ed. Orig. 1972; 2nd ed. 1973 (San Diego: ICR Publishing Co.). Also in "public school" ed. – identical except biblical references are omitted. (**Gish** 1985 is a rewritten version.) Preface by **H. Morris.** Widely distributed; one of the best-known and most-cited creation-science works. Presents standard creation-science paleontology arguments. Fossil record refutes evolution, which is a philosophical faith. Many quotes by scientists critical of evolutionary theories, or interpreted as critical of evolution (Leakey, Zuckerman, **Kerkut,** E.C. Olson, G.G. Simpson, **Grassé,** etc.). Also cites and quotes creationist scientists. Many scientific references; includes biblical arguments also – refers to Adam and Eve, the Flood, Babel. Compares evolution with Ptolemaic theory. Scientists accept evolution because it is a materialistic, naturalistic explanation, and they are unbelievers. "While the *theories* and *opinions* of some scientists may contradict the Bible, there is no contradiction between the *facts* of science and the Bible." "Evolution is indeed no less religious nor more scientific than creation." Creation itself can't be studied scientifically, "for God used processes which are not now operating anywhere in the natural universe." Strong emphasis on fossil gaps; lack of transitional fossils; lack of "indisputable" Precambrian fossils. Such evidence is predicted by creation "model." Quotes various scientists on inadequacy of fossil record for evolutionary origins of each major group. Focuses on gaps between fish, amphibians, reptiles and mammals, *Archaeopteryx,* the horse series, and fossil hominids. Relies on **O'Connell** for *Homo erectus* criticism. Refers to Tasaday in arguing "cave-men" are really degenerates from original civilized condition. Drawings, photos; index.

616 _______. *Speculations and Experiments Related to Theories on the Origin of Life: A Critique* (booklet). c1972. San Diego: Institute for Creation Research. ICR Technical Monograph No. 1. Foreword by **H. Morris.** Criticizes origin-of-life theories and experiments of Oparin, Miller, Urey, Fox, Abelson, Wald, Sagan and others. Fairly technical bio- and geo-chemistry, with math calculations, chem. equations, many scientific citations. Compares evolutionists' desire to create life with pagan idolatry (Foreword). Emphasizes evidence against assumption of reducing atmosphere in early earth; difficulty of polymerization (precursor substances tend to break down, not aggregate), improbability of attaining resultant biological structure by chance, inability of proposed precursors to replicate.

617 _______. *Have You Been Brainwashed?* (pamphlet). 1974. Seattle, Wash.: Life Messengers. Comic-book format pamphlet of Univ. Calif. Davis lecture by Gish. Over two million copies distributed. Basic creation-science; standard arguments. No (indisputable) Precambrian fossils. "All paleontologists acknowledge that Archaeopteryx was a true bird." No short-necked giraffes found—and how could they have survived anyway? Lack of transitional fossils; hominid hoaxes. Notes that **J.N. Moore** stated chromosome numbers contradict evolution at AAAS meeting. Earth just right for life; coastal cities would be flooded if climate were warmer. (ICR has on occasion denied that Gish wrote this pamphlet, but usually credits him as author.)

618 _______. *A Decade of Creationist Research* (reprint). 1975. Ann Arbor, Mich.: Creation Research Society. Popular reprint from *CRSQ* (vol. 12, no. 1), distributed at creation conferences, etc. Cities research done on geological overthrusts, fossil anomalies (Paluxy prints, Meister's trilobite-in-shoeprint, Moab skeleton, **Burdick**'s Precambrian pollen, Flood origin of Ararat, **Coffin**'s study of petrified trees, **Northrup**'s study of Lompoc diatoms, fossil reefs, **Barnes**'s magnetic decay, genetic studies by **Lammerts** and **Tinkle,** Galapagos Island research by **Klotz,** seed survival in floods by **Howe, Frair**'s molecular taxonomy, **E. Williams**'s thermodynamics analyses, etc. Suggests areas for further research.

619 _______. *Gish Answers Faculty* (pamphlet). 1975. Seattle, Wash.: Life Messengers. Comic-book format pamphlet of Gish debating college professors.

620 _______. *Dinosaurs: Those Terrible Lizards.* 1977. San Diego: Master Books. Marvin Ross: illustrator. Children's book. Descriptions of dinosaurs with heavy creationist emphasis. Paintings on every page. Paluxy prints prove dinosaurs lived same time as humans. Mentions pre-Flood Water Canopy; demise of dinosaurs result of climate change after Canopy collapse. Special section on bombardier beetle: its chemical defense system suggests that dinosaurs actually breathed fire—dragon legends are true. Index.

621 _______. *The Challenge of the Fossil Record.* 1985. El Cajon, Calif.: Creation-Life. Revised, expanded and rewritten version of *Evolution? The Fossils Say No!* Acknowledges assistance of John Woodmorappe (pseudonym of Jan Peczkis; frequent contributor to *CRSQ*) in section on mammal-like reptiles. Section on fossil hominids in particular much expanded—discusses new disputes arising from rival evolutionist interpretations and proposed phylogenies, further misidentifications of bones, presents more quotes. Photos, drawings; index.

622 _______. *The Non-Evolution of the Bombardier Beetle* (pamphlet). Undated? Catholic Center for Creation. (Listed in SOR CREVO/IMS.) Creationists claim that the bombardier beetle's chemical defense system is a particularly good refutation of evolution, which requires that all stages of gradual development be adaptive. **A.J.M. White** distributes another tract (undated) reprinting Gish's argument: *Bombardier Beetle Explodes Evolution Myth.*

623 _______. *Scientific Evidence Demands a Verdict: Evolution vs Creation* (booklet). Undated. New Delhi, India: J.K. Gupta, c/o Sabina Press. Pub. and distr. by J.K. Gupta (ed.?). Also distr. at ICR. Standard creation-science arguments and quotes against biological evolution. Inc. biog. note on Gish.

624 _______**, and Arthur C. Cunningham (eds.).** *The Christian World View of Science and Technology* (report). 1986. Mountain View, Calif.: Coalition on Revival. A Coalition on Revival "Sphere Document." COR seeks to present and encourage fulfillment of Christian (biblical inerrantist) worldview in all areas of life; Steering Committee includes over a hundred prominent evangelists, creationists, fundamentalist authors and activists. Jay Grimstead: general editor. *Gish:* Chairman of COR Science and Technology Committee. *Cunningham:* Co-Chair. With contributions from other Committee members. Presented as series of propositions: affirmations and denials. Affirms inerrancy of Bible, strict creation, young earth, Adam and Eve as first humans, worldwide Flood of Noah, Flood Geology, death and disease the result of the Fall. Denies evolution, the Big Bang, that pollution results from Genesis Mandate, that mankind can be saved through science or any human endeavor. Supports military spending. Declares that science must be biblically centered.

625 _______**, and Donald Rohrer (eds).** *Up with Creation.* c1978. San Diego: Creation-Life. Compilation of Institute for Creation Research "Acts & Facts" newsletter and "Impact" articles: 1976–1977. Includes "Impact" series #31–54. "Acts & Facts" items include reports on debates, lectures, seminars, educational programs, special interest stories, and **H. Morris**'s Director's Column.

626 _______**, and Clifford Wilson.** *Manipulating Life: Where Does It Stop?* San Diego: Master Books. A creationist consideration of genetic engineering, cloning, recombinant DNA research, test-tube babies, surrogate mothers, and abortion. Authors cautiously endorse genetic engineering if used to improve health, but condemn any tampering with "divine limits" of species. Artificial insemination approved if it enables births otherwise impossible, but abortion strongly condemned as against God's will, as is any destruction of embryos with potential to achieve live births. Presentation of different techniques and topics largely straightforward reporting. Authors say Rorvik's cloning account is probably fictional, but agree with its presentation of the dangers of cloning. Includes a biblical warning that humans are trying to mimic God's creative powers in these End Times. But also hope: in the coming Kingdom there will be no disease, and we will be immortal. Diagrams.

627 Gitt, Werner. *The Flight of Migratory Birds* (pamphlet). 1986. El Cajon, Calif.: Institute for Creation Research. ICR "Impact" series #159. *Gitt:* director and prof. at Physikalisch-Technische Bundesanstalt (Braunschweig, Germany). Migration of golden plover from Alaska to Hawaii proves creationism.

628 Gladstone, William E. *The Impregnable Rock of Holy Scripture.* 1896 [1890]. Philadelphia: Henry Altemus. *Gladstone:* British Prime Minster (1892; d. 1898). A reconciliation of Genesis with science. Dismisses literal six-day creation but also notion that creation days refer to vast geological ages. Favors instead a 'literary' or 'framework' hypothesis: Bible uses phenomenological language; describes events in terms comprehensible to "relators" and audience. Creation Days are like chapters about important events and topics—may overlap chronologically. Discusses reconciliation of various events in biblical account with scientific evidence and theories. Accepts Nebular Hypothesis of solar system formation. Cites many scientific works. Also cites a few creationists—relies heavily on **J. Dawson.** Refutes at length Huxley's attacks on scientific truth of Bible. Bible account does prove to be inspired by Author of Creation: must be Divine Revelation. Also discusses recent discoveries of Mesopotamian Creation and Flood stories; sees these as flawed derivatives of original Bible versions. (Gladstone also wrote book on Homeric Greece, suggesting ancients had different color perception based on analysis of color terms in the *Iliad.*)

629 Glashouwer, Willem J.J., and Willem J. Ouweneel. *Het Onstaan van der Wereld.* 1980. The Netherlands: Stichting "De Evangelische Omroep." Dutch. *Glashouwer* and **Ouweneel:** scriptwriters for well-known Films for Christ *Origins* film series (co-produced with Dutch group); Glashouwer was also researcher for film series; Ouweneel was featured scientist in series. **McDowell** and Stewart's 1984 book (which contains much material from the film series) is "adapted from" this book.

630 Gloag, Paton J. *The Primeval World: A Treatise on the Relations of Geology to Theology.* 1859. Edinburgh: T.&T. Clark. *Gloag:* minister; also wrote religious books. Concordist approach. Earth is ancient, but evolution is impossible. Flood wasn't worldwide—but it did cover entire region inhabited by man. Denies that presence of death before sin (the Fall) is anti-scriptural. Genesis creation was "not the original creation ... out of nothing, but a new arrangement or remodelling of previously existing materials"—i.e. a Gap Theory interpretation. Quotes and discusses **Buckland**'s Bridgewater presentation of Gap Theory at length; also **Pye Smith**'s modified Gap Theory requiring only regional re-creation. Gloag proposes instead that the pre-Adamic destruction was worldwide but only partial: not all life became extinct.

631 Good, Mrs. Marvin. *How God Made the World.* 1978. Rod and Staff.

632 Good, Ronald D'Oyley. *Features of Evolution in the Flowering Plants.* 1956. London/New York: Longmans, Green. **J.N. Moore** (in 1964 *CRSQ*) describes this as an examination of problems of evolution, particularly as demonstrated by plants. Good demonstrates that some of the "best-known speculations about organic evolution are seen to have a less general applicability than is usually claimed." In another work, Good further criticized principle of natural selection; said it was not appropriate to the present moral, social and educational climate. Illustrations; bibliog.

633 Goodman, Jeffrey. *American Genesis.* 1982. New York: Berkley. "Startling New Evidence that the First Americans were also the First Modern Humans!" *Goodman:* anthropology Ph.D.—Calif. Western Univ.; director of Archaeological Research Associates (Tucson, Ariz.). Goodman got anthropology M.A. at Univ. Ariz., and claimed breakthrough in "psychic archeology" using ESP at his Flagstaff dig (wrote 1977 book *Psychic Archaeology*). He claims here that humans inhabited New World 70,000, maybe as early as 500,000 years ago. These were fully modern in form (earliest modern Old World humans: 35,000 B.P.). They spread from Southern California to the rest of the world. Suggests that this American Indian evidence is not accepted for racist reasons. In 1977 book, Goodman explicitly argued that these first humans came from the lost continents of Lemuria (Pacific) and Atlantis. Goodman also wrote *We Are the Earthquake Generation* (c1978; New York: Seaview Books), a "Psychic-Scientific Prediction" about imminent catastrophes which relies on the *I Ching,* **Cayce, Velikovsky,** astrology, and Bible prophecy (esp. Hal Lindsey's End Times books).

634 ________. *The Genesis Mystery: A Startling New Theory of Outside Intervention in the Development of Modern Man.* c1983. New York: Times Books. Pre-human hominid ancestors suddenly got full human mental ability when zapped by a "superior intelligence" ("God? Spaceman? Hitchhiking spirits?"). They then had fantastic psychic and shamanic powers, altered states of consciousness, and performed incredible paranormal feats—controlling wildlife, healing magically, and seeing into the past and future. Rejects Darwinian gradualism; asserts that neo-Darwinian theories of human evolution and standard interpretations of fossil hominids are completely wrong. Darwin was a fraud who stole **Wallace**'s idea of evolution (relies on **Brackman** here). Like Wallace, Goodman argues that man's development must have occurred outside of the evolutionary process—which Darwin was completely wrong about anyway. Goodman says the sudden zapping by outside Intelligence occurred at the Neanderthal/Cro-Magnon transition. Suggests that "man was set apart from the evolved fang-and-claw instincts of the animal world with much nobler origins..." Discusses various interventionist scenarios: a Creator God, extraterrestrial seeding or visits by spacemen, and "spirits from other realities" who took joy-rides in hominid bodies but eventually became trapped in these physical vehicles. Much of book is reasonably accurate critical review of history of paleoanthropology, and description of competing theories of man's evolution. Many scientific references and quotes. Cites evidence of New World humans much earlier than traditionally assumed. Presents his "interventionary" theory as alternative to materialist evolution and biblical creationism. Stresses body/mind duality (appealing to Wallace, Eccles, Wigner, etc.). Very readable style. Drawings, maps, charts; bibliog., index.

635 Gordon, W.R. *The Science of Revealed Truth Impregnable as Shown by the Argumentative Failures of Infidelity and Theoretical Geology.* 1878. New York: Board of Publishers—Reformed Churches of America. Based on 1877 Vedden Lectures at Theol. Sem. of Rutgers College (New Brunswick, N.J.). Defends literal Mosaic Cosmogony. Rails against atheistic and liberal theologians and geologists who deny Bible or try to bend Scripture to fit geological theories. Denounces "theoretical" geology as opposed to that based on solid fact. Attacks Darwinists; also old-earth geologists such as **Hitchcock, Pye Smith,** etc.

636 Gosse, Philip Henry. *Omphalos: An Attempt to Untie the Geological Knot.* 1857. London: Van Voorst. *Gosse:* member of Plymouth Brethren; respected naturalist, marine biologist (popularized the aquarium). Gosse tried to accommodate his strict belief in literal creation with mounting evidence for old earth and succession of fossil organisms. He expected this book to be a triumphant breakthrough, a reconciliation embraced by atheist geologist and fundamentalist alike. Gosse argued for creation with appearance of age: just as Adam was created with a belly-button ('omphalos'), so too all creatures and the earth itself were created with apparent age—illusory evidence of previous existence. God created all things in cycles of existence, so His creations showed signs of previous stages of the cycle from the moment of creation. He created trees with tree-rings (from non-existent previous growth), and animals with signs of earlier growth and wear—even with excrement in their intestines. The living world had to be created as an ongoing process in order to function. When God created organisms, he necessarily created them at some point in the cycle of existence. He gave each an arbitrary beginning, "but one which involved all previous rotations of the circle, though only as an ideal, or, in another phrase, prochronic." (Gosse contrasts such "prochronic," created events with ordinary "diachronic" events which occur "during" time.) Geological strata were created with fossils already in them. "The past conditions or stages of existence in question, can indeed be as triumphantly inferred by legitimate deduction from the present...; they rest on the very same evidences; they are identically the same in every respect, except this one, they were *unreal.* They exist only in their results; they are effects which never had causes." Gosse's theory was rejected by all sides. Most creationists today are embarrassed by the bold totality of Gosse's concept, but continue to rely on creation with appearance of age for specific cases of refractory evidence. (Millhauser says Gosse's theory was anticipated by **Mellor Brown** and **G. Penn.)**

637 Gower, D.B. *Radiometric Dating Methods* (pamphlet). Undated. England: Evolution Protest Movement. EPM pamphlet #207. *Gower:* biochem. reader at

Guy's Hospital (London). Also wrote EPM pamphlet #220: *Chemical Evolution—Theories of the Origin of Life.*

638 Grace Theological Seminary. *Biblical Creationism* (tract). Undated [1979]. Winona Lake, Ind.: Grace Theological Seminary. Tract affirming strict creationism as doctrinal position of Grace Theol. Sem. faculty. Adopted by faculty in 1979. Begins by declaring that what one believes about creation affects what one believes about Bible and God. "And since no human being observed God's work of original creation, we are totally dependent upon God's revelation" in understanding origins. Insists on literal six-day creation and Flood Geology; explicitly rejects evolution. Argues against Day-Age and Gap Theory creationism (Fall of Satan was after creation week). "Heavy demand" claimed for this tract.

639 Graebner, Theodore C. *Evolution: An Investigation and a Criticism.* 1929 [1921]. Milwaukee, Wisc.: Northwestern Publishing House. 4th ed. Orig. 1921. Also pub. by Concordia (St. Louis). *Graebner:* Missouri Synod Lutheran; philosophy and New Testament prof. at Concordia Theol. Sem. (St. Louis, Mo.). Graebner presents "the argument against evolution . . . derived from the study of religion." Vigorous denunciation of evolution as anti-Christian and unscientific. Mentions that reading **Mivart** confirmed his reservations about evolution; also **Dana,** LeConte, **Wallace, H. Miller, Fairhurst.** Many scientific quotes. "It is evident that the evolutionary theory not only contradicts the Bible story of creation but, if true, deprives Christianity of every claim of being the true religion." Evolution excludes divine fiat, revelation, supernaturalism, immortality of soul, and any absolute standard of morals. The Resurrection—the best-confirmed and most tremendous fact in history—stands four-square against evolution. "The evolutionary hypothesis is contradicted by the facts of religion, of history, and of natural science." It does not account for origin of the universe, of life, of species, or of man. Evolutionist paleontology is based on "reasoning in a circle": the geologist dates rocks by their fossils, assuming they have evolved from simple to complex; then the evolutionist says this confirms evolution (says Tingelstad pointed this out in 1908). Declares that "species are *fixed.*" Discusses vestigial organs, fossil evidence, instinct, evidences of design, alleged evolution of religion.

640 ________. *Essays on Evolution.* 1925. St. Louis, Mo.: Concordia. Essays (rev.) orig. pub. in *Theol. Quart., Luth. Witness,* and *Walther League Messenger.* 1925 Scopes Trial allowed evolutionists to ridicule creationist fundamentalists. Graebner refutes the arguments of the various scientific witnesses at the trial: "Never has the hollowness of evolutionistic claims become so apparent" as in the statements of these scientists. Predicts that Scopes appeal will be decided on technical issues (rights of parents vs. teachers) rather than deciding merits of evolution. Quotes Nobelist Millikan as saying that no scientist can prove evolution. Describes various evolutionist textbooks. Denounces H.G. Wells' 1921 *Outline of History* as evolutionist propaganda; also T. Roosevelt's 1916 *Nat'l Geog.* article on the antiquity of man (itself based on book by H.F. Osborn); inc. letter from Roosevelt to Graebner. Asserts that evolutionists will want to destroy epileptics and other handicapped persons. Uses example of ichneumon wasp paralyzing caterpillar for its offspring's food as evidence of instinct which refutes evolution. Explains that disease germs invaded man after the Fall; previously, they were harmless. Discusses Einstein's relativity as demonstrating the ephemeral nature of scientific theories; later notes that it renders geocentrism as plausible as heliocentrism. Shows that savages are merely degenerate humans, and that advanced humans are evident in the earliest reaches of history. Presents drawing of "dinosaur" by Ariz. Indians as conclusive disproof of man's descent from apes. Quotes many anti-Darwinians and anti-evolutionists; includes several standard creation-science arguments, though stressing anti-biblical nature of evolution. "The severest indictment that must be brought against the God-dishonoring theory of evolution is that it denies that there was a fall; therefore there is no need of the plan of redemption or of the Savior . . ." Photos, drawings.

641 ______. *God and the Cosmos: A Critical Analysis of Atheism, Materialism and Evolution.* 1943 [1932]. Grand Rapids, Mich.: William B. Eerdmans. 2nd ed.; rev. and enlarged. Orig. 1932. 439 pages; largely an expansion of arguments in previous books. Refutes atheism, materialism, and evolution in separate sections, but includes many anti-evolution arguments in all three. Concedes book is work of a theologian, but denies it is theology. Declares that science proves theism true and evolution false. Describes militant atheist efforts and communist affiliations. Many examples of design in nature. Full of quotes by anti-Darwinian, anti-evolutionist, and Bible-believing scientists and writers (**Jeans, Driesch,** Millikan, **Smuts, Wallace, Boodin,** Sir Oliver Lodge, Austin Clark, many others). Includes references to many works not cited elsewhere. Denounces H.G. Wells' 1929 *Science of Life* (written with J. Huxley); also **G.B. Shaw,** Haeckel, **Bergson,** many others. Says that Heisenberg's indeterminacy principle rules out materialism (mechanistic interpretation). Points out there are dozens of competing theories of evolution. Argues that evolution is logically atheistic. Asserts fixity of species, but says many "species" are actually mere varieties. Discusses the "barrier of heredity" (Mendel's laws, failure of Lamarckian theories), fossil evidence (gaps, persistence of forms, and evidence only of degeneration), discredited recapitulation theory, alleged fossil ape-men, difference between humans and apes. Argues that evolving organs would be selected against in their rudimentary stages; also other standard creation-science arguments proving the collapse of Darwinism. Christianity elevates humanity, but evolution "unquestionably has degrading, demoralizing, brutalizing influences." "Their rivalry is at the bottom of all human affairs." Bibliog., index.

642 Grant, Frederick William. *Creation in Genesis and Geology* (booklet). Undated. New York: Loizeaux Bros. Treasury of Truth series: no. 10. *Grant:* d. 1902.

643 Grassé, Pierre-Paul. *The Evolution of Living Organisms.* 1977 [1973]. New York: Academic Press. Orig. French, 1973 (*L'Evolution du Vivant;* Paris: Editions Albin Michel). *Grassé:* prof. of evolutionary studies at the Sorbonne; editor of 28-volume *Traite de Zoologie;* ex-president of French Acad. Science. Highly critical of neo-Darwinian assumptions and complacency. Facts first, then theories! "Our duty is to destroy the myth of evolution, considered as a simple, understood and explained phenomenon which keeps rapidly unfolding before us." "Ultra-Darwinism" is wrongly assumed same as "evolution." But Darwinism "inspires fallacious interpretations." Darwinists assume that fossil data accords with evolutionary theories—but it doesn't. Severely criticizes neo-Darwinian notion that pure chance can account for evolution; especially critical of J. Monod. Different sense organs of mammals all evolved at same time: ridiculous to suppose that so many simultaneous, harmonious and fortunate mutations could have occurred by chance. Says the dogma of the one-way flow of information from DNA was decisively refuted with discovery of ability of some viruses to transfer genetic information into other organisms. Dismisses story of gradual fish-amphibian transition—says there are mud-skippers today which haven't changed in millions of years. Questions evolutionary role of mitochondrial DNA; suggests that cytoplasm influences DNA. Argues that complex instincts could not have been programmed by chance processes. "Absurd" to believe that evolution of human brain could have occurred so quickly as a result of mutations alone. Much cited by creationists.

644 ______. *L'Homme en Accusation.* 1980. Paris: Albin Michel. "Man Stands Accused." Quoted and discussed in **Bucaille** 1984, who says it is "extremely critical of today's neo-Darwinism." Grassé does not question the fact of evolution, but argues that we have no valid explanation for how it works. Argues that random mutations are insufficient to play a determinative role in evolution. Lamarck was the true father of evolution. Bucaille says Grassé is "extremely critical of Darwin for having drawn his inspiration from Malthus and for the unfortunate influence he created." *Grassé:* "Due to its basic precepts and final conclusions, Darwinism is the most anti-religious and most materialistic doctrine in existence." Expresses amazement that Christian scientists do not seem aware of this; points out that Marx was. Notes that often the "fitter" animals may be killed; considers this an argument against effectiveness of natural selection. Many organisms are constantly adapting for millions of years, but never evolve into different forms.

645 Gray, Bennison, and [Mrs.] Gray. *Evolution and the Revolution that Failed: The Semiotics of Taxonomy.* 1983(?). *B. Gray:* Ph.D.—USC. Cited in 1982 *Kronos* article by Gray, in which he champions **H. Nilsson**'s "secular creationism."

646 Gray, James. *The Earth's Antiquity in Harmony with the Mosaic Record of Creation.* 1849. London: Parker. 2nd ed. (1851) titled *Harmony of Scripture and Geology; or, The Earth's Antiquity [etc.].* Gap Theory creationism. (Cited in **Custance** 1970, via Hoare 1860 reference.)

647 Gray, James M. *Modernism: A Revolt Against Christianity; A Revolt Against Good Government.* 1924. Chicago: Moody Bib!e Institute Colportage Assoc. *Gray:* president of Moody Bible Inst. (Chicago). Evolution is a threat to American democracy. Modernism in general and evolution in particular result in spread of communism and overthrow of our government. (Quoted in Gatewood.)

648 ________. *Why a Christian Cannot Be an Evolutionst* (pamphlet). 1925. Chicago: Moody Press.

649 Green, Melody, and Sharon Bennett. *The Crime of Being Alive* (tract). 1984. Lindale, Texas: Last Days Ministries. *Green:* very well-known anti-abortion speaker and rally-leader; often seen with Reagan; widowed wife of Christian rock musician Keith Green. One of many anti-abortion tracts from Last Days Ministries by Green; reprinted from *Last Days Newsletter.* **Hitler** built on foundations already accepted by society. German schools taught survival of the fittest —that helping the handicapped, therefore, went against nature. The "grandfather" of their philosophy of natural selection is the same Darwin honored in our own educational system. "If man evolved, then he is a mere animal. His value is determined strictly by what he can offer society. If man is created in the image and likeness of God, his value is determined by his Creator."

650 Greenberger, Josh. *Theories and Fantasies.* 1986 [1983]. Priv. pub. (Brooklyn, N.Y.: Systematic Science). Updated version of *A Thin Line Between Theory and Fantasy* (1983) written under pseudonym **David Balsam.** *Greenberger:* computer consultant—software for NASA Goddard, Bell Labs, Western Electric; also newspaper columnist (gives New Canaan, Conn. address). Sarcastic book; attempts to be humorous. No references cited. Belief in intelligent Creator is scientific, not religious. Naively refutes evolution by appealing to genetic information, harmfulness of mutations, lack of life on other planets, falsity of dating methods, success of venereal diseases. Scientists can't yet even explain our own solar system.

651 Greenman, W.W. *Evidences of the Flood* (booklet). 1974. Gospel Standard Baptist. (Listed in SOR CREVO/IMS.)

652 Greenough, George Bellas. *A Critical Examination of the First Principles of Geology.* 1819. London: Longman, Hurst, Rees, Orme and Brown. Suggests that a comet caused the universal Flood, and that a single Deluge was preferable for explanatory purposes to several by the principle of parsimony.

653 Greenwood, George. *The Tree-Lifter.* 1876 [1844]. London: Longmans, Green. Concerns a new method for transplanting trees. Cited by Millhauser as claiming man is coeval with oldest fossils: Scriptural Geology.

654 ________. *Rain and Rivers, or Hutton and Playfair Against Lyell and All Comers.* 1876 [1857]. London: Longman, Brown, Green, Longmans, & Roberts. 3rd ed.

655 Greenwood, William Osborne. *Biology and Christian Belief.* 1939. New York: Macmillan. The reappearance of indeterminism in modern science. Mechanism is now just an "as if" assumption. Biology need not conflict with Christian belief. Purpose in evolution (cites **D'Arcy Thompson**); design in universe (cites **J. Jeans**); spirit exists apart from matter (cites **Smuts**). We see pattern everywhere. Mechanism can't explain Mind. Immortality of the soul; also of the body of unicellular organisms. The psychical is real, and demonstrates separateness of spirit and mind from physical body.

656 Gridley, Albert L. *The First Chapter of Genesis as the Rock Foundation for Science and Religion.* c1913. Boston, Mass.: The Goreham Press. Also pub. by R.G. Badger (Boston). Apparently progressive creationist harmonization of science and Bible. Accepts theory that earth was initially molten, and sequential appearance of life-forms, but says the beginning was recent. Also wrote *Jesus Only, Suborganic Evolution,* and *Organic Evolution.* (In Schadewald files.)

657 Griggs, Jolly F. *Science Says No!* (booklet). Priv. pub. (Ventura, Calif.). *Griggs:* teaches at Ventura College; heads Creation Science Assoc. of California. By promoting evolution, public schools unjustly attack religious beliefs of creationists, thereby violating students' constitutional rights. Schools also suppress evidence unfavorable to evolution and indoctrinate students in anti-Christian dogma. Griggs presents the standard creation-science arguments and examples, and includes many anti-evolution quotes. "Evolutionists have closed their minds to all evidences that negate the sacred theory and this is not the correct approach to science." Asserts that evolution violates many scientific laws. Argues for young earth; rejects Big Bang scenario. Denies that Christians are seeking to teach the Bible in public schools, but argues that since schools censor scientific criticism of evolution, all churches ought to establish private schools to avoid evolutionist indoctrination. The public school "systematically steals traditional family values by indoctrinating our children with the one-sided, unfounded, false theory of evolution." Drawings.

658 Grumley, Michael. *There Are Giants in the Earth: Survivors Since Genesis.* 1974. Garden City, N.Y.: Doubleday. *Grumley:* described as "graphic artist as well as naturalist"; lives in New York City; also co-authored book on Atlantis. This book about Bigfoot (Sasquatch) of the Northwest, Yeti of the Himalayas, and Mono Grande of the Andes. Suggests all these anthropoid giants are living descendants of *Gigantopithecus* (a fossil primate which survived as recently as a million years ago in China). Lively account of various reported sightings. In this "evolutionary and legendary history," Grumley further suggests that a "chakra"—a center of physical and psychic energy—was set up by extraterrestrial beings in the Andes in order to spread religion and culture to the pre-human primates then living. Later, the chakra site was relocated to the Himalayas, due to a shift of the earth's energy field. The alien "gods" or spacemen vastly increased the consciousness and abilities of the pre-humans, setting up repositories of arcane priestly knowledge. Other chakras also existed at various times in Atlantis and Lemuria. The reference to "Nephilim" (giants) in Genesis reflects an age when gigantopithecines mated with humans, a practice which came to be abhorred by humans. Photos, drawings by the author; bibliog., index.

659 Grunlan, Stephen A., and Marvin K. Mayers. *Cultural Anthropology: A Christian Perspective.* 1979. Grand Rapids, Mich.: Academie Books (Zondervan). *Grunlan:* student of Mayers; on St. Paul Bible College faculty; formerly at Moody Bible Inst. *Mayers:* Ph.D.—Univ. Chicago; Dean of Intercultural Studies and World Missions at Biola Univ.; director of Summer Inst. of Linguistics (Dallas); linguistics prof. at Univ. Texas Arlington; fieldwork among Pocomchi (Guatemala) with Wycliffe Bible Translators. Authors seek to combine "biblical absolutism" with "cultural (not ethical) relativism." Includes many Bible references; chapters on "Missions," "Anthropology and the Bible." Discusses creationism favorably as alternative to evolution for origin of man. Reviews development of anthropological theory; notes it passed through evolutionist stage. Refers to Chomsky's linguistic theory as supporting creationism. Includes many standard anthropology topics (kinship, etc.). Diagrams; bibliog., index.

660 Gunkel, Herman. *Schöpfung und Chaos in Urzeit und Endzeit.* 1921 [1895]. Gottingen, Germany: Vandenhoeck & Ruprecht. German. *Gunkel:* Berlin theology prof. Cosmogonic warfare between God and evil Chaos at Creation. Cited in

Bixler 1986, who calls it a "landmark" work.

661 Gutzke, Manford George. *Plain Talk on Genesis.* 1975. Grand Rapids, Mich.: Zondervan. Vol. in Gutzke's "Plain Talk" Bible commentary series. *Gutzke:* grad. of Bible Inst. Los Angeles; psych. M.S.—S.M.U.; D.D.—Austin Coll.; Ph.D.—Teachers Coll. (Columbia Univ.); Presbyterian minister. Theological rather than scientific arguments, but insists on historical and scientific reliability of the Bible. "In the whole account of creation the amazing thing is that the order in which God created the universe, as it is written, exactly fits the best thinking of scientists today. . ." Compares creationism with the "harmful theory of evolution." Asserts that the Genesis account is "plain as day"; it proclaims strict fiat creationism, and flatly rejects evolution and development by natural processes. Continuity of biblical 'kinds' of organisms is also "basic to all morality." Doctrine of "after its kind" insures that good comes from good, and evil from evil. "For a man to think of himself as an animal is debasing." It leads to "unchecked immorality." Insists there is not "a single bit of proof or shred of evidence" to support evolution. Literal interpretation of Adam and Eve, Satan, Eden, the Flood. Implies young-earth creationism.

662 Guyénot, Emile. *The Origin of Species.* 1964 [1944]. New York: Walker (Sun Book). Orig. 1944; French (Presses Universitaires de France). *Guyénot:* prof. at Univ. Geneva. Guyénot laments that evolution has descended from scientific study to political and religious argument, and regrets that evolution is now linked with monism and materialism. Evolutionists have unfairly extrapolated from their theory the assertion that the only reality is matter "governed finally by the laws of chance." Urges a return to truly scientific study of evolution. Suspicious of notion life arose spontaneously; if it did, there were probably many separate origins of life. Agrees with **Caullery** that different groups of organisms evolve at different rates, and that this indicates "intrinsic properties of the organisms" are more significant for evolution than external environment. Evidence from anatomy, embryology and paleontology does not prove the evolutionary hypothesis but makes it fairly certain. Contrary to Darwinism, paleontology shows that evolution is discontinuous, without transitional forms. Species evolve suddenly, then remain stable. Guyénot suggests that large mutational jumps have played decisive roles in producing new genera, orders, and even classes. Mutation theory, however, is "powerless" to explain the general adaptation of organisms. Guyénot stresses that chance mutations cannot account for complex structures such as the eye: "Here the mutationist explanation comes up against a genuine impossibility." Some other factor, "the nature of which is unknown," must come into play. Sympathetic to idea of orthogenesis; notes it has resulted in unfavorable exaggeration of some traits (though he also suggests these are related to general size increase). Drawings; index. Guyénot also wrote *La Variation et L'Evolution* (1930; Paris). The 2nd ed. of *La Variation* (same work?—1950; Paris: G. Doin) was pub. in the series "Encycl. Scientifique: Bibliotheque de Biologie Generale" (Caullery: Directeur).

663 Guyot, Arnold H. *Creation; or The Biblical Cosmology in the Light of Modern Science.* 1884. New York: Charles Scribner's Sons. *Guyot:* Ph.D.; geology and physical geography prof.—College of New Jersey (Princeton); Swiss-born. Rejects higher criticism of Bible. The Bible teaches "that this universe has a beginning; that it was created—and that God was its Creator. The central idea is creation." Appeals to modern physics in describing the gaseous state of the original chaos, and the subsequent creative days. True "creation" *(bara)* occurs three times in Genesis: creation of matter, animal life, and humans. These creative acts demand direct divine intervention and preclude evolution, though microevolution may operate within animal kinds. "The Bible narrative, by its simplicity, its chaste, positive, historical character, is in perfect contrast with the fanciful, allegorical, intricate cosmogonies of all heathen religions. . . . By its sublime grandeur, by its symmetrical plan, by the profoundly philosophical disposition of its parts, and, perhaps, quite so much by its wonderful caution in the statement of facts, which leaves room for all scientific discoveries, it betrays the supreme guidance which directed the pen of the writer, and kept it throughout within the limits of truth." Revelatory theory of creationism: "The same divine hand which lifted for Daniel and Isaiah the veil which covered the tableaux of the time to come, unveiled to the eyes of the author of Genesis by a series of graphic visions and pictures the earliest ages of the creation. Thus Moses was the prophet of the past as Daniel and Isaiah and many others were the prophets of the future." May also endorse Day-Age creationism. Drawings. Guyot also wrote *The Earth and Man,* about geography in relation to the history of mankind, in print from 1849 to 1906. He introduced the study of scientific geography to America. Flat-topped seamounts—"guyots"—are named after him.

664 Hadd, John R. *Evolution: Reconciling the Controversy.* 1979. Glassboro, N.J.: Kronos Press. Pub. in association with the Center for Interdisciplinary Studies at Glassboro College (a Velikovskian operation). Champions **Velikovsky;** cites **Dewar, Wysong,** and other creation-scientists favorably.

665 Hagin, Kenneth E. *The Origin and Operation of Demons* (booklet). c1983. Tulsa, Okla.: RHEMA Bible Church (Hagin Ministries). Volume 1 of the "Satan, Demons, and Demon Possession Series." RHEMA Bible Church: a.k.a. Kenneth Hagin Ministries (Tulsa, Okla.). Gap Theory creationism: Satan had kingdom on earth which God destroyed; God then created Adam and Eve to repopulate the earth. "Scientists tell us there is a vacant spot in the North" where throne of God is located. Towns and cities can have spirits ruling them, whose presence Hagin can feel.

666 Haigh, Paula. *What's Wrong with Evolution?* (booklet). Undated [1975]. Caldwell, Idaho: Bible-Science Association. Also pub. by Catholic Center for Creation Research (Louisville, Ky.). *Haigh:* reference librarian at Spalding College (Louisville, Ky.); heads CCCR. Presents many standard creation-science arguments (fossil record, thermodynamics, transitional forms, geologic column, etc.). Discusses St. Thomas vs. Augustine, but mostly relies on Protestant creationists. Cites **Morris, Whitcomb, Gish, Dewar, Acworth, Macbeth,** other creationists. Offers proof of God and of creation. Says Protestants have been more vigilant in protecting Bible than Catholics. Theology is the best science. Disputes Nogar's "process" argument (Catholic). Were evolution true, species would be indistinguishable—nature would be one big mess. **Teilhard** and others overcompensated for the Church's treatment of Galileo, and now accept all that science claims. Totalitarianism is not the result of certainty of possession of truth, as critics claim; oppression and tyranny can only issue from vice.

667 ________. *Thirty Theses Against Theistic Evolution.* 1976. Louisville, Ky.: Catholic Center for Creation Research. Strict creationism.

668 Hall, Christopher. *The Christian Teacher and the Law: Rights and Opportunities* (booklets). 1975. Oak Park, Ill.: Christian Legal Society. 1975, with 1978 Supplement by **J.W. Whitehead.** *Hall:* B.S.—UCLA; J.D.—Loyola Univ. (Los Angeles); practiced law in Newport Beach; lives in San Bernardino. Booklets deal with prayer, proselytizing, religious freedom, etc.; not specifically with evolution.

669 Hall, George M. *Farewell to Darwin: The Unified Field Theory of Physics, the Genetic Process, and Psychology.* c1977. St. Louis, Mo.: Warren H. Green. *Hall:* attended West Point; hopes for M.D. 500 pp. book; plans 7-vol. expanded version. Eclectic, witty, cleverly-written pseudo-science. Physical/psychic dichotomy: each has separate axioms and laws. Man evolved from animals but transcendental psychic domain took over. Creator God, First Cause, supreme design, absolute time and space from Big Bang. Genetic force, disease entropy, cure for cancer, laws of psychology, sociology, economics, ethics and philosophy. Full of scientific and literary references and quotes. Diagrams, cartoons.

670 Hall, Marshall, and Sandra Hall. *The Truth: God or Evolution?* 1975 [c1974]. Grand Rapids, Mich.: Baker Book House. Orig. pub. by Craig Press (Nutley, N.J.). *M. Hall:* Ph.D. student—

Center for Advanced Int'l Studies, Univ. Miami. *S. Hall:* attended schools in China, Philippines, Mexico; M.A.–Middlebury College (Ver.). The Halls now head the Fair Education Foundation (Murphy, N.C.). "Original publishers of Darwin agree: '. . . the book does undermine the theory of evolution'!" Endorsements include William McCall (emer. prof.–Columbia Teacher's College). Strict creationism; includes most standard creation-science arguments. Emphasizes communist implications of evolution. Crudely argued; quotes often out of context. Many scientific references, but shows little understanding of science. Ridicules evolutionist proposals; frequently attempts humor. Darwin, like Marx and Freud, is accepted because he provides Godless explanations, but all three are unscientific frauds. The Halls believe they have utterly demolished evolution, which is "not even a theory," by rigorous scientific and logical analysis, and they proclaim God's Word as the only alternative. Bibliog.

671 _______, and _______. *The Connection Between Evolution Theory and One's Philosophy of Life* (booklet). c1977. Lakeland, Fla.: P/R Publishers. The Connection Papers series: Monograph #1. "Each of these [Connection Papers] is interconnected with the others by one common theme, the theme that evolution is a lie, that man is not an evolved animal. Given this common theme, our intention has been to demonstrate that exposing the evolution lie to the world is certain to bring about the exposure of every major false teaching and anti-Bible force in the world today."

672 _______, and _______. *The Connection Between Evolution Theory and the Sure Knowledge that Miracles Are Real* (booklet). c1977. Lakeland, Fla.: P/R. The Connection Papers series: Monograph #2.

673 _______, and _______. *The Connection Between Evolution Theory and a Great Turning to the Bible as the Source of All Truth* (booklet). c1977. Lakeland, Fla.: P/R. The Connection Papers series: Monograph #3.

674 _______, and _______. *The Connection Between Evolution Theory and the Separation of Preachers into Supporters or Opposers of the Bible* (booklet). c1977. Lakeland, Fla.: P/R. The Connection Papers series: Monograph #4.

675 _______, and _______. *The Connection Between Evolution Theory and Theistic Evolution–A Destroyer of the Bible & of True Christianity* (booklet). c1977. Lakeland, Fla.: P/R. The Connection Papers series: Monograph #5.

676 _______, and _______. *The Connection Between Evolution Theory and the Collapse of the Twin Doctrines of Salvation by Faith Only & Once Saved, Always Saved* (booklet). c1977. Lakeland, Fla.: P/R. The Connection Papers series: Monograph #6.

677 _______, and _______. *The Connection Between Evolution Theory and the Errors in the Calvinist Doctrine of Predestination* (booklet). c1977. Lakeland, Fla.: P/R. The Connection Papers series: Monograph #7.

678 _______, and _______. *The Connection Between Evolution Theory and Roman Catholicism–Coming: An Open Confrontation with the Bible Utter Destruction of the Roman Catholic Church* (booklet). c1977. Lakeland, Fla.: P/R. The Connection Papers series: Monograph #8.

679 _______, and _______. *The Connection Between Evolution Theory and the Realization that Denominational Protestant Churches Teach Against the Bible* (booklet). c1977. Lakeland, Fla.: P/R. The Connection Papers series: Monograph #9.

680 _______, and _______. *The Connection Between Evolution Theory and the Revelation of the Un-Biblical Origins of the Two-Stage "Secret Rapture" Doctrine and the Subsequent Collapse of Premillenialist Fundamentalism* (booklet). c1977. Lakeland, Fla.: P/R. The Connection Papers series: Monograph #10. Explains and refutes premillennial doctrine (the belief, held by a majority of fundamentalists, in the "rapture" of the faithful to heaven, the coming of Anti-Christ and Armageddon, and the Second Coming of Christ to rule on earth at the Millennium) as unscriptural–a Satanic counterfeit which is soft on evolution.

681 _______, and _______. *The Connection Between Evolution Theory and the Doctrine of a Future Millenium–A*

Demonstration That This "Jewish Fable" Will Not Stand After the Exposure of the Evolution Lie (booklet). c1977. Lakeland, Fla.: P/R. The Connection Papers series: Monograph #11. The biblical "Millennium" is not intended literally. Literal interpretation results in neglect of other, vital Bible doctrines.

682 _______, **and** _______. *The Connection Between Evolution Theory and the Charismatic Movement—The Big Quesion: Back to the Bible or Against the Bible??* (booklet). c1977. Lakeland, Fla.: P/R. The Connection Papers series: Monograph #12.

683 _______, **and** _______. *The Connection Between Evolution Theory and Religious Cults—Discussed Are: Mormonism, Jehovah's Witnesses & Seventh Day Adventism, and Others* (booklet). c1977. Lakeland, Fla.: P/R. The Connection Papers series: Monograph #13. All are Satanic deceptions to lure us away from genuine biblical faith (even though Seventh-day Adventists in particular have argued strenuously against evolution).

684 _______, **and** _______. *The Connection Between Evolution Theory and Exposing the Real Source of Power Behind Astrology & the Occult* (booklet). c1977. Lakeland, Fla.: P/R. The Connection Papers series: Monograph #14. God and Satan both have supernatural powers. The occult, which is truly supernatural, derives from Satan. UFOs are Satanic deceptions.

685 _______, **and** _______. *The Connection Between Evolution Theory and the Discrediting of All the World's Non-Christian Religions (Islam, Judaism, Hinduism, Buddhism, and Others)* (booklet). c1977. Lakeland, Fla.: P/R. The Connection Papers series: Monograph #15.

686 _______, **and** _______. *The Connection Between Evolution Theory and Some Unusual News for Those Who Call Themselves Jews* (booklet). c1977. Lakeland, Fla.: P/R. The Connection Papers series: Monograph #16. Present-day Jews are no longer heirs to God's biblical promises, which He already fulfilled; they are not descendants of Israelites anyway (endorses **Koestler**'s theory of Khazar ancestry of Ashkenazim). Satan is the real god of Talmudic religion.

687 _______, **and** _______. *The Connection Between Evolution Theory and Racism* (booklet). c1977. Lakeland, Fla.: P/R. The Connection Papers series: Monograph #17. The Halls, whose anti-Semitism is fairly blatant, here fervently repeat the charges that evolution is the basis (i.e. the cause) of Nazi and other racist ideologies. The Communist-Talmudic revolution was led by non-Semitic Jews. They plan to take over the world and implement their racist ideology.

688 _______, **and** _______. *The Connection Between Evolution Theory and a World-Wide Awakening to the Fact that There Are Numerous and Very Real Conspiracies that Have as Their Central Goal the Destruction of the Bible and Christianity* (booklet). c1977. Lakeland, Fla.: P/R. The Connection Papers series: Monograph #18. Any anti-Bible organization is carrying out Satan's strategy, and is therefore part of the great Satanic Conspiracy. Includes communism, Catholicism, Masonry, British-Israelism, Zionism, Ecumenicism, Millennialism, the U.N., and the usual assortment of groups (political and economic organizations, secret societies) named by conspiracy theorists.

689 _______, **and** _______. *The Connection Between Evolution Theory and the Exposure of the Anti-Bible Nature of Masonry* (booklet). c1977. Lakeland, Fla.: P/R. The Connection Papers series: Monograph #19.

690 _______, **and** _______. *The Connection Between Evolution Theory and Communism—Destroying Evolution Theory Will Destroy the Basis of Communism, Namely, Its "Scientific Materialism;" Moreover, the Plot of the Talmudic Jews Who Control Communism Will Be Revealed* (booklet). c1977. Lakeland, Fla.: P/R. The Connection Papers series: Monograph #20. The Communist Revolution was entirely carried out by Jews as part of Satan's plan to destroy true Christianity.

691 _______, **and** _______. *The Connection Between Evolution Theory and the Failure of the John Birch Society and Other Anti-Communist Organizations to Stop the Spread of Communist Ideology in the USA and the World* (booklet). c1977. Lakeland, Fla.: P/R. The Connection Papers series: Monograph #21. The

Birch Society has not been effective enough in combating communism because it is soft on evolution. Also discusses Billy James Hargis (Christian Crusade), Carl McIntire, Fred Schwarz (Christian Anti-Communism Crusade), Gerald L.K. Smith (Christian Nationalist Crusade). Praises anti-Zionism.

692 _______, and _______. *The Connection Between Evolution Theory and the Need for a New Definition of Anti-Intellectualism* (booklet). c1977. Lakeland, Fla.: P/R. The Connection Papers series: Monograph #22. Rails against promulgation of cultural relativism, Freud's "Talmudic sex cult worship," and theorizing in general. All built upon evolution theory.

693 _______, and _______. *The Connection Between Evolution Theory and the Dethroning of the Religion Which Deifies Man, viz., Humanism* (booklet). c1977. Lakeland, Fla.: P/R. The Connection Papers series: Monograph #23.

694 _______, and _______. *The Connection Between Evolution Theory and Agnosticism—A Logical Demonstration That—with the Destruction of Evolution Theory —Agnosticism Will Become Just as Untenable as Atheism* (booklet). c1977. Lakeland, Fla.: P/R. The Connection Papers series: Monograph #24.

695 _______, and _______. *The Connection Between Evolution Theory and the Economy—There Are Several Powerful Economic Reasons for Destroying Evolution Theory; Two Specific Areas—the "Military" and "Crime"—Are Examined Here* (booklet). c1977. Lakeland, Fla.: P/R. The Connection Papers series: Monograph #25.

696 _______, and _______. *The Connection Between Evolution Theory and the Space Program—Pure Evolution Propaganda from NASA Scientists!* (booklet). c1977. Lakeland, Fla.: P/R. The Connection Papers series: Monograph #26. "(Contains also a brief consideration of President Carter's message to 'galactic civilizations')." Space program is built on assumption of extraterrestrial life, a result of evolution propaganda. NASA's goal is to find ET life to prove evolution. This is an enormous swindle of tax money, since evolution is proved false.

697 _______, and _______. *The Connection Between Evolution Theory and the Metric Swindle* (booklet). c1977. Lakeland, Fla.: P/R. The Connection Papers series: Monograph #27.

698 _______, and _______. *The Connection Between Evolution Theory and Education —the Rationale and the Formula for Driving Materialist, Evolution-Based Humanism from the Schools.* (booklet). c1977. Lakeland, Fla.: P/R. The Connection Papers series: Monograph #28.

699 _______, and _______. *The Connection Between Evolution Theory and Homosexuality—(And Other Biblically Prohibited Sexual Behavior)* (booklet). c1977. Lakeland, Fla.: P/R. The Connection Papers series: Monograph #29.

700 _______, and _______. *The Connection Between Evolution Theory and Women's "Equal Rights" Movement* (booklet). c1977. Lakeland, Fla.: P/R. The Connection Papers series: Monograph #30.

701 _______, and _______. *The Connection Between Evolution and the Church-State Question—The President and Any Member of Congress Who Opposes a Congressional Investigation into Evolution Theory Will Be Guilty of Treason According to the Constitution* (booklet). c1977. Lakeland, Fla.: P/R. The Connection Papers series: Monograph #31. The Halls means this quite literally. They shrilly demand a full-scale Congressional investigation of evolution. Since evolution is the basis of communism (as well as every other evil doctrine), anyone who supports it is a traitor.

702 _______, and _______. *The Connection Between Evolution Theory and the Coming Together of the True Church* (booklet). c1977. Lakeland, Fla.: P/R. The Connection Papers series: Monograph #32. Includes a call to defend the biblical teaching of geocentrism. Criticizes unbiblical doctrines of various groups, including fundamentalists and "Restoration" churches (Reconstructionists), but seems to suggest the latter are closest to true Christianity.

703 _______, and _______. *The Connection Between Evolution Theory and the End of Time?—A Golden Age for Bible Christianity? Or a Division of the*

World into God's Camp & Satan's Camp? The Consequences of Exposing the Evolution Lie and All the Deceptions Tied to It Point to One or the Other (booklet). c1977. Lakeland, Fla.: P/R. The Connection Papers series: Monograph #33.

704 _______, and _______. *The Great Evolution Deception: The Error, the Fruit, & the Solution* (booklet). c1980. Murphy, N.C.: Citizens Against Federal Establishment of Evolutionary Dogma (CAFEED). CAFEED is the Hall's own organization. Booklet contains one-page summaries of many topics in "Connection Papers" series. Also includes section on proposed national bill to prevent federal censorship of creationism: all federal funds for scientific research to be split equally between evolutionists and creationists. Bill requires Congressional hearings to determine scientific evidence against evolution; an equal number of creation-scientists and evolutionists would testify. "Any member of Congress who would seek to prevent the exposure and destruction of evolution, the central and necessary tenet of this country's officially recognized enemy, would obviously be giving 'aid and comfort' to the enemy and would therefore be a traitor as defined by the Constitution. This is no hollow, academic threat. It is exactly as strong as the Constitution." Drawings.

705 [Hallonquist, Earl]. *The Bankruptcy of Evolution* (pamphlet). 1985 [1973]. Vancouver: Creation Science Association of Canada. Pamphlet does not name author (Hallonquist). Orig. 1973. Rev. and enlarged eds. 1974, 1983, 1985. Brief statement of standard creation-science arguments. Many quotes of scientists. "All statements in this Booklet Fully Documented and Referenced."

706 _______. *Continental Split* (tract). Undated. Syracuse, N.Y.: Book Fellowship. Acceptance of continental drift was a complete scientific turnaround. Similarly, evolution may be overthrown soon. Proposes catastrophic drift to solve problems of Noah's Ark theory.

707 Ham, Ken. *The Relevance of Creation* (booklet). 1983. Sunnybank, Australia: Creation Science Foundation. Casebook II (magazine format special supplemental issue) from *Ex Nihilo,* the CSF magazine. Widely distributed. *Ham:* director of ministry for Creation Science Foundation (Australia); now works full-time in U.S. at ICR to promote "creation evangelism." Emphasizes that evolution is a belief system, "and you don't need to be a scientist to combat it." Evolution says that everything, including truth, is relative and evolving. Creation is the foundation of Christian belief: explains origin of life, of man, of sin, government, marriage, different cultures, nations, death, clothes, etc. "Many Christian girls go bra-less and wear clingy T-shirts or wear clingy clothes to show off their breast or sexual parts." Lists creation-science resources. Drawings.

708 _______. *An Introduction to Creation Science for the Layman* (tapes). Undated [1985?]. Geebung, Australia: Calvary Communications. Set of three audiotape cassettes: "The Six Days of Creation," "Creation, Humanism and Christian Education," and "Creation: Facts and Bias." Largely same "creation evangelism" messages Ham delivers in speaking tours, such as 1986 traveling Creation Festival (sponsored by Films for Christ) at the start of Ham's U.S. ministry.

709 _______. *The Lie: Evolution.* 1987. El Cajon, Calif.: Master Books (Creation-Life). Foreword by **L. Sunderland** ("If there is no Creator, there is no purpose in life"). Genesis, literally interpreted, is "foundational" to the rest of the Bible and all of Christianity. Substitution of evolution for creationism results in the collapse of Christianity. Creationism is thus of vital importance. Ham urges "Creation evangelism" in order to restore biblical foundations. "If God did not mean what He said in Genesis, then how could one trust Him in the rest of the Scriptures?" People have been "deceived" into believing evolution is science. Everyone is biased, and holds religious positions based on faith; evolutionists are biased against God and creationism. But creationism is the "best" bias to have, since it is based on infallible divine revelation. Schools have eliminated Christianity and replaced it with humanism, an anti-God religion. Ham describes teaching creationism in public schools using presuppositionalist approach – identifying various assumptions used to interpret evidence. Declares that if something disagrees with the Bible it is wrong, no matter what the evidence. Discusses evils of evolution (lawlessness, homosexuality, drugs, abortion, Nazism, racism, etc.). "Evolution is an anti-God religion held by many people today as justification for their continued pursuit of self-gratification and their rejection of God as Creator." "An all-out attack on evolutionist thinking is possibly the only real hope our nations have of rescuing themselves from an inevitable social and moral catastrophe." Final chapter "Creation, Flood and Coming Fire" on prophecy of II Peter as referring to "willingly ignorant" uniformitarian evolutionists who deny Creation and the worldwide Flood; warns that God is about to destroy the world again. Includes listing of creationist resources. Many cartoon-style drawings; index.

710 Hamilton, Floyd E. *The Basis of Evolutionary Faith: A Critique of the Theory of Evolution.* 1946 [1931]. London: James Clarke. Some chapters orig. pub. in 1926 *Princeton Theol. Rev.;* revised and updated. *Hamilton:* Presbyterian minister; Bible prof. at Union Christian Coll. (Korea); also wrote *The Basis of Christian Faith* and *The Basis of Millennial Faith.* Hamilton declares that evolution must be examined carefully and fairly; if it is true, then Christians must accommodate it. The facts, however, force him to side with anti-evolutionists. "It is the hope of the writer that many readers whose faith in Christ and the Bible has been weakened or destroyed by anti-Christian evolutionary teaching, may be led to see the weakness of the evidence in favour of evolution, and may recover their faith..." Asserts that scientists "became convinced of the truth of evolution thirty years ago, and, using evolution as an assumption, proceeded to construct a number of sciences such as biology, geology, anthropology, etc., which today can hardly be taught on any other basis than that of the assumption of the truth of evolution." Anti-evolutionist scientists must conceal their beliefs or suffer persecution. Only religious and non-scientific publications criticize evolution, because scientists are too closeminded. Though evolution is no longer criticized, the Darwinian theory has been abandoned as explanation. Says that acknowledgment of degenerative evolution is tantamount to conceding creationism. Evolution does not admit the reality of sin, and denies need for a Divine Saviour and redemption. If evolution is true, then "religion and morality and ethics might as well be cast on the scrap heap," as there is no life after death. Cites J. Bose's experiments showing sensitivity of plants as evidence of transcendence and immanence of God. Maintains that God's supernatural activities are not miracles, because "it is the way God *ordinarily* acts." Many problems such as origin of life and instinct are insoluble to the materialist evolutionist. Even if man did create life, it would only prove God used intelligence. Claims that if evolution were true, the genes for all traits of all organisms "would have to be hidden in the first chromosomes of the first cells..." Chromosome numbers refute common ancestry. Chapters on genetics, taxonomy; comparative anatomy, embryology, vestigial organs, G. Nuttall's tests comparing blood antiserums of different species, biogeography, paleontology. "The whole classification system is a pure deduction from the theory of evolution itself." Convergence disproves evolution. Nuttall's evidence demonstrates biochemical similarity, but since blood is not the carrier of heredity, it cannot prove common descent. Discusses various creationist theories, inc. Xenogenesis theory of H.W. Magoun (Johns Hopkins prof.)—a type of progressive creationism—but prefers youngearth creationism. Accepts **G.M. Price**'s Flood Geology. Suggests a planetary collision caused tilt of earth's axis, and vast tidal currents (the Flood).

711 Hampden, John. *The Earth in Its Creation, Its Chronology, Its Physical Features, and the One Alone Portion of the Universe Adapted to Man's Occupation and Service* (pamphlet). 1880. London: W.H. Guest. Orig. series of letters to the *Christian Journal. Hampden:* Oxford grad.; flat-earth advocate. "If the Mosaic records of Creation are provably false, our Saviour himself wilfully and persistently condoned the fraud..." Creation: "nearly six thousand years ago" in six solar days, with Wednesday falling on autumnal equinox. (In Schadewald files.) In 1870, Hampden offered reward to any scientist who could prove earth was not flat; **A.R. Wallace** accepted the challenge, but the results of his test were not accepted by Hampden.

712 Hand, John Raymond. *Why I Accept the Genesis Record* (booklet). c1972 [c1953]. Lincoln, Neb.: Back to the Bible. 1972 ed. revised by **Bolton Davidheiser.** Orig. 1953 (Van Kampen: Wheaton, Ill.); pub. by Back to the Bible in 1959. 733,000 printed as of 1979. 1972 Foreword by **J.C. Whitcomb.** *Hand:* Sc.D.; physics teacher. Standard creation-science arguments against evolution. "Until science can provide me with one [answer] that will satisfy me better than the one I have found in God's word, I shall stand my ground." If evolution is true, then there was no created Adam and Eve, no Fall, no Atonement, no Christ as Redeemer and Saviour. **T. Dwight** and Paul Shorey (Univ. Chicago) quoted on tyranny of evolutionists and pressure put on non-believers. Many scientists quoted, but no dates or references given (many—such as the two just mentioned—are from the 1920s). Stephen Langdon, Oxford Assyriologist, quoted in support of original monotheism. Cites contemporary human and dinosaur tracks, human and trilobite fossil, Cambrian pollen. Presents population increase argument. 1959 Appendix mostly quotes from **Morris** 1951. Davidheiser excises Hand's 1959 declaration that Darwin "repudiated most of his original conclusions before he died."

713 Handrich, Theodore L. *Everyday Science for the Christian.* 1947 [1938]. St. Louis, Mo.: Concordia. *Handrich:* Minnesota high school teacher; Missouri Synod Lutheran.

714 ________. *The Creation: Facts, Theories, and Faith.* 1953. Chicago: Moody Press. "Darwin's outdated theory of Gradual Ascent is being gradually abandoned even by evolutionists." They are replacing it with Mutation Theory, then Explosive Evolution. Handrich presents Day-Age creationism, Gap Theory, and recent creationism all as plausible—but suggests creation was 6–35,000 years ago. Catastrophic origin of coal. Glaciation was short-term or result of Flood. Discusses radioactivity, including helium escape and pleiochroic halos. The "strongest proof" against evolution "is that acquired traits are not transmissible"—the environment can't make new genes. Population increase argument.

715 Hansen, Julia. *God Makes a World: Bible Study Guide.* 1973 [1971]. Denver: Baptist Publications. Rev. ed.; orig. 1971. "Fourth Grade Teacher." (Sold at CHC [ICR] bookstore.)

716 Hanson, James. *A New Interest in Geocentricity* (report). Undated [1979?]. Minneapolis: Bible-Science Association. Edited transcription of 1979 talk given at Assoc. for Christian Schools (Houston). *Hanson:* computer science prof.—Cleveland State Univ.; contributes to *CRSQ* and *Bull. Tychonian Soc.* (geocentrist journal). Scientific and biblical arguments for geocentricity. "I sincerely believe that evolution and heliocentricity go together.... To me it appears as inconsistent for people to accept creation and then to oppose geocentricity." Cites Varshni *(J. Astrophysics and Space Science),* Vera Ruben (Caltech). Approves Canadian geocentrist Walter van der Kamp's view that planets and stars are in a shell 60 light-years distant. Cites objection by C.L. Poor (Columbia Univ. astron.) to Einstein and bending of light. Praises Oliver Heaviside's 1890 book which uses classical physics. Advocates Lesage's idea that gravity is a bombardment from exterior of sphere of universe as compatible with geocentricity.

717 Hardie, Alexander. *Evolution: Is It Philosophical, Scientific or Scriptural?* 1924. Los Angeles: Times-Mirror Press. *Hardie:* Methodist Episcopal Church (S. Calif. Conf.). Book was "written to show that Evolution—which is the most common and degrading manifestation of materialism—...is utterly unphilosophical, unscientific, and unscriptural. The further purpose is to save our dear school children and students from the mental and moral defilements of the 'Mud Philosophy.'" Evolution is "proof of the mental and moral insanity of man" and "the unregenerate wish to avoid the Divine Presence..." Evolutionists have simply plagiarized the old heathen materialists. Praises Plato and Aristotle. Darwin abandoned science by indulging in reckless and vague speculation, and by casting aside biblical truth. "Thus, it is seen that this disgraceful craze for an animal ancestry is built upon nothing but *supposition*.... All unbelievers are making frantic efforts to propagate this man-dishonoring and God-denying, Satanic explanation of the Cosmos, to the mental and moral debasement of the rising generation." Criticizes Spencer's view of evolution primarily. Quotes **William Dawson** and many other anti-Darwinian and anti-evolutionist scientists. Presents examples of the wonders and mysteries of science, and of design in nature as proof of divine creation. "This recent evolution rage, which has cursed colleges and schools, is the outstanding proof of the insanity of sin." "The moral mental mortality among college professors is one of the most humiliating and appalling facts of our times." Describes evil effects of evolution; proclaims it is destructive to Christian faith. The most vociferous followers of evolution include "demented atheistic Germany" and France, various "undesirable parasites on society," anarchists and other troublemakers. Blasts apostate science: "Now this treacherous Science declares,/ That the beasts are our deities great,/ And that life is a doom of despairs,/ Because darkness forever's our fate." Evolutionists "deify mud and monkeys." Reprints Lady Hope story (Darwin's deathbed confession), quoted from **L.T. Townsend.** Advocates both Day-Age creationism and a Gap Theory interpretation. First two verses of Genesis "describe events and conditions anterior to the work of the seven periods occupied in transforming chaos into cosmos." Earth was originally molten. The creation "days" were long periods, and thus can accommodate the "vast geological strata of our earth which must have required long aeons..." Accepts date of Flood as 1,656 years after creation of Adam.

718 Harding, Judith Tarr. *Establishing Scientific Guidelines for Origins-Instruction in Public Education* (pamphlet). 1981. El Cajon, Calif.: Institute for Creation Research. ICR "Impact" series #93. How to convince educators that creation-science is as scientific as evolution.

719 Harris, John. *The Pre-Adamite Earth: Contributions to Theological Science.* 1870 [1846?]. Boston: Gould and Lincoln. 6th ed. Orig. pub. 1846(?) by Ward (London). Gap Theory creationism. "My firm persuasion is that the first verse of Genesis was designed, by the divine Spirit, to announce the absolute origination of the material universe by the Almighty Creator; and that it is so understood in the other parts of holy writ; that, passing by an indefinite interval, the second verse describes the state of our planet immediately prior to the Adamic creation, and that the third verse begins the account of the six days' work." Millhauser says that Harris criticized Chambers' *Vestiges* for scientific errors, but felt its first fault was theological. *Harris:* "After the primary act, according to [Chambers'] view, the Creator might have ceased to be"—as far as the created universe was concerned. Urges instead a God who intervenes directly and repeatedly in affairs of the world and of man. Does not repudiate science or geology; simply sees no conflict.

720 ________. *Primeval Man.* 1849. London: Ward. Gap Theory creationism.

721 Harris, Lester E., Jr. *Galápagos: A Creationist Looks at Darwin's Islands.* 1976. Nashville, Tenn.: Southern Publishing Association. Harris called here "one of the nation's leading authorities on herpetology." Descriptions of unusual adaptations of various Galapagos animals and plants. Contains no real argument, but final paragraph in each section usually praises God for His wisdom and design. Notes merely that Darwin "jumped to the wrong conclusions," thereby leading many away from God's Word, and that "no new types have evolved." Presence of sin confuses God's original plan for nature; results in tooth-and-claw struggle. Mentions belief in Noah's Flood and suggests young earth.

722 Harris, R. Laird. *Man: God's Eternal Creation: Old Testament Teaching on Man and His Culture.* 1971. Chicago: Moody Press. *Harris:* Bible prof. at Faith Theol. Sem. (Del.). Quoted and discussed in **Whitcomb** 1973. Argues against world-wide Flood in this book (apparently a change of mind). Discusses Israeli religious painter and architect Meir Ben Uri's theory of diamond-shaped Ark (cross-section); accommodation of animals, deck space, tools. Harris also wrote *Inspiration and Canonicity of the Bible* (c1969;

Grand Rapids, Mich.: Zondervan).

723 Harvey, Jeff, and Charles Pallaghy. *The Bible and Science.* 1985. Melbourne, Australia: Acacia Press. Foreword by **Clifford Wilson.** Strict creationism, Flood Geology. All geology and fossils are the result of the Flood; also the Ice Age. Discusses polystrate fossils, rapid coal formation, design and content of Ark, antediluvian giants, Pre-Flood Water Canopy, co-existence of man and fire-breathing dinosaurs. Viper and crocodile biochemistry less similar than crocodile or viper to chicken: this disproves evolution. (Quoted in **C. Wilson** 1985.)

724 Hasskarl, Gottlieb C.H. *The Terrible Catastrophe, or, Biblical Deluge; illustrated and corroborated by mythology, tradition, and geology, to which is added a brief interpretation of the creation, with notes from theologians, philosophers, and scientists.* 1885. Philadelphia: C. Henry.

725 _______. *The Missing Link, or, the Negro's Ethnological Status.* 1898. Chambersburg, Penn.: Democratic News.

726 Hathaway, Henry W. *Evolution, Science and the Bible.* 1955. London: The Dawn Book Supply. 3rd ed.

727 Haughton, Samuel. *Principles of Animal Mechanics.* 1873. London: Longmans, Green. *Haughton:* physiologist; geology prof. – Dublin Univ. Quoted and discussed in Hull. Hull says Haughton is extremely ill-informed about evolutionary theory, but typical of physical scientists trained in math and classics – an influential group highly critical of Darwin. These scientists felt that evolution could be disproved by reasoning from simple math principles.

728 Haun, Delton. *Evolution and the Bible* (booklet). c1977. Pasadena, Texas: Delton Haun Tract. Standard anti-evolution arguments.

729 Hausmann, William John. *Science and the Bible in Lutheran Theology from Luther to [?].* 1978. Washington, D.C.: University Press of America. Exposition of views of leading Missouri Synod Lutheran scholars. (Missouri Synod Lutherans have been among the most active and influential creationists.) The author does not advocate any particular view.

730 Hayhoe, Douglas. *The Creation Psalms of David: Meditations from Science and Scripture.* 1980. Niagara, Ontario, Canada/Sudbury, Penn.: Believer's Book-shelf. Bible-science essay. Discusses meteorology, agriculture, botany, cosmology, pre-natal growth, etc.

731 Hayward, Alan. *God's Truth.* 1977 [1973]. London: Lakeland. Rev. ed. Orig. 1973 (London: Marshall, Morgan & Scott). *Hayward:* physicist; research and development advisor at Redwood Int'l.; was "principal scientific officer in a gov't. research lab" until 1977. Includes discussion of Genesis and geology; refutes young-earth creationism. Declares that a Creator is necessary. Book is largely concerned with why this Creator is the Christian God and not Allah or some other god. Most creatures now regarded as pre-human hominids were really highly-developed animals now extinct. Adam was the first real man.

732 _______. *God Is: A Scientist Shows Why It Makes Sense to Believe in God.* c1978. Nashville, Tenn./New York: Thomas Nelson. Mostly concerned with anti-evolution. Evolution is only a hypothesis, and has become a religious dogma. Refutes J. Monod's naturalistic "chance and necessity" view. Endorses Anthropic Principle. Presents riddles inexplicable by evolution: instinct, migration, sex, the bombardier beetle, etc. Man's brain too large to be result of adaptation. Psychic sense is proof of the spirit; endorses Uri Geller. Jesus was either a lunatic or the Son of God. Bible prophecies have been fulfilled. Refutes Flood Geology creationism. Proposes successive creation: millions of creative acts. Favors Day-Age or Revelatory creationism. Adam was first New Stone Age man, 10,000 years ago. Index.

733 _______. *Creation and Evolution: The Facts and Fallacies.* 1985. London: Triangle. Urges "middle position" of "ancient creationism." Darwinism is "contrary to the evidence, and . . . evolution is therefore nothing more than an unsupported speculation." But "succession" of fossil types is undeniable. It is not due to evolution, though, but to "successive acts of creation over a long period." Includes a thorough, carefully researched refutation of young-earth creation-science; demolishes Flood Geology. Hayward insists that recent-creationists are "friends and allies," though. They "share a belief in an inspired Bible. We agree that Darwin was mistaken, and that God is the Creator of every living thing. Compared with this, the question of the age of the earth pales into insignificance." Notes that recent-creationists are obliged to rely on **Gosse**'s argument. Comprehensive review of recent non-religious anti-evolutionist and anti-Darwinist scientists and writers, especially British and French. Cites many of these as exposing weaknesses of evolution theory. The new physics proves that the universe is carefully designed to support life—**Paley** is being "rehabilitated." Hayward proposes creationist theory of "Days of Divine Fiat": creation was *declared* in six days, but the process of creation was manifested over ages, and the six ages can overlap. Genesis contains parentheses, which describe the intervening processes between God's creation fiats. Presents "repunctuated" version of Genesis creation account. Says this theory was first proposed by F.H. Capron in 1902 (in *The Conflict of Truth;* London: Hodder & Stoughton). Adam and Eve are real; the basis of Christianity. Well-written. Index.

734 Heard, Gerald (Henry Fitzgerald). *Is God in History?: An Inquiry into Human and Prehuman History in Terms of the Doctrine of Creation, Fall, and Redemption.* 1950. New York: Harper. *Heard:* apparently claims to be an anthropologist. The same year (1950), he published one of the earliest flying saucer books, *The Riddle of the Flying Saucers* (London: Carroll & Nicholson), reprinted in U.S. as *Is Another World Watching?* (1951; Harper and Bros.). Heard concluded that the saucers were piloted by superintelligent bees from Mars. (From Schadewald files.)

735 Hedtke, Randall. *The Secret of the Sixth Edition.* 1983. New York/Los Angeles: Vantage Press. *Hedtke:* biology teacher at St. Cloud Tech. High School (St. Cloud, Minn.). Hedtke has taught "applied creationism" in his classes for 17 years, but is now embroiled in conflict with Board of Educ. regarding his curriculum. Book consists of six parts: nos. 2-4 originally appeared in *CRSQ;* Hedtke also contributes to *Bible-Science Newsletter.* The "Secret" is that "Darwin, in his old age, abandoned natural selection, the mechanism by which evolution was believed possible." Hedtke claims that Darwin's *Origin of Species* is rarely actually read—especially the sixth (last) edition, in which Darwin allegedly discards natural selection—but that evolution is simply accepted as truth. (Darwin was, in fact, troubled by arguments against the sufficiency of natural selection, and he did—reluctantly —come to rely more on Lamarckian mechanisms in later editions.) Darwin's chronic illness was psychoneurosis caused by anxiety—it was because of this debilitating anxiety that he gave up natural selection. **Mivart**'s objections were a cause for this change. Basically a conspiracy theory view. Hedtke argues that Darwin followed an obsolete pre-Baconian, pre-Cartesian scientific method, "overloading" facts in order to fit philosophies, and shunning experiment. Darwin was not a true scientist, but a biased propagandist whose style and arguments are very obscure and deliberately devious. Lyell, Huxley and others promoted Darwinism for selfish reasons. Accepts the Lady Hope story that Darwin privately renounced evolution on his deathbed. "It all comes down to this: Evolutionary theory, allegedly one of the greatest scientific theories of all times, the foundation for many philosophies, religions, and political systems, is merely a metaphor 'proved' by an analogy, an abomination of science. Those who believe it have been over-influenced by the clever persuasion tactics of a natural philosopher." The purpose of evolutionary theory is not scientific; it is a philosophical attempt to "ungod the universe."

736 Heffren, H.C. *Who "Mythed" the Boat?* (tract). Undated. Priv. pub. Reprinted from Marantha [sic] Messenger (March 1965).

737 Hefley, James C. *Lift Off!: Astronauts and Space Scientists Speak Their Faith.* 1970. Grand Rapids, Mich.: Zondervan. "Adventures with God": fifteen astronauts share their testimony. Hefley: writer (lives in Tennessee). Has also written *God Goes to High School, Adventurers with God: Scientists Who Are*

Christian, and many other books on Christian athletes, Christians in government, Wycliffe Bible translators, and other religious subjects.

738 ________. *Textbooks on Trial.* 1976. Wheaton, Ill.: Victor Books. Newer printings titled *Are Textbooks Harming Your Children?* Foreword by John Conlan (ex-Congressman, Ariz.; co-author with **J. Whitehead** of creation-science and anti-humanist law article). Laudatory account of the efforts of Texas textbook monitors **Mel** and **Norma Gabler**'s "ongoing battle to oust objectionable textbooks from public schools." Includes the Gabler's battle against the BSCS biology textbooks; also against the MACOS series. Appendix lists creation-science organizations, and information on the CRS creation-science biology textbook.

739 Heim, Karl. *The World: Its Creation and Consummation.* 1962 [1958]. Edinburgh/London: Oliver and Boyd. 2nd ed. Orig. German; 1958. Directed evolution. Divine Will and the culmination of man's Salvation.

740 Heindel, Max. *The Rosicrucian Cosmo-Conception; or Mystic Christianity.* c1973 [c1909]. Oceanside, Calif.: Rosicrucian Fellowship. 28th ed. "An elementary treatise upon man's past evolution, present constitution and future development." Rosicrucian occultism: reincarnation, astrology, vibrations, the various heavens and spirit worlds, "cosmogenesis and anthropogenesis." Cosmic evolution, like human evolution, spirals upward in cycles of rest and activity. "It will be noted that the modern evolutionary theory . . . would, if it were completely reversed, be in almost perfect accord with the knowledge of occult science." (I.e., within each great period, forms degenerate.) Seven Worlds and Seven Periods (rebirths). We are now in Fourth Period. The purpose of evolution, which is guided by advanced beings and spirits, is to awaken consciousness. Polarian, Hyperborean, Lemurian, Atlantean and Aryan Epochs. Seven Atlantean races: Rmoahals, Tlavatlis, Toltecs, Turanians, Semites, Akkadians, Mongolians. Semites were "seed-race" for seven Aryan races: Indian Aryan, Babylonian, Persian-Graeco-Latin, Celtic, Teutonic-Anglo-Saxon ("to which we belong"). Slavs will produce two more races. Seed-race for Sixth Epoch will arise in U.S. Also describes physical and mental differences between modern races. Includes chapters "Back to the Bible," and "The Occult Analysis of Genesis." Interprets Genesis creation in terms of Rosicrucian scheme; occult epochs correspond to creation days. Urges racial purity. The Lucifers are degenerate Angels. "Science merely states the fact, the occult scientist gives the reason." Diagrams; index.

741 Heinze, Thomas F. *Creation Vs. Evolution Handbook.* c1973 [c1970]. Grand Rapids, Mich.: Baker Book House. 2nd rev. ed. Orig. c1970. *Heinze:* Dallas Theol. Sem. grad.; missionary in Italy for Conservative Baptist Foreign Mission Society. "An evaluation of the theory of evolution in the light of scientific research." Foreword comments by **R. Whitelaw** and **D. Gish.** Standard creation-science arguments; recent creation, Flood Geology. Some Bible references (especially at end), but relies primarily on standard scientific quotes and references, popular literature, and creation-science sources. Discusses evil effects of evolution, and concludes by exhorting readers to turn to Christ and the Bible before God again destroys the world.

742 Henning, Willard L. *How Valid Is the Theory of Evolution?* (booklet). 1962. Dayton, Tenn.: Bryan College Press. **J.N. Moore** (1965 *CRSQ*) says Henning "critically examines common evidences offered in support of evolution with due mention of assumptions, circumstantial data, and postulates. After mention of archaeological evidences of the Genesis account, he itemizes briefly ten degrading effects of teaching evolution for the origin of man." (Bryan College was established in Dayton, Tenn. – site of the Scopes Trial – in commemoration of **W.J. Bryan.**)

743 Henry, Carl F.H. *Science and the Supernatural* (tract). Undated. Garland, Texas: American Tract Society. Abridged from *Christianity Today. Henry:* very influential conservative theologian; founder of *Christianity Today* magazine (evangelical). Quotes scientists from RCA, Harvard Dental Sch., MIT, Cambridge Univ., etc., as denying that science is anti-Christian or anti-supernatural. Mentions

H. Enock (sic–**Enoch**]. Henry also wrote *Remaking the Modern Mind* (1948; Eerdmans) and other major works.

744 Hepp, Valentine. *Calvinism and the Philosophy of Nature.* c1930. Grand Rapids, Mich.: William B. Eerdmans. The 1930 Stone lectures at Princeton. Calvinist; strict creationism. **D. Young** 1982 says Hepp "vigorously denounced efforts by Christians to find millions of years in earth history; to him this was a compromise with evolutionism. Hepp called for a total rethinking of geology in terms of recent creation and a global flood."

745 Here's Life. *Creation vs. Evolution* (pamphlet). Undated. Priv. pub. Published transcript of *Here's Life* T.V. show: a debate between "Kelly Seagraves" [sic–**Segraves**] of CSRC and Daniel Osmond (theistic evolutionist), with moderator Poland. Here's Life is apparently same group that publishes Campus Crusade for Christ books.

746 Herget, John F. *Questions Evolution Does Not Answer.* c1923. Cincinnati: Standard Publishing. *Herget:* Baptist minister (Cincinnati). "This book is the result of a purpose to find out what facts have been discovered by scientists to support the theory of the evolution of organic life. I have tried to distinguish between the facts which they present and their deductions from those facts." Facts must be accepted–but not "philosophical opinions." Consists largely of quotes by scientists. Evolution provides "no explanation of the origin of life," or of the unbridgeable gap between man and animals, or evidence of transitions between major groups. "Just how He made the worlds; the aeons of time He took to do it; through what processes He created and developed organic life as we know it to-day; in just what way He made man of the dust of the ground, whether directly or indirectly. . ." are questions we cannot answer until we enter the next world.

747 Herrmann, Kenneth C., Robert A. Ginskey, and William Stenger. *Probability–What It Says About Evolution* (pamphlet). c1974 [1971]. Pasadena, Calif.: Ambassador College. Articles by Herrmann, Ginskey, and Stenger (presumably reprinted from **H.W. Armstrong**'s *Plain Truth*). Standard anti-evolution arguments. Pictures, diagrams.

748 Herrmann, Robert A. *The Reasonableness of Metaphysical Evidence; and Nature: The Work of a Supreme Mathematical Logician* (reports). [1982, 1983]. Annapolis, Md.: Institute for Mathematical Philosophy. Spiral-bound reports. The first orig. 1982 *(J. Amer. Scientific Affil.);* the second was invited paper for 1983 Baltimore Creation Conference. *Herrmann:* in math dept. at U.S. Naval Acad. (Annapolis); director of Inst. for Math. Philos. (IMP); has advised Congress (including anti-evolution congressman Dannemeyer); contributes to *CRSQ.* Mathematical philosophy proves that Marxism, humanism, and Supreme Court decisions are illogical; proves that Christian doctrines are logical. Section on "Creation" is largely about quantum mechanics–no creation-science references or arguments. Also includes sections on fulfilled Bible prophecy and revelation. Offers $25,000 for proof of any math error in his work.

749 ________. *Mathematical Philosophy and Evolutionary Processes* (report). c1983. Annapolis, Md.: Institute for Mathematical Philosophy. IMP Monograph #130. Mathematical philosoophy and logic demonstrates that evolutionary processes are impossible. No empirical data; no creation-science works cited. Abstruse. Herrmann's other IMP monographs are on topics such as *Mathematics and the Word, The Miraculous Model,* and sub-atomic particles.

750 Hertwig, Oskar. *Das Werden der Organismen: Eine Widerlegung von Darwins Zufallstheorie.* 1916. Germany. *Hertwig:* Berlin cytologist and anatomy prof.; student of Haeckel; studied differentiation of cell layers in embryo; established nature of sperm-egg fertilization from sea urchin experiments; described process of meiosis. Criticism of Darwinian doctrine of chance variations and natural selection. **Nordenskiöld** says Hertwig "devoted the latter part of his life" to attacking it, emphasizing that it is based on false analogies from human social existence and from artificial selection, and that it is misinterpreted to boot. Says that Hertwig "most emphatically maintains the heredity of acquired characters. . ." Hertwig also deplored the ethical and political consequences of using natural selection to justify ruthless competition and war.

751 Hiebert, Henry. *Evolution: Its Collapse in View?* 1979. Beaverlodge, Alberta, Canada: Horizon House. "To those high school and college students who fear that science has discredited the Bible." Standard creation-science arguments. Supports Flood Geology, and asserts that geological uniformitarianism "represents the most incredible misinterpretation of scientific data ever made in human history," and will "undoubtedly take its place alongside the Flat Earth concept of the Middle Ages." Repeatedly refers to creationist **M. Cook** as "Nobel Prize Medalist." Pre-Flood Vapor Canopy. Paluxy dinosaur tracks also include prints that are "unmistakable human." Includes population increase argument, probability, ape-men hoaxes, and the rest of the regulars. Section on answers to common questions evolutionists might ask creationists; also sections emphasizing biblical implications. Mentions that family member was miraculously healed by Kathryn Kuhlman. Many references to creationist works. Index.

752 Higgins, W. Mullinger. *The Mosaical and Mineral Geologies Illustrated and Compared.* 1833 [1832]. London: Simpkin and Marshall. Mentioned by Millhauser as Gap Theory advocate. Higgins was among those who accused non-Scriptural geologists of being infidels (especially in atheist France, where geological theories were pushed to extremes). Higgins also wrote *The Book of Geology* (1842; London), and *The Earth* (1836; New York).

753 Highton, Henry. *The History of Creation as Deduced from Scripture and Science.* 1860. Cited by Millhauser as Day-Age creationism.

754 Higley, L. Allen. *Science and Truth.* c1940. New York: Fleming H. Revell. *Higley:* Ph.D., D.Sc.; chemistry and geology prof. at Wheaton College (Ill.); first president of Religion and Science Assoc. (1930s creationist group). Very strong presentation and defense of Gap Theory creationism. No scientific references despite frequent appeals to science; discussion mostly theological. "It is our purpose to disprove evolution and many other false speculations which dishonor the Creator..." "The Bible is the one foundation on which all true science must finally rest, because it is the one book of ultimate origins." Facts of science must be biblically standardized, then classified, so that ultimate truth can be distinguished from passing speculation. Any scheme which contradicts the Bible is "necessarily false." God could not have created the world imperfect, incomplete and void—this condition arose only after sin entered the world. Earth was created perfect to be Lucifer's throne; angels already lived on other planets. God destroyed this world in a great cataclysmic judgment after Satan corrupted it with evil, then He re-created the world in six literal days, and created Adam. Evidence for the pre-Adamic cataclysm is everywhere, but scientists refuse to recognize it because of belief in evolution. "Evolution is purely speculation. It is pseudo-science, because it is directly opposed to the clearly observed facts and definitely established laws of science as well as directly opposed to the definite statements of the Bible.... Evolution and cataclysms are mutually opposed. If one is true the other must be false. Cataclysms are based directly on sin, the very thing that evolution evades or seeks to set aside by denying its existence." Higley refutes Flood Geology; explains that it cannot account for fossils and geological strata. Young-earth creationists fail to realize there was an even greater cataclysm (and Flood) before Adam. Noah's Flood was due to man's sin; the pre-Adamic Flood was due to sin of fallen angels, and thus was much greater. The destruction was spread over long periods of time, and involved a "series of disasters" in different places. Higley describes the six-day re-creation in detail. Plant seeds survived. The sun and moon, knocked out of position during the pre-Adamic cataclysm, were returned to their proper positions on the Fourth Day, and the earth's axis was tilted to its present position. Citing **Dimbleby,** Higley refutes Halley's theory that the axis was tilted at the Flood. God wrote the "entire plan of salvation" in the constellations. The re-created earth was covered by a Vapor Canopy, which remained on high until the Flood, giving earth a uniformly mild tropical climate. The Ice Age followed the canopy collapse and Flood. That animals were created "after their kind" indicates they had a previous (pre-Adamic) existence. (There were no pre-Adamic humans though.) Bacteria are a "perversion of nature"; mountains are also (they are a result of sin). Man was created when everything required for his welfare was in place. Describes Eden and the Fall, and the coming Millennium. Also explains diseases as consequence of fallen man's sinful carnivorous diet. Men refuse to believe in supernatural miracles because they vainly refuse to acknowledge God's omnipotence. They advocate evolution, which is really a false religion intended to justify pleasure-seeking. Study questions; index.

755 Hill, Harold. *How to Live Like a King's Kid.* c1974. Plainfield, N.J.: Logos International. "As told to" Irene Burk Harrell. *Hill:* former president of Curtis Engine Co. (Baltimore); now writes religious books for children and heads King's Kid Korner (Baltimore). Hill has written evangelical *How to* ... book series on overcoming various problems (he is a recovered alcoholic). This book contains chapter on the widespread story of the NASA computer proof of Joshua's Long Day, which originated with Hill. NASA computer scientists were checking an orbital program prior to a satellite launch. They ran it back in time, but the computer stopped because it came across a "missing day." A scientist familiar with the Bible noted that this day occurred at the time Joshua commanded the sun to stand still. However, there were still 40 minutes missing, according to the computer. The scientists then remembered the biblical story in which Hezekiah told God to make the sun go backwards ten degrees —exactly 40 minutes. Hill's version is a modern variant of **Totten**'s math-astronomical proof (Hill cites him as authoritative proof). Hill claims to have witnessed this event as NASA, but he provides no names, dates, or other details. Says his "inability to furnish documentation of the "Missing Day" incident in no way detracts from its authenticity. 'God said it—I believe it—that settles it.'" Hill's story was picked up by news services and appeared in many newspapers and tracts. NASA was deluged with inquiries.

756 ________. *From Goo to You by Way of the Zoo.* 1985 [c1976]. Old Tappan, N.J.: Fleming H. Revell. Written "with" Irene Harrell. Rev. and updated ed. Orig. c1976 (Plainfield, N.J.: Logos). Earlier printings titled *How Did It All Begin?* Foreword by **Wernher von Braun.** 1985 ed. has second foreword by astronaut **James Irwin.** Intended for children. Remarkably ignorant and derogatory attack on evolution. Hill strives to be cute and humorous throughout; relies mostly on sarcasm and misrepresentation. Follower of **H. Rimmer:** "one of the foremost scientists of this century"; states that Rimmer's widow asked Hill to carry on his work. Hill reprints full creationist bibliography from **Morris** (ed.) 1974, and adds Rimmer's works. Asserts that evolutionists must rely on inheritance of acquired characteristics. Evolutionists are deceitful propagandists who try to hide all the evidence which destroys evolution. Einstein's relativity, the strong nuclear force, etc., are among the scientific truths predicted by the Bible. Standard anti-evolution quotes. Section on Donald Liebman's "glory meter," a device which measures human brain energy: born-again Christians shoot off the scale. Says Jodrell Bank Observatory has been receiving messages from outer space, probably from ETs telling us to "turn on to Jesus." Declares "There's nothing unscientific about resurrection"—plant seeds "die" before they sprout. Lots of jokes about primeval slime creatures and "reeking, hairy jungle baboons" as our supposed ancestors. Professors ignore the overwhelming evidence against evolution because they don't want to give up their "fat royalty incomes" and "pretty coeds." Engineers—like Hill—are the real scientists, because they have to provide results. Darwin was only a theologian—not a scientist. Humorous drawings, charts.

757 Hills, Edward F. *Space Age Science* (booklet). c1979, c1964. Des Moines, Iowa: Christian Research Press. "The Bible Is True!" 2nd ed. Orig. c1964. *Hills:* Yale BA; Westminster Theol. Sem. and Columbia Sem. grad; Th.D.—Harvard. We must begin not with facts of nature and history, but with God and His revelation. Criticizes search for life in space and attempts to colonize space as wasteful—Bible says only earth has life. Most of booklet is criticism of Einstein's relativity (a "religious philosophy"), which Hills rejects in favor of Newtonian absolute space. Includes section on "Why Only Bible Believing Students Can Meet the Challenge of Einstein." Also criticizes Einstein's anti-military political stance. States that the Bible teaches geocentricity; also appeals to Michelson-Morley experiment. Drawings; index. Hills also wrote other booklets in New Space Age Christian Library series: *The King James Version Defended: Space Age Edition,* and *Believing Bible Study: Key to the Space Age.*

758 ________. *Evolution in the Space Age* (booklet). c1967. Des Moines, Iowa: Christian Research Press. Book dedicated to **N.G. Moore,** Hills's grandfather. Well-written; many scientific references; amusingly illustrated. Strict creationism; Flood Geology. Standard creation-science arguments. Unabashedly appeals to the Bible for proof of assertions, but also includes many scientific references. Drawings.

759 Himmelfarb, Gertrude. *Darwin and the Darwinian Revolution.* 1962 [1959]. New York: W.W. Norton. "A biographical, historical, and philosophical study of the impact of Darwinism on the intellectual climate of the nineteenth century." Very literate and important but quite unsympathetic critique of Darwin. A. West characterized this book as "an advanced case of Darwinitis, a complaint that afflicts those of a literary bent and with strong attachments to pre-scientific culture, who find in the theory of evolution a disturbing and mysterious challenge to their values." Himmelfarb, who seems to get carried away by her own prose, denigrates Darwin's training, writing style, personality, philosophy, and the logic of his arguments for evolution. Attempts to refute many of his scientific arguments; assails his objectivity, his originality, and his motives. Sees him as a product of a materialistic age. Argues that Lyell was a secret evolutionist. Claims that Darwin very deviously turned weaknesses of his argument into supposed proofs. Chastises Darwin—and scientists in general—for unfeeling reductionism and materialism. Gives a very misleading account of the Piltdown affair; implies evolutionists depended on it all along as central proof, and that it renders all of evolution suspicious. Well-researched; an excellent source of Darwiniana. Much cited by creationists. Bibliog., index. Himmelfarb's husband, editor and publisher Irving Kristol, has also criticized Darwinism.

760 Hinderliter, Hilton. *The Earthquake Story* (booklet). 1979. Minneapolis: Bible-Science Association. Personal, chatty account of "bizarre" 1977 journey by Hinderliter to attend conference in Bulgaria on "space-time absoluteness" organized by **S. Marinov.** Heard about conference via H.C. Dudley (Univ. Illinois Med. Ctr.), who was similarly interested in **Velikovsky**'s "neutrino sea" theory of non-random radioactivity (with implications for radiometric dating). Conference "patron" was to have been A. Sakharov. Hinderliter received messages of expected earthquakes at conference site, which he realized were sent by communist authorities to deter attendees. Unable to locate Marinov in Sofia; Bulgarian Acad. of Sciences denied his presence. Hinderliter describes ominous intervention, by both communist authorities and supernatural agencies, intended to disrupt the conference, which Hinderliter never finds. On return, gets letter from Marinov describing his frequent imprisonment and psychiatric incarceration by communists, who are determined to thwart his scientific and spiritual message.

761 Hinn, Benny. *War in the Heavenlies.* 1984. Winter Park, Fla.: Benny Hinn Ministries. *Hinn:* Israeli-born; grew up in Canada; attended French Catholic schools; experienced visions of God; stuttering miraculously cured by Kathryn Kuhlman; founder-pastor of Orlando Christian Center (Fla.); televangelist. Advocate of Gap Theory creationism: the "war in the heavenlies" refers to pre-Adamic cosmic battle between Satan and God (the war still rages). Describes Lucifer's Rebellion and Fall, followed by the six-day re-creation of Genesis. Much of the book concerned with theological classification of various kinds of angels, demons, levels of heaven, the five underworlds, etc. Genesis "sons of God" were angels who co-habited with women. Demons are not fallen angels, but evil spirits of pre-Adamic creatures who seek to inhabit human bodies. Explains how evil drum beats attract demons. Affirms reality of global sin-destroying Flood. No scientific references.

762 Hitchcock, Edward. *The Connection Between Geology and the Mosaic Account of the Creation.* 1836. Edinburgh: T. Clark. *Hitchcock:* natural theology and geology prof. at Amherst College (Mass.), also president of Amherst; studied Conn. Valley dinosaur tracks; charter member of Nat'l Acad. Sci. Sought to prove compatibility of Bible and modern geology. Advocated ancient earth and endorsed Gap Theory creationism as a reconciliation. Admired Lyell, but emphasized changing intensities of forces in geology (e.g. glaciation, flooding); thus, uniformitarians and catastrophists both claimed him as a supporter.

763 ________. *The Historical and Geological Deluges Compared.* 1837. Edinburgh: T. Clark. Noah's Flood affected upper geological strata only – not responsible for deposition of lower sediment layers. The Bible uses phenomenological language, not modern scientific terminology; things and events are described as they appear to the observer or narrator, not as modern science knows them to be.

764 ________. *Elementary Geology.* 1841. New York. 2nd ed. Various eds. up to 1871. Best-selling textbook. Introduction by **Pye Smith.** Many sedimentary layers were deposited in quiet waters, not a violent Flood. Most fossils are of prehuman forms not existing now, but the Flood occurred in human times; therefore these fossils were deposited before the Flood. Sedimentary deposits are far too deep to have been all caused by the Flood. (Quoted and discussed in **Hayward** 1985.)

765 ________. *The Religion of Geology and Its Connected Sciences.* 1851. Boston: Phillips, Sampson. Also pub. 1861. Tremendously popular book. Hitchcock eloquently argues the unity of truth of both science and theology, as opposed to the view that there are separate domains of truth. Hitchcock stresses that religion should have nothing to fear from modern science – that the truths of science must harmonize with biblical truth, since God is the author of both nature and scripture. "Scientific truth is religious truth"; it is a perversion of science to use it against religion. Praises and quotes scientists who harmonize Genesis with modern geology, especially **Pye Smith;** also **John Harris, Buckland, Sedgwick, Whewell, Hugh Miller.** Chides theologians and Scriptural Geologists **(Penn, Fairholme, Young, Cole)** who resist or denounce modern science. Praises Congregationalists (his own denomination) and **Chalmers'** Free Church of Scotland for enlightened attitudes in this regard. Quotes many of the Bridgewater Treatises often and at length (Buckland, **Bell, Kirby,** Chalmers); also Babbage's so-called *Ninth Bridgewater Treatise.* Says that attempt to find modern scientific discoveries anticipated in the Bible is misguided, though, since the Bible is not intended to explain the natural world. The Bible describes things "as they appear to the common eye, and not in their real nature" – according to "optical" rather than physical truth, and it employs many of the "erroneous notions which prevailed" in its time. Rocks are result of secondary causes: they show signs of having undergone long processes and changes, and were not created in their present state. The continents have undergone tremendous change: high land was once ocean bottom, and vice-versa. Fossils are not promiscuously thrown together, as in a single, violent catastrophe; fossilization is a "quiet and slow process." Fossils are generally found near where the organisms lived, "arranged, for the most part, in as much order as the drawers of a well-regulated cabinet." There have been at least five distinct periods of life on earth, each characterized by distinct and independent sets of organisms. There were also different centers of creation for different animal types, which radiated only as far as their adaptations permitted. Ecological conditions changed over time; organisms were adapted only for their particular period, then died off, to be replaced by the next set. These successive changes rendered the earth progressively more fit for man, the final creation: each system was "most beautifully adapted to its place in the great chain, and yet each successive link becoming more and more perfect." Hitchcock emphasizes various scientific proofs of the enormous antiquity of the earth. Argues for Gap Theory creationism. Cites many early Gap Theory advocates: Dathe, Rosenmuller, Chalmers, **Sharon Turner;** also claims Augustine and other Church fathers, Milton. The Bible says man was created 6000 years ago, but geology proves the earth is vastly older. Genesis allows, however, long ages prior to the six day creation of the present environment and its inhabitants: "admit of a long period between the first creative act and the six days, and all difficulties vanish." Argues that death existed long before Adam; what his Fall caused was a new manner of human death: fear and decay—death's "sting." The Flood left no mark in the geological record. Refutes theories (**Burnet, Woodward, Scheuchzer,** Catcott) that the Flood dissolved the surface of the earth, and that it can explain fossil deposits. The Flood may have been regional—limited to western Asia. Ark did not land on Mt. Ararat, but probably in Babylonia. Refutes materialism as a naive attempt to escape God; also atheist notion of eternalism. Refutes the "hypothesis of creation by law" promoted by Oken, Lamarck, and in Chambers' *Vestiges*—the notion that life could have originated from powers inherent in matter, and that different forms can evolve. Discusses claims that life can be produced by electrical discharges—says bodies of organisms may be so produced, but they are not alive. Hitchcock stresses the uniformity of law and of natural processes: the "same general laws appear to have always prevailed upon the globe, and to have controlled the changes which have taken place upon and within it." But also argues that geology, more than any other science, proves that God has intervened directly to guide and alter earth history. "No other science presents us with such repeated examples of special miraculous intervention in nature." God's providential interference operates via secondary, natural phenomena. Divine benevolence is proved by beneficial constitution of the earth's crust and the distribution of water and minerals. The successive sets of organisms were separate miraculous creations, not metamorphoses from previous species. Volcanos, glaciers and other destructive phenomena cause short-term damage but are necessary to render earth productive for mankind. These beneficial long-term processes have caused the earth to be perfectly adapted to man. Suggests our glorified bodies at the Resurrection will not contain the same atoms which constituted our living bodies, but rather atoms of the same kind and proportions. The Resurrection world will be physical, but of an "etheric" type. Uses speed of light analogy (viewing the past on distant worlds) to suggest how all our actions may reverberate through eternity.

766 Hitching, Francis. *Earth Magic.* 1977. New York: William Morrow. *Hitching:* member Royal Inst. Archaeol.; T.V. documentary and musical producer; has worked for **Landsburg** and Nimoy's *In Search Of* series. Megaliths were built to harness mysterious paranormal force fields of the earth. They were sited along "ley lines"—lines of power connecting ancient monuments (cites A. Thom). Churches were later built on these same sites because of this power; thus truth of miraculous healing legends, etc. Dowsing is vestige of this power. The great symbols of myth are taken from observed patterns and manifestations of the force fields. Cosmic cataclysm 15,000 years ago: associated with an evolutionary jump, UFOs, Atlantis. Man then learned to build megaliths over spots emitting this force, developed incredible ESP and psychic powers (now largely lost). Also includes pyramidology, astrology, etc. Drawings, maps; index.

767 ________. *The Mysterious World: An Atlas of the Unexplained.* 1978. New York: Holt, Rinehart and Winston. Illustrated compendium of anomalies. Section "Emergence of Man" includes accounts of surviving Neanderthals. Sections on "The Deluge," "Noah's Ark," "Old Testament Truths"—presents all as credible. Champions **Velikovsky**'s interpretation of history, especially of Exodus and the Ipuwer papyrus.

768 ________. *The Neck of the Giraffe:*

Darwin, Evolution, and the New Biology. 1982. New York: Mentor. Orig. pub. by Ticknor & Fields (New York). Lively attack on Darwinian evolution; very widely cited and endorsed by creationists because of its assault on evolutionary theory. Good presentation of various anti-Darwinian arguments. Uses many creation-science arguments and examples (**M. Bowden** in fact accused him of plagiarism), and quotes many creationists as authorities (**R.E.D. Clark, R. Daly, Davidheiser, D. Dewar, Gish, Macbeth, Salisbury, Wysong,** and others). Though he ridicules creationism itself, he suggests many of its criticisms of evolution are valid. Favors Lamarck, and **E.J. Steele.** Strongly influenced by **Koestler** and **Velikovsky.** Proposes catastrophist evolution to replace discarded Darwinian theory. Combines various anti- and non-Darwinian hypotheses into colorful hodgepodge: "severe environmental crisis accelerates embryonic restructuring, and isolated mutants survive." Suggests that global catastrophes led to extreme biological stress and resultant genetic pressure, causing many "hopeful monsters"—organisms with drastically altered genetic structure—to be born. Some of these survived as dramatically different creatures in the new environment. Asserts this scenario is "profoundly un-Darwinian"—e.g. it is *absence* of [direct interspecies] competition following catastrophe which allows for new forms to emerge. Drawings; bibliog., index. This book was excerpted in both *Life* magazine (1982) and *Reader's Digest* (1982).

769 Hitler, Adolf. *Mein Kampf*. 1939 [1925]. New York: Reynal & Hitchcock. Orig. 1925 (vol. 1) and 1927 (vol. 2) (Frz. Eher Nachf.). 1939 ed. first complete English trans. (copyright Houghton Mifflin). Written during Hitler's imprisonment following unsuccessful coup attempt. Because of his frequent and vociferous appeals to evolutionary struggle for existence between nations and races, fundamentalists consider Hitler the epitome of evolutionism. However, the "evolution" he uses to justify Nazi militarism and aggression is a crude caricature of biological evolution (though promoted by many German scientists). Hitler —like other tyrants—rationalized his actions by appeal to "scientific law." He also appealed to Christian and biblical beliefs, though he eventually sought to replace Christian worship with a mystical pseudo-scientific Nordic neo-pagan Nazi religion which supposedly originated in Atlantis. In *Mein Kampf* Hitler describes his mission of destroying the Jews as the "Lord's work": the will of the Creator. Hitler begins his chapter "Nation and Race" by declaring that the outstanding law of nature is that all species reproduce only after their own kind. Any blending of types is a sin against Nature's will. Such race poisoning is the "original sin" against the will of the Creator, who intended the races to remain "as He Himself created them." "*The most believing Protestant* could stand in the ranks of our [Nazi] movement next to *the most believing Catholic,* without ever having to come into the slightest conflict of conscience with his religious convictions." Hitler expressed admiration for Christianity's "fanatical intolerance"—its compulsion to destroy all opposing creeds which is its "absolute presupposition." (He especially admired the Jesuits.) Hitler was considered the Messiah; he accused those who opposed his myth of Aryan destiny of "sins against the benevolent Creator," and warned of "expulsion from Paradise" of his "thousand year Reich" (the Millennium). Hitler even uses the analogy of a few members of a favored culture-bearing race surviving a global Flood catastrophe. All men are "fundamentally deeply religious," but it is the German State and the Aryan race which deserve the highest worship. Jewish religion—unlike Hitler's Aryan conception—"lacks the conviction of the continuation of life after death. . . ." Hitler's "evolutionism," with its stress on "racial purity"—the constant invocation of the "sacred obligation" of purity of the blood—is profoundly un-Darwinian. Race-mixing "necessarily" degrades the "higher" race, which is conceived of as an ideal Type or essence. As with Stalin's notorious support of Lysenko, Hitler relies on a Lamarckian notion of evolution as Progress—the striving of the Will. American neo-Nazi groups such as the Church of the Creator (N. Carolina) explicitly advocate these doctrines, demanding that Aryans forcibly assert their superiority over Jews and other "mud races." The Church of the Creator denies being atheist (founder Ben Klassen has written books such as *Nature's Eternal Religion* and *The White Man's Bible*), but ridicules the fundamentalist interpretation of Genesis. Klassen says that Hitler, the greatest man who ever lived, erred only by worshipping the German State as much as the White Race.

770 Ho, Mae-Wan, and Peter T. Saunders (eds.). *Beyond Neo-Darwinism: An Introduction to the New Evolutionary Paradigm.* 1984. London: Academic Press. *Ho:* biologist at Open Univ. (England). *Saunders:* mathematician at Univ. London. Saunders and Ho are legitimate scientists who argue for non-Darwinian evolution. In a 1976 *J. Theor. Biol.* article they proposed that "complexity, rather than fitness or organization, is the quantity whose increase gives a direction to evolution." Complexity (apparently of information) is a cause, not a result, of increase in fitness—a conclusion which is in "direct conflict" with Darwinian evolution. In this book they suggest a catastrophist (after R. Thom) hypothesis of evolution of new forms: external environmental influences and increased informational load somehow catapulting the genome into a new state. They argue that random changes cannot account for evolution: "The neo-Darwinian concept of random variation carries with it the major fallacy that everything conceivable is possible." "All changes are held [by neo-Darwinists] to be possible and equally likely." This book contains chapters by distinguished biologists such as origin-of-life researcher Sidney Fox. Also includes chapter by "transformed cladists" G. Nelson and N. Platnick (Amer. Mus. Nat. Hist.—N.Y.), who insist that Darwinism is "a theory that has been put to the test and found false." ("Transformed cladists"—unlike cladists—somehow feel the need not only to ignore evolutionist preconceptions but also to seemingly repudiate the very existence of descent relationships: ironically, since cladism itself is the construction of taxonomies exclusively on the basis of genealogical connections.) Index.

771 Hoare, W.W. *Geology and Its Reference to Religion.* 1856. Cited by Millhauser as advocating Gap Theory creationism.

772 Hoare, William H. *The Veracity of the Book of Genesis.* 1860. London: Longman, Green. *Hoare:* minister. Mentioned in Custance; presumably Gap Theory creationism.

773 Hobbs, Herschel H. *The Origin of All Things: Studies in Genesis.* 1975. Waco, Texas: Word Books. *Hobbs:* presents Baptist Radio Hour broadcasts. Earth may or may not be billions of years old, but Genesis is a factual account. Doubts Gap Theory; prefers Day-Age creationism. Quotes Apollo 17 astronaut Ron Evans that it is clear the world is "too beautiful and perfect for it to have happened by accident, that there is someone greater than us all." Discusses California textbook situation: evolutionists tried to ban creation-science.

774 Hodge, Charles. *What Is Darwinism?* 1874. New York: Scribner, Armstrong. *Hodge:* famed Presbyterian preacher at Princeton Theol. Sem.; author of *Systematic Theology;* editor of *Princeton Review;* key figure in development of doctrine of biblical inerrancy. "A more absolutely incredible theory was never propounded for acceptance among men" (quoted in Zimmerman 1959). C. Hummel says that "Hodge provided documentation to show how the theory held that all the organs of plants and animals, as well as instincts and mental capacities, could be accounted for without reference to divine purpose and guidance." He considered the Design argument incontrovertible, and argued that either Darwin was wrong or that God did not exist. "Is development an intellectual process guided by God, or is it a blind process of unintelligible, unconscious force, which knows no end and adopts no means?" According to Marsden, Hodge concluded that biblical supernaturalism was "utterly incompatible with the naturalism that he saw as essential to Darwin's position." (In his *Syst. Theol.,* Hodge wrote: "If natural science be concerned with the facts and laws of nature, theology is concerned with the facts and the principles of the Bible." The proper task of science is to arrange and systemize the facts of nature; general truths then emerge. The theologian does the same with the facts of the Bible.)

775 Hodgman, Stephen Alexander. *Moses and the Philosophers, in Three Parts . . . The Whole Together Giving a View of the Universe, as Written by Moses, the Servant of God.* 1886 [1884]. Philadelphia: Ferguson Bros. Also 1881 edition, titled *Plain Facts in Plain Words.* Quoted in Cavanaugh 1983 as a strict Calvinist. Infidelity is now rife; Darwin, Huxley and Spencer teach we developed from inferior animals; this is taught to our sons at college. But Moses still lives; the Bible will spread—this is pre-ordained. "Moses wrote the true and philosophical account of the origin of things," and all the facts of science now confirm the truth of Mosaic science. The absurd fictions of false science will disappear.

776 ________. *Fallacies and Follies of Science, Falsely So-Called* (pamphlet). c1882. Philadelphia: J.T. Bryce.

777 ________. *The Miracle of Creation as Contrasted with the Theory of Evolution* (pamphlet). 1884(?). Philadelphia: J.T. Bryce.

778 ________. *A Discourse on the Days of Creation.* Undated. Philadelphia: J.T. Bryce. "They are not geological periods, but natural days of twenty four hours each."

779 Hoeksema, Homer C. *"In the Beginning God . . .".* 1966. Grand Rapids, Mich.: Reformed Free Publishing Association. *Hoeksema:* prof. at Theological School of Protestant Reformed Churches in America. Strict creation. Mostly theology—much discussion on Dutch Reformed doctrine and teachings; but includes chapter on "Genesis and Science."

780 Hoen, Reu E. *The Creator and His Workshop.* c1951. Mountain View, Calif.: Pacific Press. *Hoen:* Seventh-Day Adventist. Mostly biblical rather than scientific creationism. "Every detail of the creation reveals a plan and design in its structure." The creation of the universe by God is the second greatest truth of religion, after the existence of God; the third is redemption. "There is only one significant difference between idolatry as practiced by the heathen and the reasoning of evolutionists. The heathen make their idols of material substances . . ., while present-day philosophers select evolutionist theories, which are mere figments of the imagination. Both are worshipped with equal ardor and zeal . . ." Affirms creation *ex nihilo* and Flood Geology; describes the six literal creation days, Adam and Eve, Eden, Satan, the Sabbath, the Genesis dominion mandate. Quotes theologians (esp. **E.G. White**), some scientists (including Austin Clark). Marine deposits at high elevations attest to the Flood; fossils prove a worldwide catastrophe. Anti-biblical evolutionists constitute a "literal and tragic fulfillment of Christ's prophecy that the attitudes of Noah's day would be repeated." "Doubtless evolutionism is the most clearly defined antithesis of the Biblical doctrine on creation and redemption." Describes coming restoration of God's Kingdom on earth.

781 Hoggan, David L. *The Myth of the 'New History': The Techniques and Tactics of the New Mythologists of American History.* 1965. Nutley, N.J.: Craig Press. 1985 reprint edition pub. by Institute for Historical Review (Torrance, Calif.). *Hoggan:* history Ph.D.—Harvard; taught at MIT, Berkeley, Amerika Inst. (Munich). Orig. edition dedicated to **R. Rushdoony,** who Hoggan obviously admires greatly and cites frequently. Largely an iconoclastic de-mythologizing of world history analyzing all the wars America has fought in plus theoretical issues. By "New History" Hoggan means the present dominant historiographic tradition: anti-Christian humanist propaganda advocating the perfectability of man. Hoggan and other "revisionist" historians seek to oppose this biased view. He advocates Christian values and nationalism as defense against communism and the threat of internationalism. The Civil War was an avoidable disaster; blacks were not worse off under slavery—and the South was about to abolish it anyway. The U.S. was manipulated into intervening in WWI, and shouldn't have. Likewise for WWII. Hoggan, strongly anti-communist and pro-German, denies German guilt and claims U.S. treated the Germans cruelly after WWII. Darwin, Freud and Marx were all humanist-communist pseudo-scientists beloved by New Historians because they destroy faith in Christianity, American institutions, and morality. Evolution is an unproven, anti-Christian propaganda ploy. Bibliog., index. The Inst. for Historical Review is pro-Nazi revisionist group which denies systematic killing of Jews in WWII. IHR also published Hoggan's book *Der Erzwungene Krieg* ("The Imposed War"), and Hoggan has written for the IHR journal, stating in a 1985 article that he "fully accept[ed] the verdict" that "**Hitler** was a better leader than ever the German people or any other people had ever observed."

782 Holbrook, David L. *The Panorama of Creation: As Presented in Genesis Considered in Relation with the Autographic Record as Deciphered by Scientists.* c1908. Philadelphia: The Sunday School Times Co. Based on lectures given 1906. Holbrook says that Willis J. Beecher holds the same view expressed here. First chapter of Genesis is literature rather than science. It is not a narrative of origins; the beginnings it describes are of "appearance rather than essence." Its propositions, however, are factual rather than merely poetic, and there is a profound harmony between science and Genesis. Genesis deals with terrestrial matters, in a pictorial fashion. It portrays a panorama of creation in six divisions, like a series of paintings of geological landscapes. Genesis is "phenomenal": it presents a plain account of the visible progress of creation (after a general announcement of God's initial creation)—the preparation of the earth for man—as it would appear to an ordinary human observer. God's successive fiats and anthropomorphic actions in the creation "week" are "rhetorical" devices to give vividness to the account. Cites the Day-Age interpretations of **Guyot** and **Dana** as admirable scientific efforts; also **J.W. Dawson** and **Winchell.** Also cites the revelatory and literary interpretations of **Miller** and **Gladstone,** which seek to avoid chronological difficulties of strict Day-Age creationism. Holbrook argues that his phenomenal interpretation avoids these problems. By using the language of appearances, the Bible avoids dependence on particular scientific theories. Holbrook tends to agree with nebular hypothesis: earth originally molten; waters suspended above surface in vapor form. The land and seas have risen and fallen over long ages. Describes the concordance between the stages in Genesis and in the geological record. "Creation" of light: first penetration of sunlight through the dense clouds. Not till much later (Mesozoic) were the stars visible. Land vegetation created during the Paleozoic. "Sea-monsters" of fifth "day" are dinosaurs: marine and shore-dwellers. Sixth "day": land animals of the Cenozoic. Presents descriptions of both nebular hypothesis and planetesimal hypothesis side-by-side with parallel Genesis verses to demonstrate harmony. That the order of eight events in Genesis corresponds to the scientific order statistically demonstrates its truth. No other creation accounts preserve this correct order.

783 Hooper, E. Ralph. *Does Science Support Evolution?* c1947 [c1931]. Grand Rapids, Mich.: Zondervan. Orig. c1931 in pamphlet form by Defender Pub. (Toronto/Wichita, Kan.) (Defenders of the Christian Faith is **G. Winrod**'s organization.)

784 Hoover, Arlie J. *Fallacies of Evolution: The Case for Creation.* c1977. Grand Rapids, Mich.: Baker Book House. *Hoover:* Ph.D.—Univ. Texas; history prof. at Abilene Christian Univ.; was history and philos. prof. 1964–1977 at Pepperdine Univ. (also Church of Christ affil.); later acad. dean at Columbia Christian College.

785 _______. *The Case for Teaching Creation.* 1981. Joplin, Mo.: College Press (Know the Truth). Standard review of creation-science arguments. Urges that creation-science be taught in all schools; advocates two-model approach. "I feel that the present policy of teaching only evolution is wrong because the problem of origins is still an open question." Not allowing creationism constitutes rule by force, comparable to Nazi or Soviet policies. Argues that teaching creation is perfectly legal, and that evolution is religious. Strict creation; praises ICR and other creation-scientists. Scientific references. Hoover concedes he is advocating biblical Christianity, but insists that he is "primarily arguing a case for science..." Includes section on answers to objections by evolutionists regarding teaching of creation. Hoover explicitly bases his case on apocryphal quote by Darrow at Scopes Trial: "It is bigotry for public schools to teach only one theory of origins." (This very widely repeated quote was popularized by **W. Bird** 1978, and later **Geisler;** it apparently originated in a 1974 article by **J. Griggs** in *Science and Scripture.*) Accuses evolutionists of reductionist fallacy by allowing only naturalistic, mechanistic explanations in science. Includes creation-science book list.

786 Hooykaas, R[eijer]. *Natural Law and Divine Miracle: A Historical Critical Study of the Principle of Uniformity in Geology, Biology, and Theology.* 1963. Leiden, Holland: E.J. Brill. 2nd ed. *Hooykaas:* history of science prof. at Univ. Utrecht (The Netherlands); has been visiting prof. in Germany, Great Britain, Poland, Portugal, and U.S.

787 ________. *Religion and the Rise of Modern Science.* 1972. Grand Rapids, Mich.: William B. Eerdmans. A serious scholarly study by a legitimate historian of science. Hooykaas argues that modern science is largely a result of the Judeo-Christian tradition in Western thought; that the original structure of science may have been Greek but its impetus and spirit is biblical. The Bible "de-deifies" nature —removes divinity from nature, depersonalizes it and makes nature a creation of a God who exists apart from His creation —thus opening up nature to scientific investigation but avoiding a mechanistic world view. Stresses Protestant and especially Puritan contribution to science. "Science is more a consequence than a cause of a certain religious world view." This book is often cited by creationists as validation of Bible-science and creation-science, though Hooykaas stresses Creation rather than "creation-science."

788 Hopkins, Evan. *Cosmogony, or the Principles of Terrestrial Physics.* 1865. London: Longman, Green, Longman, Roberts, & Green. *Hopkins:* Fellow—Geol. Soc. "There are no 'demonstrable scientific facts' yet found that can legitimately upset the Mosaic Cosmogony." Refutes the theory that earth was originally molten, and shows how earth history can be explained in terms of scientific law without violating the "plain meaning" of Genesis: recent creation in six literal days. "If we only grant the geological 'speculators' a few millions of centuries, and a command over the agencies of nature, to be brought into operation when and how they please, they think they may form a world with every variety of rocks and vegetation, and even transform a worm into a man!" Hopkins explains that the semi-aqueous earth is constantly changing. The land drifts continuously towards the North Pole at the rate of twenty seconds of a degree per year (the apparent shift of the polar star—precession of the equinox); this accounts for fossils of tropical organisms found in northern latitudes. Referring to principles elucidated in his earlier book *On the Connexion of Geology and Terrestrial Magnetism,* Hopkins explains that ocean and magnetic currents spiral from the South to the North Pole: the South acts as a fountain, the North as a whirlpool. Hopkins dismisses other theories as mere speculations unsupported by direct observations, and declares that nothing but "absolute prejudice" could oppose his theory, because it is based on the "real facts of science," and explains so much. On the fifth day of creation, this earth movement was tremendously accelerated—organic remains from the Southern Hemisphere were transported to the Arctic in a day. Adam was created near the equator; the earth's land mass was then a single continent. Noah's Flood was another intensification and acceleration of the northward spiralling. "If the Mosaic record is a myth, how can we believe in the Gospel?"

789 Horigan, James E. *Chance or Design?* 1979. New York: Philosophical Library. *Horigan:* a "science-oriented lawyer"; lives in Denver; specializes in natural resources. Argument from apparent design in nature. Evidences of obvious intelligent design of the universe "far too bewildering to attribute to chance." The materialist view of creation by pure chance is "unsupportable, if not irrational." Even parapsychology is better explained in terms of the view Horigan proposes. Argues against J. Monod, J. Bronowski, G.G. Simpson; in favor of **Paley** and **J. Monsma** (whose 1958 book he admiringly compares to the Bridgewater Treatises). Evolution is guided by conscious behavior; cites A. Hardy. The Designer seems to be the Christian God: Horigan says "the biblical account of Genesis fits quite comfortably" with his interpretation. Bibliog., index.

790 Howard, Philip. *The Scriptural History of the Earth and of Mankind, Compared with the Cosmogonies, Chronologies, and Original Traditions of Ancient Nations.* 1797. London: R. Faulder. Subtitle: "An abstract and review of several modern systems; with an attempt to explain philosophically, the Mosaical account of the creation and deluge, and to deduce from this last event the causes of the actual structure of the earth..." Based on an earlier (1786) work pub. in French.

791 Howard, Warren R. *Jesus Christ and Modern Science: A Biologist's View of Christ, Creation, and Evolution.* 1967. Manchester, N.H.: Counselor Publications. *Howard:* assoc. ed. of *North Amer. Christian Magazine.* Admits that "evolution may have been the symptom of atheism rather than the cause," and that "belief in evolution does not automatically result in communism or fascism." Non-scientific presentation, though Howard has contributed to *CRSQ.*

792 Howe, George F. *Carbon-14 and Other Radioactive Dating Methods* (pamphlet). 1970. Caldwell, Idaho: Bible-Science Association. *Howe:* botany Ph.D. —Ohio State; formerly natural science prof. at Westmont College (Santa Barbara, Calif.), now at Los Angeles Baptist College; on *CRSQ* ed. board.

793 _______ (ed.). *Speak to the Earth: Creation Studies in Geoscience.* 1975. Nutley, N.J.: Presbyterian and Reformed. Selected articles from the *Creation Research Society Quarterly:* 1969–74. The Creation Res. Soc. was formed in 1963 (largely as a result of the influence of **Whitcomb** and **Morris**'s 1961 book) by ASA members dissatisfied with that group's increasing accommodation of evolution. Founding members were **Lammerts, Tinkle, E. Monsma, Rusch, Klotz,** J. Grebe, **R.L. Harris, Gish,** and **H. Morris.** The CRS now has some 700 members with advanced degrees in science (broadly defined). Though it now tries to sponsor some creation-science research, and published **Moore** and **Slusher**'s 1970 creation-science textbook and other books, its primary activity remains the publication of its *Quarterly:* a slick, impressively scientific-looking magazine which began in 1964 and is still the most prestigious creation-science journal. The CRS Statement of Belief, featured in every *CRSQ* issue, requires adherence to strict young-earth creationism, Flood Geology, belief in biblical inerrancy and a born-again commitment to Christ.

794 Howison, G.H. *The Limits of Evolution: and Other Essays Illustrating the Metaphysical Theory of Personal Idealism.* 1901. New York: Macmillan. *Howison:* philosophy prof. at Berkeley. Mind cannot have evolved from matter; minds "have no origin at all—no source in time whatever." "The World of Spirits ... can therefore neither be the product of evolution nor in any way subject to evolution...." If evolutionary philosophy is used to explain mind it is "destructive of the reality of the human person, and therefore of that entire world of moral good, of beauty, and of unqualified truth...."

795 Howitt, John R. *Karl Marx As an Evolutionist* (pamphlet). Undated [1964]. Hants, England: Evolution Protest Movement. EPM pamphlet #111. *Howitt:* Canadian psychiatrist and hospital superintendent. "The unholy alliance of Darwinism and Marxism has, therefore, undermined and destroyed the basic concept of a Christian society."

796 _______. *A Biblical Cosmology* (pamphlet). 1976. Toronto: International Christian Crusade. Howitt not credited in pamphlet; no author listed. Gap Theory creationism. Argues against evolution and also against young-earth theory. **Ussher**'s chronology defended as pertaining to events since the six-day 're-creation.' Cautions that the (7th) Millennium may not begin exactly in A.D. 2000, although it has been about 6,000 years since the first six. Succinct statement.

797 _______. *Evolution: "Science Falsely So-called"* (booklet). 1981. Toronto: International Christian Crusade. 20th ed. (Orig. date?) 205,000 copies by 1981 ed. Author Howitt not credited in booklet. 96 pages; 230 scientific references listed (evolutionist and "creation-science"). "A Handbook for Students." Requests students to show booklet to evolutionist teachers and challenge them to point out any scientific error. Standard creation-science arguments; full of quotes. **D. Gish** says this booklet converted him to creationism. Convenient source of creationist quotes and examples. Includes section on "fruits of evolution": jungle ethics, militarism, and atheism. Complains that the "dead hand of Darwinism" has severely retarded progress in biology. Also includes biblical references (such as title).

798 Howorth, Henry H. *The Mammoth and the Flood: An Attempt to Confront the Theory of Uniformity with the Facts of Recent Geology.* 1887. London: S. Low, Marston, Searle, & Rivington. Also photocopy ed. from **Corliss**'s Sourcebook Project (Glen Arm, Md.). Asserts the Bible is "absolutely valueless in geological discussion, and has no authority whatever," except as a collection of cosmological tales, myths and traditions. But argues that the Flood tradition, as reflected in legends from around the world, including the Bible, is evidence of a real catastrophe. Flood was widespread but not worldwide; emphasizes gap between the paleo- and the neolithic. Sudden severe climate change froze Siberian mammoths during Flood. Equates mammoths with biblical Behemoth. Describes Flood myths from different cultures. Despite Howorth's rejection of scientific authority of Bible, he is often cited by creationists for his catastrophic Flood interpretation.

799 _______. *The Glacial Nightmare and the Flood: A Second Appeal to Common Sense from the Extravagance of Some Recent Geology.* 1893. London: Sampson, Low, Marston. 2 vols. The glacial (Ice Age) theory is a "nightmare" because it is illogical and false. Argues for Flood. Earth's history is of "intermittent violence and repose." Unlike the speculative "religion" of Hutton and Lyell, Howorth says that what he proposes here is of "no school of thought . . . merely an inductive argument from the facts." Does not mention C. Darwin. Flood accounts for Pleistocene extinctions, including paleolithic man; there were also earlier catastrophes. Appeals to **Sedgwick.** Declares Bible to be myth.

800 Hoyle, Sir Fred. *The Intelligent Universe.* 1984. New York: Holt, Rinehart and Winston. *Hoyle:* astronomer and mathematician, mostly at Cambridge Univ.; head of Inst. Theor. Astron. at Cambridge until 1972; former pres. of Royal Astron. Soc.; former VP of Royal Soc. "A sickly pall now hangs over the big bang theory." (Hoyle was one of the originators and champions of the rival steady-state theory.) "The Darwinian theory of evolution is shown to be plainly wrong." Life was purposefully created by some cosmic "Intelligence"; it is far too complex to have originated by the random processes of evolution. "I am haunted by a conviction that the nihilistic philosophy which so-called educated opinion chose to adopt following the publication of the *Origin of Species* committed mankind to a course of automatic self destruction. A doomsday was then set ticking." "Personally, I have little doubt that scientific historians of the future will find it mysterious that a theory which could be seen unworkable came to be so widely believed."

801 Hoyle, Fred, and N. Chandra Wickramasinghe. *Lifecloud: The Origin of Life in the Universe.* 1978. New York: Harper and Row. Also pub. 1978 by J.M. Dent (London). *Wickramasinghe:* native of Sri Lanka; head of Dept. Applied Math. and Astronomy at Univ. College (Cardiff, Wales); Buddhist. Organic molecules present in outer space produce complex prebiotic products. Comets absorbed many of these, including "polysaccharides and related organic polymers" (cellulose). Earth's atmosphere and oceans were formed from material brought in by comets. "Our argument is that life arrived eventually on Earth by being showered already as living cells from comet-type bodies." The authors claim that infrared astronomy indicates that interstellar dust is largely composed of cellulose—a complex organic polysaccharide. Photos, charts, diagrams; index. (Hoye anticipated this thesis in his 1957 science fiction book *The Black Cloud,* about a living interstellar cloud.)

802 _______. *Diseases from Space.*

1979. New York: Harper and Row. In *Lifecloud,* the authors argued that although life arrived from space, it developed on earth by Darwinian evolution. In this book they argue that not merely prebiotic ingredients but viruses, bacteria, and even more complex organisms formed inside comets and were brought to earth by comets. Earth was—and is—being showered by living organisms. Each new arrival may stimulate evolution by providing new lifeforms: the influx of cometary genes is responsible for the increasing complexity of the paleontological record. Diseases, especially epidemics, result from organisms brought in by comets from space. Cancer was intended as a genetic program for yeast, but was inadvertently inserted into other organisms. The course of human history has been determined largely by these diseases from space. Maps, drawings, charts, diagrams; index.

803 ________. *Evolution from Space.* 1981. New York: Simon and Schuster. Also pub. 1981 by J.M. Dent (London). Here Hoyle and Wickramasinghe assert the impossibility of "chance" origin of life, and declare their belief in a Creator. "These conclusions dispose of Darwinism, which cannot produce genetic changes quickly.... The speculations of *The Origin of Species* turned out to be wrong as we have seen.... Nobody seems prepared to blow the whistle on Darwinian evolution. If Darwinism were not considered socially desirable and even essential to the peace of mind of the body politic, it would of course be otherwise." They also claim there are intelligences existing between this Creator and humans. The one responsible for our development was an "extremely complex silicon chip" which designed programs for bacteria, in order that they develop into humans who could then produce more computer chips. The probability that life could be formed by random naturalistic processes are no better than the possibility that a tornado blowing through a junkyard would assemble a Boeing 747. Wickramasinghe testified as a witness for creation-science in the 1981 Arkansas trial (and was scheduled to appear again in the Louisiana trial) because of his anti-evolutionism—though many felt that his testimony of "seeding" of genetic material and organisms by comets did little to support the creation-science case. (At the trial he declared that the earth must be very old—contrary to the creation-science bill he was testifying for—and also conceded he believed that insects might be smarter than people.) In 1982 Hoyle and Wickramasinghe wrote a booklet *Why Neo-Darwinism Does Not Work* (U.C. Cardiff Pr./Longwood). Hoyle has endorsed and popularized the accusation that *Archaeopteryx,* the celebrated half-reptile, half-bird fossil, is a fake. (This claim was earlier made by Jewish creationist Lee Spetner, and endorsed by **M. Trop.**) Hoyle and Wickramasinghe coauthored a 1987 book, *Archaeopteryx, the Primordial Bird: A Case of Fossil Forgery* (Longwood).

804 Hudson, Thomas Jay. *The Divine Pedigree of Man: Or, the Testimony of Evolution and Psychology to the Fatherhood of God.* 1899. Chicago: A.C. McClurg. A reconciliation. "...I have shown that every fact and every argument that sutains the theory of evolution also proves, with stronger reason, the divine origin of life and mind." Hudson also wrote *The Law of Psychic Phenomena,* and *A Scientific Demonstration of the Future Life.*

805 Hughes, Philip Edgcumbe. *Christianity and the Problem of Origins* (booklet). 1964. Philadelphia: Presbyterian and Reformed.

806 Hull, Marion McH. *Evolution: What It Is and What It Does* (pamphlet). *Hull:* M.Sc., M.D. Address delivered at Atlanta (Ga.) Bible School. Praises **G.M. Price** as "scientist of the highest repute." Discussed in 1926 *Science* article by E. Linton, who says that Hull is "unmindful of the dates of the authorities whom he quotes"; he quotes no scientist opposed to evolution more recent than **Agassiz** and **W. Dawson.**

807 Humberd, Russel I. *Are Evolutionists Intelligent??* (tract). Undated. Randleman, N.C.: Pilgrim Tract Society. As did Nebuchadnezzar, evolutionists crawl on the ground and look down at bugs instead of up at God. Rev. Humberd, who lives in Indiana, says he will provide more information on evolution on request.

808 ________. *Evolution* (booklet). Undated. Priv. pub. (Flora, Ind.). (Listed in Ehlert files. CREVO/IMS lists *Evolution and Creation.*)

809 Hunt, Dave, and T.A. McMahon. *The Seduction of Christianity: Spiritual Discernment in the Last Days.* 1985. Eugene, Ore.: Harvest House. A warning about the satanic temptations of the New Age Movement. We are in the midst of the apostasy prophesied in the Bible, and many are succumbing to Satanic deceptions on the eve of Christ's Second Coming. A new world religion is being established, which will result in the worship of Antichrist. The old lie of Satan—that we can "become as gods"—is now widespread. Hunt and McMahon warn about the dangers of New Age mind science in its myriad forms; various cults; and doctrines of mental, psychic and metaphysical evolution (instigated largely by **Teilhard**) which are opposed to true Christian creation. "Far from being scientific, evolution has been an integral part of occultism/mysticism for thouands of years, where it has always been understood to be the mechanism behind reincarnation." I.e., evolution (and "scientism" in general) remains primarily a means by which to spread false and dangerous New Age beliefs in reincarnation, mysticism, and occultism. This book is a Christian bestseller, though controversial for its hard-line criticism of many popular Christian leaders (especially advocates of the "Wealth and Prosperity Gospel," plus advocates of Dominion Theology, post-millennialists, and the Christian Reconstructionists), who Hunt and McMahon accuse of heretical New Age practices. They claim that "we are in the midst of an unprecedented revival of sorcery worldwide that is deeply affecting not only every level and sector of modern society, but the church as well." They argue that post-millennial eschatology, which rejects the doctrines of the Rapture and Armageddon, and claims that man will triumphantly establish God's Kingom here on earth before Christ returns, is a dangerous apostasy leading to adoption of satanic techniques in the attempt to gain wealth and power. Hunt, who has been featured by **Swaggart** on T.V. and in print, has since written a sequel, *Beyond Seduction.*

810 Hurlbut, Jeffrey A. *Evolution, Creation, Scriptures and My Two Cents Worth* (report). Undated. Unpub. MS. *Hurlbut:* Ph.D.—Univ. Calif. Santa Barbara; chemistry prof. at Metropolitan State College. Long unpublished(?) MS in ICR Library. Several sections on scientific evidences for creation and truth of Bible; evidences from archeology, history, etc.; truth of Bible prophecy.

811 Hurley, Morris E., III. *The Doctrine of Biological Evolution: A Critical Study of Contemporary Acceptance of a Theory* (report). 1968 [1965]. Unpub. MS. *Lieut. Hurley:* then stationed at Fort Lee AFS, Vir. Orig. a 1965 paper for Berkeley philosophy course under Paul Feyerabend (the radical philosopher of science who criticizes reliance upon scientific authority as dogmatic and oppressive). Preface dated 1968. Evolution as religion, myth, metaphysics and dogma. Well-written. Discusses E.L.G. Watson's criticisms (in 1961 *Sat. Eve. Post*) of the inadequacies of evolution. Includes statement that it is not to be cited as a published work. In ICR Library.

812 Huse, Scott M. *The Collapse of Evolution.* 1983. Salisbury, N.Y.: Pinecrest Publications. *Huse:* Th.D.; with Pinecrest Bible Training Center (Salisbury, N.Y.). Repeats the usual creation-science arguments. Borrows heavily from **Morris,** other ICR writers, **Wysong, Bowden,** etc. Crude and direct attack on evolution: "the very concept of organic evolution is completely absurd and impossible." Dedicated to Jesus Christ, "our Creator and Saviour." Scientific and biblical references. Cites criticism of evolution by Nobel science laureates D. Gabor and E. Chain. Presents many standard anti-evolution quotes. "The collapse of evolution, then, is already a reality in the minds and hearts of the well-informed. The ever-accumulating weight of scientific and Biblical fact has finally crushed the superficial fraud of organic evolution. ... Alleged intellectual difficulties with Bibical Christianity are usually nothing more than a smoke screen for moral rebellion against God and His Word. The erroneous concept of organic evolution is ephemeral, ... destined to pass into obscurity. Therefore, place your faith this very day in your Creator and Saviour..." Includes Appendix of all scientific facts supporting evolution: a blank page. Attractive drawings. Eight-page list of creationist organizations. Index. This book was recently excerpted in **J. Chick**'s *Battle Cry* newspaper.

813 Hutchinson, John. *Moses's Principia.* 1748. London: J. Hodges. *Hutchinson:* prof. at Cambridge Univ. Constructs a complete philosophical system of nature from the Bible. Assaults the Newtonian system as "atheistic"; argues mystically against Newton's gravity. Says the Deluge was caused by an "expansive force" of fires from the earth's center. Hutchinson apparently not concerned with evolution directly. Also wrote *Glory or Gravity* (London, 1749 [1738]); *Power Essential and Mechanical* (London, 1749); *The Religion of Satan, or Anti-Christ* (1736).

814 Hutton, James Laurence. *Acts: Deluge—The Other Cheek and the Dragon* (booklet). 1969. Priv. pub. (Victorville, Calif.). *Hutton:* mechanical engineer; head of Hutton Safety Hinge Co. Deluge was 4625 years ago. Cain forsook the way of simplicity, and invented things. Cain's evil race mixed with Seth's (the Sons of God), corrupting their blood. There were six times as many people and forty times as many animals before the Flood as now; earth then had better climate, better soil, and people lived a lot longer. "The Bible proves that when races mix, it is a perfect set-up for Satan." Includes anecdote about a UCLA student whose professor tells him that students aren't educated if they believe in God.

815 Hyma, Albert, and Mary Stanton. *Streams of Civilization: Ancient History to 1600 AD.* 1976. San Diego: Creation-Life/Milford, Ill.: Mott Media (co-pub.). Vol. I of creationist history textbook (Vol. II concerns history after 1600). Foreword by **H. Morris.** *Hyma:* former history prof. at Univ. Michigan. Intended for junior and senior high school. From the Foreword: "Certain sections, especially in chapter 1 [which deals with origins of earth, life and man, and the Flood], were written by the Director of the Institute for Creation Research [i.e. by Morris himself]."

816 Ikenberry, Larry D. *Noah's Ark: Mystery of Ararat.* 1976. Olympia, Wash.: Cascade Photographics. *Ikenberry:* professional photographer; member of 1973 ICR Ararat expedition.

817 Institute for Creation Research. *21 Scientists Who Believe in Creation* (booklet). 1977 [1973]. San Diego: Creation-Life. Booklet version of ICR "Impact" series #9 (1973). All twenty-one scientists listed are either members of ICR Technical Advisory Board or ICR/CHC staff. **T. Barnes; E. Blick;** D. Boylan (Ph.D.—Iowa State; prof. and dean of engin. at Iowa State); L. Butler (Ph.D.—UCLA; biochem. prof. at Purdue); **K. Cumming;** Malcolm Cutchins (Ph.D.—Virginia Tech.; aerospace engin. prof. at Auburn Univ.); Donald Hamann (Ph.D.—Virginia Tech; food technol. prof. at N. Carolina State); Charles W. Harrison (Ph.D.—Harvard; taught engineering at Harvard, Princeton; electromagn. research at Sandia Lab.); Harold R. Henry (Ph.D.—Columbia Univ.; prof. of civil and mining engin. at Univ. Alabama—Tuscaloosa); Joseph Henson (entomol. Ph.D.—Clemson; head of Science Div. at Bob Jones Univ.); John R. Meyer (Ph.D.—State Univ. Iowa; physiol. and biophysics prof. at Univ. Louisville Med. Sch.); **J.N. Moore; C. Ryrie** (a theologian); **J.C. Whitcomb** (theologian); **D. Gish; H. Morris; H. Slusher; G. Parker;** Wm. A. Beckman (chem. Ph.D.—Western Reserve; science prof. at CHC and at S.W. Jr. Coll. in San Diego); **R. Bliss; C. Wilson.** Gives biog. notes and quotes an anti-evolution passage from each.

818 _______. *Publications of Staff Scientists* (booklet). 1983. El Cajon, Calif.: Inst. for Creation Res. Lists all publications, secular and religious, by ICR scientists up to May 1983. Lists books, monographs and theses, subdividing these into scientific, creation-science, and apologetics & Bible categories; and articles, sub-divided into refereed scientific journals, misc. science journals, refereed creation-science, non-refereed creation-science, and misc. creationism & apologetics categories. Intended to counter the "widespread but erroneous charge" that creation-scientists are not bona fide scientists. Their list of publications "compare favorably" with those of secular scientists. Listings in refereed creation-science category mostly from *CRSQ* but also some legitimate science journals. M.S. and Ph.D. theses included. Useful bibliography; comprehensive.

819 _______. *ICR Summer Institute on Scientific Creationism: 1983.* Undated [1983]. El Cajon, Calif.; Institute for Creation Research. Large loose-leaf binder on course outlines, notes, and supplementary material for week-long ICR course on creation-science offered to the public and taught by ICR faculty. (Available for college and graduate school credit). Includes lectures by **H. Morris, D. Gish,** D. McQueen, **S. Austin, K. Cumming, G. Parker, R. Bliss, H. Slusher, L. Vardiman.** Many lecture sections very comprehensive and complete; others sketchy. Standard creation-science arguments as presented to sympathetic paying audience; many explicit biblical references. Some lectures videotaped for sale and rental from ICR.

820 _______. *ICR Summer Institute on Scientific Christian Evidences: 1984.* Undated [1984]. El Cajon, Calif.: Institute for Creation Research. Course outlines, notes, and supplementary material for 1984 ICR Summer Institute on creation-science. Includes lectures by **H. Morris, D. Gish, R. Bliss,** D. McQueen, **K. Cumming, H. Slusher, G. Parker, L. Vardiman, J. Morris, S. Austin, R. Niessen.**

821 _______. *ICR Summer Institute on Scientific Christian Evidences: 1985.* Undated [1985]. El Cajon, Calif.: Institute for Creation Research. Course outlines, notes, and supplementary material for 1985 ICR Summer Institute on creation-science. Includes lectures by **H. Morris, D. Gish, R. Bliss, G. Parker, K. Cumming,** D. McQueen, **S. Austin, J. Morris, R. Niessen, W. Bird.** (Bird's lecture notes on legal aspects of creationism are comprehensive and exhaustively footnoted.)

822 _______. *ICR 1982–83 Graduate School Catalog.* Undated [1982]. El Cajon, Calif.: Institute for Creation Research. ICR, which began its graduate program in 1981, offers M.S. degrees in astro/geophysics, biology, geology, science education, and general science. Catalog includes section on "ICR Educational Philosophy" (also reprinted in **H. Morris** 1984) presenting the explicitly creationist tenets the school is based upon: personal Creator-God, Bible as infallible revelation of scientific as well as moral truth, etc. Presents tenets of both "scientific" and "biblical creationism"—the former containing no overt religious references, the latter explicitly Christian fundamentalist. ICR Graduate School has been approved by Calif. State Dept. of Educ. to grant M.S. degrees. Catalog also pub. in 1984–85, and 1986–87 editions.

823 _______. *ICR Summer Workshop: Science Curriculum for the Christian School.* Undated [1985]. El Cajon, Calif.: Institute for Creation Research. Two loose-leaf binders of course outlines, class and lab exercises, and supplementary material, for ICR summer workshop led by **Richard Bliss.** How to teach science from the creationist perspective in Christian schools. Also contains much ICR promotional material: brochures, reprints, etc. For teachers grades K–12. Includes computer-assisted teaching program.

824 _______. *Institute for Creation Research Grand Canyon Field Study Course: Biology/Geology 537: April & June 1985.* Undated [1985]. El Cajon, Calif.: Institute for Creation Research. Guidebook and course material for ICR Grand Canyon trips, which can be taken for ICR graduate school credit. Includes geological maps, photos, biology and geology guides, reprints of various creation-science articles, a biblical creationist introduction and study guide, and supplementary materials. Most of the geological maps and descriptions are standard (non-creationist) presentations. Assembled and written by **S. Austin** of ICR. Several other ICR faculty members have participated in these Grand Canyon camping and hiking field courses also.

825 Institute for Creation Research (ed.?). *Collected Essays on Evolution/Creation* (reports). Undated. Collected papers; in ICR Library. Includes **R.L. Whitelaw** ("Creation or Evolution?: The Christian Faith Today"); also non-creationist Stephen B. Darby ("Creation Controversy in California"; a chapter from Princeton B.A. biology thesis).

826 Institute for Space-Time Studies. *The Einstein Myth and the Ives Papers:*

A Counter-Revolution. 1979. Old Greenwich, Conn.: Devin-Adair. "Production of the Institute for Space-Time Studies." Herbert Ives, a physicist at Bell Labs who developed the Snipascope and other inventions, rejected Einstein's relativity theory. (He is much cited by **H. Slusher** and other creationists.) On Einstein's relativity: "Clearly, this would lead to a complete moral relativism in the universe. ... In fact, I encounter several students in my classes every year who invoke Einstein's theories to justify their moral relativism."

827 International Bible Association. *The Missing Day (As Discovered by Astronauts)* (tract). Undated. Dallas: International Bible Association. IBA tract #6. Joshua's Long Day proved true by NASA space scientists' computer, which discovers a missing day at that date in history. This account here attributed to a Longview, Texas newspaper; credits **H. Hill** for the original story. (This tale, originated by Hill, is a space-age version of **Totten**'s math-astronomical proof that the sun did stand still for Joshua. It is very widely reprinted.)

828 International Christian Crusade *The Monkey Trial* (pamphlet). 1975. Toronto: International Christian Crusade. 3rd ed. Answers the Bible questions **W.J. Bryan** couldn't handle successfully at the 1925 Scopes "Monkey" Trial. Snakes used to walk—the evolutionists say so themselves. Sun was appointed light-bearer on the fourth day but was created on the first.

829 _______. *Is Evolution a Fact?* (pamphlet). 1976. Toronto: International Christian Crusade. 5th ed., rev. Standard creation-science arguments. "If evolution were true then man is only an animal and life is cheap. ... If we are just reconstructed monkeys or apes, and if survival is all that matters, then life would be futile and there would be no hope beyond the grave."

830 International Council on Biblical Inerrancy. *Summit II: Hermeneutics Papers.* 1982. Oakland, Calif.: International Council on Biblical Inerrancy. Papers from 1982 ICBI conference. Includes several papers advocating old-earth creationism; only one contributor defends recent creationism. Walter L. Bradley [**Thaxton**'s co-author]: "Trustworthiness of Scripture in Areas Relating to Natural Science"—urges old-earth creationism. **N. Geisler:** Appendix I: "Yom of Genesis I: Several Views"—outlines all possible variations of creationism [favors old-earth himself]. **G. Archer:** "A Response to the Trustworthiness of Scripture in Areas [etc.]"—also posits old-earth creationism. **H. Morris:** "A Response to the Trustworthiness [etc.]"—alone holds out for recent creation.

831 _______. *Inerrancy: Does It Matter?* (tract). Undated [1980s]. Oakland, Calif.: International Council on Biblical Inerrancy. ICBI was founded in 1977. ICBI Council includes **G. Archer, J. Boice, N. Geisler,** J. Grimstead, M. Rosen **[Jews for Jesus];** advisory board includes **H. Blocher,** W. Bright [Campus Crusade for Christ], **W. Criswell, D.J. Kennedy, F. Schaeffer,** and many other well-known evangelists and theologians. Tract includes excerpts from the "Chicago Statement on Biblical Inerrancy," a set of affirmations and denials, which was signed by 250 Christian leaders. Strongly creationist. "We ... deny that scientific hypotheses about earth history may properly be used to overturn the teaching of Scripture on creation and the flood." ICBI also issued a similar "Statement on Hermeneutics," also strongly and explicitly creationist. Both "Statements" are reprinted in full in the ICBI Catalog.

832 Iowa Campaigns for Christ. *Creation or Evolution?* (booklet). Undated. Des Moines, Iowa: Iowa Campaigns for Christ. Quotes **G. Kerkut** and **W.R. Thompson.** Iowa Campaigns for Christ pub. the weekly *Truth.*

833 Irwin, James B. *More Than an Ark on Ararat.* 1985. Nashville, Tenn.: Broadman Press. "Spiritual lessons learned while searching for Noah's Ark." "Written with" Monte Unger. *Irwin:* one of the Apollo astronauts who walked on the moon; now heads evangelistic High Flight Foundation (Colorado Springs, Colo.). Book chronicles Irwin's 1982 High Flight expedition searching for Noah's Ark on Mt. Ararat. Mostly photos; brief text. Does not include history or other tales of Ark-eology; only Irwin's journey. Includes photo of Irwin bloody and battered after near-fatal climbing accident.

Also photos of Turkish military presence in Ararat region.

834 Jackson, Wayne. *Fortify Your Faith . . . In an Age of Doubt* (booklet). c1974. Priv. pub. (Stockton, Calif.). *Jackson:* Church of Christ minister (Stockton, Calif.); co-leader (with **B. Thompson**) of Apologetics Press. Much of booklet consists of standard anti-evolution arguments. Introduction by **J. Bales.** Suffering is largely the result of collapse of the water canopy before the Flood. Also includes proofs of Bible prophecy, man's need for religion, historical proof of the New Testament, and the evil fruits of belief in evolution.

835 ________. *A Fly in the Ointment of Evolution.* 1976. World Evangelists. [Listed in SOR CREVO/IMS.]

836 ________. *The Mythology of Modern Geology: A Refutation of Evolution's Most Influential Argument* (booklet). 1980. Stockton, Calif.: Apologetics Press. Foreword by **B. Thompson.** Standard creation-science arguments. The "geologic time table is nothing more than a graphic conglomeration of assumptions, wistfully conceived and arbitrarily thrown together, in an attempt to prove the unprovable hypothesis of organic evolution." Scientific and biblical references; also cites creation-scientists. Missing and out-of-order strata; lack of transitional fossils; living "fossil" organisms; human fossils from all strata (includes Paluxy, Moab skeleton, trilobite-in-footprint, etc.); Flood Geology and recent creation. Scripture clearly states the earth was created recently in six literal days.

837 ________**, and Bert Thompson.** *Evolutionary Creationism: A Review of the Teachings of John Clayton* (booklet). c1979. Montgomery, Ala.: Apologetics Press. Jackson and Thompson refute the old-earth creationism of fellow Church of Christ member **John Clayton,** accusing him of compromising with evolution by advocating a mix of Gap Theory creationism, Day-Age creationism, and even theistic evolution. They attribute Clayton's false and heretical old-earth beliefs to his training in earth sciences rather than the Bible. Clayton, they claim, averaged $2,800 a month in 1978 from contributions for his anti-evolution presentations.

838 Jacobsen, Louis A. *The Evolution Myth: An Indictment of the [. . .]* (pamphlet). 1982. Evangel Missionary Fellowship. [Listed in SOR CREVO/IMS.]

839 Jacobson, Steven. *Mind Control in the United States.* c1985. Santa Rosa, Calif.: Critique Publishing. Critique is publisher of *Critique: A Journal of Conspiracies & Metaphysics,* some of whose contributors (Otto Scott, Anthony Sutton) are also affiliated with **Rushdoony** and **G. North** and publish in Christian Reconstructionist journals; it also features **J. Keel** and followers of **E.C. Prophet.** Jacobson's book argues that Darwin was wrong. Mankind is the result of mating of celestial beings – angels – with apes, as the Bible states. Sodom is allegory. Man can regain the angelic Force and transcend his flesh by using the subconscious mind. Evil mind control is real also – the satanic influence of rock music, especially backwards masking; subliminal perception techniques used by the media; government propaganda; hypnosis; the film *Reefer Madness* which was really made to promote marijuana use by means of hidden messages; the financial elite; the threat of world government; communism.

840 Jaki, Stanley L. *Science and Creation: From Eternal Cycles to an Oscillating Universe.* 1974. New York: Science History Publications. *Jaki:* Hungarian-born Benedictine monk; theologian and physics prof. at Seton Hall Univ. (S. Orange, N.J.); visiting prof. at Edinburgh Univ. and Oxford; respected historian and philosopher of science; winner of **LeComte du Noüy** prize; d. 1987. The fallacy of eternal cyclical or oscillating universe, as opposed to the Christian view of a created historicist universe. Marxism and Nazism have explicit ties to the dogma of cyclic recurrences. Equates belief in "oscillating worlds" with "wavering minds."

841 ________. *The Road of Science and the Ways to God.* 1978. Chicago: Univ. of Chicago Press. Orig. Gifford Lectures (natural theology) at Univ. Edinburgh, 1974–75 and 1975–76. We must follow both the Road and the Ways. Argues against materialists and those who deny Purpose – particularly Darwin, Spencer, Huxley, Monod. Darwin is vague, logically contradictory, and scientifically weak. Chapters 18 and 19 are strongly anti-Darwinist; Jaki vigorously promotes the necessity of teleology. People exhibit Purpose, so the universe and life must also have it. Dense but well-written. Index.

842 _______. *The Origin of Science and the Science of Its Origin*. 1979. South Bend, Ind.: Regnery/Gateway. Orig. 1977 Fremantle Lectures at Oxford. Serious history and philosophy of science. Christianity created the intellectual climate that allowed science to flourish. Modern science resulted from the Christian rejection of the mythologized worldview of the Greeks and others; Christianity de-personified nature. All science is cosmology. The Gospel and belief in God's creation offers the ultimate perspective for origins. Objects to K. Popper's rejection of "historicist" dogmas and revealed or Platonic absolutes, and to L. White's attribution of modern anti-ecological attitudes to the Christian coercive approach to nature. Argues against T. Kuhn, but absolves Einstein of accusation of moral relativism. Index. Jaki has also written a biography of P. Duhem, whom he praises in this book.

843 _______. *Cosmos and Creator*. 1980. Chicago: Regnery Gateway. Orig. pub. by Scottish Academic Press. Universe began billions of years ago, but its evolution has been quite constrained. This "startling specificity" proves its contingency and leads to a Christian theology. Includes chapter on extraterrestrial life, which Jaki feels is "unlikely," though fervently believed in by Darwinists. Agrees that the "true" message of Darwinism is rule of the strong and advocacy of a master race. "Darwinism is a belief in the meaninglessness of existence." Distinguishes Darwinism from the "conviction fully shared by this author in the factual though still very imperfectly understood instrumentality of a species in the rise of another" (i.e., Jaki accepts at least some degree of evolution). Strong criticism of the "almost mystical faith" of Darwinists.

844 James, Constantin. *On Darwinism, or the Man-Ape*. 1892 [1877]. Paris: E. Plon. Orig. French. *James:* French Catholic physician. Refutation of Darwin's *Descent of Man*. Uses scientific arguments, but also highly contemptuous: called *Descent* a fairly tale so fantastic and burlesque it must be a joke. According to A. White, Pope Pius IX thanked him for this book refuting Darwinist aberrations masquerading as true science, and made him a member of the Papal Order. The Archbishop of Paris urged James to write a revised ed. stressing the scientific truth of Genesis; James did so (1892 [1882?]). Title of rev. ed.: *Moses and Darwin: The Man of Genesis Compared with the Man-Ape, or Religious Education Opposed to Atheistic*.

845 Jamieson, Robert, A.R. Fausset, and David Brown. *Critical and Explanatory Commentary on the Old and New Testaments*. 1948 [1871]. Grand Rapids, Mich.: William B. Eerdmans. Orig. 1871 (England). Also pub. in one-vol. abridged ed. (Zondervan) and 3-vol. ed. *Jamieson:* Presbyterian minister in Glasgow. **Wonderly 1977** quotes and praises Jamieson's "Introduction to the Mosaic Account of Creation" in Vol. 1 for accepting the findings of geologists regarding the age of the earth and of life, rather than condemning science for being anti-religious. Jamieson accommodates the truths of science without, however, elevating them above Scripture. The Bible is concerned with religion; the province of science is "to deal with the facts drawn exclusively from the volume of nature: and these facts . . . will be found to prove the truth, and give strong confirmation to the statements contained in the Mosaic account of Creation." Facts discovered by science must always agree with the Bible, as God's Divine Word cannot be contradicted by true science. Jamieson cautions that some scientific interpretations may be wrong, though; he shuns the full uniformitarian view. We should beware of arraying certain immature scientific speculations against Scripture. The "*thoroughly* established principles of Geological Science" are "in perfect unison" with the Mosaic account. Gap Theory creationism: the original creation was in "remote and unknown antiquity." Then the earth was "convulsed and broken up . . . in confusion and emptiness" ("without form and void"). Out of this chaotic state, God created the present world in six days. This process is "described in the natural way an onlooker would have done, who beheld the changes that successively took place." Cites **Hitchcock.** Describes Noah's Ark, but advocates regional (not global) Flood. In Noah's time human habitation did not extend beyond the Near East; the Flood covered this region only. Agrees with **H. Miller** that this section of land subsided, allowing a rise of waters. (One-vol. ed. does not discuss extent of Flood, but states that there are only 300 species of birds and beasts.) This Commentary is still highly regarded by many fundamentalists.

846 Jansma, Sidney J. *UFO's and Evolution.* c1981. Priv. pub. Distributed by Master Books (ICR). Foreword by **H. Morris.** *Jansma:* president of Wolverine Oil (Michigan). UFOs are "diabolic manifestations, one and the same with demonic occult phenomena." They are really "ISAs": "*I*dentifiable as *S*atanic paranormal *A*pparitions by their violation of the laws of nature in speed and motion and to materialize and dematerialize at will. The coldness of UFOnauts, their sulfuric stench, and their lying also testify to Hellish origin." Satan originated the evil lie of evolution; he is also spreading the supernatural deceptions—false revelations mimicking God's powers—of psychic phenomena, ESP, astrology, spontaneous human combustion, Buddhism, Hinduism, Islam, yoga, karate, ouija boards, Uri Geller, Marian visions, the Kabala, etc.—all discussed by Jansma, who considers them very real manifestations of demonic power. All paranormal phenomena stem from the satanic cultist concept of continuous evolution; the ascent of man to godhood. Jansma denounces Ruth Carter Stapleton's charismatic healing as demonic. Book includes a chapter by **R. Niessen** (CHC) refuting theistic evolution, a section by **R. Whitelaw,** and chapter endorsements by **D. Gish** and **R. Bliss.** Jansma, who is vehemently anti-evolution (Gish dedicated his 1972 book to him), believes the most outrageous claims of paranormal activity, and recounts many of them here. Cites **Steiger, von Däniken, Berlitz,** Adamski, Vallee, and many other UFOlogists and psychics.

847 ________. *Six Days.* c1985. Priv. pub. (Grand Rapids, Mich.). Distr. by Master Books (ICR). Foreword by **H. Morris.** Chapter introduction by **Gish.** "Man did not evolve from animal ancestors, in spite of what illusions Satanic deceptions may have inspired." Insists on recent literal six-day creation and Flood Geology; castigates rival creationist theories as well as evolutionism: "Theistic Evolutionists, Progressive Creationists, Process Creationists and the like, deny the Genesis story of creation and of a world wide Flood, and do violence to the laws of nature." This book presents "scriptural evidence that Satan is the author of the notion that God used a naturalistic, mechanistic process of gradual evolution over a vast stretch of time as His method of 'creation,' and that the theory of evolution is scientifically unsound as well." Includes standard creation-science arguments for fossil record, criticism of old-earth dating methods and geological uniformitarianism. Warns against evils of genetic engineering, and protests against "delusions" of "godless scientists" such as **Jastrow**'s claims regarding extraterrestrial life. "God Says Creation—Satan Says Evolution... I do not see evolution simply as the beastly actions and language of Satan but as his lying and blasphemy as spoken through the possessed." Man cannot be "an animal made from slime, mud, and muck!"

848 Jastrow, Robert. *God and the Astronomers.* 1978. New York: Warner Books. Orig. pub. by W.W. Norton (New York); based on 1978 AAAS lecture. *Jastrow:* physicist and astronomer; director of NASA's Goddard Institute for Space Studies. Big Bang cosmogony, expansion of the universe, and thermodynamics prove the universe began at a definite moment 20 billion years ago. Beyond this moment of creation, we cannot know. "For the scientist who has lived by his faith in the power of reason, the story ends like a bad dream. He has scaled the mountains of ignorance; he is about to conquer the highest peak; as he pulls himself over the final rock, he is greeted by a band of theologians who have been sitting there for centuries." Two Afterwords: by John O'Keefe of NASA (Catholic), who discusses the anthropic principle—creation of universe for man to inhabit; and by Steven Katz (Dartmouth religion dept.; Jewish), who says that evolution by chance is true, but that the whole universe shows cause and purpose. Photos; index.

849 ________. *The Enchanted Loom: Mind in the Universe.* c1981. New York: Simon and Schuster. Third vol. in trilogy consisting of *Red Giants and White Dwarfs: Man's Descent from the Stars* (1967), and *Until the Sun Dies.* In the first, which concerns stellar evolution and the origin of life, Jastrow presents standard evolutionary views. In this book he strongly affirms biological evolution ("The fact of evolution is not in doubt"), beautifully illustrating and describing many transitional and ancestral forms. But he argues that scientific proof of the universe's Beginning (the Big Bang) suggests divine Creation, and states that "Scientists have no proof that life was not the result of an act of creation..." In chapter "A Guiding Hand," Jastrow suggests that naturalistic, random Darwinian evolution may be insufficient to account for organs such as the eye and the human brain, which seem to demand a Designer. "My own views on this question remain agnostic, and close to those of Darwin." Evolutionary theory seems to account for life adequately without invoking any supernatural forces, yet the history of life, with its "clear direction" leading to man, makes him wonder if there is something else. Science can tell us nothing about the "larger questions of plan and purpose" in the Universe. Jastrow also states that organic evolution is virtually complete, while the evolution of artificial intelligence is just beginning. He predicts that computers will soon become "an emergent form of life, competitive with man," and that we will transfer our brains into computers, as extraterrestrial life-forms probably already have. Among sources relied on are Jerison (brain evolution), Monod and his critics (chance evolution vs. plan and purpose), Popper and Eccles, and **MacKay** (mind and brain). Photos, drawings; index. In a chapter in **Varghese** 1984, Jastrow says that naturalistic origin of life is "plausible," and naturalistic evolution of man from primitive organisms is "fairly convincing"—but not certain.

850 Jauncey, James H. *Science Returns to God.* 1961. Grand Rapids, Mich.: Zondervan. *Jauncey:* claims ten academic degrees, including Ph.D.s in math and religion; former principal of Kenmore Christian College in Australia. Foreword by **J.N. Moore** (whose direct held Jauncey acknowledges). Compatibility of science and Christian belief. Jesus affirmed divine accuracy of Genesis. The scientific truth in the Bible is mostly in "embryonic" form—simplified non-technical language. Eventually God will disclose how all biblical miracles harmonize with natural law. Discusses literal vs. figurative intepretation; says we should not employ figurative merely to avoid apparent scientific difficulties (Adam and Eve, the Flood). Urges reconsideration of **Gosse**'s theory. Includes chapter on "The Origin of Man"—states that "the evolutionary theory is at a dead end." Evolution is an unproved theory based on philosophical presuppositions. "Nature is crowded with features which cannot be explained in terms of usefulness in the survival struggle." Argues that lack of consistent "progression" in the fossil record refutes evolution; also lack of surviving transitional hominid forms and presence of modern man in ancient strata. Cites **Adler, Fleming, Rendle-Short, Velikovsky,** Millikan, other scientists. In chapter on biblical eschatology, says that prophesied destruction of the world refers to atomic power. Jauncey claims that even if evolution were proved true this would not weaken the evidence for God. Creation of life in the lab would merely prove the necessity of the Creator. Bibliog.

851 Javor, George T. *Once Upon a Molecule* (booklet). 1979. Nashville, Tenn.: Southern Publishing Association. *Javor:* Ph.D.—Columbia Univ.; teaches biochemistry at Andrews Univ. (Berrien Springs, Mich.; Seventh-day Adventist—affil. with Loma Linda Univ.). Chemical arguments against evolution and the origin of life. Distributed free by Seventh-day Adventist Information Ministry (Berrien Springs).

852 Jeans, Sir James H. *Eos, or the Wider Aspects of Cosmology.* 1928. London. *Jeans:* D.Sc., Sc.D., F.R.S.; theoretical physicist, mathematician and astronomer; British; taught at Princeton Univ.; demonstrated untenability of original nebular hypothesis of planetary formation; anticipated continuous creation theory later advocated by **F. Hoyle** and others. "Everything points with overwhelming force to a definite event, or a series of events, of creation at some time or other, not infinitely remote. The universe cannot have originated by chance out of its present ingredients, and neither can it have always been the same as now." Compares universe to wound-up clock. Often cited by creationists as refuting naturalistic explanations in cosmology and cosmogony.

853 ________. *The Mysterious Universe.* 1930. New York: Macmillan. Argues against mechanistic interpretation of universe as purposeless aggregation of blind random forces. Rather, "the universe begins to look more like a great thought than like a great machine. Mind no longer appears as an accidental intruder into the realm of matter; we are beginning to suspect that we ought rather to hail it as the creator and governor of the realm of matter... We discover that the universe shews evidence of a designing or controlling power that has something in common with our own individual minds..." "If the universe is a universe of thought, then its creation must have been an act of thought." The Universal Mind is a Supreme Mathematician. Diagrams; index.

854 Jenkins, John E., and George Mulfinger. *Basic Science for Christian Schools.* 1983. Greenville, S.C.: Bob Jones Univ. Press. Science for Christian Schools textbook series; this book for grade 9. Series is "written from an entirely Christian perspective" stressing scientific infallibility of the Bible. "All distortions of evolution and humanism have been eliminated..." Teacher's edition and lab manuals also available.

855 Jennings, Abraham Gould. *The Mosaic Record of the Creation Explained, Scripture Truth Verified* (booklet). 1893. New York: Fleming H. Revell.

856 ________. *The Earth and the World: How Formed?* c1900. New York: Fleming H. Revell.

857 Jensen, Karen G. *Subject Index to Creation Research Society Quarterly 1964–1976* (report). 1977. Unpub. MS. Cumulative index for vols. 1-14 of *CRSQ.* Topic headings include Animal Design; Ararat, Ark and Flood; Archaeology; Astronomy; Bible Exposition; Climate Change (Pre-Flood, Ice-Age); Cosmology, Cosmogony and Extraterrestrial Life; Biochemical Evolution; Biology; Botany and Paleobotany; Biblical Creation and Science; Creationism—History; Darwinism; Dating Methods; Earth Science; Fossil Record; Human Body—Design; Human Origins; Microbiology; Molecular Biology; Philosophy of Science; Physics; Solar System; Taxonomy/Phylogeny; Teaching Creation/Evolution. (SOR CREVO/IMS also has *CRSQ* cumulative index through 1984—*CRSQ.RPT,* listing author, title and date. *CRSQ* also indexes each yearly volume in last volume issue.)

858 Jepson, J.W. *The Social Consequences of Evolution* (pamphlet). 1978. England: Evolution Protest Movement. EPM pamphlet #218. Edited by **C.E.A. Turner.** Orig. in *The Evangelical Beacon;* later in *The Prophetic Witness* (1977). *Jepson:* M.Sc. Major inconsistency in our culture between biblical principles affirming human value and materialistic tenet that "Man is just an evolved animal, a machine." Society must choose one or the other. Presents ominous picture of future society if evolution is chosen. According to evolutionist tenet, there is no such thing as crime, nothing deserving of punishment. Man has freedom and intrinsic worth "only because the Biblical accounts of Creation and origin of man are true!"

859 Jesus People USA. *Monkey's Uncle* (tract). c1979. Chicago: Jesus People USA. Standard creation-science arguments, naively presented. Presents usual anti-evolution examples; urges special creation. Another Jesus People tract explains that UFOs are caused by Satan's fallen angels confined to atmospheric heavens by God.

860 Jews for Jesus. *Evolution* (tract). Undated [1980s]. San Francisco: Jews for Jesus. Jews for Jesus is headed by Moishe Rosen. Evolution says we are getting better—but we're not. It's really "devilution."

861 Jochmans, Joey R. *Strange Relics*

from the Depths of the Earth (booklet). 1979. Lincoln, Neb.: Forgotten Ages Research Society. Reprinted 1981 by Bible-Science Assoc. Anomalies. Includes Paluxy manprints and other creation-science evidences which refute orthodox geology, paleontology and anthropology. Jochmans also publishes the journal *Ooparchist* ("Out-of-Place Archeology and History").

862 Johannsen, E. *Nest-Making Inexplicable by Evolution* (pamphlet). Undated. England: Evolution Protest Movement. EPM pamphlet #43.

863 Johnson, J. Wallace G. *The Case Against Evolution* (booklet). 1976. Louisville, Ky.: Catholic Center for Creation Research. Rev. ed. distributed by Keep the Faith. *Johnson:* Australian Catholic; his books have Nihil Obstat and Imprimatur, signifying they are free of doctrinal error.

864 ________. *The Crumbling Theory of Evolution.* 1982. Priv. pub. (Brisbane, Australia). Inspired by **P. O'Connell** 1959. Orig. lectures, printed by **P. Haigh,** then reworked into present form. Strict creationism; presents all the standard creation-science arguments. Argues at length against **Teilhard.** Appendices on Church versus Teilhard; on Church's position supporting creationism (cites Papal encyclicals; other Catholic documents). Infiltration of Church by Marxists. "Demolishing the theory of evolution is the essential first step in manning the barricades against the modern anti-God assaults." "Definitive remedy" from heaven given at Fatima in 1917 (Marian apparition). Many scientific references; quotes many creationists extensively. Argues against continental drift. Reports on 1980 Chicago macro-evolution conference as "demise of Darwinism." Lists creation-science resources. Photos, drawings, charts.

865 ________. *Miracles for Moderns* (audiotapes). North Haledon, N.J.: Keep the Faith. Two audio-cassettes. Evolution results in materialism and the rejection of supernaturalism. Lourdes, Shroud of Turin, Fatima and Virgin of Guadalupe. Keep the Faith also distributes another anti-evolution tape set by Johnson, *A Wolf in Sheep's Clothing.*

866 Johnson, Paul S.L. *Creation.* 1938. Philadelphia: Paul S.L. Johnson. Epiphany Studies in the Scriptures: Vol. II. Jehovah's Witnesses or related Russellite sect. A 585-page tome in which the author explains Everything. God set matter (gases) in motion ages ago; the six creation days occurred much later. Denies Gap Theory, however—says the earth was created gradually, and Satan did not recruit fallen angels til just before the Flood; there was no sin before Adam. Each creation "day" is 7,000 years. Employs **Vail**'s Ring/Canopy theory: early earth was molten, and surrounded by seven great rings or canopy layers arranged by density. These eventually all collapsed, one at the end of each creation "day," the lowest (densest) layers first, thus forming all geological deposits. The six great layers which comprise the Grand Canyon are "irrefutable and factual proof" of this. The seventh canopy layer, composed of water, caused Noah's Flood when it collapsed. Light was visible after collapse of first canopy; the sun and moon during collapse of third. Discusses classification of angels and demons, space ether, planet Vulcan near sun; states that heaven is Alcyone (466 light-years away). Satan ruled earth until 1799, when End Times began. Spread of God's Truth began in 1874—the Millenium; the Tribulation will follow. Concludes with chapter refuting evolution; cites and quotes many scientists. "Evolution is built on the most extravagant *guesses*..." Mendelism and biometry disprove evolution. Includes rate of population increase argument, shrinking sun, degeneration of man in history, and many other standard arguments. "Degeneration is the trend, not evolution." "The earth is not old enough" for evolution. "The distribution of plants and certain animal life disproves evolution." "Against evolution the fact tells that some insects have more knowledge and practical ability than apes." "If evolution were true..., why have not all lower species passed out of existence through evolving into higher species until by this time there would be but one species—man...?" "Everything in the earth shows design, and that for the most part in anticipation of man's arrival on earth..." Johnson's book is now distributed by **Laymen's Home Mission-**

ary, which has reproduced many of his arguments in their tracts.

867 Jones, J. Cynddylan. *Primeval Revelation: Studies in Genesis I–VIII.* 1897. New York: American Tract Society. Evolution is not proved, but a "super-naturalistic" version of evolution is OK for Christians. "Man may, if the phrase be allowed, be a supernatural development from a prior animal; he cannot be the product of natural evolution." This "modified" evolution allows for "immediate Divine intervention at certain critical moments in the creation and development of life..." But these interventions are few and far between; evolution may proceed for millions of years between interventions. Jones advocated regional Flood, but warned against attempts to explain the Flood and Noah's Ark in too naturalistic a fashion. (Quoted in **Ramm** 1954; **Whitcomb** and Morris.)

868 Jones, Orson P., Jr. *Science and Christian: A Review of Scientific Creationism* (booklet). 1986. Priv. pub. (El Cajon, Calif.). Privately-published critique of young-earth creationism and Flood Geology—especially of ICR's strict creationism. (Jones lives in El Cajon, home of ICR.) Many quotes from **H. Morris**'s *Scientific Creationism.* Jones says that universe and life were created by God, but that Genesis does not tell us when, how, or why. "Creation science" is based on mistaken English translations of the Bible. Jones seems to favor a modified Day-Age creationist interpretation: Genesis "days" are not in strict chronological order and vary in duration. Affirms fundamentalism and creationism, but says God may have used some evolution.

869 Kachur, Victor. *The World That Was* (booklet). 1983. Priv. pub. (Dublin, Ohio). "Story of the world destroyed by a deluge, based upon the information preserved in the ancient sources." The story of Noah and the Flood in Old Slavonic, from the Eastern European *Book of Vles.* Vles is Japheth, one of Noah's sons. Kachur has also published *The Book of Vles* ("Vles Kinyha"), his own translation, and *Vlessiana* newsletter. He believes the Ark is still on Ararat, inscribed with the names of the craftsmen who built it. Elephants helped with the construction, but Noah left the mammoths behind to freeze in the Flood. He also left the 600-pound beavers behind, who had helped control the rivers in the relatively rainless pre-Flood climate. Distributed by Donald Cyr, ed., *Stonehenge Viewpoint,* a journal inspired by **Vail**'s Canopy Theory which interprets myths, artifacts and megaliths as reflections of the ancient (pre-Flood) atmospheric conditions and cosmic catastrophes.

870 Kang, C.H., and Ethel R. Nelson. *The Discovery of Genesis: How the Truths of Genesis Were Found Hidden in the Chinese Language.* 1979. St. Louis, Mo.: Concordia. *Kang:* former missionary to China; now lives in Singapore. This book is based upon his 1950 book *Genesis and the Chinese* (listed under **Khang**). Foreword by **P. Zimmerman.** Analysis of Chinese characters demonstrates that they contain and depict the Genesis account of Creation and Noah's Flood, and that the ancient Chinese were monotheistic. Equates ShangTi ("heavenly ruler") with Creator-God. Many Chinese characters illustrated and analyzed. Authors select characters from ancient oracular inscriptions, later seal inscriptions, and more modern forms. Various radicals are interpreted as depicting the Trinity, God creating man out of the earth, Garden of Eden, Noah's Ark, creation of Eve out of Adam, etc. Bibliog. (C. Chui, an active young-earth creationist who grew up in mainland China, conceded to me that many of the interpretations were doubtful, but emphasized there were enough valid ones left to be highly significant statistically.)

871 Kappeler, Max. *The Seven Days of Creation.* 1951. Kappeler Institute. Compendium for the Study of Christian Science: No. 2.

872 Katter, Rueben Luther. *The History of Creation and Origin of the Species: A Scientific Theological Viewpoint.* 1984 [1967]. Minneapolis: Theotes Logos Research. 3rd ed., rev. Orig. 1967. *Katter:* businessman and religious college administrator; claims to be a specialist in various pseudo-science and theological fields. Theotes Logos Research apparently Katter's one-man group. "An exhaustive treatise reconciling the theological and scientific viewpoints fo the CREATION of the Universe, and its forms of life." 450-page book; densely and obscurely written, intricate and bizarre. Explicitly Christian (though quite idiosyncratic); claims to be "interdenominational." Reveals God's colossal plan for the future and explains how entire history of the world and of life is part of divine conception. Variant of Gap Theory creationism, plus some Day-Age creationism. Katter accepts standard geological ages but interprets them as God's carefully prepared stages. Earth was created twenty billion years ago. Earth then under Lucifer's management, who turned evil. Ice Ages began about 20,000 B.C., then God precipitated worldwide Catastrophe by shifting earth's axis. (II Peter refers to this event – not to Flood.) God re-created world 6–8,000 B.C. as told in Genesis. Adam created Oct. 23, 4004 B.C. Flood: on Halloween, 2348 B.C. Discusses dispensational scheme of history as exhibited and prophesied in Great Pyramid; also other Bible prophecy and numerology. Also explains twelve vast energy systems and dimensional levels of the cosmos, and proposes a new atomic force. Charts, diagrams; index.

873 ________. *Creationism: The Scientific Evidence of Creator Plan and Purpose.* 1979. Minneapolis: Theotes-Logos Research.

874 Kautz, Darrel. *Man's History from Adam to Abraham* (booklet). c1979. Priv. pub. (Milwaukee, Wisc.). The Contemporary Bible-Study Guides (OT) series: part 1 of 15. For Christian schools, Bible classes, survey and teacher-training courses, etc. *Kautz:* M.A. – Lutheran School of Theology. Standard creation-science arguments; cites creationist works; scientific and biblical references. Discusses the Fall; Cainites and Sethites (Daughters of Men; Sons of God); the advanced pre-Flood civilization of two billion people; **Navarra**'s Ark discovery; other Ark-eology; descent of human races from Noah's sons; Flood Geology; trilobite-in-shoeprint. Inc. study question, exercises. Photos; bibliog.

875 Kearley, F. Furman. *The Relation of Evolution to Modern Behavior Problems* (pamphlet). 1974. Lubbock, Texas: World Mission Publishing. *Kearley:* former director of graduate Bible studies at Abilene Christian Univ. (Texas); editor of *Gospel Advocate;* Church of Christ. Olney (in #1339) says this is "in-depth essay of the basis of evolution in Nazism, communism, human eugenics and experimentation, crime, family decline, etc."

876 ________. *The Effect of Evolution on Modern Behavior* (booklet). Undated. Montgomery, Ala.: Apologetics Press. (May be same text as Kearley 1974.) Subjects evolution to "fruit test": examine the fruit it bears. Evolution promotes selfishness, and is the basis for militarism, Nazism, Fascism, racism, communism, religious liberalism, eugenics, etc. Discusses C. Darrow's deterministic view of crime. Reprints early Soviet "decree" detailing forced collectivization of women for sexual purposes (but omits bibliog. reference for cited source).

877 Keel, John A. *Our Haunted Planet.* 1971. London: Neville Spearman. Excerpt reprinted in 1986 *Critique* conspiracy journal: "Origins: Creation, Evolution and Extraterrestrial." Proposes combination of all three theories: religious creation, evolution, and descent from extraterrestrials. Evolution is a terribly weak conjecture – the worst of the three. Modern man co-existed with primitive types, and appeared suddenly. Nonphysical beings, who may have been fallen angels, lived on earth with animals and primitive hominids. Adam was an androgynous being linked with supermind of universe. Endorses **Le Poer Trench**'s theory that other "ultraterrestrials" wanted earth, so battle raged between original spirit inhabitants and newcomers. They created physical armies by descending and materializing into physical form, using Neanderthals, etc., for bodies. Different human races arise from utilization by different ET overlords for materialization. Genesis is true but symbolic. Eventually the ultraterrestrials became trapped in these bodies, and became the first real people. Keel also wrote *Why UFOs,* and *UFOs: Operation Trojan Horse.*

878 Keiser, B.E. *Can the Scientist of Today Believe Genesis 1?* (pamphlet). 1962. Priv. pub. (Trenton, N.J.). *Keiser:* engineer. **J.N. Moore** (in 1965 *CRSQ*) quotes Keiser as concerned with conflict "which appears to exist between the Bible and modern science relative to the age of the earth and the origin of life," and says he also discusses "changing character of limited findings of men of science."

879 Keith, Alexander. *Demonstration of the Truth of the Christian Religion.* 1838. Edinburgh: W. Whyte. Attempted to make geological eras literally equal days.

880 Keith, Bill. *Scopes II: The Great Debate.* c1982. Shreveport, La.: Huntington House. *Keith:* former Louisiana state senator (recently defeated); previously journalist and city editor of Shreveport newspaper; now president of Creation Science Legal Defense Fund. Keith wrote the Louisiana creation-science bill struck down in 1987 by the U.S. Supreme Court. Book is primarily aggressive defense of the Louisiana creation-science case. Vehemently anti-evolution: Darwinian evolution is the "greatest hoax of the 20th century." It is religion not science; voodoo logic; a fairy tale like the Easter Bunny, Santa, and the Tooth Fairy. Includes usual creation-science arguments. Many quotes of creationist and anti-Darwinian scientists. Descriptions of all creationist witnesses; including E. Boudreaux (chemist–Univ. New Orleans), C. Harlow (engineer–Louisiana State), and lengthy quotes from **L. Sunderland**'s presentations. Reprints many editorials, newspaper and journal coverage. Blasts ACLU and humanists; says their master plan is to destroy our society and promote communism. Declares that abortion is worse than Nazi holocaust. Keith objects to being labelled believer in 6,000 year old earth–he is an old-earth creationist. Also long section on Arkansas trial: attacks creation-science defense team as inept and ill-prepared and chastizes them for not accepting help from **CSLDF, W. Bird, J. Whitehead,** and **D. Gish.** Concludes with advice on how to influence legislative and educational process; urges creationist lobbyists to stress scientific evidences and avoid discussion of religious implications of creationism. "Creation-science is pure science and it belongs in the public school classrooms." Keith also wrote Foreword to J. Aranza's 1983 book *Backward Masking Unmasked: Backward Satanic Messages of Rock and Roll Exposed.*

881 Kellogg, Howard W. *The Canopied Earth.* Undated [1936]. Los Angeles: American Prophetic League. Early advocate of pre-Flood Water Canopy theory. **Ramm** 1954 says Kellogg apparently follows **I. Vail** closely.

882 ________. *The Coming Kingdom and the Re-Canopied Earth.* c1936. Los Angeles: American Prophetic League. The pre-Flood Canopy Theory. God will re-erect the canopy at the Millennium, and the Edenic, pre-Flood conditions will once again prevail.

883 Kennedy, D. James. *New Evidences for Creation* (pamphlet). Undated [1977]. Fort Lauderdale, Fla.: Coral Ridge Ministries. Text of sermon preached 5-29-77. *Kennedy:* M.Div.–Columbia Theol. Sem.; M.Th.–Chicago Grad. Sch. Theol.; D.D.–Trinity Evangelical Coll.; Ph.D.–NYU; pastor Coral Ridge Presbyterian Church (Ft. Lauderdale, Fla.); televangelist broadcasting in 50 states and 22 countries. Many scientists quoted; expanding universe, divergent planetary rotations, complexity of cells, left-handed amino acids. Old and new quotes mixed without dates; creationists quoted alongside scientists. Implies A. Keith advocated evolutionary morality. "I say that evolution is based upon no evidence or proof at all!" It is "the big lie . . . the most destructive, pernicious lie that has ever come down the pike. . . . It has already resulted in the deaths of more people than have been killed in all of the wars in the history of mankind." Coral Ridge Ministries distributes free transcripts or audiotape copies of Kennedy's sermons from his weekly telecast. Kennedy has also urged support for Judge Bork and the Shroud of Turin.

884 ________. *Why I Believe.* c1980. Waco, Texas: Word Books. Includes chapter "Why I Believe in Creation": many scientific and creationist references. Usual creation-science arguments presented in glib, smooth style. Scientists believe in evolution only in order to deny God. Evolution is the basis of every false and evil doctrine around. Quotes **Gish, Clark** and **Bales, Enoch, Coppedge, Wakely, Fleming.** There are "only two religions. . . . One . . . is Christianity; the other religion is evolution. Anyone who does not realize that evolution is a religion does not know much about evolution." Another chapter on archeological confirmation of Bible; accuracy of Bible prophecy; also design argument. Other chapters doctrinal.

885 _______. *The Collapse of Evolution* (pamphlet). Undated [1981]. Fort Lauderdale, Fla.: Coral Ridge Ministries. Text of sermon preached 9-27-81 at Kennedy's Coral Ridge Presbyterian Church. Quotes many scientists. Standard creation-science arguments: discusses lack of transitional forms, "hopeful monsters," punctuated equilibria. Many have rejected the Bible, assuming it has been "disproved on its very first page by the theory of evolution. This pernicious theory which has supported atheistic secular humanism is nothing other than communism waiting to receive its political rights and powers."

886 _______. *The Crumbling of Evolution* (pamphlet). Undated [1983]. Fort Lauderdale, Fla.: Coral Ridge Ministries. Text of sermon preached 1-23-83 at Kennedy's Coral Ridge Presbyterian Church. Also reprinted in **H. Duncan** 1979. Quotes many scientists. Big Bang cosmogony, non-reducing early atmosphere, complexity of living cell, lack of transitional fossil forms: all prove creation. Evolution is the "pseudo-scientific foundation" of "every single anti- Christian" system—especially Nazism, communism and secular humanism. Yet "The whole of evolution is in absolute chaos today and the public does not know it. Students are still being taught the same old lies." Anti-evolution facts are "repressed."

887 _______. *Evolution's Bloopers and Blunders* (pamphlet). Undated [1986]. Fort Lauderdale, Fla.: Coral Ridge Ministries. Text of sermon preached August 1986 at Kennedy's Coral Ridge Presbyterian Church. Haeckel was a liar and cheat; Piltdown hoax. *Archaeopteryx:* "another case of deliberate fraud and deceit"—cites **F. Hoyle**'s accusations. Orgueil meteorite hoax (contained organic material). At Scopes Trial G. Osborn claimed a "whole race of men" lived a million years ago in Nebraska—based on pig's tooth [actually this was never mentioned at trial]. Zuckerman and Oxnard prove australopithecines were not ancestral. Mussolini and Marx endorsed evolution. UNESCO head J. Huxley liked evolution because idea of God interfered with free sex. Like Kennedy's other anti-evolution sermons, this is delivered with great confidence and authority, yet is filled with highly misleading distortions and outright falsehoods. Kennedy states that "one of the disheartening things about the advent of Darwinism is that it introduced into science an element of dishonesty which had not been seen before." This sermon also reprinted in *Bible-Science Newsletter.* In 1986 Kennedy circulated a petition in California protesting that state's textbook selection process. He also delivered the keynote address at the 1986 Int'l Creation Conf. in Pittsburgh, *Origins: Creation or Evolution?*, (later broadcast on his T.V. show) and distr. in pamphlet form.

888 _______. *Creationism: Science or Religion?* (pamphlet). Undated [1987]. Fort Lauderdale, Fla.: Coral Ridge Ministries. Text of sermon preached 1987 at Kennedy's Coral Ridge Presbyterian Church. Christians need to fight their enemies where the battle is hottest: the evolutionist attack. Denounces legal action against Ark. creation-science bill. Suppression of creation-science is censorship. Evolution "flies in the face of established scientific laws." It is "not only religious, it is more religious than creation." On the other hand, "Creationists invented science! Without creationists there wouldn't be any science." "Now has come the time to tell the public: *There are no intermediate links.*" To the evolutionist, "life has no meaning. It is just haphazard chance. . . . Life has no values and there is no basis for any moral standards."

889 Kennedy, J.G. *Evolution Discredited by Evolutionists, By Facts [. . .].* 1925. Los Angeles: Biola Book Room. [Listed in SOR CREVO/IMS.]

890 Kerkut, G.A. *Implications of Evolution.* 1960. New York/Oxford/London/Paris: Pergamon Press. Vol. 4: International series of Monographs of Pure and Applied Biology. *Kerkut:* physiology and biochemistry prof. at Univ. Southampton (England). Legitimate non-creationist critique of evolutionary theory and its excesses. Chides evolutionists for thoughtless and uncritical acceptance of Darwinism as Revelation; apparently an attempt to jolt them out of their complacency and to force his students to think independently. "It is my thesis that many of the Church's worst features are still left embedded in present-day studies." Education still emphasizes faith in and quoting of theological authorities rather than critical thinking. **Paley**'s *Evidences of Christianity* was required reading at Cambridge until 1927. Accepts proof of "special" evolution theory—species variation and change (microevolution), but says "general" theory—common ancestry of all life—is unproved. It is believed by appeal to Book (Darwin's *Origin*)—just as others appeal to Bible or *Das Kapital*. Kerkut discusses seven "basic assumptions" of the general theory of evolution, claiming that none of them are proven. These assumptions are: the origin of life from non-life, and descent links between various major groups of organisms. Kerkut advocates polyphyletic origin of life and evolution instead of evolution from a unique source. Presents evidence that "there are many discrete groups of animals and that we do not know how they have evolved nor how they are interrelated." Very widely quoted by creationists.

891 Kester, Phyllis. *What's All This Monkey Business?: Scientific Creation Vs Humanistic Evolution*. 1981. Harrison, Ark.: New Leaf Press. Written with Monty Kester. *P. Kester:* math M.S. *M. Kester:* math Ed.D. Part I is account of the Kesters' involvement in the 1977 Texas textbook hearings (with the **Gablers**) after discovering son was being taught evolution. Exhorts Christians to assert their rights to prevent bullying by humanists. Written in breathless, melodramatic style; very personalized narrative: "His strong arms swept my trembling body close to him" (referring to husband Monty). Describes many supernatural events: family lived through 75 mph head-on crash with truck; battled hurricanes sent by Satan who was trying to prevent them from attending Texas hearings. Sinister "spiritual warfare" between Christians and forces of Satan. On high school texts: "The entire chapter on human sexuality was so explicit with descriptions and pictures that a person probably couldn't read it without becoming sexually aroused." Part II gives standard creation-science arguments, presented in debate format by creationist arguing against a humanist evolutionist and a fence-straddler. Relies heavily on, and borrows extensively from, **R. Wysong** and **H. Morris** 1144; also other creationists. Paluxy manprints, Nebraska Man, bombardier beetle, and the other usual examples. Claims that unnamed university "ditched" a research project which showed geological column was completely wrong, then "did all it could to squelch any of the information leaking out." Stresses importance of origins debate: "it means you must choose between being just another animal that is nothing more than the end result of millions of chance happenings, or you are the highest creation of a Creator who planned and designed our universe and uses the order and complexity of it to reveal Himself and His power to us." The Kesters' book is distributed by **B. Keith**'s CSLDF, which featured it on the cover of its journal (1984).

892 Keyser, Leander S. *The Problem of Origins: Whence Came the Universe? Whence Came Life and Species? Whence Came Man?: A Frank Discussion of the Doctrines of Creation and Evolution*. 1926. New York: Macmillan. Also (orig.?) pub. 1925 by Wartburg Press (Ohio). *Keyser:* Lutheran theologian at Hamma Divinity Sch. of Wittenberg Col. (Columbus, OH). Strongly anti-evolutionist book. The fact that man alone among animals "naturally stands and walks upright" is proof of "the doctrine of special creation in the divine image" (quoted in Gatewood).

893 Khalifa, Rashad A. *Creation: Why We Must Teach It in the Schools*. 1982. Denver: Islamic Productions. *Khalifa:* Egyptian-born chemist with biochemistry Ph.D. from U.S.; directs Arizona pesticide residue section; now heads Tucson Mosque; editor of *Muslim Perspective*. Khalifa claims that his computer study of the Koran reveals mysterious number code, proving its divine origin, and has written several works since 1972 promoting this idea. Khalifa strenuously defends his own translation of the Koran against heretical versions. In 1984 he filed a $38 million lawsuit against the National Acad. of Sciences for their anti-creationist booklet, stating it was spreading "deliberately distorted information" and suppressing scientific freedom. He said the NAS damaged his business and claimed that he had verifiable proof of creationism. (The suit was later dropped.) Khalifa has since produced a video *Creation or Evolution: The Final Argument* which is directed at the Supreme Court in relation to the Louisiana case.

894 Khang, Kiat Tien. *Genesis and the Chinese.* 1950. Hong Kong. 1985 reprint ed. by Leaves-of-Autumn Books (Payson, Ariz.). *Khang:* born Catholic, in Singapore; converted and became Seventh-day Adventist minister; "alias **Kang Chong Heng.**" This book was the basis for Kang and **Nelson**'s *Discovery of Genesis.* Chinese writing characters have retained depictions of the whole Genesis story—Creation, Adam and Eve, the Flood—which Khang uncovers. He analyzes characters for "God" ('Shangti'), "forbidden fruit," "flood," "seventh-day Sabbath," "in the beginning God," "creation," "devil," "soul," "serpent," "sin," "naked," and others. "These ancient Chinese certainly seemed to know the creation story more thoroughly than we do, though we have the Bible record. Yet how few believe it! They choose rather to place credence in evolutionary theories which certainly require more faith to believe than the Bible testimony."

895 Kidd, John. *On the Adaptation of External Nature to the Moral and Intellectual Constitution of Man.* 1833. London: Pickering. 2 vols. One of the Bridgewater Treatises. Also wrote *A Geological Essay on the Imperfect Evidences in Support of a Theory of the Earth* (1815).

896 Kilgore, Charles. *The Mainline Chronology Table* (scroll). c1986. Stafford, Conn.: D.L. Cooper. *Kilgore:* English and theology B.A.s; aerospace technical writer. This work is a computer printout (several yards long) of Bible chronology from 5567 B.C.—Creation—to A.D. 3075. No gaps in biblical genealogies. (Displayed at 1986 Int'l. Creation Conf.)

897 Kindell, Thomas J. *How to Introduce Scientific Creationism into [. . .]* (booklet). 1981. Priv. pub. [Listed in SOR CREVO/IMS.]

898 ________. *Questions and Answers on Scientific Creationism* (booklet). 1981. Priv. pub. [Listed in SOR CREVO/IMS.]

899 ________. *Evolution on Trial: With Evolutionists at the Witness Stand* (booklet). 1985. Priv. pub. (Pomona, Calif.). *Kindell:* received creation-science training at Christ for the Nations Inst. (Dallas) and at ICR; has been giving creation-science seminars for the past ten years; VP of Creation Concern (Portland, Ore.); has gotten creation-science into several public school curricula; creation-science specialist for Oregon Moral Majority; now lives in Pomona, Calif. Foreword by **D. Gish.** Usual creation-science arguments; many scientific quotes and references (religious references at conclusion). Courtroom trial conceit. Demands verdict against evolution and for creationism; quotes evolutionists as "hostile witnesses." Says he uses these witnesses "exclusively" to support his case—but cites many creationists and anti-Darwinists also. Strict young-earth creationism. Standard anti-evolution examples. Paluxy human/dinosaur prints; live dinosaurs in Africa. Cites **Jochmans** on out-of-place human remains found in all geological strata since Cambrian. Asserts proof of Bible prophecy and Resurrection.

900 King, Jesse. *The Mosaic Account of the Creation Affirmed, and Silent Monitors of the Past Described and Illustrated with Object Lessons of Each Day's Part of the Creation.* 1892. Philadelphia: J. King. "Together with the Formation and Coloring of Rocks, and the Formation of Different Kinds of Coal Supported by Sciences as Taught and Understood at the Present Day, from Creation to the Present Time."

901 ________. *Elementary Geology; an Endorsement of Genesis in Stone, Studied in the Fields from Nature.* 1899. Philadelphia: American.

902 Kingsley, Charles. *Glaucus.* 1855. London. *Kingsley:* Anglican clergyman. Book based on reviews of natural history books; became a general critique of transmutation, which Kingsley then thought impossible. According to Millhauser, it was largely a reaction to Chambers' *Vestiges,* expanded to become a popular book in reaction to the notoriety of *Vestiges.* Later, Kingsley became a leading clerical supporter of (Darwin's) evolution. He received an advance copy of *Origin* as a suspected sympathizer. Later elected to Geological Society. Wrote *Water Babies* (1863), which creationists consider evolutionist propaganda for children because of its advocacy of transformism; also *Hypatia* (novel about the burning of the great library at Alexandria and its pagan heroine).

903 Kingston, Lyle H. *A Return to Creationism* (pamphlet). Undated. Priv. pub. (Wisc.). Wonders and design of nature.

904 Kinney, LeBaron W. *Acres of Rubies: Hebrew Word Studies for the English Reader.* 1946 [1942]. New York: Loizeaux Bros. Bible Truth Depot. 2nd ed. Orig. 1942. Each chapter an essay on a Hebrew word or phrase from the Bible, exploring its meaning and context. The chapter "A Precious Unfathomable Mystery" concerns *rachaph* ('flutter' or 'move'), from Gen. 1:2: "And the earth was without form and void; and darkness was upon the face of the deep, and the Spirit of God 'moved' upon the waters." Kinney presents in this chapter a Gap Theory interpretation of creationism. Earth had been created perfect ages ago in Gen. 1:1; but it became corrupt, a formless ruin. God then re-created it for man.

905 Kinns, Samuel. *Moses and Geology: or, The Harmony of the Bible with Science.* 1895 [1892]. London: Cassell. Rev. ed. Orig. 1882. *Kinns:* fellow of Royal Astronomical Soc.; member of Soc. Biblical Archaeology Soc.; vicar of Holy Trinity. Copy of this book officially accepted by the Queen. Prime Minister **Gladstone** also liked it. Day-Age creationism; Kinns thought his interpretation could be reconciled with Huxley's scheme.

906 Kipp, Peter E. *Is Moses Scientific?: First Chapter of Genesis Tested by Latest Discoveries of Science.* 1893. New York: Fleming H. Revell. Day-Age creationism. God's "method of creation" also involved evolution. Man appeared about 10,000 years ago.

907 Kirby, William. *On the Power, Wisdom, and Goodness of God As Manifested in the Creation of Animals, and in Their History, Habits, and Instincts.* 1837 [1835]. London: Pickering. Also pub. 1837 in Philadelphia. One of the Bridgewater Treatises. Millhauser says Kirby "subscribed unexpectedly to theories of the diluvial origin of strata and of a literal Abyss of Waters" and that this work was similar to **Fairholme** – and **Burnet.** Praised, quoted and discussed in **B. Nelson** 1931 for its exposition of the Flood and the Ark. Flood was worldwide; Noah's Ark settled on Mt. Ararat. The Flood dispersed fossils widely – they are not always found in their native areas; it also altered the climate and atmosphere.

908 Kircher, Athanasius. *Arca Noe.* Amsterdam: Apud J. Janssonium a Waesberge. "The Ark of Noah." In Latin. "I: de rebus quae ante diluvium. II: de iis, quae ipso diluvio ejusque duratione. III: de iis, quae post diluvium a Noemo gesta sunt, quae omnia nova methodo, nec non summa argumentorum varietate, explicantur, & demonstrantur." *Kircher:* German Jesuit priest; polymath who researched and wrote about virtually all fields of knowledge of his day. Made important contributions to optics, microscopy and telescopy, astronomy, magnetism, and decipherment of hieroglyphics. His 1665 *Mundus Subterraneus* dealt with various geophysical processes. In this book, Kircher attempted a "scientific" study of the Ark, calculating how it must have been constructed. The Ark had 300 stalls on the first deck, a granary on the second, and 2,000 cages for birds and space for Noah's family on the third deck. Admitted some animals didn't make it to the Ark – they became extinct. He also tried to deduce the nature of the Flood, suggesting that elements of air combined to form water for the Flood. Illustrations, maps.

909 Kirchmaier, George Kaspar. *De Diluvii Universalitate Dissertatio Prolusoria.* 1667. Geneva. Advocated regional rather than global Flood. (Cited in **Whitcomb** and Morris.)

910 Kirkpatrick, Foy. *God's Wonderful World* (booklet). c1970. Dallas: Gospel Teachers Publications. Teenage Teachers Manual. Comes with Pupil's Workbook and Visual Aid Packet (includes chart of Grand Canyon and geological time table, items created on each of the six creation days, Bible chart with prophecies, and pictures of miracles). Manual contains many creation-science arguments; scientific as well as biblical references. "The Bible Teaches a Creation which is Absolutely Inconsistent with Evolution." Evolution is uncritically taken for granted, or accepted in order to get rid of God. Drawings, charts, NASA photos.

911 Kirwan, Richard. *Geological Essays.* 1799. London: T. Bensley. Also pub. in Arno Press reprint ed. *Kirwan:* mineralogist (Irish Inspector of Mines); chemist (advocate of phlogiston theory); president of Royal Irish Acad. Kirwan defended the Neptunist theory of geology; this book was written as a rebuttal to Hutton's uniformitarianism and the Vulcanist or Plutonic theory. Insisted on literal adherence to Genesis account; asserted that geology corroborated its divine authority. Earth originally covered by an aqueous miasma; then a great evaporation and deposition of solids; then vulcanism and formation of atmosphere; then appearance of land, and creation of animals (Kirwan apparently allowed for long creation "days.") Some species left out of Noah's Ark; others created since the Flood. Noah settled in high mountains near China. Kirwan viciously attacked Hutton's non-scriptural interpretation, and also criticized **Burnet, Whiston, Woodward** and **Deluc** for not being literal enough in their scriptural geology. Discussed and quoted in Gillispie.

912 Klaaren, Eugene M. *Religious Origins of Modern Science: Belief in Creation in Seventeenth-Century Thought.* 1977. Grand Rapids, Mich.: William B. Eerdmans. *Klaaren:* grad. of Western Theol. Sem.; Ph.D. – Harvard; religion prof. at Wesleyan Univ. (Middletown, Conn.). "While it is a truism that the rise of modern science occurred within Christendom, I have sought to show ... that belief in creation constituted a definitive context within which the basic questions ... of the new science were raised, pursued, and developed." Serious scholarly study of cultural and intellectual significance of Christian belief in divine creation. Acknowledges **Hooykaas.** Focuses on Robert Boyle; also on Johan Baptist van Helmont. Contains no creation-science, but often cited by creationists as supporting Bible-science.

913 Kline, Theodore. *Cosmic Patterns and the Bible.* c1983. Winona, Minn.: Justin Books (Trinity Pub.). *Rev. Kline:* advocates Hebrew Christianity; apparently a "Christian Jew." This book is a "biblical application" of Kline's earlier work *The Theory of Universal Trichotomy.* Kline discovered the "utter literalness" of the Bible. Creation was in six days – but these were "universal days" of a billion years each. The Fifth Day was the age of dinosaurs. Lucifer and his followers were apparently active at that time, in this reptilian form, before God destroyed them. There were pre-Adamic men at the start of the Sixth Day; some survived in Adam's race. The "sons of God" (godly descendants) mated with "daughters of men" (other, evil races); then all were wiped out by the Flood.

914 Klingman, George. *God Is.* 1929. Nashville, Tenn.: Gospel Advocate. Also pub. by F.L. Rowe (Cincinnati). "An antedote [sic] for the poisonous propaganda of 'The American Association for the Advancement of Atheism'."

915 Klotz, John W. *Genes, Genesis, and Evolution.* c1970 [1955]. St. Louis, Mo.: Concordia. 2nd rev. ed. Orig. 1955. *Klotz:* biology Ph.D. – Univ. Pittsburgh; natural science prof. at Concordia Senior College (Ft. Wayne, Ind.); Missouri Synod Lutheran. 544-page biology textbook with biblical as well as scientific references. Comprehensive treatment of biology (especially genetics) and evolution; Klotz has solid training in biology. Provides fairly thorough coverage of evidences for evolution but emphasizes problems of evolution and clearly advocates creation. Includes standard creation-science arguments; discusses the "species problem," homology, vestigial organs, comparative physiology and biochemistry, embryology, mimicry, biogeography, paleontology, selection and isolation, genetics and mutation, human evolution and alleged hominid ancestors, and various theories and proposed mechanisms for evolution; also criticizes radiometric dating. Advocates strict recent creation (but allows for genealogical gaps in Genesis). "As Christians we know that in the Bible we do not have a theory which is subject to all sorts of changes, a theory which has come about as a result of the restricted reasoning abilities of human beings, but we have the inspired account of the only Being who was present at Creation." Unlike science, the Bible is infallible. "Certainly evolution is by no means proved, and it is not the only explanation for the organic diversity that we find. It is not unreasonable, then, to assume that the changes which have occurred have been finite and limited and that they have occurred within closed systems, the 'kinds' of creation." Photos, drawings; index.

916 _______. *Challenge of the Space Age.* c1961. St. Louis, Mo.: Concordia. Discusses satellite launches and nuclear power. Does not address evolution specifically. Urges a Christian approach to science and to scientific problems. Index.

917 _______. *Modern Science in the Christian Life.* 1961. St. Louis, Mo.: Concordia. "Written to provide an understanding of the scientific method and the relativity of scientific truth, this volume deals with past and current conflicts between science and religion" (**J.N. Moore,** in 1965 *CRSQ*). Includes section on evolution.

918 _______. *Ecology Crisis: God's Creation and Man's Pollution.* c1971. St. Louis, Mo.: Concordia. Knowledgeable discussion of pollution and ecology. Urges recognition that God owns the whole earth, and that mankind must care for it properly (biblical "stewardship"). Denies Lynn White's theory (proposed in 1967 *Science* article) that exploitative attitude towards nature arose from Christian concept of asserting "dominion over the earth." Index.

919 _______. *Studies in Creation.* 1985. St. Louis, Mo.: Concordia. "A General Introduction to the Creation/Evolution Debate." Includes sections on history of science, science and religion, history of evolutionary theory, scientific method, and confessional statements of various denominations (especially Lutheran) regarding creation. Also concedes and discusses "Problems for the Creationist"—areas of creation-science vulnerable to criticism: biogeography, extinction theories (post-Flood), continental drift. Admits evidence of biogeography fits evolution theory better than creationism, but adds that there are problems with evolutionist theory also. Discusses "Problems for the Evolutionist" in long final chapter. Includes standard creation-science arguments; emphasizes animal language, molecular biology, origin-of-life experiments, prehuman fossil hominids. Includes many recent scientific references. Emphasizes that evolution is based on "assumptions and presuppositions [that] are as much matters of faith as is acceptance of the Biblical creation account." Though the evolutionist seeks to exclude supernaturalism from science, he "does indeed have a god ... the god of chance." Deplores warfare analogy for science/religion relations; points out polemical nature of A. White's 1896 book. Comparing Linnean taxonomy with Genesis: "There is nothing right or wrong about a system of classification, for ... any system is arbitrary." Admits it is theoretically possible to reconcile an old earth with creationism, but says that if we accept the historicity of Creation and the Fall it makes no sense to wait so long for the Redeemer. "While it is true that there are observations which fit better with the theory of evolution than they do with the theory of special creation, there are also areas ... which fit better with the concept of special creation. One of these is the study of the evolution of man himself. ... Another ... is the suggested mechanism for evolution. ... Still another ... is the whole question of the complexity of living things." Attractive presentation; Klotz is more knowledgeable and accurate than most creationists.

920 Knaub, Clete. *A Critique of Molecular Homology.* 1983. Unpub. thesis (Institute for Creation Research). ICR biology M.S. thesis. Committee: **Parker** (chair), **Cumming, Gish.** *Knaub:* in Army Security Agency; then fish and wildlife management B.S.—Montana State. Standard creation-science arguments against validity of molecular taxonomy, in expanded form. (Evolutionist W. Thwaites of San Diego State had earlier suggested to Knaub that he use published data on known DNA sequences to try to falsify evolutionist phylogenies. He didn't.)

921 Knaub, Clete, and Gary Parker. *Molecular Evolution?* (pamphlet). 1982. El Cajon, Calif.: Institute for Creation Research. ICR "Impact" series #114. Synopsis of Knaub's ICR thesis: phylogenies constructed from "molecular clocks" are unreliable and contradictory.

922 Knight, Charles Spurgeon. *Both Sides of Evolution: A Debate.* c1925. San Jose, Calif.: Arthur H. Field. Apparently a Rosicrucian. Also wrote *The Mystery and Prophecy of the Great Pyramid* (1933; San Jose: Rosicrucian Press AMORC), with Introduction by **Arthur I. Brown**.

923 Knoche, Keith. *Incredible Voyage* (pamphlet). c1975 [c1963]. Mountain View, Calif.: Pacific Press. Reprinted from *HIS* magazine (InterVarsity Fellowship; c1963). Describes a voyage from the realm of the very small to the very large: demonstration of Design and creation. Seventh-day Adventist.

924 Koch, Kurt E. *Occult ABC.* 1978. Germany: Literature Mission Aglasterhausen. Enlarged ed. Orig. German. Distributed by Grand Rapids International. A compendium of occult, psychic, paranormal and demonic beliefs. Stresses reality of Satan and demons (Koch is an exorcist); argues that psychic and paranormal events are real but demonic—satanic counterfeits of Christian belief. Includes faith healing, Jehovah's Witnesses, other cults, rock music, UFOs, the *700 Club* (Pat Robertson's healing is not of Christ), and belief in "Descent from the Ape." Quotes **Bettex** in evolution section. If you believe you came from an ape, then go back to the jungle. Very naive; doubts fossils are very old—says we should C-14 date them.

925 Koestler, Arthur. *The Case of the Midwife Toad.* 1971. New York: Random House. *Koestler:* Hungarian-born writer; became British citizen. A vindication of Austrian biologist Paul Kammerer, who shot himself in 1926 after his experiment demonstrating the inheritance of acquired characteristics was shown to be faked. Koestler says Kammerer was innocent, and that his results may yet prove to be true. Argues that Kammerer's rival, Bateson (the English biologist who championed Mendelian genetics) was unfair to him. Koestler has criticized Darwinian evolution and urged reconsideration of Lamarckism in many books: *Ghost in the Machine, Act of Creation, Beyond Reductionism* (co-ed.), *Janus.* Koestler also argues for the validity of ESP and psychic phenomena, and is well-known for his eloquent renunciation of communism.

926 Kofahl, Robert E. *Handy Dandy Evolution Refuter.* c1980 [1977]. San Diego: Beta Books. Rev. and expanded ed. *Kofahl:* chemistry Ph.D.—Caltech (founded Caltech Christian Fellowship as undergrad); 21 years as teacher and president of Highland College (Christian school in Pasadena, Calif.); since 1972, science coordinator of Creation Science Research Center. Preface by **K. Segraves.** Standard creation-science arguments in question-and-answer format. Explicitly biblical—evolution is bad because it contradicts the Bible and hinders people from accepting Christ—but packed with scientific references; also cites many creationists. Includes most of the popular creationist quotes. Strict recent creationism; Flood Geology. Neither evolution nor creation are testable scientific theories, since they deal with origins; both are accepted on faith. Design in nature, origin of life, 2nd Law of thermodynamics, mutations, fossil gaps, radiometric dating, evidences for young earth, earth specially designed for man. Fossil hominids are all frauds or just apes. Accepts Paluxy manprints, Calaveras skull. Genetics shows Middle East origin of man. Creation of Eve from Adam: God just doubled one of Adam's X chromosomes. Virgin birth of Jesus: created as hybrid being (man/God)—Y chromosome came from God. Scopes Trial was ACLU plot to discredit Christianity. God created Eden, and Adam, with "false appearance of age," but this wasn't deceptive because He told us what he was doing. "Supposedly scientific theories such as evolution which contradict the Bible can cause some people to doubt the Bible and thus hinder them from coming in humble faith to Jesus Christ for salvation." Lists creation-science resources. Index.

927 _______, and Kelly L. Segraves. *The Creation Explanation: A Scientific Alternative to Evolution.* c1975. Wheaton, Ill.: Harold Shaw. Creation-science textbook; includes biblical as well as scientific references—explicit appeals to reader to turn to Christ. One of the most cited creation-science books. Comprehensive presentation of all the standard creation-science arguments, examples, and sources. Includes bombardier beetle, other examples of Design; Second Law and degeneration; human/trilobite fossil; **Burdick**'s Precambrian pollen; Darwin's use of speculative phrases; uniqueness of Man; rate of human population increase; critique of radiometric dating, other young-earth proofs; Noah's Flood; dispersal of humanity at Babel; fossil hominids either apes, modern man, or hoaxes. "Life—Miracle, Not Accident." Acknowledgments include Victor Oliver (sections on Man), R. Koontz (entomology), **J.N. Moore,** Carole Barklow (astronomy), and Everett Purcell, **John Read** and James Honeyman (all CSRC Board of Trustees, for technical information). "The most powerful evidence for creation and against evolution is . . . to be found in specific evidences of intelligent, purposeful design." Purpose of book is to show "true science in its proper perspective, as a searching out of the handiwork of the Creator for the glory of the triune God. . ." It was "written to demonstrate how the facts of the sciences support what the Bible says about creation and the providential rule over the world by Jesus Christ." "The attributes and powers of man cannot be explained on the basis of a purely materialistic process of development from chemicals to cells to animals to man. Man is a spiritual and personal being who must have had a spiritual and personal source." The Christian scientist has the advantage of divine revelation, and is thus "confident that the final judgments of true science and history will entirely concur with the biblical record." The Bible tells us that those who reject the God of creation are "without excuse." Presents a "Creation Model" matching earth and life history with Genesis verses. Photos, drawings, diagrams, charts; bibliog., index.

928 Kölliker, Rudolf Albert von. *Über die Darwinische Schöpfungstheorie.* 1904. *Kölliker:* M.D.—Heidelberg; Swiss; student of Muller and Oken; prof. at Univ. Wurzburg; eminent embryologist, microscopist; did important research on cell division, cell nucleus, and sex cells; also in histology and neurology. Kölliker praises Darwin's demonstration of descent, but criticizes his theory of natural selection and became one of the leading opponents of Darwinian evolution in Germany. According to **Nordenskiöld,** he rejected special creationism, but suggested hypotheses for non-Darwinian development —"either that all organisms have arisen each out of its own primary form, or that the species have come into existence through one primary form or through a few." If the latter, there must be a common law governing formation for development of hgher forms from lower, involving changes in embryological development. He believed evolution proceeded by jumps. Kölliker criticized Darwinism by noting lack of transitional forms, lack of proof or experimental evidence for species formation and natural selection, and that variations within species may be more sterile than the originals. He also accused Darwinian selection of embodying a "teleological" notion of finality. "The development theory of Darwin is not needed to enable us to understand the regular harmonious progress of the complete series of organic forms from the simpler to the more perfect. The existence of general laws of nature explains this harmony, even if we assume that all beings have arisen separately and independent of one another. Darwin forgets that inorganic nature, in which there can be no thought of a genetic connection of forms [as in crystal growths], exhibits the same regular plan, as the organic world (of plants and animals). . ." (Quoted in **Graebner** 1921).

929 Kreischer, Louise. *The Stars Tell the Story Too: An Interpretation of the Bible Prophecies as They Are Illustrated by the Constellations and Depicted in Stone by the Great Pyramid of Gizeh.* 1942. New York: Maranatha.

930 Kronos. *Evolution, Extinction, and Catastrophism* (special issue). 1982. Wynnewood, Penn.: Kronos Press. Special issue of *Kronos,* a pro–**Velikovsky** journal. Authors include **N. Macbeth; Bennison Gray** ("Alternatives in Science: The Secular Creationism of **Heribert Nilsson**"); Peter J. James ("Ever Since Darwin: A Review"); Tom Bethell ("Darwin's Unfalsifiable Theory") [Bethell has written several anti-Darwinist articles for *Harpers* and has been working on a book about the failure of evolutionary theory]; Lynn E. Rose ("On Velikovsky and Darwin"); Velilovsky (on the Ten Lost Tribes of Israel).

931 Kurtz, Johann Heinrich. *The Bible and Astronomy.* 1857 [1842]. Philadelphia: Lindsay & Blackiston. 3rd ed. Orig. 1842; German; pub. by Wohlgemuth (Berlin). 2nd ed. 1849. *Kurtz:* theology prof. at Dorpat. Kurtz presents a classic version of Gap Theory creationism, and discusses scriptural evidence of the angels whose Fall caused the earth to become desolate and waste, prior to God's re-creation. He then presents, and also advocates, a pictorial version of the Revelatory Theory of creationism. **H. Miller** 1857 discusses and endorses Kurtz's Revelatory Theory. God reveals to Moses (or possibly an earlier seer) the history of creation in a series of visions. These visions of these "prophetic days" are analogous to the prophetic visions of the seer, but concern the past rather than the future. "Before the eye of the seer, scene after scene is unfolded, until at length, in the seven of them, the course of creation, in its main *momenta,* has been fully represented." These revelatory visions of creation were described by Moses as they appeared to him. The "prophetic days" were real days: the events of the re-creation (but not necessarily the revelation of the visions) actually occurred in six literal days (which may, however, have been of unusual and indeterminate length).

932 _______. *A History of the Old Covenant.* 1859. Edinburgh: Clark. Mentioned in **Weston Fields** 1976 as advocating Gap Theory creation.

933 Labrentz, Arnold J. *The Young Christian and Science.* 1975. Kisumu, Kenya: Evangel Publishing House. *Labrentz:* teacher, editor and minister for eight years in Kenya. Book is addressed to Kenyans. Mainly concerned with creation-science arguments, largely taken from *Bible-Science Newsletter.* Young-earth creationism; Flood Geology. Explicitly religious; many creationist, some scientific references. Describes the great faith of NASA scientists and astronauts. Claims that, since dates were obtained for all samples, "radiocarbon has shown that there are no remnants of life tested and found to be older than forty thousand." Darwin's theory was accepted because it was "non-sacred"—anti-religious. It depended on inheritance of acquired characteristics, now known to be false. "This purposeless philosophy supported by the theory of evolution is uninspiring, degrading and downright depressing. . . . How much healthier and happier to believe in ... an all-wise Creator who knows our weaknesses and problems..." Cartoon-style drawings.

934 LaHaye, Tim. *The Battle for the Mind.* c1980. Old Tappan, N.J.: Fleming H. Revell. *LaHaye:* pastor of Scott Memorial Baptist Church (San Diego); founder-president of Family Life Seminars; co-founder (with **H. Morris**) and president of CHC (creationism-centered college, home base of ICR); now works in Wash. D.C. as chair of Amer. Coalition for Traditional Values (group with broad fundamentalist support lobbying for Bible-based society). (Creation-Life Publishers was named for ICR and for Family Life.) Christianity vs. atheistic humanism. "Most of the evils in the world today can be traced to humanism, which has taken over our government, the UN, education, TV..." etc. Includes much creationism. Declares that evolution is the foundation of all secular education. Evolution is based on faith, but science is based on fact. Cites ICR scientists. "No humanist is qualified to hold any governmental office..." A call to political action. LaHaye has also written many other fundamentalist books, including *The Battle for the Public Schools, The Battle for the Family, The Hidden Censors* (the anti-Christian humanist media), *The Beginning of the End* (Bible prophecy).

935 _______**, and John D. Morris.** *The Ark on Ararat.* c1976. San Diego: Creation-Life/Nashville: Thomas Nelson (co-pub.). Account of 1972 ICR expedition to Mt. Ararat to find Noah's Ark, led by Morris (son of **H. Morris**). Reviews Ark sightings, Ark-eology tales. Skeptical of many claims—**Roskovitsky, Navarra,** LandSat, Holy Ground Mission—but certain nonetheless the Ark is there. Comprehensive and relatively critical summary of Ark-eology efforts.

Discusses **Burdick**'s geological survey of Ararat, which proves Flood. Demons have attacked Christian investigators to prevent them from proving Bible right and evolution wrong. Morris himself was struck by lightning and seriously injured. Includes chapters on Flood traditions around the world; the design, construction, and cargo of the Ark. "The real significance of the Ark in the days of Noah should be analyzed, since world conditions indicate we are living in the last days." God destroyed the world in the Flood; the world will shortly be destroyed by fire. Unless God reveals it sooner, the intense heat of the Great Tribulation will expose the Ark by melting Ararat's glacial ice, finally convincing the skeptical. "Climb on board the Ark then. Receive Jesus Christ as your personal Savior..." Unlike many other Arkeologists, who often boast of illegal Ark searches, Morris urges friendship and complete cooperation with Turkish authorities. (Morris wrote most of the book; LaHaye wrote the opening and closing chapters about the Ark and the Bible.) Photos, drawings, maps; index.

936 Laing, Sidney Herbert. *Darwinism Refuted: An Essay on Mr. Darwin's Theory of "the Descent of Man"* (booklet). 1871. London: Elliot Stock.

937 Lambert, Edmond. *La Deluge Mosaique: L'Histoire et la Geologie.* 1870. Paris: Victor Palme. 2nd ed., enlarged. *Abbe Lambert:* Th.D.; member of French Geological Soc., and Linnaean Soc. of Bordeaux. "Accord de la Science et de la Religion."

938 Lammerts, Walter E. *Discoveries Since 1859 Which Invalidate the Evolution Theory* (reprint). 1964. Ann Arbor, Mich.: Creation Research Society. Reprint from *CRSQ* Annual. Also reprinted in **Lammerts** (ed.) 1973. *Lammerts:* genetics Ph.D.—Berkeley; prize-winning plant breeder (especially roses), acknowledged as a leader in the field; former ornamental horticulture prof. at UCLA; later director of Germain's Seed Co., then in private business and research; key figure in formation of CRS; first CRS president and *CRSQ* editor. Discusses molecular evolution, species variation, genetics, mutations, chromosome numbers, hybrids, the Genesis Flood; cites some of his own technical work. Lammerts read **G.M. Price** in college, and still writes on "wrong-order" geological formations and other subjects for *CRSQ*.

939 ________. *The Challenge of Creation* (booklet). 1965. Caldwell, Idaho: Bible-Science Association.

940 ________ **(ed.).** *Why Not Creation?* 1973 [c1970]. Grand Rapids, Mich.: Baker Book House. Selected articles from the *Creation Research Society Quarterly:* Vols. I through V (1965–1968). Orig. c1970 by Presbyterian and Reformed Pub. Includes articles by John Grebe (director of research at Dow Chemical); William Meister (on fossil proving man contemporary with trilobites); R.H. Brown (physics prof. at Walla Walla College, Wash., on radiocarbon dating); Harold Armstrong (physics—Queens Univ., Canada, on DNA); N.A. Rupke (State Univ. Groningen, The Netherlands; doing grad. study in geology at Princeton Univ.); Arthur F. Williams (Cedarville College, Ohio); **J.N. Moore** (on social influences of Darwinism); **E. Williams** (entropy); **Rusch; G. Howe** (paleobotany); **Klotz; Mulfinger** (theories of the origin of the universe); **Whitelaw** (on radiometric dating); **H. Morris; Gentry; M. Cook** (on the Meister fossil); **Lammerts** (on mutations; also other topics); **Frair** (on DNA biochemistry); **Tinkle** (on genetic stability of Genesis kinds); **Custance; Gish.** Basically strict young-earth creationism and Flood Geology (though some *CRSQ* contributors, such as Custance, are Gap Theory believers, or allow for other creationist interpretations). Covers standard creation-science topics: philosophy of science, cosmogony, thermodynamics, radiometric dating, catastrophist geology (Flood), fossil evidence (anomalies, out-of-order fossils), biochemical evolution, criticism of evidences for and influence of evolution. Lammerts analyzes Galapagos finches from Calif. Acad. Sci. collection. Rupke discusses many classical creationists and Flood geologists **(Steno, Ray, Woodward, Scheuchzer, Deluc, Fairholme),** and stresses importance of polystrate fossils. Custance summarizes his arguments on human evolution. Lammerts: "Our aim is a rather audacious one, namely the complete re-evaluation of science from a theistic viewpoint." Photos, diagrams.

941 ________ (ed.). *Scientific Studies in Special Creation.* 1971. Nutley, N.J.: Presbyterian and Reformed. Selected articles from the *Creation Research Society Quarterly:* Vols. I through V (1964–1968). Companion volume to *Why Not Creation?* Includes articles by Donald Acrey (geophysicist; with Graham Plow Co. in Amarillo, Texas); Harold Armstrong; N.A. Rupke; P.W. Davis (**Frair**'s co-author); Richard Korthals (Lt. Col. USAF [ret.]; former astronautics prof. at AF Acad.; now physics prof. at Concordia Jr. Coll., Ann Arbor, Mich.); **R. Reymond; Rita Ward; T. Barnes; Burdick; H. Clark** (on Flood Geology interpretation of red bed deposits); **H. Coffin; M. Cook** (summarizes criticisms of radiometric dating from his 1966 book); **Davidheiser; Gish; G. Howe** (on survival of plant seeds during Flood); **F. Marsh** (on the Genesis 'kinds'); **Lammerts; H. Morris** (on population increase and Bible chronology); **Rushdoony; Shute** (on "remarkable adaptations"); **Tinkle; J. Whitcomb; P. Zimmerman.** Covers standard creation-science topics. Korthals article also reprinted in **Zimmerman** 1972. Rushdoony on premises of evolutionism: "God, clearly, is an inescapable premise of human thought. Man either faces a world of total chance and brute factuality, a world in which fact has no meaning and no fact has any relationship to any other fact, or else he accepts the world of God's creation and sovereign law." Lammerts on "Immorality in Natural Selection": "If we feel that man is an animal, we are in danger of losing our sense of responsibility, for animals do not and cannot have this trait." It is "dangerous to believe that man is the product of struggle among selfish, irresponsible lower organisms." Lammerts also discusses artificial mutations and rose breeding. Reymond affirms that creation in Genesis is *ex nihilo,* as does Whitcomb, who also refutes Gap Theory. Includes biog. notes. Photos, drawings, diagrams.

942 Land, George T. Lock. *Grow or Die: The Unifying Principle of Transformation.* 1973. New York: Random House. Also pub. by Dell (Delta). Blurbs by Karl Menninger, Margaret Mead, S. Dillon Ripley and others. *Land:* "creativity expert." Explains all phenomena in the universe – physical, biological, cultural, psychological and behavioral – by "transformation theory": three-stage growth processes ("accretive," "replicative," and "mutual") involving new levels of information and organization. Darwinian theory cannot account for the obvious directionality and progress of evolution. Points out tautological nature of Darwinian theory. Praises **Koestler** and Kammerer; also **Teilhard.** Entropy is opposed by evolutionary syntropy. Lamarckian theory sees evolution as cumulative (organisms changing their environment); Darwinian theory sees it as repetitive (organisms changed by environment); Land advocates "directed selection." Information is a form of energy which allows organisms to transform their environment so they can adapt to it. "Organisms create the conditions in their environment that favor the selection of particular kinds of mutations." Cites Heisenberg – effect of observer on phenomena. Origin-of-life experiments show that chance alone "cannot explain such consistent results"; there is an "inevitability." Man now controls his own evolution. Growth is irreversible and essential: "Not to grow is to die." Charts, diagrams; bibliog., index.

943 Landsburg, Alan. *In Search of Extraterrestrials.* 1977 [c1976]. New York: Bantam. Foreword by Leonard Nimoy. *Landsburg:* T.V. and film producer (has worked on Cousteau and Nat'l. Geog. programs); producer of NBC *In Search Of...* T.V. series. UFOlogy; standard flying saucer sources, examples, and authorities. Alien spacemen are visitng earth. They were also responsible for man's "cosmic and psychic leap" from brutish stone-age existence to civilization. Dogon lore of Sirius, ancient carvings and tales of UFOs, messages from outer space, etc. Photos; bibliog. Landsburg's weekly *In Search Of...* series (aired since 1976), with Nimoy as host, has prominently featured ICR and other creationists on several episodes, especially "In Search of Noah's Flood" (1979) and "Creation vs. Evolution" (1981; another version titled "In Search of the Missing Link"). These episodes treat creationism as of equal scientific validity as evolution, and seem to endorse Flood Geology and Ark-eology.

944 _______, and Sally Landsburg. *In Search of Ancient Mysteries.* 1974. New York: Bantam. Also NBC-TV special (done with **F. Warshofsky**). Foreword by Rod Serling. *S. Landsburg:* wife of Alan. "Did Man Begin on Earth – Or Was He Sent Here from Other Worlds?" Spacemen colonized earth and interbred with subhumans. Standard examples of extraterrestrial influence in ancient history: Nazca lines, various monuments and megaliths, prehistoric 'aircraft' and 'computers,' etc. – all supposedly baffling mysteries only explicable by ET hypothesis. Chapter "Back to the Bible": discusses NASA designer Joseph Blumrich's idea of spaceships of Ezekiel. Chapter "How Did Man Get His Brain?": Darwin was wrong; **A.R. Wallace** was right. Evolution occurred, but cannot account for man's brain. Cro-Magnon man was result of infusion from another planet. Photos; bibliog., index.

945 _______, and _______. *The Outer Space Connection.* 1975. New York: Bantam. Also a feature film. Foreword by Rod Serling. Extraterrestrials seeded earth by sending body cells and cloning them here. Accounts for sudden rise of Sumerian and other civilizations, which we know possessed computers, spaceships, lasers and anti-gravity devices. Genesis Flood is historically true; includes summary of usual Ark-eology tales. Pre-Flood patriarchs really had immense life-spans: lost secrets of immortality. Ancient neurosurgery (trepanning). Mayans were genetically identical – clones; also genetic experiments to modify body forms. ETs built the ancient pyramids to harness incredible pyramid power to preserve cells for cloning. Egyptian mummification is crude mimicry of this; Aztec heart removal is vestige of advanced surgery. ETs arrived in 3113 B.C., are observing us now, and will return on Dec. 24, 2011.

946 Lang, Walter. *The Mythology of Evolution* (pamphlet). 1968. Caldwell, Idaho: Bible-Science Association. *Lang:* Missouri Synod Lutheran minister; began *Bible-Science Newsletter* in 1963 and organized **Bible-Science Association** in 1964; has toured, lectured and campaigned ceaselessly for creationism since then. Now emeritus director of BSA, and deprived of direct control, Lang concentrates on his new Genesis Institute and **Faulstich**'s Chronology-History Research Inst. with which it is affiliated. Most of the BSA tracts (see under BSA) were "edited" or written by Lang, as were most of the BSA-sponsored creation conferences, though many of these BSA publications do not list any author. "In Scripture we find the absolutes which are not found in science." Quotes **A. Schnabel.** "Our social life demands the creation approach, for socialism and communism destroy the individual. All fields of science need the creation approach for it is the simplest and best explanation of origins and agrees best with the latest scientific data. Government needs the creation approach – the alternative is a **Hitler** or a Stalin."

947 _______. *Was God an Astronaut?* (tract). 1975. Minn.: Bible-Science Assoc. Refutation of **von Däniken** – links him with evolutionism. Asserts evolutionists want truth by majority vote, and that von Daniken is popular but false. "Evolutionists need to find life in space! If life began by accident on earth, it must have begun earlier by accident in space."

948 _______. *Genesis and Science.* 1982. Minneapolis: Bible-Science Association. A study of Genesis 1–11. Revision of "Five Minutes with the Bible and Science" series of articles (BSA publications orig. coordinated with radio program; now included in *Bible-Science Newsletter*).

949 _______. *History of Recent Creationism and Its Future* (audiotape). 1985. San Fernando Valley Bible-Science Association. Tape cassette of 1985 lecture by Lang to San Fernando Valley (Calif.) chapter of BSA. Lang has known over 5,000 people who have converted from evolution to creationism, but didn't know more than three or four who switched because of the scientific evidence; almost all changed not because of any evidence, but because they became Christians first. Creationism is in a new stage now. Why rely on evidence so much if so few people are changed by it? Mentions that creationist **John Read** is a geocentrist.

950 _______. *Creation Scientists at*

Work (booklet). Undated. Minneapolis: Bible-Science Association. Includes names and descriptions of many creationist scientists. Useful biographical information on contemporary creationists.

951 ______. *Evangelism Program* (tract). Undated. Bible-Science Association. Lang and **C. Burdick** realized a Bible-science program was an effective evangelism tool. Recommends evangelism effort using BSA tracts and materials. Account of witnessing to a Jewish girl on 1971 BSA Middle East tour. "There are no absolutes in nature due to contamination of sin throughout the universe..." Mildly critical of charismatic evangelism: a pursuit of saved "feeling" which he likens to pietism prior to French Revolution. Lists 1973–76 BSA tracts.

952 ______ **(ed.).** *The Challenge of Creation* (booklet). c1965. Caldwell, Idaho: Bible-Science Association. Essays presented at 1964 Creation Seminar in L.A. area. Authors: **Lammerts; Rusch; Burdick; Howe; Davidheiser; Lang.**

953 ______ **(ed.).** *Theology and Science.* 1973. Minneapolis: Bible-Science Assoc. Essays presented at Lutheran Research Society creation symposium at Faith Lutheran Sem. (Tacoma, Wash.).

954 ______ **(ed.).** *A Challenge to Education II: Popular Level Essays on Creationism* (booklet). 1974a. Caldwell, Idaho: Bible-Science Association. Essays from 1974 Creation Conference in Milwaukee.

955 ______ **(ed.).** *A Challenge to Education II: Technical Essays on Creationism* (booklet). 1974b. Caldwell, Idaho: Bible-Science Association. Essays from 1974 Creation Conference in Milwaukee. Authors: **T. Barnes; Slusher; H. Morris; Lammerts;** John Grebe (Dow Chemical); **H. Coffin;** Charles Roessger (Milwaukee Museum); **Gish;** Walter G. Peters (science and geology teacher); **Overn; E. Williams;** Joseph L. Henson (entomol. Ph.D. – Clemson; science prof. at Bob Jones Univ.); **Mulfinger;** Ray Hefferlin (Ph.D. – Caltech; physics prof. at Southern Missionary Coll. in Collegedale, Tenn.) and Hans Boksberger (Swiss-born physics student at Southern Missionary Coll.). Biog. notes included.

956 ______ **(ed.).** *A Challenge to Biology.* 1975. Minneapolis: Bible-Science Assoc. Essays from 1977 Creation Conference in Philadelphia. Speakers included **Lang; E. Williams; Ouweneel;** David Christensen, M.D.; David Kaufmann, Ph.D. (Creation Health Foundation: Taylors, S.C.); **D. Crowley, Jr.;** Albert Anderson (M.D.; Creation Health Foundation); William James, M.D.; **R. Bliss; D. Gish;** George Koshy; **L. Lester;** Paul Hooley, M.D.; Henry Grimm; David Livingston (director of Associates for Biblical Research, Huntingdon Valley, Penn.); Chris Hummer; **Elmendorf;** Micah Leo, Ph.D.; Rush Acton, M.D.

957 ______ **(ed.).** *Unidentified Flying Objects* (pamphlet). 1975. Caldwell, Id.: Bible-Science Assoc. *Five Minutes with the Bible & Science* series. Primarily a critical discussion of **F. Salisbury**'s book *The Utah UFO Display.* Lang accepts the reality of UFO accounts, but insists that they are supernatural manifestations: either of angels, or (most likely) Satanic deceptions. Urges acceptance of Christ as protection against this and other evil lies of Satan.

958 ______ **(ed.).** *The Challenge of Design.* 1978. Essays from 1978 Creation Conference in Wichita, Kansas (presumably edited by Lang). Speakers included **J. Whitcomb; P. Ackerman; Clifford Wilson; G. Parker; R. Bliss; T. Barnes; Davidheiser; D. Gish; H. Slusher; Burdick;** D. Kaufmann; A. Anderson; George Graf, M.D.; **H. Morris;** Edward Coleson, Ph.D.; **Lang;** Rev. Robert Ingram; Rev. Greg Bahnsen; **Mulfinger; Lubenow;** Mike Wilson; Verne Bigelow.

959 ______ **(ed.).** *1980 Bible-Science Association Convention.* 1980. Minneapolis: Bible-Science Association. Essays from 1980 conference in River Forest, Ill. (presumably edited by Lang). Speakers included **Lang; Overn; Bartz;** Rev. Vernon Harley; **J.N. Moore;** Carl Mueller; **Kofahl;** Paul Freeman; Dick Caster; **Arndts;** Anne Driessnack, Ed.D.; Wesley Chase; **B. Deen; Rusch; Rev. Herman Otten** (editor of *Christian News;* active in opposing modernism in Lutheran Missouri Synod; involved in formation of CRS and BSA); **L. Lester; Klotz;** R. Surburg; Dean Griffith; David Nelson (son of **Byron Nelson**); Woodmorappe; Albert Anderson, M.D.; Charles Roessger; Lane Anderson; David Kaufmann, Ph.D.;

David Watson; Paul Leithart (M.D.; Creation Health Foundation); **Beierle;** Robert Shaibley; Robert Preuss, Ph.D.; Gerald Mallmann; Verne Bigelow; George Saxenmeyer; William Strube; Ron Schuchard; Edith Mitchell; **James Hanson;** Prof. Paul Boehlke; **G. Bouw.** (See **Bible-Science Association** for 1979 and 1983 convention essays.)

960 _______ (ed.). *Genesis and Science* (booklet). 1982. Priv. pub. (Richfield, Minn.). "A Study Course Prepared for The Genesis Institute." Discusses Ebla tablets, OT commentaries and interpretations of Genesis, the "science commission of Genesis, the "waste of space research," biblical interpretation of the constellations, Genesis 'kinds,' the Flood and Noah's Ark, *Nephilim* of Genesis, plus other creation-science topics.

961 _______, and Valeria Lang. *Two Decades of Creationism.* 1984. Priv. pub. (Minneapolis). *V. Lang:* wife of Walter. Disjointed, apparently hastily-written history of Bible-Science Assoc. and Lang's 20-year, full-time crusade against evolution. Useful source of information on recent creation-science activity; index is helpful. Lists all speakers at BSA-sponsored creation conventions, and others who have assisted with BSA activities—tours, seminars, debates, publications. Includes comments on geocentricity (Lang is sympathetic), relativity (Lang is not), alleged population problem, creationism in Lutheran Missouri Synod, Stanley Taylor's film *Footprints in Stone,* Nobel Prize (awarded for pro-evolution work proved false by creationists), **S. Marinov,** and many other creationists. Lang describes his "100 Year Plan" for creationism: evolution has had 100 years to take over, so creationists should be patient and see what creation-science can do given time. Also includes many standard creation-science arguments. "The revolution taking place in creation/evolution is a revolution demanded by the apostle in Romans 12:2. It means turning the world upside-down for Christ and His Gospel." Index.

962 _______, and Vic Lockman. *Have You Been Brainwashed About Ape-Men?* (tract). 1975. Minneapolis: Bible-Science Association. Alleged ape-men refuted—usual cast of characters. Peking Man was human; ape parts mixed in. Middle East origin of man—fossils of primitive humans are all found far away. "Very primitive people still exist today, e.g., Bolivia and the Philippines."

963 Lanz-Liebenfels, Joerg. *Theozoologie oder die Kunde von dem Sodoms-Äfflingen und dem Götter-Elektron.* 1905. ("Theo-zoology or knowledge of Sodom's monkeys and electron of the gods.") Peter A.J. Lanz was a Benedictine monk who changed his name to von Liebenfels. In 1900 he founded the Order of the New Temple in Vienna and the journal *Ostara* to promote "heroic racialism" and "race science." Lanz describes the prehistoric Ario-heroes, who lived like gods in prehistoric times. But the Aryan women mix with dark ape-like beings, producing the inferior Chandala race. The Ario-heroes lose their racial purity and their electro-magnetic-radiological powers. Lanz-Liebenfels urged extermination of Jews and other "ape-races." His occultic pseudo-scientific racism apparently had some influence on the Nazis who developed **Hitler**'s racist and eugenics policies. Hitler was influenced by *Ostara.* (Discussed in **Zepp-Larouche** 1984. Zepp-LaRouche claims that Kitchener and Strindberg were members of Lanz's New Temple Order.)

964 Laramore, John. *The Creator of this World and Universe.* Undated. Port Washington, N.Y.: Ashley Books.

965 Larkin, Clarence. *Dispensational Truth, or God's Plan and Purpose in the Ages.* 1920 [1918]. Orig. 1918. Philadelphia: Rev. Clarence Larkin. 3rd ed. rev. and enlarged. Based on 1918 pamphlet. This book was influential in popularizing Gap Theory creationism. Includes 90 detailed chronological charts (some reproduced in Marsden). Teaches about pre–Adamic destruction of earth due to activity of Satan and his fallen angels. The original creation was in the "dateless past." "It was not at the beginning of the first day as described in Gen. 1:3–5. The six days' work as described in Gen. 1:3–31 was the *restoration of the earth* (not the heavens or the starry space), to its original condition before it was made 'formless and void,' and submerged in water and darkness." The inhabitants of the pre-Adamic world (probably humans, dwelling in cities) became sinful, causing God to destroy this original world. II Peter ("the world that then was, being overflowed with water, perished") refers to this awful catastrophe – not to Noah's Flood. Noah's Flood was not worldwide. Demons are the disembodied spirits of the pre-Adamites. Dispensational scheme of history is encoded in the Great Pyramid.

966 Larson, Muriel. *God's Fantastic Creation.* c1975. Chicago: Moody Press. Acknowledges **G. Mulfinger** for verifying scientific facts in book. Paean to exquisite design in nature as unarguable proof of Designer. Planets don't bump into each other; they follow strict orbits. Bible says the earth goes around the sun; Columbus, because of this, discovered America. If evolution were tested in a law court, it would be thrown out. Includes sections on plants, animals, man, etc.; each section also used to illustrate Bible teachings. A few scientific references; many Bible quotes. Several creation-science books in short bibliog.

967 Lasater, B.E. *The Scientific Proof of the Soul!* (booklet). Undated. Priv. pub. (Whittier, Calif.). Rev. Lasater's account of Satan's battle for his faith. Attended Univ. Arkansas ("devil's territory") where he was exposed to evolution. These teachings on origin of life and of man blinded him to God. Before he entered medical school, his infected hand was miraculously healed. Describes debate there between C. Darrow and a rabbi, on immortality. Rabbi proved immortality by appealing to "first law of Chemistry which states: Nothing can be created, nothing can be destroyed!" His surgical training proved existence of man's soul: the heart feels no physical pain (no nerves) yet experiences great emotions. Accepted Christ forthwith.

968 Lavallee, Louis. *The Early Church Defended Creation Science* (pamphlet). 1986. El Cajon, Calif.: Institute for Creation Research. ICR "Impact" series #160. *Lavallee:* M.S. – Harvard; M.Div. – Reformed Theol. Sem. (Jackson, Miss.); environmental engineer at Miss. Dept. Natural Resources. Early Church Fathers – Paul, Justin, Theophilus, Basil, Origen, Augustine – accepted and defended young-earth creation and Flood. By contrast, Greeks and other pagans all had evolutionist theories.

969 Laymen's Home Missionary Movement. *The Bible Vs. Evolution* (tract). Undated. Chester Springs, Penn.: Laymen's Home Missionary Movement. *Laymen's Home Missionary Movement:* Jehovah's Witnesses or similar Russellite sect; publishes *The Bible Standard and Herald of Christ's Kingdom* (mostly millennialist prophecy); distributes many books and tracts. This tract largely based on **Paul Johnson**'s 1938 book (distr. by LHMM). Day-Age creationism: each 'day' 7,000 years long; but Spirit of God "energized" the waters long before the six creation days (i.e. also proposes a Gap Theory variant). Earth "brought forth" plants and animals in Genesis – each 'kind' was not created separately. But man is a separate creation. Infertility of hybrids proves species can't change – "fixity of species." Man's brain is no larger now than in ancient civilizations, though great increase of knowledge shows we are in End Times. Evolution denies central tenets of the Bible: original perfection, etc.

970 ________. *The Evolution Theory Examined* (pamphlet). Undated. Chester Springs, Penn.: Laymen's Home Missionary Movement. Includes citations up to 1967. Old references cited without dates; many quotes out of context. Many standard creation-science arguments presented, but very naively. Ignorant of science, relies on obsolete sources. Gap and Day-Age theories of creationism both endorsed; Genesis 'days' 7,000 years each. Serological argument: author misunderstands antibody reactions (apparently taken from **W. Williams** 1925). Mendelism and Biometry disprove evolution. **Rimmer** quoted (no date). Nebraska Man presented at Scopes Trial [false; from Rimmer]. Lady Hope tale quoted and approved. Evolution contradicts the Bible. Much of text taken verbatim from **P. Johnson** 1938. "The pessimism and indulgence of brute instincts inculcated by evolutionists have in large part produced the moral collapse seen today." Pronounces a "Solemn Indictment" of evolution for causing the "direst of evils" and turning our youth into atheists and infidels.

971 Lee, Francis Nigel. *Communism*

Versus Creation. 1969. Nutley, N.J.: Craig Press. "This book is an extended discussion of the evolutionary views of Marx, Engels, and Lenin" (**N. Pearcey,** in 1986 *Bible-Science Newsletter*). Lee quotes Lenin: "we may regard the material and cosmic world as the supreme being, as the cause of all causes, as the creator of heaven and earth"—communism makes matter its God.

972 _______. *The Origin and Destiny of Man*. Memphis, Tenn.: Christian Studies Center. Recommended in **R. Walton** 1984 as creationist book.

973 Lefevre, J. *Manual Critique de Biologie*. 1938. Paris.

974 Lehman, Chester Kindig. *The Inadequacy of Evolution as a World View*. 1933. Scottdale, Penn.: Mennonite Publishing House. *Lehman:* Dean of Eastern Mennonite School (Harrisonburg, Vir.). Introduction by **L. Keyser.** Facts speak for creation. Evolution is supported only by theories and arguments. Cites and quotes many anti-evolutionist sources. Follows **G.M. Price.** Index.

975 Lemoine, Paul, R. Jeannee, and P. Allorge (eds.). *Encyclopedia Francaise, Tome 5: Les Etres Vivants*. 1937. Paris: Societe de Gestion de l'Encycopedie Francaise. *Lemoine:* president of Geological Soc. of France; director of Museum d'Histoire. Volume on natural history from the prestigious French encyclopedia; critical of neo-Darwinian theory of evolution. Widely cited by creationists. Lemoine concludes volume with article "How Valid Are the Theories of Evolution?" "The theories of evolution with which our studious youth have been deceived, constitute actually a dogma that all the world continues to teach; but each in his specialty, the zoologist or the botanist ascertains that none of the explanations furnished are adequate." Natural selection is not adequate for evolution. "The result of this consideration is that the theory of evolution is impossible." He describes evolution as a "fairy tale for adults," and as "a sort of dogma in which the priests no longer believe, but which they uphold and maintain for the people." Few French scientists believe the theory of evolution. "It is necessary to have the courage to say this in order that men of a future generation may orient their research in a different manner." (Quoted in **P. Zimmerman** 1959, 1963.)

976 LePelletier, Jean. *Dissertations sur l'Arche de Noe*. 1700. Rouen, France: Jean B. Besongne. 520-page "discussion of the size, arrangement, construction, and logistics of the Ark. Apologetic refutation of objections to the historicity of the flood and the Noahic voyage. Author holds to a universal deluge." (Described in **J.W. Montgomery** 1974.)

977 Lester, Lane P. *Mimicry* (pamphlet). 1974. El Cajon, Calif.: Institute for Creation Research. ICR "Impact" series #18. *Lester:* genetics Ph.D.—Purdue Univ.; ICR "extension scientist" and S.E. representative; formerly worked for BSCS; has taught at Univ. Tenn. (Chattanooga) and at CHC (ICR); now biology prof. at Falwell's Liberty Univ. (Liberty Baptist) and director of its Center for Creation Studies and its creation-science museum. Batesian mimicry in butterflies as support for creationist biology.

978 _______**, and Raymond G. Bohlin.** *The Natural Limits to Biological Change*. 1984. Grand Rapids, Mich.: Zondervan/Dallas: Probe Ministries International (co-pub.). Christian Free Univ. Curriculum series. *Bohlin:* population genetics M.S.—Texas State Univ.; now in mol. biology doctoral program at Univ. Texas (Dallas); Research Projects Manager of Probe Ministries International (Dallas). Includes a brief "Response" by V. Elving Anderson (genetics and biology prof. at Univ. Minn.; former pres. of Sigma Xi) which is mildly critical. Lester and Bohlin have a better understanding of biology than most creationists. Several chapters consist of a fairly accurate review of biological, genetic, and evolutionary concepts. They stress that evolutionists as well as creationists can and do defend their metaphysical models by adapting facts to fit theories, and point out cultural and historical biases influencing scientific views on origins. Evidence can be found to support various evolutionary theories or creation; individuals choose according to their world views. Two chapters are criticisms of neo-Darwinism and punctuated equilibrium theory. Authors propose criteria for defining "prototypes"—organisms descended from same created 'kinds'—as alternative to evolutionary theories and differing from criteria of **Marsh** and other creationists. They argue that organisms cannot transcend limits of "prototype." Many scientific references, including creation-science works. Calls *H. erectus* "fully human" (but degenerative). Good drawings, diagrams; index.

979 Leupold, Herbert C. *Exposition of Genesis.* 1949 [c1942]. Grand Rapids, Mich.: Baker Book House. Two vols. Orig. pub. c1942 by Wartburg Press (Columbus, Ohio). *Leupold:* D.Div.—Capital Univ.; Old Testament prof. at Evangelical Lutheran Theol. Sem. of Capital Univ. (Columbus, Ohio). A theological, not scientific, work—though Leupold mentions geology, anthropology, astronomy and other sciences in order to criticize theories which contradict the Bible. Says **G.M. Price** and **B. Nelson** are better scientists than those whose theories contradict fundamentalist belief. Refutes "higher criticism" of the Bible. Praised by **H. Morris** for rigorously defending literal six-day creationism and Flood Geology as the only legitimate interpretation of Genesis.

980 Lever, Jan. *Creation and Evolution.* 1962, Grand Rapids, Mich.: William B. Eerdmans. Orig. Dutch. 1958 ed. (English) by International Publications (Grand Rapids, Mich.). *Lever:* pupil of **Dooyeweerd;** zoology prof. at Free Univ. of Amsterdam. Concerned more with theology and philosophy than straight creation-science. Creation 'days' not necessarily 24-hour days. **H. Taylor** 1967 criticized Lever's book as too "accommodating"; **Rushdoony** has similarly attacked Lever for abandoning strict creationism. Other creationists have dismissed it as theistic evolution.

981 Levi. *The Aquarian Gospel of Jesus the Christ.* c1964 [c1907]. Marina del Rey, Calif.: DeVorss. "The Philosophical and Practical Basis of the Religion of the Aquarian Age of the World. Transcribed from the Akashic Records by Levi." *Levi:* Levi H. Dowling; raised as Disciples of Christ; served in Civil War as army chaplain; graduate of two medical colleges. Biblical style and format. Zodiacal Ages or Dispensations. Jesus travels to the East, studies in Greece, Egypt, Persia, India, and Tibet. He repudiates Buddhist doctrine of evolution ("protoplast evolved, becoming worm, then reptile, bird and beast, and then at last it reached the form of man"). Jesus reveals true origin of man: the Triune god created ether planes, each of which had its own form of life. Their vibrations slowed down, and these forms became physical.

982 Levitt, Zola. *Creation: A Scientist's Choice.* 1976. San Diego: Creation-Life. Orig. pub. by Victor Books (Wheaton, Ill.). *Levitt:* M.A.—Indiana Univ.; formerly performing musician; a "Hebrew Christian" (born-again Jew); has nationwide radio/T.V. ministry (stresses Jewish heritage of Christianity). Book is extended interview with creationist **John N. Moore** (natural science prof. at Michigan State Univ.) and an exposition of creation-science. Levitt says "he secretly believed in evolution" until the day he met Moore. Includes standard creation-science arguments. Charts, diagrams, tables. Levitt has written over 40 Christian books (on Bible prophecy, Israel, Pentecost, Jewish traditions, etc.). He co-authored *UFOs: What on Earth Is Happening?* (c1975; Irvine, Calif.: Harvest House) with John Weldon. That book, which explains UFOs as Satanic deceptions aimed at luring us away from Christ in these End Times, denies the possibility of extraterrestrial life. Since life did not evolve on earth, it cannot evolve elsewhere. Acknowledges assistance of **Coppedge, Slusher,** Hal Lindsey and Walter Martin.

983 Lewis, C.S. *Christian Reflections.* 1967. Grand Rapids, Mich.: William B. Eerdmans. Edited by Walter Hooper. *Lewis:* prof. of Medieval and Renaissance literature at Cambridge Univ. Lewis underwent a conversion experience and wrote many influential and very popular works advocating and defending Christianity. **Geisler** categorizes Lewis as a "liberal-evangelical" who believes Genesis includes "myths" and who subscribes to theistic evolution. This book includes a chapter "The Funeral of a Great Myth," in which he distinguishes between the science and the myth of evolution, criticizing the latter. "In the science, Evolution is a theory about changes; in the Myth, it is a fact about improvements." (It is not clear here what Lewis thinks about evolution as science.) Stresses that evolution is a hypothesis, subject to change, which is accepted largely for metaphysical reasons and desires. Elsewhere, in his poem "Evolutionary Hymn," he satirizes belief in evolutionary progress: "Goodness = what comes next." Among Lewis's most famous books: the *Narnia Chronicles* (children's books with Christian themes); the *Perelandra* trilogy (science fiction with Christian themes); *The Screwtape Letters* (1943), a modern account of the Devil; *Mere Christianity* (1943–1945), a highly regarded defense of Christianity; *Miracles* (1947), arguing for the reasonableness of believing that God intervenes directly in nature and human affairs, transcending natural law; *The Discarded Image* (1964), a study of the medieval concept of the universe which survived in Renaissance literature and beyond—the closed hierarchy of immutable Heaven and imperfect, changing earth (Lewis admits to a nostalgic yearning for the security and comfort of this obsolete cosmology). **E. Myers,** in 1985 *CSSHQ,* claims that Lewis really opposed evolution: that he was "unalterably opposed to emergent evolutionism as a philosophy or myth; that he maintained a good deal of dry skepticism about the biological hypothesis of evolutionism as not necessarily scientifically true; and that he would have gladly welcomed the rise of creation science in our own generation."

984 Lewis, Tayler. *The Six Days of Creation; or, The Scriptural Cosmology, with the Ancient Idea of Time-Worlds, in Distinction from Worlds in Space.* 1880 [1855]. Schenectady, N.Y.: G.Y. van Debogert. *Lewis:* prof. of Greek at Union College; member Reformed Church. Defends biblical account against evolution, and lambasts the pretensions of science. Lewis criticized Chambers' *Vestiges* for scientific errors, but primarily for theological reasons. As Millhauser notes, Lewis felt it showed "impudence, arrogance, and profound ignorance of Revelation"; it reduced God to a remote logical principle or an epitome of natural law. Lewis accepts pre-Genesis ages of chaos (before six creation days); also a Day-Age interpretation of creation days. Argues that Bible uses phenomenal language. Suggests infinite cycles of worlds ("time-worlds"). "Lewis' writings without question provide the most exhaustive exegetical study of Genesis 1 from the point of view that the days cannot possibly be twenty-four-hour days. What makes Lewis' work all the more compelling is his desire *not* to be influenced by current scientific findings" (**D. Young** 1982).

985 Life Messengers. *The Lonely Cabin on the 40-Mile* (pamphlet). c1982. Seattle, Wash.: Life Messengers. "Adapted script" and art by Daryl Skelton. Comic-book format pamphlet about Joe Conlee, narrated by Charles S. Price; a "true story." Conlee studied for the ministry, but then read Darwin, Huxley, Renan, etc., and doubt battled faith. He quit the ministry, became an editor in Los Angeles and then an alcoholic gambler. He was born again while holed up in a cabin in the Yukon gold rush.

986 _______. *Who Am I?* (pamphlet). Seattle, Wash.: Life Messengers. "Small creationist booklet"—"highly recommended" by **Malik** 1981, who quotes its scientific and probability arguments against evolutionary origin of life from non-life.

987 Lightfoot, John. *A Few, and New Observations, Upon the Book of Genesis. The Most of Them Certain, the Rest Probable, All Harmless, Strange, and Rarely Heard of Before* (booklet). 1642. London: T. Badger. *Lightfoot:* distinguished biblical scholar; later Vice-Chancellor of Univ. Cambridge. It is commonly but erroneously asserted that Lightfoot "improved upon" Archbishop **Ussher**'s famous calculation of the date of creation (Oct. 23, 4004 B.C.) with calculations of even greater precision, determining it occurred at 9 AM. This misconception spread after inclusion in A. White's 1895 book. Lightfoot's work in fact preceded Ussher's, and his 9 AM time refers to the creation of man—not the world. In later works (all prior to Ussher's)—*Harmony of the Four Evangelists Among Themselves and the Old Testament, with an Explanation of the Chiefest Difficulties both in Language and Sense* (1642), and *Harmony, Chron-*

icle, and Order of the Old Testament (1647)—Lightfoot demonstrates his system of determining chronology by correlating the lifespans of the patriarchs. He arrived at a date for creation similar to Ussher's: 3928 B.C. Lightfoot assumed creation occurred on the autumnal equinox (that is, September).

988 Lindsay, Gordon. *Evolution—the Incredible Hoax* (booklet). 1963. Dallas: Voice of Healing. 2nd ed. Gordon also wrote *The Mystery of the Flying Saucers in the Light of the Bible* (1953), and *Strange Facts About Adolph Hitler* (c1942), a pamphlet pub. by the Anglo-Saxon Christian Assoc. (Portland, Ore.) in its "Blueprints of God" series.

989 _______. *The Creation.* Undated(?). Dallas: Christ for the Nations.

990 Lindsell, Harold. *The World the Flesh & the Devil.* c1973. Minneapolis: World Wide Publications. "Published for the Billy Graham Evangelistic Assoc." *Lindsell:* then editor-publisher of *Christianity Today* (founded by Graham). Includes discussion of science in the chapter "The Creator and the Redeemer." Science can only postulate theories, not ultimates; e.g. evolution. Science cannot deal with first causes, so both creationists and evolutionists enter metaphysical realm when dealing with ultimate origins. The Christian must turn to the Bible to answer these questions that science cannot. Lindsell mentions that he is a special creationist, but does not defend this in detail here. Scripture is not clear whether Satan fell before or after creation of Adam.

991 Lindsey, Arthur W. *The Problems of Evolution.* 1931. New York: Macmillan. *Lindsey:* prof. at Denison Univ. "All of the selection theories, all ideas of isolation, all of our knowledge of mutations, serve only to show that the characters which make up a species may be reassorted, re-distributed, preserved in part and destroyed in part, or modified to some degree" (quoted in **Field** 1941). Though apparently a critic, Lindsey did not deny evolution; he later wrote textbooks on evolution and on genetics.

992 Lindsey, George. *Evolution—Useful or Useless?* (pamphlet). 1985. El Cajon, Calif.: Institute for Creation Research. ICR "Impact" series #148. *Lindsey:* Ed.D.; on ICR graduate faculty. Evolution retards scientific research and teaching. "Creation with plan and purpose, on the other hand, provides an excellent basis for solving problems in the real world of scientific research."

993 Linnaeus, Carolus [Karl von Linné]. *Systema Naturae.* 1758 [1735]. Stockholm/Leiden. *Linnaeus:* the Swedish botanist who devised the modern system of biological taxonomy—"Linnaean" classification. "The concept of design by God lay at the heart of Linnaeus' philosophy of nature. The very fact that we can classify species into an orderly system bespeaks the existence of a rational Creator. The relationships of similarity between species that today are seen as a sign of common evolutionary ancestry were for Linnaeus elements in the divine plan..." (Bowler). Initially, he believed 'species' (the second, more "specific" term in his famous binomial nomenclature) were the originally created biblical "kinds"—thus he advocated strict "fixity of species." A brilliant classifier, Linnaeus adopted **Ray**'s theory of species as clearly defined immutable units created from the very beginning: "We count as many species now as were created in the beginning." Later, he came to believe (largely from his experiments in plant hybridization) that the fixed biblical "kinds" were really genera, and that species could change within genera. Linnaeus believed the 'genus' was a "natural" category definable by essentialist (typological) characteristics, and strictly delineated from other genera, but he considered higher-level taxa of what we call the "Linnaean hierarchy" more or less artificial and conventionalist. **Nordenskiöld** describes Linnaeus's conception of God in Nature as "half-pantheistic." Though resolutely creationist, Linnaeus's hierarchical classification helped later biologists conceive of taxa in terms of descent relationships.

994 Little, Paul E. *Know Why You Believe.* 1978 [1967]. Downers Grove, Ill.: InterVarsity Press. Orig. c1967 by Victor Books (Wheaton, Ill.). *Little:* on staff of InterVarsity Christian Fellowship; prof. of evangelism at Trinity Evangelical Divinity Sch. (Deerfield, Ill.). Apologetics: book provides rational answers to questions asked by skeptics and critics of Christianity. Chapter on science and the Bible stresses dogmatic nature of most evolutionist belief. Long excerpts from **Kerkut.** Critics falsely assume all Christians believe **Ussher**'s date for creation—but this is not in the Bible. Evolution is not proved (quotes **Mixter** and **Ramm**), but notes that Bible allows for change between species (quotes **H. Morris**). Also discusses archeology, miracles, etc.

995 Lohmann, E. *Descent from the Monkey.* According to **Graebner** 1925, Lohmann, a German, "cites a number of cases which prove the insufficient character of the evidence being used by evolutionists." This is presumably Ernst Lohmann, who wrote various religious books.

996 Lord, David N. *Geognosy; or the Facts and Principles of Geology Against Theories.* 1855. New York: Franklin Knight. Denounces doctrines "openly hostile to revelation" featured in popular education which are "masked under the form of facts or truths of natural science..." "The Geological Theory Contradicts the Sacred History of the Creation." Various chapters refuting the "false theories of geologists." Geology "is not a science—it has no laws"; it cannot account for the formation of strata or the origin of animals and man. Quotes the false old-earth views of **Buckland** 1836, **Sedgwick** 1833, **Hitchcock,** Lyell. If the earth and life were not created six thousand years ago, as God told Moses, then not only the Genesis creation account, but the whole Bible is disproved, and loses its status as "heaven-descended reality." To suppose that the earth formed out of hot gases millions of years ago is a blatant contradiction of scientific laws. Present processes cannot account for the facts of geology.

997 Lord, Eleazar. *Geological Cosmogony; or, An Examination of the Geological Theory of the Origin and Antiquity of the Earth, and of the Causes and the Object of the Changes It Has Undergone.* 1843. New York: R. Carter.

998 ________. *The Epoch of Creation.* 1851. New York: C. Scribner. "The Scripture doctrine contrasted with the geological theory." Mentioned in Millhauser as part of the reaction by the orthodox against **H. Miller,** whose "reconciliation was as scandalous as transmutation." Old-style Scriptural geology; fossils prove the historicity of the Flood.

999 Lotsy, Johannes Paulus. *Evolution, By Means of Hybridization.* 1916. The Hague, The Netherlands: M. Nijhof. *Lotsy:* Dutch botanist; lecturer at Leiden Univ.; founded journal *Genetica;* friend of W. Bateson and **H. Nilsson.** Moore (1964 *CRSQ*) says: "Contains extensive discussion of his work in producing 'new species' from crosses of existing species. Lotsy felt that new classes and phyla have occurred *suddenly* from recombinations of factors brought together in crosses." Lotsy's theories were largely a response to De Vries' mutation theory, which he rejected. He became convinced that species originated and evolved by hybridization—the "only known source" of new species. 'Linneons'—ordinary Linnean species—are the homozygous remnants of previous crossings. Evolutionary development results from the crossing of different Linneons. (Discussed in **Fothergill** 1953.) Lotsy developed experimental approaches to study the genetics and mechanisms of evolution.

1000 Løvtrup, Søren. *Darwinism: The Refutation of a Myth.* 1987. England: Croom Helm. *Løvtrup:* zoophysiology prof. at Univ. Umea, Sweden. Løvtrup is a macromutationist. Quoted in **W. Bird** 1987, who says Løvtrup "poses a new [evolutionary] theory while brilliantly repudiating Darwin." Løvtrup: "After this step-wise elimination, only one possibility remains: the Darwinian theory of natural selection, whether or not coupled with Mendelism, is false. ... Hence, to all intents and purposes the theory has been falsified, so why has it not been abandoned? I think the answer to this question is that current Evolutionists follow Darwin's example—they refuse to accept falsifying evidence."

1001 Lubenow, Marvin L. *Does a Proper Interpretation of Scripture Require a Recent Creation?* (pamphlet). 1978. El Cajon, Calif.: Institute for Creation Research. ICR "Impact" series #65 and 66 (in two parts). *Lubenow:* B.A.—Bob Jones Univ.; science M.S.—E. Mich. Univ.; Th.M.—Dallas Theol. Sem.; pastor of First Baptist Church (Ft. Collins, Col.); active in BSA and CRS.

Account of Wheaton College panel discussion: **Gish** and Lubenow arguing for affirmative; Kaiser and Willis against.

1002 ________. *"From Fish to Gish."* c1983. San Diego: CLP (Creation-Life). "Morris and Gish Confront the Evolutionary Establishment." Chronicle of a decade of creation-evolution debates. This is a strongly biased creationist account, but Lubenow does report on the contents of several dozen debates. He is more interested, however, in affirming and applauding the creationist arguments, and in pointing out the fallacies of the evolutionist debaters, than in real analysis or comprehensive coverage of the debates. Concludes with R. Doolittle's televised debate with **Gish** sponsored by J. Falwell. Light, journalistic style. Covers most standard creation-science arguments. Includes detailed appendices listing debates by city and by debate participants. In Introduction, Lubenow tells of attending 1975 lecture by evolutionist J.W. Schopf. He hides his creationist identity, but later asks Schopf some questions damaging to evolution. A young man overhears this and recognizes Lubenow as a fellow creationist. This man, not identified by name, is D. McQueen (now of ICR; then a grad student at Michigan). Since this book, Lubenow has continued an intensive study of paleoanthropology; at the 1987 Creation Conf. he argued that *H. erectus* was a true human.

1003 Lucas, Charles William, Jr. *Soli Deo Gloria: A Renewed Call for Reformation* (booklet). c1985. Temple Hills, Md.: Church Computer Services. *Lucas:* M.S.—Univ. Maryland; theor. physics Ph.D.—William & Mary College; has presented at creation conferences. Call for reform of science and theology. Moral permissiveness is the result of materialist humanist view. Problems of theoretical physics stem from idealization of point particles, which should be eliminated. Special relativity, then quantum mechanics, was invented to circumvent these problems. Physics really consists solely of electrodynamics; electro-magnetism is the only force (cites Ives **[Inst. Space-Time]**, Poincare). Says three other physicists have come to similar conclusions: **Barredo** (a Catholic); J. Kenny at Bradley Univ.; and **T. Barnes** (ICR). Contains Bible references. Includes section "No Universal Scientific Theories of Evolution" (dated 1984).

1004 Lucas, Jerry, and Del Washburn. *Theomatics: God's Best Kept Secret Revealed.* 1977. New York: Stein and Day. *Lucas:* former All-American basketball player with New York Knicks; now a memory expert (wrote *the Memory Book* and *Remember the Word*); founder of Memory Ministries (Calif). "Theomatics scientifically proves that God wrote the Bible." Numerology: each Hebrew and Greek letter assigned number value; analysis of words and phrases using multiples and clusters proves intricate mathematical design. Authors assert this demonstrates God—not the devil—still rules on earth, though sin is rampant. God is now revealing His Truth via theomatics in these End Times prophesied in the Bible (significance of 666). This refutes prevalent view which denies that "God created all things by personal and direct action," and the assumption that everything is governed by impersonal laws of nature. Does not deal directly with evolution.

1005 Lunn, Arnold H. *The Revolt Against Reason.* 1950 [1930]. London: Eyre & Spottiswoode. Orig. pub. by Greenwood in 1930. *Lunn:* Catholic. Lunn quotes a Fellow of the Royal Soc. as expressing gratitude that Lunn is tackling evolution, since the professional scientists' "hands are tied." Those in authority "regard Darwin as a Messiah"; "no jobs are going except to those who worship at the Darwin shrine." Quotes anatomy prof. **T. Dwight** on "tyranny" of evolutionist control, of which outsiders have no idea—"oppression as in the days of the Terror. How very few of the leaders of science dare tell the truth concerning their own state of mind!" "Faith is the substance of fossils hoped for, the evidence of links unseen." Since natural selection still operates, "if Darwinism is true we should expect to find that the world was full of transitorial forms; but the world is full of fixed types, and the five thousand years of recorded history are eloquent in their witness, not to transitorial forms fading into each other, but to the stability of type." (Quoted in **Field** 1941; **P. Zimmerman** 1972; **Graebner** 1932.)

1006 ________. *The Flight from Reason*. 1932 [1931]. Rev. ed. 1932. M. Gardner says Lunn praises **G.M. Price** for pouring "well-deserved ridicule on the arbitrary rearrangement of strata."

1007 ________, **and J.B.S. Haldane.** *Science and the Supernatural: A Correspondence Between Arnold Lunn and J.B.S. Haldane*. 1935. New York: Sheed & Ward. A written debate (31 letters, from 1931–34), largely concerning evolution. *Haldane:* famous British geneticist (Univ. London); played important role in bringing genetics into harmony with evolutionary theory in the 1930s (the neo-Darwinian synthesis); also known for outspoken advocacy of Marxism. Lunn presents many standard creation-science arguments: unoriginality of Darwin; *Archaeopteryx;* Paley's Design argument; Mendel was right and Darwin was wrong; Nebraska Man; persecution of anti-evolutionists; fossil gaps; etc. Relies on Thomas Aquinas's proofs. Argues for truth of psychic phenomena as proof of supernatural; endorses Lourdes. The Darwinian controversy demonstrates the anti-supernatural bias of most scientists. There is "no real evidence in support of Darwinism." Darwin was a mere "fact collector" and a "weak reasoner." Quotes **Mivart, Belloc, Berg, Fabre, O'Toole, Vialleton, Macfie,** other anti-Darwinians. Haldane observes that all people are affected by science, but few realize the nature of scientific thought. Index.

1008 Lutheran Research Forum. *Theology and Science* (booklet). 1973. Lutheran Research Forum. Essays presented at Lutheran Res. Forum (Tacoma, Wash.), 1973. Missouri Synod Lutherans. Authors: **W. Overn,** on Bible-Science research; **O. Friedrich,** on God's grace; Rev. Vernon Harley, on Lutheran theology in science teaching; **W. Lang,** theology and science; Marcus Braun (credit agency owner), on the new Arianism.

1009 Lutzer, Michael. *Exploding the Myths that Could Destroy America*. 1986. Chicago: Moody Press. One of these twelve myths is that life is the result of Godless chance. Also excerpted in *The Rebirth of America* (1986; Arthur S. DeMoss Found.), written and compiled by R. Flood.

1010 Lysons, R.W. *Viruses No Proof of Evolution* (pamphlet). Undated. Evolution Protest Movement. EPM pamphlet #85. Also wrote EPM pamphlet #101: *More Recent Opinions on Evolution*.

1011 Maatman, Russell W. *The Bible, Natural Science, and Evolution*. 1970. Grand Rapids, Mich.: Reformed Fellowship. *Maatman:* chemistry prof. and natural sciences chairman at Dordt College (Sioux Center, Iowa); Dutch Reformed. Arguments are primarily philosophical and theological; few scientific references. Contrasts the "diametrically opposed" assumptions and views of the Christian and the "natural man," who supposes there is nothing beyond nature. Demonstrates that science has flourished when and where the Bible has been honored. The Bible commands man to study nature in order to exercise dominion. Affirms absolute truth and inerrancy of the Bible, but dismisses Bible-science claims of allusions to scientific discoveries and inventions (e.g. of atomic war); says science should not be used to prove the Bible. States that quantum uncertainty, and origin and extent of the universe, present irresolvable dilemmas for "natural man." "Throughout the evolution discussion in this book, an attempt is made to discover *first* what the Bible teaches. This procedure should be followed in the study of any question upon which the Bible sheds light. If this is not done, and one uses another source first, he might sin by contradicting what God states in the Bible." Argues that all or most of the six Creation 'days' were long periods, though "there is no doubt that *each creation event* was instantaneous" and *ex nihilo*. Supports scientific evidence that universe is billions of years old. Declares that "the Bible nowhere teaches that Adam's sin brought death to nature," though it did affect it — allowing bacteria to harm man and thorns to annoy him. Denies that evolution would contradict 2nd Law or that refutation of spontaneous generation refutes it; also denies creationist accusation that evolutionists conspire against them. But insists evolution is "Not proved." Discusses natural law and miracles. Vanishingly small probabilities of natural origin of life from non-life make this no less miraculous than divine creation. Quotes **Kerkut** and Austin Clark on lack of evidence for monophyletic origin of life; also **J. Rosin.** Origin of different kinds of life is equally a miracle whether God created them de novo or out of existing organisms. Denies human evolution: "The Bible may allow for a miraculous conversion of one animal or plant into another animal or plant, but it does not allow for the miraculous conversion of an animal into man." Criticizes **Lever** and **du Noüy;** refutes theistic evolution. Evolution requires racial differences, which implies inequality. Evolutionists can provide no good argument to deny this expected inequality, thus they cannot validly oppose the "Hitler Fallacy." "Accepting the idea of human evolution encourges men to accept evil ideologies," though the fact that evil ideologies are based on evolution does not itself disprove it. However: "The Nazi, the racist, and the Marxist have merely carried the evolutionary concept to its logical conclusion." Quotes Carleton Putnam's 1961 racist book *Race and Reason: A Yankee View.* (**S. Wolfe** says that staunchly Calvinist Dordt Coll. is the "leading center" in the U.S. of the Cosmonomic movement.)

1012 ________. *The Unity in Creation.* 1978. Sioux City, Iowa: Dordt College Press. Theology and philosophy of science. Does not deal with arguments against evolution specifically, but objects to evolution as "positivistic."

1013 MacAlister, John. *The Scientific Proof of Origins by Creation* (booklet). Undated. Priv. pub. (Midland, Texas). Distributed free by Servants for the Coming New Order (a group apparently consisting of MacAlister alone). Exceptionally incoherent diatribe. Full of references to the insidious designs of Satan, atheists and communists. Elaborate charts illustrating correlation of geologic column and fossils with "creative sequence" of God revealed in Bible. Day-Age creationism. Insists sun was not created until fourth "Process-of-Time" ('day'). Says God played "cosmic jokes" such as this in order to ridicule false scientists—evolutionists—who vainly attempt to dispute His word. Man created about 4000 B.C. Refutes C-14 dating (acknowledges **R. Gentet** for information). Exhorts "brainwashed" students to resist Satan's theory of evolution. Presents questions to challenge evolutionists with: e.g., Why don't we see life emerging from non-life today? If apes evolved into man, why are there still apes? How can order come from disorder? Rails against Darwin and against "Satan's commUNist-controlled media and so-called textbooks" emanating from New York. Triumphantly proclaims tale of Darwin's deathbed repudiation of evolution: "Fortunate, indeed, in these last days of Choice-Making, are we to have this account of Darwin's closing hours, with which to counteract the life-losing brainwash by the atheist-'evolutionists'. . . ." Also rambles on about Russian-Zionist psycho-politics.

1014 Macbeth, Norman. *Darwin Retried: An Appeal to Reason.* 1973 [c1971]. New York: Dell. Orig. pub. c1971 by Gambit (Ipswich, Mass.). Based on 1967 *Yale Review* article. *Macbeth:* B.A.—Stanford; J.D.—Harvard Law; attends N.Y. Mus. Nat. Hist. Systematics Group meetings. Book has blurbs—elliptical endorsements—by K. Popper and **Koestler.** Retired lawyer Macbeth puts Darwinian evolution on trial, proves it illogical and unconfirmed. Concludes that all aspects of Darwinism are "sadly decayed." Most so-called evolutionary explanations are "hardly worthy of being called hypotheses." Fatal deficiencies of evolution are "open secret" among scientists who try to keep this secret from public. Urges evolutionists to "confess." Conspiracy-theory attitude. Adopts naive falsificationist view (invoking Popper): the scientist must abandon his theory the moment it is found incompatible with any fact. Burden of proof lies on evolutionists, whereas critic can freely snipe at shortcomings and gaps in theory without offering better alternative. Macbeth identifies "best-in-field fallacy": the mistaken notion that evolutionary theory must be true just because there is no better theory. Darwin was guilty of unjustified extrapolation in assuming variation within species accumulates indefinitely resulting in macro-evolution. Stresses tautology of natural selection (the 'fittest' are those that survive)—it is devoid of explanatory value. Selection is not creative. Ridicules R. Fisher for excessive and incomprehensible use of math in genetics. Says **Paley**'s design argument has not been defeated; Chance vs. Design issue is still open. Points out that selectionists often make their theories unfalsifiable. Argues that "Darwinism itself has become a religion." Suggests greater role for catastrophism, citing **Velikovsky** and Goldschmidt. Persuasive and well-written; many scientific references (quotes heavily from Simpson, Mayr, Huxley; also other leading evolutionists). Praises **Standen.** Macbeth admits to no religious bias or belief (though he says some creationists have good criticisms, and he was given a warm welcome at ICR). Bibliog., index. This book is widely cited and very highly praised by creationists for being devastating attack on evolutionary theory.

1015 MacBrair, Robert M. *Geology and Geologists; or, Visions of Philosophers in the Nineteenth Century* (booklet). 1843. London: Simpkin, Marshall. *MacBrair:* missionary in Africa and linguist. MacBrair assailed geology for lending support to the wicked and nonsensical notion of biological "development" (evolution), and he presented a signed copy to Chambers—prior to his *Vestiges.*

1016 McCabe, Joseph, and George McCready Price. *Is Evolution True?* Girard, Kansas: Haldeman-Julius. McCabe is pro-evolution; Haldeman-Julius is the famous atheist publisher. Presumably a debate between McCabe and **Price.**

1017 McCann, Alfred Watterson. *God—Or Gorilla: How the Monkey Theory of Evolution Exposes Its Own Methods, Denies Its Own Inferences, Disproves Its Own Case.* c1922. New York: Devin-Adair. Satirical, amusing and cleverly-written attack on evolution. Denounces the "ape-man hoax." Expresses amazement and outrage at the distortions, pretensions, and bad science of the evolutionists. Includes many standard creation-science arguments, especially alleged pre-human fossils. Opens with ridicule of Piltdown hoax, then Trinil (Java) Man and many other proposed ancestors or missing links. The scientist knows there are no missing links and "admits there is no evidence in favor of any such ascending evolution." Blood anti-serum reactions and other biochemical analyses give many results contradictory to evolution. Discusses embryology, convergence; refutes the touted horse series. Objects particularly to "ape-manologists" but also argues there is no evidence of descent between any major group. Assails "chemic creed" of evolutionists, which denies the soul, as reducing man to brutish, apish "things." "Psychical activity" and music not explainable by materialist evolution. Praises Bateson's genetics as antidote to evolution. DeVries, a Mendelist, showed that natural selection is a sieve, not a creative force. Also commends **Fabre**'s study of insect instinct. Blasts H.G. Wells for his evolution propaganda; Haeckel for his embryo frauds and other lies. Dismisses materialist theories for failure to explain mind and genius. Many scientific quotes and references. Quotes and cites many anti-Darwinist scientists. McCann, a Catholic, says Augustine, Aquinas, and other Catholics were vulnerable to bad evolutionist science. Argues for Day-Age creationism; demonstrates its correspondence with Genesis. Accuses evolution of fostering German militarism. "That there should be no weakening of the fascination of 'Darwinism,' as the theory of man's ape-origin, is ... the most disquieting and at the same time most inexplicable phenomenon of the twentieth century, for the simple reason that the preponderance of scientific evidence, including all the established data and all the opinions based on truth as it has been stripped of error, have come into court solidly against the ape, whereas, on the other hand, there remains on the side of the ape nothing but the old inferences and assumptions, nothing but the old hypotheses and unsupported theories based on erroneous or deliberately fabricated premises, nothing but the old conflicts and contradictions, nothing but the old falsifications and exposures. In their choice the nations have the alternative of chaos or Christ." Photos; index.

1018 McCann, Lester J. *Blowing the Whistle on Darwinism.* c1986. Priv. pub. (St. Paul, Minn.). *McCann:* biology Ph.D.—Univ. Utah; prof. at Coll. St. Thomas (St. Paul); has done research and written on wildlife management. Seeks "third mode" of theoretics, "Scientific Biogenesis," in place of Creationist and Darwinist positions, both of which are based on faith. Describes Darwinism in flamboyant military metaphors as anti-religious campaign. Darwinism is an evangelistic belief system which has "invasively" infiltrated and insinuated itself in social, political, educational and religious institutions. Non-Darwinists are shunned and discriminated against; intolerant Darwinists pugnaciously intimidate and suppress all dissent. "Any challenging of the tenets of Darwinism all too often evokes a bristling, bellicose reaction, which makes one realize that a personal set of beliefs is involved." But now non-biologists are challenging the ridiculous things that Darwinists are saying, pointing out "direct conflicts with fundamental scientific principles." Denigrates Darwin's scientific training, method, and logic. Much of Darwinian theory is "superficial conjecture" pretending to be real science. Chance origin of life would require an "impossibly gigantic abridgement of the Law of Entropy." "Chance Is Not a Force ... Only Controlled Forces Construct." Random forces cause only degeneration. Criticizes Miller's origin-of-life experiment. Declares that "Science Is Not History": since science depends on verifiable facts, "questions regarding the past history of organisms are of questionable scientific merit." The concept of natural selection is tautological – circular reasoning – though Darwinists ascribe magical powers to it. Mutations cannot generate evolution; genes cannot be changed constructively; chance cannot be a cause. "For Darwinists to keep saying that chance produced life and that chance produces species, goes against the most fundamental of scientific principles. Moreover, for Darwinists to maintain their position despite these clashes with the realities of science reduces Darwinism to a belief sytem – a blind creed." Darwinists are preoccupied with chemistry: "They stand with arms akimbo and sternly insist that the activities of living systems by decree must be explained only on physico-chemical, mechanistic grounds." Darwinism is not only false, it has a "baneful" influence on science, since it violates so many scientific principles. Discusses "embarrassing history" of hoaxes and deceptions in field of human evolution (Piltdown, Nebraska Man). Cites creationists, occultic evolutionists and other anti-Darwinians – **Wysong, Himmelfarb, Hitching, F. Hoyle, I. Cohen, Wilder-Smith,** others. Warns against Secular Humanism, which has maintained Darwinism in power. Intelligence is necessary for any development of order: McCann proposes that individual cells (not genes) possess intelligence. Ascribes to protoplasm itself "the ability to perform meaningfully by virtue of its own built-in intelligence." Urges a "determined movement away from Darwinism, with its menacing presence and ill-boding effects philosophically..."

1019 McCaul, Alexander. *Some Notes on the First Chapter of Genesis.* 1861. *Rev. McCaul:* Hebrew and Old Testament prof. at King's Coll., London. Day-Age creationism. Also endorses Gap Theory as complemetary to Day-Age Theory. Defends historicity of Genesis account; demonstrates its harmony with modern science. Hebrew exegesis indicates that "In the beginning" refers to "pre-mundane existence" (time prior to God's creation), and does not specify when creation occurred. It allows for all the geological ages. The creation 'days' are of indefinite length – mostly long ages. Additionally, the first day was unique – indefinitely longer than the second. (**Wonderly** 1977 quotes and discusses an 1861 essay, "The Mosaic Record of Creation.")

1020 McCone, R. Clyde (et al.). *Evolutionary Time: A Moral Issue* (pamphlet). Houston, Texas: St. Thomas Press. (Listed in Ehlert bibliog. Presumably same as article in **Patten** 1970.) *McCone:* anthropology and sociology Ph.D. – Mich. State Univ.; anthropology prof. at Calif. State Long Beach; on CSRC Board.

1021 ________. *Culture and Controversy: An Investigation of the Tongues of Pentecost.* 1978. Philadelphia/Ardmore, Penn.: Dorrance & Co. Disputes assumption the Disciples spoke in tongues they didn't understand at original Pentecost. Galilean Disciples spoke in Gentile languages: Aramaic, Greek, Latin. Listeners were amazed because the Law was only spoken in sacred Hebrew – not vernacular or local dialect. Jesus poured Spirit onto Disciples to free them from Jewish traditions, to bear witness to Christ.

1022 McCulloch, Kenneth C. *Mankind: Citizen of the Galaxy.* Advertised in Wm. Moore's UFO catalog. Mankind is descended from a colony of extraterrestrials, which was almost destroyed by a natural disaster. All that remains of previous high civilization are legends of past Golden Age and some unidentified artifacts. Rebuilding process is still underway. Once a world community is established, we can meet our ET neighbors and relatives. Includes references; index.

1023 McDaniel, Timothy R. *The Creational Theory of Man and the Universe.* 1980. Classical College.

1024 McDougal, William. *Science of Life; and Other Essays on Allied Topics.* 1934. London: Methuen. *McDougal:* psychologist at Duke Univ. Advocates a "group mind" theory. Endorses Rhine's psychical experiments (had collaborated with him at Duke). Deplores rejection of Lamarckism, which he blames on scientific materialism. Admires the moral influence of Lamarckism. Includes chapter "Was Darwin Wrong?" "I have long held that we should go boldly back to Lamarck and assume with him that the essential factor to be investigated is the effort, the more or less intelligent striving, of the organism to adapt itself to new conditions. . ." In 1920 he designed an experiment to test this, which he felt vindicated Lamarckism and refuted materialism.

1025 McDowell, Josh, and Don Stewart. *Reasons Why Skeptics Ought to Consider Christianity.* c1981. San Bernardino, Calif.: Here's Life Publishers. *McDowell:* grad. of Talbot Theol. Sem.; Campus Crusade for Christ speaker and leader; teacher at Julian Center, Int'l. School of Theol. *Stewart:* grad. of Talbot and Int'l. Seminar in Theol. and Law (Strasbourg, France); pastor in Costa Mesa, Calif.; founder-president of Answers in Christian Theology. Book is mostly on creation/evolution, but also has sections on the Bible (defense of supernatural inspiration; comparison of various editions and translations), and on Noah's Ark (usual Ark-eology tales). Covers standard creation-science arguments. **Glenn Morton** was ghost-writer for creation-science sections. Technical issues explained well. Attractive production, but many typos and errors. Full of technical and scientific references. Includes an annotated bibliog. on creation/evolution (also on other topics). Like McDowell's many other apologetics books, this is framed as series of questions and answers: a kind of faith-building catechism.

1026 ________ and ________. *The Creation.* 1984. San Bernardino, Calif.: Here's Life Publishers. Campus Crusade for Christ: Family Handbook of Christian Knowledge series. McDowell is series editor; Stewart is author of this volume. "Adapted from" *Het Onstaan van der Wereld* (**Glashouwer** and **Ouweneel** 1980), the Dutch book by the scriptwriters for the Films for Christ *Origins* series. Lavishly illustrated: color photos, paintings, drawings. Format same as that of Time-Life evolution books. Many photos are from film (of creationist John Cuozzo, e.g.). The Genesis account "is an accurate, historical summary of the original creation of this universe and all that is in it. . . . We will show the harmony between true scientific evidence (not just scientific interpretation or speculation) and the Genesis account." Deplores move away from God by science and society. Presents standard creation-science arguments; emphasizes Flood catastrophe and need to accept Christ. "The heart of our argument, and the total defect in the evolutionary argument, is that Christians have an *adequate source* for the development and complexity in the universe, while evolutionists have *no adequate source*. . ." Life "could not have evolved spontaneously or by chance. . . . Evolution is incapable of accounting for the complexity and design everywhere evident in living organisms." There is "no direct evidence of the macroevolution proposed by scientists who presuppose a naturalistic and mechanistic world." Rather, the evidence supports the creation model. Stresses need for defense of Christianity by reason and evidence. Includes thermodynamics argument; coexistence of dinosaurs and man (Paluxy, folk-art). Presents some Mendelian genetics; illustrates harmfulness of mutations. Endorses Flood Geology: fossils and geological strata attest to great catastrophe and lack of transitional forms. Author "leans towards" young-earth creation, but considers old-earth creationism a possible option. Rejects **Ussher**'s 4004 B.C. date. "The Bible clearly states ... that the entire earth was cataclysmically altered by a great flood in the time of Noah. The Bible also clearly states that life was created by God in its various 'kinds,' and that humans were God's special creation, in His own image." Describes world Flood myths; Arkeology. "We, too, can be part of 'Noah's family,' and know that God will rescue us from the coming judgment." Some scientific references; text consists primarily of quotations from creationists: **Davidheiser, Rehwinkel, H. Morris, Rimmer, Anderson** and **Coffin, Gish, Klotz,** others. Recommended Reading list consists entirely of creationist books. Index.

1027 McFarlane, P. *Exposure of the Principles of Modern Geology*. Edinburgh: Thomas Grant. Pub. around 1850. Discussed in **H. Miller** 1857 as an especially amusing example of attacks by "anti-geologists." Says McFarlane is a Scotsman who "has been printing ingenious little books and letters against the infidel geologists" for several years. Earth prior to the Fall was more than twice present size. It was hollow, supported inside by an intricate metal framework, with granite overlying the internal framework, and a shell of soil and water spread on top of that. This outer surface was tastefully and neatly arranged to produce a "perfectly paradisaical" habitat; because the atmosphere was spread much thinner, storms did not exist. There was a great fire in the earth's center; flames shot out openings in the poles, warming the earth and lighting it with magnificent auroras. The metal "firmament" collapsed at the Fall, and the granite crust settled into compacted segments or plates at different levels, separated by great cracks, thus isolating various groups of animals. During the Flood, these groups were separately submerged: this accounts for the different fossil forms, and the fact that man is found in the deeper layers. The Flood caused the total collapse of the granitic dome, extinguishing the inner fire. Our present earth is a nasty, ugly ruin—an appropriate prison for fallen mankind.

1028 MacFie, Ronald Campbell. *Science Rediscovers God*. 1930. Edinburgh. "The mud is still here; it has been here under competent scientific observation for hundreds of years; ... and yet scientific men admit that life never occurs unless as progeny of former life—that when all life is destroyed and the medium sterilized—and sterilized the earth's crust certainly was—life never occurs." (Quoted in **Graebner** 1943.)

1029 _______. *The Theology of Evolution*. 1933. London: Unicorn Press. "Too often [evolution] has degenerated into a confession of faith to be maintained by the faithful against the infidels," and its proponents ignore the mounting difficulties of the theory. Evolution is not proven. Living things prove a "prescient Mind" at work. Darwinism is a "spiritual tragedy"—a very bad influence. Cites many scientists as anti-Darwinian: includes Bateson and Haldane, plus several creationists. Discusses limitations of variation. *If* evolution occurred, evolutionary leaps were directed by Mind; Mind is God. Basically a strong anthropic principle argument. Discusses biological homology versus heterology. Includes Nebraska Man, Java Man. Index.

1030 McGee, J. Vernon. *Evolution and You* (booklet?). 1964. Los Angeles: Church of the Open Door. (In Ehlert files.)

1031 _______. *Genesis: Volume I*. 1980 [1975]. Pasadena, Calif.: Thru the Bible Books. "Messages given on the 5-Year Program of Thru the Bible Radio Network." *McGee:* Th.D.—Dallas Theol. Sem.; Presbyterian pastor; former pastor of Church of the Open Door (Los Angeles); heads Thru the Bible Radio program. This book part of set on whole Bible. Chatty style. "...I reject evolution because it rejects God and it rejects revelation." Predicts it won't survive this century. Ridicules science theories. Says "some of the most reputable scientists of the past, as well as of the present, reject evolution." Some scientists quoted: **Nilsson, Kerkut,** etc. Recommends **H. Morris** and ICR. The only approaches to questions of origins are "by faith or by speculation—and speculation is very unscientific, by the way." Mentions Paluxy evidence favorably. Scientists won't talk about moon rocks because they disprove their theories. Says biblical 'kind' equivalent to phylum. Advocates Gap Theory creationism. Describes original creation of the universe billions of years ago. Then there was a great catastrophe, followed by the reconstruction of earth by God (the six-day creation). The pre-Adamic catastrophe convulsed the entire universe. It involved some pre-Adamic creature and the fall of Lucifer (Satan). Describes the six literal days of the re-creation. Affirms worldwide Flood; mentions pre-Flood water canopy. Emphasizes parallels between conditions of Noah's time and imminent Rapture, Tribulation and Christ's Millennium. Recommends **Scofield** Bible. Does not want creationism taught in school (as recently recommended by Calif. Board of Educ.) because most teachers not qualified to teach it properly. Bibliog.

1032 McGowen, Charles H. *In Six Days.* c1976. New Wilmington, Penn.: Son-Rise. Also pub. by Bible Voice (Van Nuys, Calif.). *McGowen:* M.D.—Ohio State Univ.; USAF School Aerospace Medicine grad.; now has internal medicine and endocrinology practice in Youngstown, Ohio. Converted to Christ after wife's illness; was evolutionist until age 34. Standard creation-science arguments; recommends ICR and CRS, and follows their arguments very closely. Urges everyone "to seek ways to influence his school system so that these truths might become a regular part of the classroom instruction." Usual young-earth proofs; Flood Geology. Thermodynamics argument, radiometric dating, population increase rate, Paluxy evidence, probability arguments. Chapter "God's Greenhouse" on the pre-Flood Water Canopy thirty miles above the earth which created the uniform warm climate and allowed for the increased life-spans. Chapters on the worldwide Flood, description and cargo of Noah's Ark, dispersion of human races after Flood (explains in terms of genetics). Chapter on medical and scientific truths contained and predicted in Bible (recommends S. McMillen's *None of These Diseases*).

1033 McGraw, Onalee. *Secular Humanism and the Schools: An Issue Whose Time Has Come* (pamphlet). c1976. Washington, D.C.: Heritage Foundation. A Heritage Foundation report: Critical Concern Series. *McGraw:* government Ph.D.—Georgetown Univ.; coordinator of National Coalition for Children; education consultant for Heritage Foundation; now VP of the Educational Guidance List. Secular humanism is being taught and promoted in the schools. It opposes the Judeo-Christian tradition, and is responsible for the decline in education. McGraw mentions the "Creator God" as a basic tenet of the "Judeo-Christian tradition," but otherwise no overt reference to creationism or evolution. Exposes teaching of situation ethics, values clarification, MACOS, etc. Provides legal references for Secular Humanism as a state religion. Includes Conlan's 1976 anti-humanist bill.

1034 McIntyre, Melody J. *Dinosaur ABC's.* c1981. San Diego: CLP (Creation-Life). ICR's Two-Model Children's Book series. **Richard Bliss:** series Project Director. "Designed for use by children in the elementary grades in public schools, these two-model books contain creation/evolution discussions on an introductory scientific level" (ICR catalog). For 4 to 6-year-olds. Pictures and simple descriptions of dinosaurs, including Paluxy evidence (contemporary human footprints). Mentions Flood and Creation as explanations for fossil evidence; evolution as another possibility.

1035 MacKay, Donald M. *Science and the Quest for Meaning.* 1982. Grand Rapids, Mich.: William B. Eerdmans. 1979 Pascal Lecture series: Univ. Waterloo (Ontario, Canada). *McKay:* radar researcher in WWII; taught at Oxford, Cambridge, Berkeley, Johns Hopkins, etc.; then at Univ. Keele; pioneer of theoretical neurobiology—worked on communications theory, computing, brain (vision and hearing); also brain/mind philosophy. A distinguished scientist (d. 1987), MacKay was also an active Christian apologist. Compatibility of science with religion: science doesn't destroy meaning. Warns against "scientism." Says evolution is theologically neutral—but "evolutionism" is meta-scientific.

Recommends **Hooykaas;** notes that science "de-deified" nature—a biblical attitude. Urges "stewardship" rather than exploitation of nature. Criticizes Monod for "deifying" chance—making it an agent. Argues that miracles can't be ruled out; science can't disprove their existence. Often cited as supporting Bible-science, but MacKay does not deny evolution.

1036 Mackay, John, Andrew Snelling, Carl Wiegand, Dean Anderson, Ken Ham, Charles Taylor. *The Quote Book* (booklet). c1984. Sunnybank, Australia: Creation-Science Foundation. Compiled by members of the Australian Creation-Science Foundation. Collection of quotes, mostly from scientific books and journals by evolutionists or non-creationists, which are critical of evolutionary theory (or seem to be). 112 quotes, from 74 people (almost all scientists—except M. Muggeridge); arranged by topic. Includes many anti-Darwinian evolutionists, a few hostile to evolution, but no creation-scientists. Many quotes taken out of context—presented as critical of or harmful to evolution. Photos, drawings; index.

1037 McKerrow, James Clark. *Evolution Without Natural Selection.* 1937. London: Longmans, Green. Contains discussions of life as "habit"; the emergence of man; the evolution of "sapience." Lamarckian and/or metaphysical. McKerrow also wrote *The Appearance of Mind* (1923).

1038 McMurray, N. *Chiroptera: The Problem of Bats* (pamphlet). Undated. England: Evolution Protest Movement. EPM pamphlet #158. Also wrote pamphlet #171 on *Ophidia* (snakes).

1039 McWilliams, Donald A. *The Myth of Evolution.* c1973. Fullerton, Calif.: Plycon Press.

1040 Maddoux, Marlin. *America Betrayed!* c1984. Shreveport, La.: Huntington House. *Maddoux:* radio journalist; president of International Christian Media (Dallas). CSLDF says this book is a "clear presentation of how the teaching of evolution-humanism has contributed to the decline of moral values in this land." Maddoux denounces media manipulation and anti-Christian proselytizing. Quotes and endorses W.B. Key on influence of subliminals; presents former rock musician talk-show guest who now witnesses against the evils of rock music. Discusses and praises the **Gablers**' book-monitoring, and cites examples of evolutionist propaganda.

1041 Magne, Charles Lee. *The Negro and the World Crisis.* Undated [1970s]. Hollywood, Calif.: New Christian Crusade Church. Explicitly claims to be Christian; KJV Bible is inerrant and proves truth of arguments presented here. There is a Jewish conspiracy to destroy the White Nordic Israel race by using Negroes. Whites descended from Adam. Negroes are the top animals—members of the ape family. They are a pre-Adamic creation (with the other beasts), without spiritual consciousness. Negroes survived the (regional) Flood. The fallen angels mated with human (white) girls, producing monster offspring. Exodus, for similar reasons, commands the death penalty for miscegenation: that is, for mating with beasts (Negroes). Cites **Rushdoony** approvingly. Lincoln planned to send slaves away from U.S.; this was thwarted when evolutionist ideas took over. Garden of Eden was in the Himalayan Pamirs. Ararat was in the Tien Shan Mts. of China.

1042 Malaise, Rene. *A New Deal in Geology, Geography and Related Sciences.* 1969 [1951]. Priv. pub.(?) (Sweden). Orig. pub. 1951 in Swedish *(Atlantis, en Geologisk Verklighet). Malaise:* Swedish Museum Natural History (ret.). Does not discuss evolution, except of races of man. Develops "constriction theory" of N. Odhner (also of Swed. Mus. Nat. Hist.); rejects Wegener's continental drift. Cycles of ice ages and mountain-building. Atlantis existed between (and included) Azores and Iceland. Origin of Flood myths from subsidence of Atlantis. Atlanteans tried to expand their European empire: they were the "Sea People" who harried Egypt; the builders of Stonehenge. They migrated to America before the Indians. Pygmies are of direct descent from creatures of the Tertiary.

1043 Malik, Peter. *The Ape Men* (booklet). c1981. Uniontown, Penn.: Penn Highlands Publications. Comic book. "Christian Warriors" series #3. Prehistoric adventure of cave-men pursued by dinosaurs—turns out to be T.V.

series. Student viewers complain that cave men didn't co-exist with dinosaurs, but a Christian explains that they really did, citing the Bible and many creation-science sources. Explains about human prints found with dinosaur tracks at Paluxy. "There is no way you can fit evolution in with what the Bible says": shows contradictions between Genesis and evolution. The students get George E. King, Ph.D., from Teens for Christ to lecture on creation-science at their school. King demolishes evolutionist teacher with scientific arguments (standard creation-science arguments; many creationist references). "I think by now we should be able to see that Darwin's theory of evolution is simply a religious belief based upon faith and opposed to fact. If evolution is to be taught in public schools, then scientific creationism should be given time!" Evolutionist explodes with rage when King mentions God, but King continues to expound scientific creationism in local church. Urges acceptance of Christ for salvation.

1044 Mandock, Randal L.N. *Scale Time Versus Geologic Time in Radioisotope Age Determination.* 1983. San Diego: Creation-Life. Orig. 1982 Institute for Creation Research M.S. thesis.

1045 Mandock, Richard E.P. *Theoretical Thermal Calculations for Heat Distributions with Spherical Symmetry.* 1983. Unpub. thesis: Institute for Creation Research (El Cajon, Calif.). ICR M.S. thesis.

1046 Maniscalco, Joe. *Flying Saucers and Winged Serpents* (booklet). Undated. Inchelium, Wash.: The Loud Cry. Comic-book format. Satan is programming the world for his master deception, when he will delude millions by impersonating Christ, arriving in a UFO. Maniscalco, a Seventh-day Adventist, has also written nature books.

1047 Marinov, Stefan. *Eppur Si Muove.* 1977. Brussels, Belgium: C.B.D.S.—Pierre Libert. Foreword by Andrei Sakharov. *Marinov:* space scientist at Sofia Observatory, Bulgaria; now lives in Belgium (**Bouw** says he was exiled because of his beliefs); Greek Orthodox. A geocentrist and creationist, Marinov participated in a U.S. Bible-Science Creation Conference. Argues for space-time absoluteness and against Einstein's relativity. First part on theoretical aspects of relativity of motion; second part on experimental evidence (inc. many of Marinov's own experiments). (Discussed by **Bouw** in 1981 *CRSQ.*) Title ("But it does move") is sotto voce remark attributed to Galileo after he was forced to publicly renounce heliocentricity. Not to be confused with 1987 book by astronomer F. Graham of same title, which is a priv. pub. refutation of modern creationist geocentrism (Graham, with **Niessen,** debated Bouw and **Hanson** at 1986 Creation Conf.). Marinov also wrote *The Thorny Way of Truth Part II: Documents on the Invention of the Perpetuum Mobile, on the Centurial Blindness of Mankind, and on Its Frantic Perseverance in It,* about his perpetual motion machine which extracts energy from earth's motion through the aether (Marinov believes that Joe Newman's free energy machine employs the same principle).

1048 Marra, William. *Evolution and the Lordship of God* (audiotape). Distributed by Keep the Faith (Catholic), which also distributes other tapes by Marra: *The Distractions of Evolution* (against Teilhard), and *Is Science Under a Cloud? Marra:* philosophy prof. at Fordham Univ.; chairman of Roman Forum (Catholic action org.). 1988 U.S. pres. candidate. (Children of Mary, another Catholic group active in promoting the Bayside Shrine [Marian apparitions], distributes an anti-evolution video by Marra: *The Evolution Controversy.*)

1049 Mars, Ross, and Peter Havel. *And God Said "Let There Be. . ."* (pamphlet). Undated [1980s]. Guilford (?): Hamar Publications.

1050 Marsh, Frank Lewis. *Fundamental Biology.* 1941. Priv. pub. (Lincoln, Neb.). *Marsh:* zoology M.S.—Northwestern Univ.; botany Ph.D.—Univ. Nebraska; biology prof. at Union College (Lincoln, Neb.); later at Emmanuel Missionary College (now Andrews Univ.) in Berrien Springs, Mich.; Seventh-day Adventist; a founder of CRS; first chairman of Geoscience Research Inst. (Loma Linda). Includes statement of tenets of special creation, patterned after H.H. Newman's statement on what evolution is and isn't. Creationism appeals to science, and requires less faith for belief than evolution. Modern humans are not the same as originally created man—modern races are degenerated forms of first created man. The medieval doctrine of special creation, retained by **Agassiz,** is mistaken. Nature is not static; variation occurs within "Genesis kinds." Creationism brooks no acceptance of bestial behavior by man, and denies that man descended from monkeys or that half-human races can be formed by breeding with beasts. It does not teach that all indigenous species were created *in situ:* plants and animals have been redistributed since the worldwide Flood of Noah. Coins the term "baramin" for Genesis 'kind' (from "create" and "kind"). Marsh, who carried on a long correspondence with evolutionist Dobzhansky, is praised by **Frair** as a central figure in the development of modern creation-science.

1051 _______. *Evolution, Creation and Science.* c1947 [c1944]. Washington, D.C.: Review and Herald. 2nd ed. Orig. c1944. Knowledgeable about biology; avoids many of the more egregious creationist mistakes. Attempts to correct obsolete versions of creationism inherited from Medieval scholastics with modern scientific version. Darwinism triumphed because contemporary critics had only a distorted and scientifically inaccurate version of creationism. "The only authority that the scientist can accept as a scientist is the authority of the facts of natural history." Affirms, however, scientific inerrancy of Bible, and says the Genesis account of creation provides a totally satisfactory explanation of all phenomena. Special creation requires "less faith" to believe in than organic evolution. Both evolution and creation are testable, falsifiable. Considers mutation and variation sufficient to have produced all races of man from Adam. Discusses processes of variation: hybridization, mutation, etc. Distinguishes between 'variation' (occurring within 'kinds') and 'evolution' (held to occur between kinds). Genesis kinds cannot cross-breed. Expands on "baramin" concept, emphasizes it is not the same as 'species'; notes that mankind is only sure example of a baramin. All land animals destroyed by global Flood except those in Noah's Ark. Discusses variation since Flood—adaptation and biogeography. Relies on **C. Courville** in embryology chapter. Summarizes and endorses **G.M. Price**'s paleontology arguments. Many scientific references. Includes chapter "A Creationist's Creed": affirms strict young-earth creation, global Flood which produced most or all of the paleontological record. Bibliog.

1052 _______. *Evolution or Special Creation?* (booklet). c1963 [1947]. Washington, D.C.: Review and Herald. 2nd ed.; rev. Orig. 1947. Distributed free by Seventh-day Adventist Information Ministry. Emphasizes special creation, and fixity of 'kinds' (not of species, which can vary). Standard creation-science arguments. Affirms scientific accuracy of Bible. Concedes that nature does not compel belief in creationism—it is a matter of faith whether we choose to believe in evolution or creation. A summary of arguments from Marsh's books. "Surely the time is ripe for a return to the fundamentals of true science, the science of creationism."

1053 _______. *Studies in Creationism.* c1950. Washington, D.C.: Review and Herald. Is man "Bestial or Divine?" Discusses earlier creationist theories (the Scholastics, **Bonnet, Linnaeus, Cuvier, Agassiz**); contrasts these with "modern creationism." Also gives history of evolutionist concepts. Argues against uniformitarianism in refuting calculations of ancient age for earth. Accepts creation with appearance of age (not deceptive, since God explained it to Adam). Affirms Bible-science. "There is no conflict between true science, that is, natural facts, and the true Christian religion except as the student of this vitally important issue employs faulty technique." Presents standard creation-science arguments. Scientific, theological and biblical references; cites several anti-Darwinists, especially **G.M. Price, E.G. White.** Second half of book is creationist exegesis of Genesis. Long section describing Creation Week. Favorable description of "Vapor-Envelope Theory" (pre-Flood Canopy). Sun was created before fourth day, but covered by clouds. Since the Fall, Satan has caused nature to deteriorate, causing disease by "derangement" of organisms, inducing them to invade our bodies. Genesis "sons of God" and "daughters of men"—Sethites and Cainites. Most fossil hominids are really early post-Flood humans, but Heidelberg and Peking Man may be antediluvian. Discusses Noah's Ark and Flood Geology. Jacob and Laban's sheep (Bible teaches Mendelism). Clean and unclean animals (hygenic and dietary reasons). Advocates lacto-ovo-vegetarian diet. Bibliog.

1054 ________. *Life, Man, and Time.* 1967 [c1957]. Anacortes, Wash.: Outdoor Pictures. Rev. ed. Orig. pub. by Pacific Press (Mountain View, Calif.); c1957. Continuation of arguments from previous books: historical theories of origins, creationism according to Genesis; degeneration of nature and organisms since the Fall and Flood; Flood Geology; biological variation within but not between Genesis 'kinds.' Solar system created *ex nihilo,* but rest of universe already in existence. Creation: 5000 or so B.C. Flood: 3300 or so B.C. Criticizes assumptions of radiometric dating. Suggests that pre-Flood water-vapor "envelope" (canopy) throws radiometric dating off. Decries ignorant ridiculing of the Bible by those who wrongly assume, without troubling to check, that it contains scientific error. "The main reason why creationism is talked down so generally today is probably the fact that evolutionists do not take the time to read the Bible carefully for themselves." Antediluvian conditions produced a healthy balanced diet; after the Flood, man became a carnivore to supplement deficiencies of post-Flood vegetation but lowered lifespan as a result. Admits that evidence can be interpreted in terms of evolution or creation, but denies notion that God is deceptive, since His Bible tells us clearly that creationism is true. Angels brought animals to Noah; small dinosaurs were included on Ark. Objects to statement by Dobzhansky that Marsh is virtually the only scientist who rejects evolution: claims all ASA members do, and that there are many creationists in universities who are forced to remain incognito because of evolutionist intolerance. Bibliog., index.

1055 ________. *Variation and Fixity in Nature.* c1976. Mountain View, Calif./Omaha, Neb.: Pacific Press. "The Meaning of Diversity and Discontinuity in the World of Living Things, and Their Bearings on Creation and Evolution." Updated and expanded discussion of the fixity of created 'kinds.' Although Marsh refers to "Genesis kinds" throughout, and also discusses Creation Week and the Flood, other biblical references are much fewer than in his previous books. "The basic types, the created kinds, the baramins, stand so manifest and so clearly defined in nature by appearance and reproductive behavior as to constitute a delight to the student observant enough to fix his attention upon the level of the forest rather than upon the trees which constitute it." Marsh insists all alleged proof of evolution is really only evidence of micro-evolution: variation within kinds. "The more that is learned about variation, the greater the harmony between demonstrable facts and special creation." Darwin discovered species can change and thought this disproved the Bible—but all this change is within "baramins." "One of the most basic and well-demonstrated of biological principles is that of the *limitation of variation.*" Develops theory that baramins are defined by ability to hybridize. "The Bible knows nothing about organic evolution. It regards the origin of man by special creation as a historical fact. . . . In view of the subjectivity of the evidence upon which a decision on the matter of origins must be made, creationism and evolutionism should be respected as alternate viewpoints." Many scientific references. One chapter adapted from 1969 *CRSQ* article. Index.

1056 Marshner, Susan. *Man: A Course of Study (MACOS)—Prototype for Federalized Textbooks.* 1975. Washington, D.C.: Heritage Foundation. Listed in E. Lazar's 1987 priv. pub. C/E bibliog.

1057 Marston, Sir Charles. *The Bible Is True.* 1938 [1934]. London: Religious Book Club. Orig. pub. 1934 by Eyre and Spottiswoode (London). *Marston:* VP of British School of Archeology in Egypt; excavated Jericho with Garstang; Fellow of Soc. Archeol. **Field** 1941 says he rejects evolution in this book, which concerns the 1925–34 digs.

1058 Martin, James Lee. *Monkey Mileage from Amoeba to Man.* 1938. Grand Rapids, Mich.: Wm. B. Eerdmans. Purpose of book is to lead open-minded searchers of the "origin of matter and Man into those channels of Thought and Reason that lead on beyond where Science ends to where Faith in a Creator of All Things begins. . . . Our research to that end may be scientific, but Science itself can deal only with *facts* reduced to an immutable *law.*" Ridicules evolution. Opens with extended first-person account of "My Evolutional Record": "When I Was an Amoeba . . . When I Was an Amphioxus [etc.]." Asks why amoebas, the purported evolutional germ of every form of life, still exist today—why they haven't all evolved. Notes that apes cannot be our ancestors since evolutionists place them at the end of descent branches. The fact that Darwinian theory has been modified is "conclusive proof that Darwin was theorizing rather than dealing with established facts; because a *fact* once established as such remains unimpeachably a truth." Most leading evolutionists have now rejected the Darwinian theory, since no Missing Links have ever been found. Branches of science such as "geology, cosmology, anthropology, paleontology, and the most highly speculative of any science or theory yet advanced—evolution, as some of them are being taught and understood today, are in direct opposition to true religion. . . . [T]he Christian scientist alone enters upon the study of science aright." Natural selection is an imaginary process invented by a deluded atheist. Says some "present-day anthropologists" classify Pygmies with anthropoid apes, and think that the human races evolved from different ape species. Explains origin of races from Noah's sons; God changed their skin colors. The "most perfect" human skull ever found was Caucasian, from near the Garden of Eden. So-called "sub-men" are degenerate forms of once perfect humans. Evolutionist fossil restorations are "highly misleading deceptions"—hoaxes. Favors recent creation: why would God wait if He can speak things into existence? Flood was ca. 2300 B.C. "To claim that man was . . . descended from lower orders and species, would imply a belief reducing to naught man's highest ambition—to live again." "That there has been no development of plant into animal, and no evolution of brute into man has already been fully established." Some scientific references. Many arguments directed at H.G. Wells' *Outline of History;* also at Dorsey. Drawings (by J.L. Martin, Jr.).

1059 Martin, Richie (ed.). *Judgment in the Gate: A Call to Awaken the Church.* 1986. Westchester, Ill.: Crossway Books. Contributors include B. LaHaye (wife of **T. LaHaye**), B. Bright (Campus Crusade colleague of **McDowell**), D. Swaggart (son of **J. Swaggart**) on government, **Maddoux** on the media, F. Schaeffer (son of **F. Schaeffer**) on economics. **Blumenfeld** has a chapter on humanism and evolution; says evolution is "pure speculation." Includes usual creation-science arguments; Bible references.

1060 Martin, Thomas Theodore. *Hell and the High Schools: Christ or Evolution, Which?* 1923. Kansas City: Western Baptist Publishing. *Martin:* Director General of Bible Crusaders of America; Field Sec'y. of Anti-Evolution League of America; Dean of Evangelical School of Union Univ. (Tenn.); former science prof. at a Texas Baptist college. "But, if evolution, which is being taught in our high schools, is true, the Savior was not Deity, but only the bastard, illegitimate son of a fallen woman, and the world is left without a real Savior. . ." Evolution, which is neither truth nor science, says that Genesis is a lie. Evolution is drilled into high school students, who are extremely susceptible to its propaganda, and even taught in lower grades. "Ramming poison down the throats of our children is nothing compared with damning their souls with the teaching of evolution, that robs them of a revelation from God and a real Redeemer." Demands that public school boards refuse to employ "any teacher who believes in evolution"; all teachers should be required to attack evolution and "expose it every time it comes up in any textbook." "We gave our sons to save the world from being crushed by the Germans, . . . but they had already stealthily crept in," infiltrating the schools. "The soul of one high school boy or girl sent to hell by your German evolution is worth more than the bodies of all our brave boys killed in the great war in Europe. But they are being sent to hell by the thousands. . ." Arrogant evolutionist scientists sneeringly brand their opponents ignoramuses. They whine about academic freedom, but there should be no freedom to teach Bible-destroying, soul-damning evolution. (Excerpted in Gatewood.)

1061 _______. *The Evolution Issue* (booklet). 1923(?). Los Angeles. Address delivered in Los Angeles in 1923.

1062 _______. *Evolution or Creation: Christ or Hell?* Undated(?). Priv. pub. (Los Angeles). (Listed in Ehlert files.)

1063 Mason, Thomas Monck. *Creation by the Immediate Agency of God, as Opposed to Creation by Natural Law; Being a Refutation of the Work Entitled Vestiges of the Natural History of Creation.* 1845. London. A refutation of Chambers' *Vestiges* (1844).

1064 Matthews, L. Harrison. *Introduction to Darwin's Origin of Species.* 1971. London: J.M. Dent and Sons. Very critical introduction to Darwin's *Origin,* much quoted by creationists. Matthews says "belief in the theory of evolution" is "exactly parallel to belief in special creation." Evolution is merely "a satisfactory faith on which to base our interpretation of nature."

1065 Maunder, E. Walter. *The Astronomy of the Bible.* 1908. London: T. Sealey Clark. Undated, but Preface dated 1908. Rev.; orig. pub. 1904 in *Sunday at Home* magazine. *Maunder:* prominent astronomer at Royal Greenwich Observatory; founder and president of Brit. Astron. Soc. Includes section on Joshua's Long Day. Command for sun to stand still really meant to "be silent." The miracle was a relief from the noonday heat by cloud cover (the hailstorm), which allowed for successful pursuit of the enemy. Book includes detailed analysis of Joshua's military tactics.

1066 Mauro, Philip. *The World and Its God.* 1908 [1907]. London: Morgan & Scott. 2nd ed. *Mauro:* lawyer; contributor to *The Fundamentals* series. Purpose of book: a rationalistic test of the biblical account of creation, especially the origin of evil. The Darwinian theory has collapsed. Mauro's anti-evolution arguments are mostly theological and philosophical, and are aimed as much at Spencer as at Darwin.

1067 _______. *The Number of Man: The Climax of Civilization.* c1909. New York: Fleming H. Revell. Also pub. by Hamilton Bros. (Boston), and Morgan & Scott (London).

1068 _______. *Evolution at the Bar.* c1922. New York: George H. Doran. 2nd ed. Orig. pub. by Hamilton Bros. Scripture Truth Depot (Boston). Mauro tries evolution as if in a court of law. Evolution loses. Evolution is "not scientific, for science has to do only with facts. Evolution belongs wholly in the realm of speculative philosophy." If life arose from inorganic matter, we should find matter reaching towards organic existence now. If evolution were true "there would be no species or other lines of division." In fact, though, nature is composed of sharply defined groups of organisms separated by impassable barriers. "Christianity . . . destroys evolution down to the ground."

1069 _______. *The Wonders of Biblical Chronology.* 1970 [1922]. Swengel, Penn.: Reiner Publications. Also pub. 1923 by S.E. Roberts (London). Orig. pub. 1922 with the title *The Chronology of the Bible.* Mauro dates the Flood at 2390 B.C.

1070 _______. *Evolution: A Religious Dogma* (booklet). 1967 [1920s]. Malverne, N.Y.: Christian Evidence League. Orig. editorials in the *Bible Champion,* all dating from the 1920s. Also includes excerpts from Mauro's *Evolution at the Bar,* and selections from **G.M. Price, Frank Allen, Fleischmann.** Agrees with **G.M. Price** that "Mendelism has put Darwin out of business." But evolution is still popular because it is a religious faith. Ridicules the "amazing credulity of the evolutionist," and demolishes the supposed evidences for evolution. "Supreme test" of every doctrine is whether it accords with Christ; the "supreme disproof of evolution . . . is the Risen Christ." Comments at length on Robinson's 1922 *Harpers* article "Is Darwinism Dead?"; also discusses bee society, quoting from Stewart's 1925 *Atlantic Monthly* article "The Bee's Knees." Lack of transitional forms, embryological evidence. Natural selection cannot allow for half-finished organs.

1071 May, Peter. *Superhuman Message Number One: The Greatest Book of*

Universal Knowledge this World Has Ever Known. 1980. Carlton.

1072 Mehlert, Albert W. *The Student's Guide to Evolution and the Bible* (pamphlet). Undated. Bible-Science Association. *Mehlert:* lives in Lawton, Australia. "Nothing in all history has been more damaging to young people's attitudes toward life and the Bible than the so-called theory of evolution which is an unproven fairy-tale of wishful thinking and speculations, designed to do away with God and with man's responsibilities to Him." Historical proof of Jesus. Asserts total lack of Precambrian fossils. Argues for creation with appearance of age: of soil, of starlight, of Adam. Refutes apemen fossils.

1073 Meldau, Fred John. *Why We Believe in Creation Not in Evolution.* 1974 [1959]. Denver, Co.: Christian Victory Publishing. 8th ed. Orig. 1959. *Meldau:* editor of *Christian Victory* magazine. Foreword by **L.V. Cleveland.** "If Evolution Is True the Bible Is False. . . . But if the Bible is true—and we are absolutely certain it is—then evolution is merely the vain imaginings of biased men, men determined they will *not* believe in a Supreme Being, but ready to believe any kind of theory that might be a possible substitute for the evident fact of creation." "Gloating" evolutionists aim for totalitarian "scientific socialism," thus bringing in communism through the back door. Most of this 359-page book consists of densely packed examples of "witnesses against evolution": wonders and designs of nature—hundreds of animal and plant adaptations, design of earth's physical properties and ecosystem, social insects, marvels of the human body, etc. Also discusses fossil frauds, Second Law. Nature teaches moral lessons to man: e.g. the deadly cunning of the snake. Meldau mentions the "cataclysm" of Gen. 1:2 as well as the Flood, implying belief in Gap Theory creationism. Many quotes and examples from popular and scientific literature (especially *Scientific Amer.*). Bibliog., index.

1074 _______ (ed.). *Witnesses Against Evolution* (booklet). 1968, Denver, Co.: Christian Victory Publishing. Contributors include **W.B. Dawson, D. Dewar, Leander Keyser, Arthur Brown,** and **Meldau.**

1075 Metcalfe, John. *Noah and the Flood.* 1976. Penn, Bucks, England: J. Metcalfe Publishing Trust. Mostly biblical arguments rather than creation-science.

1076 Meyer, Louis (ed.). *The Fundamentals: Vol. VI.* Undated [1911?]. Chicago: Testimony Publishing. Continuation of *The Fundamentals* series. (See **A.C. Dixon** for previous vols.) *Rev. Meyer:* evangelist; a "Christian Jew." This vol. inc. "The Early Narratives of Genesis" by Prof. James Orr (D.D.; United Free Church Coll., Glasgow, Scotland). Orr stresses importance of Genesis in proving God created all things, the "moral catastrophe" of the Fall and the Flood, and archeological confirmation of its historical accuracy. Affirms that "the narratives of Creation, the Fall, the Flood, are not myths, but narratives enshrining the knowledge or memory of real transactions." Science does not contradict the Creation account, which was written in popular every-day language, but rather confirms it. Accepts some evolution, however, and does not argue for recent creation. "Certainly there would be contradiction if Darwinian theory had its way and we had to conceive of man as a slow, gradual ascent from the bestial stage, but I am convinced . . . that genuine science teaches no such doctrine. Evolution is not to be identified offhand with Darwinism. Later evolutionary theory may rather be described as a revolt against Darwinism, and leaves the story open to a conception of man quite in harmony with that of the Bible." "Man's origin can only be explained through an exercise of direct creative activity, whatever subordinate factors evolution may have contributed." Suggests the Flood covered all areas inhabited by man, but was not worldwide.

1077 _______ (ed.). *The Fundamentals: Vol. VII.* Undated [1911?]. Chicago: Testimony Publishing. This vol. inc. "The Passing of Evolution" by **G.F. Wright,** who explains that "evolution" has different meanings, and some legitimate usages. "The widely current doctrine of evolution which we are now compelled to combat is one which practically eliminates God from the whole creative process and relegates mankind to the tender mercies of a mechanical universe . . . without any Divine direction." But Darwin himself never claimed as much, Wright says, though many of his followers have. Darwin only argued that varieties can evolve into different species, which Wright accepts. Wright objects to claims that all species share a common ancestor which evolved from non-life, and that there has been no supernatural influence. Argues that simultaneous evolution of coordinated traits requires supernatural guidance. "By no stretch of legitimate reasoning can Darwinism be made to exclude design." Notes that Darwinists are now divided into many competing schools. Allows for earth millions of years old, but not the "unlimited geological time required by Darwin's original theory." Emphasizes that "The failure for evolution to account for man is conspicuous." Cites scientific critics. Points out that evolutionism is ancient, and states that hostile meta-physicians are worse foes of Christianity than scientists. "The evidence for evolution, even in its milder form, does not begin to be as strong as that for the revelation of God in the Bible."

1078 _______ (ed.). *The Fundamentals: Vol. VIII.* Undated [1912?]. Chicago: Testimony Publishing. This vol. inc. "Evolutionist in the Pulpit" by "An Occupant of the Pew" (orig. in 1911 *Herald and Presbyter*), and "Decadence of Darwinism" by Rev. Henry H. Beach (Colorado). "Occupant" heaps scorn upon evolution, the "gospel of dirt," but does not address the age of the earth, and allows for evolution (degeneration) of "fallen man." Darwin's aim was to abolish dualism. Evolution is "pure supposition"; its denial of supernaturalism discredits the entire Bible. "That the Darwinian theory of descent has in the realms of nature not a single fact to confirm it is the unequivocal testimony" of distinguished scientists: names Shaler, Etheridge, L.S. Beale, **Fleischmann.** Cites books by George Paulin, *No Struggle for Existence; No Natural Selection* (Scribners), **Townsend, Dennert.** Excoriates some ministers for basing their preaching on evolution when even most scientists realize it is dead. Those scientists who still cling to it are simply seeking a mechanistic explanation in order to deny "the hated alternative of accepting Genesis with its personal God and creative acts." Evolution, nurtured in infidelity, utterly fails to explain human history and human nature, and "offers nothing but a negative reply to that supreme question of the ages, 'If a man die, shall he live again?'." Beach notes that Darwinism has been weakened by disproof of spontaneous generation and failure to find missing links. The purely academic question of whether primitive life-forms may be our distant ancestors "does not grip us," he asserts. "But the issue between Darwinism and mankind is not a purely academic question." "We cannot depend on the Bible to show us 'how to go to heaven' if it misleads us as to 'how the heavens go' regarding the origin, nature, descent and destiny of brutes and man. Darwinists have been digging at the foundations of society and souls . . ." "Darwinism degrades God and man." It is poor morals, its assumptions are false; natural selection is "self-contradictory and impossible." Beach refutes Huxley's *Encycl. Brit.* article on evolution point by point, addressing ontogeny, morphology, biogeography, paleontology. "The teaching of Darwinism as an approved science, to the children . . . is the most deplorable feature of the whole wretched propaganda."

1079 Meyer, Nathan M. *Noah's Ark—Pitched and Parked.* 1977. Winona Lake, Ind.: BMH Books. Cited in R. Moore (*Creation/Evolution* 1981). Meyer claims that Ark was discovered by Turkish investigators in 1840, apparently referring to the team sent after the great Ararat earthquake, which allegedly exposed the site of the Ark. (Moore says that Meyer may be confusing this with accounts of a team sent after the smaller 1883 quake, which Ark-eologists say saw the Ark.)

1080 Miller, Hugh. *The Old Red Sandstone; or, New Walks in an Old Field.* 1869 [1841]. Edinburgh. 1857 ed. also pub. as Arno Press (New York; 1977) reprint ed. Went through 20 editions. *Miller:* Scottish stone-mason and self-taught geologist; first attracted attention as eloquent advocate of **Chalmers'** Free Church of Scotland; later influential as editor of Free Church's newspaper *The Witness;* became a highly respected and widely-read popularizer of geology. The Old Red Sandstone is a mid-Paleozoic (Devonian) formation which Miller first quarried, then studied. The book, a vivid depiction of its fossils (mostly fish and marine plants), became enormously popular. It was praised by leading geologists (**Agassiz, Buckland**), and fueled the Victorian passion for amateur geologizing. Miller piously and persuasively argued for a reconciliation of geology and Genesis. The book was based on Miller's articles from *The Witness.*

1081 _______. *The Foot-Prints of the Creator; or, The Asterolepis of Stromness.* 1882 [1847]. New York: Robert Carter and Bros. Orig. pub. 1847 by Johnstone & Hunter (Edinburgh/London). Also pub. 1856 by Gould & Lincoln (Boston); went through 17 editions. Includes a Memoir of Miller by **L. Agassiz.** The *Asterolepis* is a mid-Paleozoic fish (a *Coelacanth*), which Miller declared was a refutation of the "development hypothesis" of Lamarck and of Chambers's *Vestiges.* Miller argued that each geological period showed increasing development, but that within each period the fossils clearly indicated degeneration—not continual growth or progress. The *Asterolepis* was the oldest—and largest—fish of its type. Miller describes the *Asterolepis* and other fossils with unmatched charm and eloquence. He also devotes considerable space to attacks on the development hypothesis; proving, i.e., that land plants did not arise from marine plants, and showing that the fossil record indicates degradation of forms instead, followed by new creations for each geological age. Describes precursors of the development hypothesis. Says it is not itself atheistic, but it does decrease devoutness. Admits that his underlying objection to evolution is belief in immortality of the soul: if man evolved, then either other animals also have immortal souls—or we do not. Engravings.

1082 _______. *The Testimony of the Rocks; or, Geology in Its Bearings on the Two Theologies, Natural and Revealed.* 1857. Boston: Gould and Lincoln. Also pub. by Hurst (New York). Most of book orig. lectures to Edinburgh Philos. Inst., Y.M.C.A., and Brit. Assoc. (Geol. Sec.): 1852–55. Published just after Miller's death (he shot himself in a morbid fit). Some chapters are descriptive paleontology, richly illustrated with engravings, quotations of poetry, and Miller's much-admired literary style. Majority of chapters concern the relationship between the "two theologies": natural (science, especially geology) and revealed (the Bible); how they may be reconciled. Praises **Chalmers**'s attempts at reconciliation. Miller mentions he formerly believed in Gap Theory creationism, following Chalmers and **Buckland,** but argues that increased paleontological knowledge now renders this theory inadequate. He presents instead a Day-Age interpretation: each creation "day" corresponding to a geological age, with its characteristic flora and fauna. Dismisses the "development hypothesis" as "unsupported by a shadow of evidence," but argues forcefully for acceptance of geological ages with succession of organisms as fully compatible with Scripture. Devotes one chapter to refutation of the "anti-geologists"—religionists who denounce geology as devilish, atheistic undermining of Bible's authority. Describes Noachian Deluge as regional, not worldwide; points out evidence against worldwide Flood. Presents 'revelatory' creationism (quoting **Kurtz** and others): God gave Moses successive visions of each day/age of creation. Earth is cloud-covered, so sun is not fully visible until fourth "day"—Mesozoic. Day 2: early or mid-Paleozoic. Day 3: Carboniferous. Fish existed before fifth "day" (late Mesozoic), but were not visible to Moses in these visions. Demonstrates that mankind becomes progressively more degraded and savage as he migrates further from original center of creation.

1083 Miller, W. Maskelyne. *The Moon, How and Why It Came Here? A Philosophical Inquiry into Astronomical Facts: The Story of the Floods Not a Myth.* 1930. Vancouver, B.C., Canada: Woodward's Ltd. Explains the Ice Age by the Flood. (A follower of Hörbiger?)

1084 Milne, D.S. *Creation or Evolution* (pamphlet). Undated. England: Evolution Protest Movement. EPM pamphlet #35.

1085 Mitchell, H.G. *The World Before Abraham: According to Genesis I.–XI.: With an Introduction to the Pentateuch.* 1901. Westminster, England: Archibald Constable. *Mitchell:* "Prof. in Boston Univ." Accepts Documentary Hypothesis of Bible authorship. God created from original chaos – not *ex nihilo.* The creation days were probably literal 24-hour days.

1086 Mitchell, N.J. *Evolution and the Emperor's New Clothes.* 1983. Diss, Norfold, England: Royden Publications. Endorsement by Wickramasinghe [**Hoyle**'s co-author] on back cover.

1087 Mitchell, Thomas. *Cosmogony.* [ca. 1880, or earlier.] American News Co. Cited by T.W. Doane (*Bible Myths;* 1910). *Mitchell:* "if the account of creation falls, Christ and the apostles follow: if the book of Genesis is erroneous, so also are the gospels."

1088 Mivart, St. George Jackson. *On the Genesis of Species.* 1871. New York: D. Appleton. Also pub. by Macmillan (London). Mivart, a Catholic convert who was later excommunicated for his evolutionism, was a sort of "renegade Darwinian" who became one of Darwin's most persistent and troublesome critics. This book is a critique of Darwin's *Origin.* Mivart advocates "specific genesis": species are not fixed, but have an innate force capable of sudden generation of new species as "harmonic self-consistent wholes." Nature's harmony and order proves divine design and purpose; all developments and adaptations are planned – though their purpose may yet be a mystery to us. Life develops through natural law – not by special creation – but not through natural law alone. Mivart severely criticized the ability of natural selection to account for organs prior to their perfected state, arguing that in their incipient stages of development they would not have been advantageous to the organism, hence not favored by selection (an argument modern creationists still find compelling).

1089 ________. *Man and Apes: An Exposition of Structural Resemblances and Differences Bearing upon Questions of Affinity and Origin.* 1876 [1873]. London: R. Hardwicke. Innate laws of evolution; pre-ordained ends and purposes – not natural selection. Some ape species are more similar to man in some traits, other species in other traits. "But however near to apes may be the body of man, whatever the kind or number of resemblances between them, . . . it is to no one kind of ape that man has any special or exclusive affinities, . . . the resemblances between him and the lower forms are shared in not very unequal proportions by different species." Thus "it is manifest that man, the apes and the half-apes cannot be arranged in a single ascending series of which man is the term and culmination." (Quoted in **Graebner** 1932; discussed in Gillespie.)

1090 Mixter, Russell. *Creation and Evolution* (booklet). 1953 [1949]. West Lafayette, Ind.: American Scientific Affiliation. Orig. pub. 1949 by Wheaton College (Wheaton, Ill.). ASA Monograph No. 2. *Mixter:* head of Wheaton College science dept.; president of ASA (1950–54). Progressive creation or theistic evolution. Proposes that biblical 'kind' approximates biological 'order.' (Mixter later conceded that some orders may have evolved from others.)

1091 ________. *The Story of Creation.* c1955. Grand Rapids, Mich.: Zondervan.

1092 ________ **(ed.).** *Evolution and Christian Thought Today.* 1960 [1959]. Grand Rapids, Mich.: William B. Eerdmans. 2nd ed. Also pub. by Paternoster (London). Symposium vol. of the American Scientific Affiliation, a group of evangelical scientists formed in 1941. Members originally held diverse views on creation/evolution, but by 1959 few were strict creationists; those who were tried to "reform" ASA, but – failing that – formed CRS instead a few years later. Of the authors in this volume, only **Carl Henry** opposes evolution strenuously, pointing out its unpleasant theological and philosophical implications and consequences. He rejects recent creation and fixity of species, however, as faulty interpretations of Genesis. Henry cites many scientists (also Spengler) in arguing against evolution as undisputed fact, and against Darwinism. "The great truths of the revealed narrative of creation retain their pointed relevance to the scientific discussions of our decade." Other contributors (basically old-earth, progressive creationists who accept at least some evolutionary change) include V.E. Anderson (Ph.D.—Univ. Minn.; biology prof. at Bethel Coll., St. Paul; ass't. dir. of Human Genetics Inst. at Univ. Minn.; respondant in **Lester** and Bohlin 1984) on biogeography; W. Heard (*ASA Newsletter* editor and co-author of **ASA** 1986) on origin of life; J.O. Buswell III (son of **J.O. Buswell;** anthropology prof. at Wheaton Coll.) on prehistoric man; T.D.S. Key (former biology teacher at Bob Jones Univ.) on the influence of Darwinism; Cordelia Barber (M.A.—Columbia Univ.; former geology teacher at Wheaton Coll.) on fossils; and others. Photos, diagrams; index.

1093 Monsma, Edwin Y. *If Not Evolution, What Then?* 1959 [c1954]. Priv. pub.? (Grand Rapids, Mich.). *Monsma:* biology Ph.D.; biology prof. at Calvin College (Grand Rapids, Mich.); a founding member of CRS. "Contains a good chapter on history of evolutionary thought, meaning of evolution, some presuppositions, and critical analysis of supposed suppositions of evolutionists, author offers his reasoned faith in creation [sic]" (**J.N. Moore,** in 1964 *CRSQ*).

1094 Monsma, John Clover (ed.). *The Evidence of God in an Expanding Universe.* 1958. New York: G.P. Putnam's Sons. Subtitle: "Forty American Scientists Declare Their Affirmative Views on Religion." Published "in connection with the International Geophysical Year." Contributors include **F. Allen, Klotz, Mixter, Lammerts, Stoner,** and others. Some of the other contributors advocate "creational evolution," "continuous creation," and other views.

1095 ________. *Behind the Dim Unknown.* 1966. New York: G.P. Putnam's Sons. Subtitle: "Twenty-six notable scientists face a host of unsolved problems and unitedly reach a conclusion." Includes chapters by Russell Artist (Ph.D.—Univ. Minn.; head of biology dept. at David Lipscomb Coll.), **D. Gish,** Leonard Burkart (wood chemist and forestry prof.), **G. Howe,** John Grebe (D.Sc.—Case Inst. Tech.; former director of Dow Nuclear and Basic Res. Dept.), **C. Burdick, H. Slusher.** Artist, who announces he is a "creationist type of biologist," describes some adaptive wonders of life. Gish describes attempts to create life, declaring that "as a scientist," he must conclude that the living cell "demands a Planner." Physical and chemical laws are completely incapable of giving rise to life. "As for the great mystery of life, I simply and definitely accept the testimony of Genesis..." Burkart says some people prefer faith in "nature and mutation" in order to deny dependence on God and responsibility to Him. Howe describes cell structure and function. Grebe states that water is being formed continuously within the earth's magnetic field, and suggests that a reversal of the magnetic field caused Noah's Flood. The Flood may have disrupted the C-14 ratio; thus, all C-14 dates may have to be drastically revised to fit within a few thousand years after the Flood. "How wonderful it would be if further data from space and the orientation of magnetite crystals in viscous lava flows would continue to clarify the details of the biblical creation and flood accounts!" Burdick argues that fossils can only be explained by Flood Geology. Genesis provides an accurate record of these events. The Flood resulted in great and permanent climate change. Since there exists no mechanism to allow for continental drift, only the catastrophic Flood can account for mountain-building and continents. Slusher says that rate of oxidation of buried hydrocarbons may force a drastic revision of the age of the earth, and argues against uniformitarianism in geophysics.

1096 Montefiore, Hugh. *The Probability of God.* 1985. London: SCM Press. *Montefiore:* Bishop of Birmingham (England); contributor to *Nature.* Discusses physics, cosmology, and biology; for the latter, relies largely on **Koestler, F. Hoyle, G.R. Taylor,** and Popper. Believes in evolution, but considers natural selection ("random" and "meaningless") inadequate to account for complex adaptations, organs, and behaviors. (Criticized and quoted in R. Dawkins' *Blind Watchmaker.*)

1097 Montgomery, John Warwick. *The Quest for Noah's Ark.* 1974 [1972]. Minneapolis, Bethany Fellowship. 2nd ed., rev. Orig. 1972. *Montgomery:* seven earned degrees, including philosophy Ph.D.—Univ. Chicago; theology doctorate—Univ. Strasbourg; former law and theology prof. at Int'l. School of Law (Wash., D.C.); Hon. Fellow of Revelle Coll.—UCSD; prof. at Melodyland School of Theology (Anaheim, Calif.); now Dean and President of the fundamentalist Simon Greenleaf School of Law (Anaheim); Lutheran minister; author of twenty books (theology, church history, philosophy). Some chapters orig. appeared in *Christianity Today*. Other chapters are excerpts from **H. Miller** 1857 (on Flood traditions around the world); E. Yamauchi (archeology and the Flood); A. Heidel (Genesis and Gilgamesh); **H. Morris** (*CRSQ* article on design of the Ark); **Filby** 1971. Also excerpts from ancient accounts of Ark sightings; from the **Cummings'** archives; from **Navarra** 1953, 1956; **J.J.F. Parrot** 1845; other Arkeology accounts; **Burdick** on geology of Ararat (from volume ed. by L.B. Hewitt of **Vandeman**'s A.R.F. team); and long account of Montgomery's own ascent of Ararat in 1970 (he wrote seven articles about Noah's Ark for the Turkish paper *Hürriyet*). Montgomery reprints address he gave at int'l. conference on remote sensing about the use of satellites for discovery of Ark; also correspondence about ERTS [LandSat] photo Montgomery believed showed the Ark, and news release about speech by Sen. Frank Moss endorsing Montgomery's ERTS tale. Though convinced that the Ark is atop Ararat, and a believer in most Arkeology tales, Montgomery thinks the Flood was regional, not global. Maps, photos, drawings; bibliog., index.

1098 Montgomery, Ruth. *The World Before*. 1982 [c1976]. New York: Fawcett Crest (Ballantine). *Montgomery:* celebrity psychic. "Arthur Ford" and other spirit guides reveal to author secrets about the past and future. Typical mishmash of paranormal fantasies mixed with religion: creation cycles, Atlantis and Lemuria, Second Coming of Christ, the Antichrist, extraterrestrials, UFOs, ESP, Aquarian Age, E. Cayce, **von Däniken,** etc. Claims that paleoanthropology proves Guides' assertion of separate creation of man 5–7 million years ago. Also cites **Gish** of ICR to prove that creationism is better theory than Darwin's evolution; endorses evidence of man/dinosaur co-existence. Souls were all created in the beginning; then, after reincarnation, monsters arose from the mating of humans with animals, and there were five separately-created human races. Women were once superior, but Moses left us a bigoted account in Genesis. Part of book is about past lives of show-biz celebrities, with the usual vacuous predictions.

1099 Mooney, Richard. *Colony: Earth.* 1974. New York: Stein & Day. Man was transplanted on earth; humans are not even related to primates. (There was no cross-breeding or genetic manipulation of primate sub-humans.) There is only a "superficial physical resemblance" between apes and man. Mooney says his theory solves problem of man's origin without either evolution or creation. Man was put here 40,000 years ago, with clothes, weapons, fire, shelter, and modern intelligence. Implies teleology: nature wants to develop an intelligent species. (Discussed and quoted by Greenwell, in 1980 *Skeptical Inquirer.*) Mooney also wrote *Gods of Air and Darkness* (1975; Stein & Day).

1100 Moore, George R. *Bible Views on Creation*. 1895. Philadelphia: John McGill White. *Moore:* minister. "Science is the reading aright the revelations of God in nature and the Bible." "Nothing is scientific that is out of harmony with any truth from any source." Science consists of induction from raw facts observed in nature; theology is the parallel task of arranging the facts revealed in the Bible. (Quoted in M. Cavanaugh.)

1101 Moore, John N. *Should Evolution Be Taught?* (booklet). c1974 [1970]. San Diego: CLP (Creation-Life). Rev. 1974. Orig. 1970. *Moore:* biology M.S. and Ed.D.—Michigan State Univ.; natural science prof. at Mich. State; longtime *CRSQ* managing editor; science editor for Zondervan Pub. Evolution comes in two "models": "limited" and "general." "Limited" evolution is true, and should be presented in schools—but this is only genetic variation. "General" evolution should be presented only as one of two competing theories on origins, if at all. Quotes many evolutionists on lack of transitional forms. No "indisputable" Precambrian fossils. Discusses impact of evolutionary thought. Strongly recommends the Genesis creation account.

1102 _______. *Questions and Answers on Creation/Evolution.* c1976. Grand Rapids, Mich.: Baker Book House. Presentation of standard creation-science arguments in question-and-answer format. Strict young-earth creationism; Flood Geology. Ideas about origin of universe, of life, and of mankind involve unobservable, unrepeatable events, and thus are outside the domain of science. They may properly be omitted from science classes (especially lower grades); if presented, it should be in a balanced, two-model form. "Uniformitarian" geologists and biologists must rely on "supranatural" explanations. Geologic column is only "hypothetical." Evolution is "mathematically impossible and really quite irrational." Entire system of evolution rests upon (mistaken) assumption that similarity indicates descent relationship. Cautiously endorses Paluxy manprints. Illustrates the "flow" of evolutionary influence in society. Evolution is required by—and hence leads to—communism. Asserts that teaching of creation-science is legal; explains how he teaches creation-science at Mich. State. Such teaching is non-religious (no worship in class), necessary for academic freedom, and popular. Arguments mostly creation-"science," but openly advocates biblical account of creation. Includes creation-science reading list.

1103 _______. *How to Teach Origins (Without ACLU Interference).* c1983. Milford, Mich.: Mott Media. This book was prepared in 1977 at CHC (ICR) while Moore was on sabbatical from Mich. St. Moore accepted early retirement from Mich. St. in 1982. This is a detailed manual in response to the question, "How do you teach creation in a public educational institution?" "It is fully legal and constitutional to teach scientific creationism in any school, whether it be public or parochial, secular or sectarian. Because of all that has gone before in previous chapters, the science teacher should no longer have any serious doubt that the creation model is a viable, scientifically based alternative to the evolution model about first origins." It is educationally unsound and unconstitutional to promote evolution in the schools; exclusive teaching of evolution is compulsory indoctrination in state-endorsed world view. Moore presents standard creation-science arguments, with numerous "classroom teaching aids" (visual displays), side-bars on "cover words" (terms prone to semantic confusion and evolutionist exploitation), and extensive scientific footnotes. Includes discussion and evaluation questions and homework suggestions for each chapter; also final examination. Includes excerpts from **H. Morris,** and other creation-science works, and section on quotes by evolutionists. Has outline summaries of **W. Bird**'s 1978 and 1979 law articles. Many charts, diagrams, tables, maps; some photos. Annotated bibliog. of creationist books; index.

1104 _______**, and Harold S. Slusher (eds).** *Biology: A Search for Order in Complexity.* 1974 [c1970]. Grand Rapids, Mich.: Zondervan. Rev. ed. Orig. pub. 1971. Major authors: **Rita Ward, Tinkle;** also Larry Butler, **Davidheiser, G. Howe, Klotz, Lammerts, H. Morris, McCone. T. Barnes:** Chairman of CRS Textbook Committee. Other members: R. Artist (biology prof. at Lipscomb Coll., Tenn.), James A. Baker (USDA plant pathol.), H.D. Dean (biology prof. at Pepperdine Coll., Calif.), **Gish, Marsh, Rusch, Sears,** Patricia A. Wilder (botanist at San Bernardino County Mus., Calif.). A creationist-oriented high school textbook was one of the first projects undertaken by the Creation Res. Society. It was published by Zondervan after refused by secular publishers. Most of book is standard high school biology, presented in a non-evolutionary form, interspersed with strongly anti-evolutionist sections and chapters. Evolutionary explanations scoffed at as inadequate throughout. Major unit on "Theories of Biological Change" emphasizes standard creation-science arguments: evidence for young-earth, "hypothetical" nature of geological column, wrong-order fossils (endorses Paluxy and Meister prints), use of index fossils for dating (criticized as circular reasoning), sudden appearance and persistence of life-forms, genetic variation within 'kinds' different from evolution, sections on "Failures of Darwinian Theory" and the many "Problems for Evolutionists." Flood Geology is "superior because it conforms to the principles of hydrodynamics." States that "mechanistic" theories cannot explain conscience or moral behavior. Criticizes fossil hominids as products of wishful thinking. Asserts evolution violates Law of Biogenesis; says complex organs cannot arise gradually. Evolutionists reject young-earth dating methods because they require long ages "for the doctrine of evolution, which has no observable evidential basis." Includes chapter questions, and bibliog. for each unit. "[T]he creation model is a framework of interpretation and correlation which is at least as satisfactory as the evolution model. However, the two laws of thermodynamics, the apparent stability of the basic 'kinds,' the existence of great gaps between the kinds, the deteriorative nature of mutations, and the catastrophic nature of the worldwide fossil-bearing formations all may be correlated far more easily with the creation model than with the evolution model." Many photos, drawings, diagrams; index. This book was approved by many state textbook committees, but was declared unconstitutional by Indiana Supreme Court in 1977. CRS Board planned a new, less overtly religious edition to circumvent this ruling, but this proposed revision became bogged down due to differences between CRS staff and ICR.

1105 Moore, Nathan Grier. *The Theory of Evolution (An Inquiry).* 1941 [1929]. Chicago: Lakeside Press. "From a lawyer's point of view." Part I: As Applied to Man; Part II: As Applied to Nature. *Moore:* grandfather of **E.F. Hills.**

1106 Moorhead, Paul S., and Martin M. Kaplan (eds.). *Mathematical Challenges to the Neo-Darwinian Interpretation of Evolution.* 1967. Philadelphia: Wistar Institute Press. Wistar Inst. Symposium Monograph No. 5. 1966 symposium at Wistar Inst. of Anatomy and Biology, held to air and confront arguments by math and engineering critics of neo-Darwinian evolutionary theory. Critics included Murray Eden (electrical engineering prof. at MIT): "Inadequacies of Neo-Darwinian Evolution as a Scientific Theory"; Marcel Schützenburger (computer science—Univ. Paris): "Algorithms and the Neo-Darwinian Theory of Evolution"; and Stanislaw Ulam (Los Alamos Lab): "How to Formulate Mathematically Problems of Rate of Evolution?" Their objections are all based on probability arguments against evolution by random mutation. The critics (physical scientists, not biologists) are replied to, with spirited discussion, by leading evolutionists such as P. Medawar (symposium chairman), L. Eiseley, E. Mayr, G. Wald, R. Lewontin, and C.H. Waddington. This book is very widely quoted by creationists, who regard it as a rigorous refutation of evolution by non-creationist scientific experts.

1107 More, Louis Trenchard. *The Dogma of Evolution.* 1925. Princeton, N.J.: Princeton Univ. Press. *More:* physics prof. at Univ. Cincinnati. There is no proof of mechanistic evolution, so divine intervention and the spiritual world—Christianity—should not be rejected as explanations. Much history of science and of evolutionary theory. Darwin wasn't bad as a scientist, but he was a poor philosopher and theorist. Malthus was wrong. Variation is a mystery. More, a Lamarckian, presents other contemporary objections to evolutionary theory. Endorses **D'Arcy Thompson;** says other biologists are not mathematically competent, and deny the Spirit. Objects to natural selection as a guide to social policy and ethics; claims that Christianity is better. Criticizes reconciliation of Genesis with geological ages as "timid." Says that belief in myths of Eden and the Flood is naive.

1108 Morgenthau, David. *Genesis and Science.* c1927. Toronto, Canada: Genesis and Science. Revised by R.C. Palmer. *Morgenthau:* Hebrew teacher. Palmer: at Cambridge Univ. Book I of a volume which also contains sections on "Judaism and Christianity." Book I reconciles Genesis creation with astronomy and archeology by exegesis of the Masoretic (Received) translation. Genesis events are not chronological, but "permuted." Genesis 'days' are eras (day-age creationism). Stars condensed out, then the earth formed, and went through a period of constant rain. Sun wasn't visible until after plants were in existence. Dinosaurs and other "crawlers" did really come from the sea; true land animals came later. There are 50,000 years of gaps in biblical post-Flood genealogies, so these years must be added to the 5687 years in Genesis since Adam.

1109 Morowitz, Harold J. *Cosmic Joy & Local Pain: Musings of a Mystic Scientist.* 1987. New York: Scribner's. *Morowitz:* Ph.D.—Yale Univ.; biochemist and molecular biophysicist at Yale. Morowitz, a pantheist, was an anti-creationist witness in the Arkansas trial. Anthropic principle argument—if the laws of nature were just slightly different, life would be impossible. Morowitz "concludes that the universe and everything in it could not have been formed by chance but rather is the work of some design and, presumably, designer" (from review by L. Dembart). *Morowitz:* "I find it hard not to find design in a universe that works so well." "Blind chance" cannot explain this perfect fit. Admires **Teilhard.** Morowitz, who has studied mutations, theoretical limits for the simplest living things, and thermodynamics of living systems, corresponded with **Coppedge**, who cites him extensively in his 1973 book on probability of chance formation of life.

1110 Morrell, R.W. *Evolutionary Contradictions and Geological Facts* (pamphlet). Undated. England: Evolution Protest Movement. EPM pamphlet #201.

1111 Morris, Henry M. *God's Way of Salvation* (booklet). 1944. Houston: Bowman Co. *Morris:* hydraulic engineering Ph.D.—Univ. Minnesota; Baptist; a founder and former president of CRS; founder and president of Institute for Creation Research; acknowledged as chief theoretician and leading exponent of modern creation-science. This booklet written while Morris was teaching at Rice Inst. (now Univ.). Morris was a theistic evolutionist in college (at Rice), but became a creationist shortly before writing this booklet, being especially influenced by **Rimmer, Price,** and Irwin A. Moon (Moon, a founder of the **Amer. Sci. Affil.,** founded the Moody Inst. of Science in 1945, and was host scientist for most of the films in its "Sermons from Science" series).

1112 ________. *That You Might Believe.* 1978 [1946]. Chicago: Good Books (Good News). The first, and most widely-read, of Morris's dozens of creation-science books. Orig. pub. in 1946, when Morris was 28, just before starting graduate study. (Morris has said he studied hydraulics primarily to prove Flood Geology.) It was revised several times; some editions titled *The Bible and Modern Science.* Last edition under orig. title was pub. 1978 (Westchester, Ill.: Good News). Most of book concerns biblical and scientific creationism, and Flood Geology; also chapters on Bible-science (scientific discoveries anticipated in the Bible), verification of Bible history, and biblical prophecy. First edition of book allowed for Gap Theory creationism, but Morris expunged this, and has argued strenuously for strict young-earth creationism and Flood Geology ever since. In one form or another, this book has been in print continuously for over 40 years.

1113 ________. *The Bible and Modern Science.* 1968 [1951]. Chicago: Moody Press. Rev. and expanded ed. of *That You Might Believe* (1946). Orig. pub. 1951, written when Morris was working on doctoral diss. at Univ. Minn. (*A New Concept of Flow in Rough Conduits,* 1950). Moody also published another, condensed edition in 1956 for its Colportage Library series. Christian evidences: apologetics. The "atheistic and satanic character" of evolution is "evidenced in the many evil social doctrines it has spawned." Genesis 'kind' may be species, genus or family. Assuming evolution, fossils are used to date rocks, and then the rock series is taken to be proof of evolution. Says all geological strata are better explained as due to worldwide Flood. Advocates pre-Flood water canopy. Defends Joshua's Long Day as literal. Presents most of the standard creation-science arguments. "The creation was accomplished by entirely different means than God now uses in His providential sustaining of the world, which are the only processes that can now be studied. The deluge also marked a catastrophic intervention by God in the operation of the normal processes of nature. Therefore, any application of the principle of uniformity, based on measurements of present processes, could not possibly extend back earlier than the time of the flood at best. Consequently, earth history earlier than this cannot be discerned geologically with any assurance; revelation is required." Includes Recommended References (scientific and creationist). Preface: "The purpose of this book, very frankly and without apology, is to win people to a genuine faith in Jesus Christ as the eternal Son of God and their personal Saviour, and to assist in strengthening the faith of those who have already received Him in this light. It is especially addressed to young people who are finding biblical Christianity under attack in many quarters in these days, nowhere more so than in the classes and textbooks of most of our colleges and universities, and even in the public schools." Morris says (1984) that the paragraph still applies "to every book I have written since, even those which are strictly scientific in content." This book has appeared in yet another incarnation, *Science and the Bible* (1986).

1114 _______. *Biblical Catastrophism and Geology* (pamphlet). 1963. Phila.: Presbyterian and Reformed. Orig. talk given to Houston Geol. Soc. (1962). Presented while Morris was hydraulic engineering professor and head of civil engineering dept. at Virginia Tech (VPI). Also same year Morris's textbook, *Applied Hydraulics in Engineering* (New York: Ronald Press), was first published. Pamphlet is basically summary of Flood Geology arguments presented in **Whitcomb** and Morris's landmark volume *The Genesis Flood* (1961). Three great facts of history are the Creation, Fall, and Flood. This history cannot be understood by study of naturalistic processes; it "can only be known through divine revelation." Endorses "creation with appearance of age": "the primeval ocean may already have been saline, radioactive minerals may already have contained daughter elements, light from distant stars may have been visible on the earth at the instant of their creation, and so on, even as Adam was created as a full-grown man." Since the Bible says there was no sin or death before the Fall, all the earth's sedimentary layers, with their fossils, must be dated later than the Fall. The fruit of evolutionary philosophy is evil (Communism, Nazism, Freudianism, materialism, etc.); therefore evolution is false.

1115 _______. *The Twilight of Evolution.* 1963. Grand Rapids, Mich.: Baker Book House. Also pub. by Craig Press, a div. of Presbyterian and Reformed (Nutley, N.J.). (P&R was owned by C. Craig.) Based on 1962 lectures to Reformed Fellowship (Christian Reformed Church; Calvinist) in Grand Rapids. Standard creation-science and Flood Geology arguments. Emphasizes evil influence of evolution. Claims there are many scientists who reject evolution, many secretly. Stresses circular reasoning of geologic column as proof of evolution and vice-versa. Presentation of thermodynamics as refutation of evolution. Long description of pre-Flood water (vapor) canopy and pre-Flood conditions; also description of post-Judgment world after present world is destroyed. If evolution is scientifically impossible, how can we explain its great popularity? "The answer is *Satan!*" Satan fathered the "monstrous lie of evolution" to blind us to God and the Gospel. The "only real issue" in the universe is "God's sovereignty versus the asserted autonomy of his creatures." Recognizing this enables us to understand the Satanic origin of evolution. Satan deceived Adam and Eve into believing they had evolved, and that God had lied to them. They fell for this deception due to their desire for knowledge ("science"). Warns against coming world government led by Satan and the Antichrist. Scientific and biblical references; many scientists quoted. Index.

1116 _______. *Science, Scripture and Salvation.* 1971 [1965]. Denver: Baptist Publications (B/P—Accent). 2nd ed. Later rev. ed. (1977) titled *The Beginning of the World.* Though it includes scientific arguments, there are no scientific references or quotes; it is very explicitly biblical. Also in Teacher Edition. "Rather than believe that life has evolved, and to accept this philosophy without a scintilla of evidence, we prefer to accept the only source of information on the origin of the universe and all its attending parts, namely, the revelation of Holy Scripture. . . . In this study we earnestly seek to glorify our Creator God, establish adults firmly in the faith, strengthen all hands against the inroads of error, and through the adults point youth in the home to the truth of God's revealed Word."

1117 ________. *Studies in the Bible and Science: Or, Christ and Creation.* 1966. Philadelphia: Presbyterian and Reformed. Also pub. by Baker. (Ehlert files list a *Studies in Bible and Science,* priv. pub. in Blacksburg, Vir. [Morris was then at VPI], which may be version of this.) A compilation of previously published articles, plus one new one ("Christ in Creation"). Most also included in a 1963 Grace Sem. syllabus edited by **Whitcomb.** Reprinted from *His* magazine (Inter-Varsity), Good News Pub., *Christian Life, Bibliotheca Sacra* (Dallas Theol. Sem.), *Torch and Trumpet* (Reformed Fellowship), *Collegiate Challenge* (Campus Crusade for Christ), *Grace J.* (Grace Theol. Sem), *Good News Broadcaster* **(T. Epp),** and *CRSQ.* One chapter explains that "The Bible *Is* a Textbook of Science." Covers all the standard creation-science arguments, with an emphasis on Flood Geology. Also concepts of energy and power in the Bible; "Water and the Word"; "the Trinity in the Universe" (after Nathan Wood's book of that title; 10th ed. 1955); critical review of **Mixter** 1959; etc. "The very nature of Christian morality is squarely opposed to that of evolution. . . . It is well known that an evolutionary philosophy is the basis of Communism, Fascism, and the many other anti-Christian systems of the day." Evolution denies the Fall, therefore the need for our Saviour.

1118 ________. *Evolution and the Modern Christian.* 1981 [1967]. Phillipsburg, N.J.: Presbyterian and Reformed. 2nd ed. Many of Morris's books contain the same material in only slightly varied form. This one presents his standard creation-science arguments and Flood Geology in brief, non-technical form. Usual denunciation of the "flagrant" circular reasoning of paleontological evidence for evolution: "Evolution is assumed in building up the geological column, superimposing rock systems from different regions all over the world on top of each other. Then the column thus constructed, with its arbitrary geologic ages, is formally offered as the best proof of the *historic fact* of evolution." Stresses scientific arguments against evolution, but includes overtly religious sections also. Since God created with appearance of age, we must rely on divine revelation to learn true age of earth—or to learn anything at all about creation. God used supernatural processes, so science cannot tell us about origins. "If we expect to learn anything more . . . about the Creation, then God alone can tell us. And He has told us! In the Bible, which is the Word of God, He has told us everything we *need* to know about the Creation and earth's primeval history." Annotated creation-science reading list.

1119 ________. *Biblical Cosmology and Modern Science.* 1970. Nutley, N.J.: Craig Press. Also pub. by Baker (Grand Rapids, Mich.). Described in Morris 1967 as: "A Biblical and scientific study of cosmology, cosmogony, naturalism, catastrophism, and eschatology. Includes special sections on evidences from population statistics, thermodynamics, sedimentation, etc., as well as a detailed critique of the day-age theory, the gap theory, allegorical theory and other compromises with evolution."

1120 ________. *A Biblical Manual on Science and Creation* (booklet?). 1972. San Diego: Institute for Creation Research. (Listed in Ehlert files and SOR CREVO/IMS.)

1121 ________. *The Remarkable Birth of Planet Earth.* 1978 [c1972]. Minneapolis: Bethany Fellowship. Orig. pub. c1972 by Creation-Life (San Diego). Standard creation-science arguments. "The discussion is primarily approached from the Biblical point of view..." Chapters on the argument from design; inevitability of degeneration rather than increasing complexity; fossil and rock record as proof of destruction of the world by Flood; complexity of living organisms and animating spirit; early man; stars; "The Strange Delusion of Evolution"; defense of recent creation. Noah's sons: Shem was ancestor of Semites, Japheth of "Aryans and Caucasians," Ham of all others (Canaanites, Egyptians, blacks, Mongoloids). Bible prophesies that Semites would be bearers of God's word; Japhethites would prosper politically and intellectually, and that Hamites would focus on physical, material aspect of life. Claims that this is a "national" rather than "racial" division, and says "race" is strictly an evolutionist concept. Starlight was created apart from or prior to the stars themselves: "real creation necessarily involves creation of 'apparent age.'" "Therefore do not be impressed by the 'apparent age' of prehistoric formations. The 'true age' is what God says it is, and there is no other way of determining it." "The only way we can determine the true age of the earth is for God to tell us what it is. And since He *has* told us, very plainly, in the Holy Scriptures that it is several thousand years in age, and no more, that ought to settle all basic questions of terrestrial chronology." Stars may be inhabited by angels, but "definitely" not by men or similar beings. As everything was created perfect and "good," the scarred and battered features of the moon and planets is evidence "either of Satan's primeval rebellion or his continuing battle against Michael and his angels." Satan is also responsible for UFOs and occultism. Both good and bad angels influence physical and astronomical processes. Index. Jerry Falwell distributed this book free after Morris appeared several times on his Old-Time Gospel Hour in 1981 to promote creationism. At the time, Falwell was waging an intense campaign to have creationism taught in public schools.

1122 ________. *Many Infallible Proofs.* 1974. San Diego: Creation-Life. "Practical and Useful Evidences of Christianity." Foreword by **J. Whitcomb.** Acknowledges assistance of **C. Ryrie.** Apologetics; Christian evidences; review of the "many infallible proofs" of truth of Christianity, and refutation of alleged fallacies and discrepancies in the Bible. Includes chapters on Bible-science, fallacies of evolution, Creation and Flood Geology, and ancient history. Observes that "by far the most influential argument against the Bible is the widespread belief that science has proved evolution and therefore disproved the Bible account of creation." Presents summary of creation-science arguments to show that evolution is completely unscientific. "The much-maligned **Ussher** chronology ... may have been discarded too quickly." Morris believes there are some gaps in Genesis chronology, but that Ussher's method is essentially correct. Includes a section presenting Bible numerology, but cautions that not all scholars endorse this; also warns that **Panin**'s method is probably misguided. Index.

1123 ________. *Introducing Scientific Creationism into the Public Schools* (pamphlet). 1975. El Cajon, Calif.: Institute for Creation Research. ICR "Impact" series #20. One of many "Impact" articles written by Morris. Says teaching of evolution only is unconstitutional. Insists that creationism can be taught in any class, but cautions teachers to exclude overt religious references. Recommends creation-science literature, especially ICR material. Suggests ways of persuading teachers and administrators to accept creationism. Advises how to form local groups for promoting and publicizing creationism.

1124 ________. *The Troubled Waters of Evolution.* 1982 [1975]. San Diego: CLP (Creation-Life). 2nd ed. Orig. 1975. Foreword by **T. Barnes;** Introduction by **T. LaHaye.** The history of evolutionary thought and its "impact on all aspects of modern life," with a "non-technical study of the evidence for creation, especially the second law of thermodynamics." Covers usual creation-science arguments. Evolution began as a religious philosophy, which is "all that it can ever be." Reviews rise of modern creationism. Stresses that 'race' is an evolutionist concept; that it is the evolutionists who have promoted racism. If the Bible is taken to be racist, however, "the error is in the interpretation, not in the Bible itself." Satan introduced evolutionary theory to Nimrod in Babylon; it was the origin of all pagan, occult and polytheistic religions. Satan and his demons met with Nimrod in the Tower of Babel, which was an astrological shrine, to scheme against God and formulate philosophies denying His creation. Sin is due to rebellion against God, which in turn is rationalized by evolution. Derides evolutionist concern about over-population; explains how rate of population increase instead proves that the Flood occurred no earlier than 6,000 B.C. Suggests ways creationists can get creationism into the schools. Emphasizes persuasion and education rather than political or legal pressure to force teaching of creationism. Photos; index.

1125 _______. *The Genesis Record: A Scientific and Devotional Commentary on the Book of Beginnings.* 1976. Grand Rapids, Mich.: Baker Book House. Copub. by Baker and by Master Books (San Diego). Foreword by Arnold Ehlert (ICR Librarian). Book grew out of syllabus for course at Christian Heritage Coll. (ICR affil.). A 716-page exegesis and commentary on Genesis, using KJV Bible. Incorporates much creation-science. Genesis is the foundation of the whole Bible, and is neessary to understand the New Testament. There are over 100 references to Gen. 1-11 in the New Testament, including six by Christ. Gen. 2 complements and amplifies the Gen. 1 account (i.e., no contradiction); Gen. 2 was probably written by Adam, and gives a human perspective on Creation, in contrast to the Godly view of Gen. 1. Adam was extremely intelligent—he named all the animals. Includes refutations of Gap and Day-Age creationist interpretations. Argues for acceptance of the "simple literal Biblical chronology," with creation somewhere between 4–10,000 B.C. The "sons of God" (Gen. 6) were Satanic angels who controlled human bodies by demon-possession. Index.

1126 _______. *The Beginning of the World.* 1977. Denver: Accent Books. Rev. version of *Science, Scripture and Salvation* (1965). Presented as biblical exegesis of first eleven chapters of Genesis, plus II Peter 3 (chapter which refers to destruction of the world by Flood, and warns of coming destruction). No scientific references cited (intended for Bible study), but presents creation-science arguments in non-technical form. Praises creationist efforts and asserts falsity of evolution. Gives summary of Flood Geology in discussion of Flood of Noah. Reviews descent of mankind from Noah's sons: Hamites (Afro-Asians) specialized for secular service—building, inventing, discovering; Semites especially religious; Japhethites (Indo-Europeans) are intellectual—they took over Hamitic lands and developed science and philosophy, and also took over the religious function from the Semites. Emphasizes that the End Time "scoffers" foretold of in II Peter, who deny the worldwide Flood and Creation by God, are evolutionist uniformitarians; castigates their "willful ignorance."

1127 _______. *Education for the Real World.* c1977. San Diego: Creation-Life. Foreword by **J. Whitcomb.** Advocates a totally Bible-based educational program. Anything taught in any course which docs not support God's revelation in the Bible must be rejected. The "real world" is not our present physical existence, but Eternity; that is what "true education" must prepare us for. Training in physical and biological science is necessary to fulfill biblical "cultural mandate" of dominion over the earth. Study of the social sciences, humanities and arts, which has also become important, is, however, much more dangerous, and needs to be carefully controlled by biblical constraints, because these "interpretive" fields are contaminated by our sinful natures. Advocates study of nature for moral lessons, parables. "[T]eaching is essentially *indoctrination* in fixed truth..." Methods which best accomplish this indoctrination must be used. The foundation of any world view is its concept of origins. Biblical cosmogony is creationist; all others are evolutionist. Public school should teach both creation and evolution; parents who want exclusive teaching in one or the other should form private schools. "In Christian schools . . . there is no justification at all for teaching falsehood [evolution] along with truth." Summarizes standard creationist arguments. Explains how Satan invented the deception of evolution—the cornerstone of all pagan, demonic, humanist religions—as a means of rebellion against God, and how Nimrod propagated this sinful doctrine in Babylon. UFOs, occult and psychic phenomena are due in part to demonic activity. Bewails the wastefulness and folly of teaching "human" wisdom. The Bible does not mandate schools as separate institutions; schools are really just extensions of the Church and the home. Index.

1128 _______. *The Scientific Case for Creation* (booklet). 1977. San Diego: Master Books (Creation-Life). "Scientific" rather than "biblical" (or "scientific biblical") creationism; this consists, that is, of scientific arguments and references, without any overtly religious material or biblical references. Small volume presenting standard creation-science arguments. Focuses on thermodynamic argument, lack of transitional fossils, arbitrariness of geologic column. Presents numerous estimates of age of earth based on rates of various physical processes—all indicating young earth. Demands that creation "model" be taught alongside evolution "model" in science classes as equally valid. Charts, diagrams, tables; creation-science bibliog.

1129 _______. *The Inspiration of the Bible* (booklet). 1980. Atlanta, Ga.: Walk-Through-the-Bible Ministries.

1130 _______. *King of Creation*. 1980. San Diego: Creation-Life. "Places the modern creationist movement in its Biblical perspective, emphasizing Christ as Creator and Sovereign of the world. Documents the scientific strength and spiritual impact of creationism." Urges all Christians to become actively involved in creationism. Describes the "Anti-Gospel of Evolution" and "Jesus the Creationist." Foreword by **J. McDowell.** Index.

1131 _______. *Creation and Its Critics: Answers to Common Questions and Criticisms on the Creation Movement* (booklet). c1982. San Diego: Creation-Life. 28 questions. These actually are the questions and accusations most commonly directed at ICR and other proponents of creation-science. A useful and concise compendium of the best (officially recommended) and most effective answers. Categories covered: creation and religion, qualifications of creationists, creationist motives and ethics, creation and science. Questions include: isn't creationism religious? Why do only Protestant fundamentalists support it? Can't evolution be considered God's method of creation? If you want creation taught in public schools, why don't you present evolution in your churches? Aren't all real scientists evolutionists? Why don't creationists publish in standard scientific journals? Also: difference between biblical and scientific creationism; relationship of ICR to CLP (its publishing affiliate); profits from promotion of creationism; quoting out of context; testability of creation and evolution; creationist entropy argument and open systems.

1132 _______. *Evolution in Turmoil*. 1982. San Diego: Creation-Life. "A documented, but easy-reading, account of the current conflicts within the evolutionary camp and the impact of modern creationism on the humanistic establishment in science and education."

1133 _______. *Men of Science—Men of God*. 1982. San Diego: Master Books (Creation-Life). "Great Scientists Who Believed the Bible." Small book intended as a "popular-level introduction" with "evangelistic motivation." Brief profiles of 66 scientists. All categorized as "creationist," but Morris includes progressive creationists and theistic evolutionists, and some anti-Darwinists whose religious convictions are unclear. Over half are pre-Darwinian. Includes **Ray, Burnet, Whiston, Woodward, Linnaeus,** Jonathan Edwards (theologian), **Cuvier, Bell, Buckland, Agassiz, J.W. Dawson, C. Piazzi Smyth** (Pyramidologist and British-Israelite), Mendel, **Fabre, Owen, Hitchcock, Maunder, J.A. Fleming, Dewar, LeMoine, von Braun,** plus many physical scientists. Also section on modern creation-science organizations. Index.

1134 _______. *The Revelation Record: A Scientific and Devotional Commentary on the Prophetic Book of the End Times*. 1983. Wheaton, Ill.: Tyndale House/San Diego: Creation-Life (co-published). Forewords by Jerry Falwell and **T. LaHaye.** Exegesis and commentary on Book of Revelation; companion to Morris's *Genesis Record* (1976). Emphasizes the "inseparable relation of the first and last books of the Bible"; their structural and mythic symmetry. The earth was created perfect, but man's sin has corrupted and degraded it. God will destroy it again, as He did in Noah's time, but will then return it to its created perfection. The pre–Flood water canopy will be restored; the whole earth will be smooth and mild—no mountains, deep oceans, deserts, or unpleasant weather. All men and animals will return to vegetarianism. Morris presents standard premillennial, pre-tribulational eschatology: God will destroy the earth again; believers will be raptured up to Heaven before the great tribulation; Christ will return to earth to rule for a thousand years; Satan will be bound in Hell during the Millenium, but will then escape for final battle against God; God will defeat Satan, then judge all souls, casting sinners into the fire and taking the righteous to live in the Heavenly Kingdom. Morris figures 100 billion people will have been born by the end of the Millenium; of these, perhaps 20 billion will be 'saved,' and will inhabit a cubic block of the Heavenly City in their angel bodies (Morris computes average living space). Index.

1135 _______. *Science, Scripture and the Young Earth: An Answer to Current Arguments Against the Biblical Doctrine of Recent Creation* (booklet). 1983. San Diego: Creation-Life. A rebuttal of **Davis Young**'s *Christianity and the Age of the Earth* (1982), which was itself a rebuttal of Morris's young-earth creationism and Flood Geology. Reiterates arguments for recent creation with updated scientific references. Focuses on radiometric dating; also coral reefs, various geological and fossilization processes, magnetic field decay. Criticizes Young's non-literal Bible exegesis, and bristles at his charges that young-earth creationists don't know enough geology.

1136 _______. *The Wonder of It All* (booklet). 1983. San Diego: Creation-Life.

1137 _______. *The Biblical Basis for Modern Science*. c1984. Grand Rapids, Mich.: Baker Book House. 516-page compendium of Bible-science. Foreword by **J. Oller.** "The Bible is indeed a book of science, as well as a book of history, literature, psychology, economics, law, education, and every other field"—even though it doesn't use technical jargon. "How could [anyone] trust the Bible to speak truly when it speaks of salvation and heaven and eternity—doctrines which he is completely unable to verify empirically—when he is taught that Biblical data that are subject to test are fallacious?" Presents a Bible-science interpretation of each scientific discipline, all based on a strongly literalist reading of the Bible. Chapters include theology (Queen of the Sciences), cosmology, miracles, evolutionism (Science Falsely So Called), cosmogony, astronomy, thermodynamics, chemistry and physics, geophysics, hydrology and meteorology, geology, paleontology, biology, anthropology, demography and linguistics, ethnology. Includes many creationist arguments: Flood Geology, antediluvian hydrologic cycle (canopy theory), population increase argument, alleged apemen, evolution as violation of 2nd Law Thermodynamics, etc. (God instituted the 2nd Law when Adam sinned, and will repeal it when he re-creates the world.) Comprehensive bibliographies for each chapter, listing many creationist sources. Photos, diagrams, maps, charts, tables; index.

1138 _______. *History of Modern Creationism*. 1984. San Diego: Master Book (CLP). Foreword by **J. Whitcomb.** An interesting and useful history of the creation-science movement by its leading proponent. Morris knew most of the creationist leaders personally, and played a key role himself in many of the most influential creationist organizations. Morris gives a brief review of early Bible-believing scientists and opponents of Darwin, but book mostly concerns creationism since the 1925 Scopes Trial. Presents biographical and background information on creationist leaders, describes history and context of creationist organizations. Most of Morris's information on modern creationism is firsthand. Chronicles the story of *The Genesis Flood* and the founding of CRS and ICR in detail. Polite but sometimes pointed criticisms of rival creationists; mentions doctrinal errors and other shortcomings of individuals and organizations, and describes conflicts and tensions within and between various creationist organizations. Has sections on "The Gospel of Creation" and "The Coming Battle for Creation"; describes the all-out war being waged against creationists by the humanist evolutionists, and emphasizes need for Christians to actively support creationism. Unlike his other creation-science books, Morris here gives credit to the pioneers of creation-science: **Price, Rimmer, Riley, A.I. Brown, B. Nelson, H. Clark,** and many others who were active before *The Genesis Flood* and the recent popular revival of creationism. Photos; bibliography of creationist books, index.

1139 _______. *Science and the Bible.*

1986 [1946]. Chicago: Moody Press. The revised and updated version of *The Bible and Modern Science* (1946; orig. titled *That You Might Believe*).

1140 _______. *Science and the Scriptures* (pamphlet). Undated. San Diego: Institute for Creation Research.

1141 _______, and Martin Clark. *The Bible Has the Answer.* c1976 [1971]. El Cajon, Calif.: Creation-Life. Orig. ed. (1971) pub. by Craig Press (Nutley, N.J.); Morris sole author of first edition. 1976 enlarged ed. adds Clark as co-author; Preface specifies who wrote which chapters. *Clark:* Bob Jones Univ. grad.; D.Ed.—Virginia Tech.; counseling director at Cedarville College (Ohio). Sections written by Morris orig. appeared in newspaper *News Messenger* (Virginia). Doctrinal issues answered from pre-Millennialist viewpoint; otherwise responses are "non-denominational." Includes sections on Bible-science, creation/evolution, the Flood, and early man (all written by Morris); these sections present standard creation-science arguments. Index.

1142 _______, and Gary E. Parker. *What Is Creation Science?* 1982. San Diego: Master Book. "Suitable for Public Schools." Foreword by Dean Kenyon. ICR recommends this book as perhaps the best general introduction to creation-science. Part I: three chapters on "The Life Sciences" by Parker; Part II: three chapters of "The Physical Sciences" by Morris. Part I is a reprint of Parker's 1980 book *Creation: The Facts of Life,* omitting the summary chapter. Part II repeats Morris's usual arguments, but adds new references and quotes. Appendix duplicates most of Morris's 1982 *Creation and Its Critics.* Morris says he became a creationist in 1943, and that his "main motivation" for entering graduate school in 1946 was to prove Flood Geology. Concentrates on thermodynamics argument; discusses Prigogine, open systems. Documents increasing interest in catastrophism amongst geologists and paleontologists; claims this confirms creation model and refutes evolution. Cites paleontologists who admit fossil record cannot be used to prove evolution, and zoologist **Grassé** that only the fossil record can prove evolution. Emphasizes theoretical nature of geologic column, and circular reasoning in using fossils to date rocks (based on assumption of evolution)—then using these rocks to date the fossils. Declares that the age of the earth is independent of proofs of creationism, but that most evidences point to a young earth, and that a young earth is fatal to evolution. Criticizes Big Bang theory (says it would produce a uniform universe); discusses uniqueness of earth and variety of heavenly bodies. Drawings, photos, tables; index.

1143 _______ (ed.). *A Symposium on Creation.* c1968. Grand Rapids, Mich.: Baker Book House. Papers from annual conference on Christian Schooling (Houston). Morris is listed as first author and generally cited as editor, though later volumes in the *Symposium on Creation* series (II–VI) are listed under **D. Patten** (ed.). Foreword by T. Robert Ingram (actual editor?). Includes papers by Morris, **Klotz, Zimmerman,** and two each by Patten and **McCone.** Biog. notes.

1144 _______ (ed.). *Scientific Creationism.* 1974. San Diego: Master Book (Creation-Life). Published in "General" and "Public School" editions, basically identical except that the former adds a concluding chapter ("Creation According to Scripture") which "places the scientific evidence in its proper Biblical and theological context." 12th printing (1985) enlarged and updated (e.g., newer, bigger creation-science reading list). Book prepared by technical staff and consultants of the Institute for Creation Research, but Morris says he wrote the "basic text." (Apparently **Gish, Slusher,** and **Austin** were Morris's most active assistants.) This book is generally regarded as the definitive presentation of creation-science: a comprehensive statement of creation-science theory and arguments. It presents "scientific" rather than "biblical" creationism—full of scientific references and quotes, but no overt religious references (with the exception of the final chapter of the General Edition). Asserts that evolutionist humanism is now promoted in schools as the official state religion. "True education in every field should be structured around creationism, not evolutionism." "There is not the slightest possibility that the *facts* of science can contradict the Bible. . ."—the creation model will always prove superior to evolution. Offers this book as means for teachers to learn how to present scientific case for creationism in school. Creationism is "mentally satisfying" and consistent with our "innate thoughts," hence true; it also assures us of meaning and purpose in life. Extended contrast of evolution and creation "models": includes assertion that evolution predicts natural law is "constantly changing." Emphasizes thermodynamic (entropy) argument, probability against chance formation of life, lack of transitional fossils, mutations as harmful rather than beneficial, problems in geologic column, superiority of catastrophist (Flood) interpretation of geology over uniformitarian assumptions of evolution, unreliability of radiometric dating, various other dating methods giving young age for earth (including rate of population increase). Says all proposed "ape-men" are either true humans or apes (considers *Homo erectus* human—contrary to many other creationists). Quotes early evolutionists to prove evolution leads to racism. The "biblical" chapter stresses the lesson of the Flood: destruction of the world because of sin; says Second Law was the result of Adam's Fall. "The Word of God must take the first priority, and secondly, the observed facts of science. . ." Refutes old-earth creationist theories. Urges reestablishment of strict creationism in order to destroy foundations of the "vast complex of godless movements spawned by the pervasive and powerful system of evolutionary uniformitarianism." Index.

1145 ______, Duane Gish, and George M. Hillestad (eds.). *Creation: Acts/Facts/Impacts.* c1974. San Diego: Creation-Life. Articles from the Institute for Creation Research free monthly newsletter *Acts & Facts:* Vol. 1, no. 1 to Vol. 11, no. 9. Includes "Impact" series articles (newsletter inserts) #1–9, which focus on various creation-science topics. This collection includes reports on the 1972 Ararat expedition, 1973 ICR Summer Institute, 1972 Nat'l. Assoc. Biol. Tchrs. convention, 1972 Calif. textbook hearings, and the Genesis School of Graduate Studies (Gainesville, Fla.—now defunct), which was affiliated with ICR and shared its faculty. Also includes reports on the college creation/evolution debtes, which were becoming popular about this time. Enthusiastic and optimistic coverage of creationism and of ICR, aimed at building support for creationism.

1146 ______ and ______ (eds.). *The Battle for Creation: Acts/Facts/ Impacts Volume 2.* 1976. San Diego: Creation-Life. ICR *Acts & Facts* articles 1974–1975, plus "Impact" series articles #10–26. Arranged by topic (not chronological). Topics include debates, campus presentations, seminars, creation-science and the public schools, confrontations with scientists, creationist research, international creationism, ICR and the media, creationism and Christian life. (Volume 3 of this series listed under **Gish** and Rohrer, eds.)

1147 ______, and Donald H. Rohrer (eds.). *Decade of Creation: Articles on Creationism from ICR Acts & Facts.* c1981. San Diego: Creation-Life. Volume 4 of articles from ICR's *Acts & Facts* (1978–1979), plus "Impact" series articles #55–78. Morris points out it has been a decade since he left VPI to found CHC and ICR, and notes that ICR has been the dominant force in creation-science. Continues coverage of college debates, seminars, international creationism, media response, and ICR's special creationist ministry.

1148 ______, and ______ (eds.). *Creation: The Cutting Edge.* c1982. San Diego: Creation-Life. Volume 5 of articles from ICR's *Acts & Facts* (1980–1981), plus "Impact" series articles #79–102.

1149 Morris, John D. *Adventure on Ararat.* 1973. San Diego: Institute for Creation Research. Foreword by **T. LaHaye.** *Morris:* son of **Henry Morris;** geological engineering Ph.D.—Univ. Oklahoma; presently geology prof. and administrator at ICR. Morris was working as a city engineer in Los Angeles when he joined an expedition to look for Noah's Ark in 1971. (This was before his graduate study.) This is an account, in diary format, of the 1972 ICR-sponsored expedition, which Morris led. Describes difficulties with Turkish authorities and local inhabitants, attacks by dogs and robbers, treacherous terrain. Recounts tensions and disputes among five-man team—attributes these to Satan's attempts to disrupt expedition. Three of the team were seriously injured by lightning. Group attempted to check Ark-like formation in Ahora Gorge photographed earlier by **C. Burdick,** who joined them later with his own group. They also scouted other areas and examined ancient inscriptions and archeological remains. Expresses optimism that Ark will be found, and stresses that its discovery will shatter the accepted geological timescale by proving Flood Geology true and the geological ages imaginary. Discovery of the Ark would "apply the final deathblow to the already fragile philosophy of Darwinian evolution." It "could well provide the final, climactic testimony to this present evil world that God is Creator and His Word is true. Evolution and all its progeny (communism, fascism, racism, animalism, etc.) are exposed as utterly and openly false. The age-long Satanic opposition to God's purpose in creation could no longer be masqueraded as 'science,' but would have to surface as overt Satanism." Includes brief review of standard Ark-eology tales, and summary of **A. Woods'** ICR study proving the geographical center of the earth is near Ankara (and Ararat). Photos.

1150 _______. *Tracking Those Incredible Dinosaurs ... And the People Who Knew Them.* c1980. San Diego: CLP. Morris is chief ICR investigator of the Paluxy River site near Glen Rose (Texas), where "manprints" have been found along with the well-preserved dinosaur tracks in the Cretaceous limestone. Morris is more critical than most creationist Paluxy advocates, but clearly believes some of the tracks are genuine human prints. He points out that proof of the co-existence of man and dinosaur would overthrow the geological and evolutionary timescale, and admits "the purpose of this book is to provide such evidence." Discusses history of the Paluxy finds, and Stanley Taylor's Films for Christ investigations for his 1973 movie *Footprints in Stone.* Also describes recent excavations by various groups and individuals, and cites many creationist publications about Paluxy. Mentions wood samples dated at 38,000 B.P. by UCLA. Claims dinosaurs are referred to in the Bible. Describes geology of Paluxy, methods of footprint identification, location and setting of all known tracks. Morris notes that some prints are carved fakes; others are possible fakes. Includes a descriptive reconstruction of Noah's Flood: suggests that humans and dinosaurs survived first part of Flood by escaping to Llano Uplift, then walked across deep Flood-deposited sediments during temporary drop in water level as Gulf of Mexico subsided. Many photos.

1151 _______. *The Paluxy River Mystery* (pamphlet). 1986. El Cajon, Calif.: Institute for Creation Research. ICR "Impact" series #151. Morris has written several "Impact" articles on the Noah's Ark expeditions and the Paluxy manprints. This one caused a stir: after much prompting (chiefly from G. Kuban), Morris returned to Paluxy to examine new evidence which demonstrated the dinosaurian origin of the "manprints." Here, Morris cautions against use of the Paluxy tracks by creationists as evidence against evolution. However, since this article Morris has been equivocal about whether the manprints have been disproved or not. ICR said it withdrew Morris's 1980 book (but actually merely inserted this pamphlet in copies being sold).

1152 Morrison, A. Cressy. *Man Does Not Stand Alone.* c1944. Old Tappan, N.J.: Fleming H. Revell. Rev. ed. (undated). Excerpted in 1960 *Reader's Digest;* rev. ed. followed. Written in response to J. Huxley's *Man Stands Alone. Morrison:* former president of N.Y. Acad. Sciences; Fellow—Amer. Mus. Nat. Hist. The wonders and design of nature prove a Supreme Intelligence and purpose. Admits strength of Darwin's theory, but maintains that **Paley**'s argument from design has not been refuted. Suggests that new facts of science demonstrate falsity of materialist atheism, but that it is not necessary to deny evolution itself. Describes uniqueness of earth as environment for life. "Life is an instrumentality serving the purpose of the Supreme Intelligence. LIFE IS IMMORTAL." The goal of the directive purpose so evident in nature is the creation of intelligent minds. God may have allowed man to evolve from lower forms—unless He created man as is, later. "The rise of man the animal to a self-conscious reasoning being is too great a step to be taken by the process of material evolution or without creative purpose." Marvelous animal instincts are impossible to account for by materialist theories; even more so the human mind. The marvelous fitness of the earth for life disproves chance origin of life. Suggests that the Genesis creation account corresponds to stages of earth history and development of life (Day-Age creationism). Openly religious.

1153 Morton, Glenn R. *The Geology of the Flood.* Dallas, Texas: DMD Publishing. Orig. 1986 (priv. pub.; Dallas). *Morton:* geophysicist with thirteen years experience in petroleum industry; now with Arco; has written many *CRSQ* articles; ghost-writer for evolution sections in **McDowell** and Stewart 1981. Morton has described himself as a "middle-earth" creationist: he wants the earth to be as young as possible for biblical reasons, but his experience in petroleum geology has convinced him that it must be more than several thousand years old, and that strict Flood Geology contains serious errors. Sedimentary layers show clear evidence of long deposition times. But he also insists that "If evolution is true, then the Bible is wrong." In this book, Morton assumes a single miracle: at the time of the Flood, God increased the permittivity of free space. This change caused atoms to move apart, some expanding more than others. Earth's radius doubled. Earth's land mass split apart—this, rather than plate tectonics, explains the continents. Differential expansion of various materials accounts for geological features such as earthquake zones, thrust faulting, etc. Continental sediments did not settle in the deep ocean basins because these did not exist in the smaller pre-Flood earth. Cambrian boundary formed at outset of Flood; early Paleozoic strata deposited in later stages of expansion. Criticizes creationist continental drift theories and pre-Flood Water Canopy theories. Creation occurred about 125,000 years ago. The Flood began 30,000 years ago; Noah's Ark landed after a year, but effects of the Flood lasted about 5,000 years. Most of the paleontological record is the result of the thousands of years of post-Flood reinundations and other adjustments. These account for fossil sequences not adequately explained by strict Flood Geology. (Reviewed by **G. Howe** in 1987 *CRSQ.*)

1154 Morton, Harold Christopherson. *The Bankruptcy of Evolution.* Undated [1925?]. London/Edinburgh/New York: Marshall Bros. *Morton:* minister. Includes appendices of serological "blood-reaction tests" (after Nuttal).

1155 Morton, Jean Sloat. *Science in the Bible.* c1978. Chicago: Moody Press. Foreword by **D. Gish.** *Morton:* Ph.D.—George Washington Univ.; member of ICR Technical Adv. Board; biochemist, science writer and teacher; works with Alpha-Omega Publications. Bible-science: chapters on astronomy, meteorology, botany, zoology, health, other sciences. Examples of scientific wisdom and predictions contained in the Bible. Does not discuss evolution or creationism directly. Says the hare is really a 'cud-chewer' since it is a caecotroph; discusses other examples vindicating the knowledge of science in the Bible.

1156 ________. *Virgil's Aeneid, By Chance Alone?* (pamphlet). 1985. El Cajon, Calif.: Institute for Creation Research. ICR "Impact" series #146. Impossibility of chance formation of proteins: probability argument. Morton also wrote "Impact" #90, on glycolysis and fermentation.

1157 Mount, Ralph H. *Evolution, a Destructive Heresy* (pamphlet?). c1970. Mansfield, Ohio: Mount Publications. (Listed in Ehlert files.)

1158 Muggeridge, Malcolm. *The End of Christendom.* 1980. Grand Rapids, Mich.: William B. Eerdmans. Muggeridge, the well-known British journalist, author, and Christian apologist, declares that in the future, Darwinian evolution will be laughed at. "Posterity will marvel that so very flimsy and dubious an hypothesis could be accepted with the credulity that it has. I think . . . this age is one of the most credulous in history. . ." He has been a featured guest on Canadian David Mainse's "Crossroads" creationism series (part of his *100 Huntley Street* telecast from Toronto), along with **J.N. Moore, Gish, Rusch, Parker, Slusher,** and other creationists.

1159 Mulfinger, George. *How Did the Earth Get Here?* (booklet). Undated [1972]. Greenville, S.C.: Bob Jones Univ. Press. *Mulfinger:* physics M.A.—Syracuse Univ.; teaches physics at Bob Jones Univ. (S.C.). Criticizes arrogance of men who try to explain origins of earth and solar system by naturalistic science; they reject God's Word, though God was the only witness of these events. The Nebular Hypothesis—origin of the solar sytem by condensation of matter—was proposed by Swedenborg, who got the idea from occult spirits. Kant and Laplace then revived the idea. Discusses failure of other evolutionary theories. The earth's origin was supernatural; men today are "willingly ignorant" of the Creation and the Flood. Scientific and biblical references.

1160 _______. *The Flood and the Fossils—Geological Puzzles Explained by Genesis 6-9.* 1969. Greenville, S.C.: Bob Jones Univ. Press. "There is probably no area of science more filled with nonsense and unscientific guesswork than that branch which deals with fossil man. . . . whole races of men are fabricated and colorful myths are constructed concerning their [supposed] primitive beastlike mode of life." (Quoted in **Noone** 1982.) Discusses Piltdown, Nebraska Man, Java Man, *Zinjanthropus.* All either frauds or pure ape.

1161 _______, and Donald E. Snyder. *Earth Science for Christian Schools.* 1979. Greenville, S.C.: Bob Jones Univ. Press. Bob Jones Univ. "Science for Christian Schools Series": this book for grade 8. Accompanying teacher's edition and lab manual. Openly creationist; stresses infallibility of Bible on scientific matters. "All distortions of evolution and humanism have been eliminated, making this series fully consistent with God's Word while accurately presenting what man has learned about God's creation."

1162 _______ (ed.). *Design and Origins in Astronomy.* 1983. Norcross, Ga.: Creation Research Society Books. Creation Research Soc. Monograph Series: No. 2. Authors: Mulfinger; **D. DeYoung; J. Whitcomb; Emmett Williams; Hinderliter; P. Wilt; P. Steidl.** Many scientific references; some sections overtly religious. Mulfinger urges a "Teleological Study of the Universe." DeYoung and Whitcomb argue that the stellar red shift is not evidence of a Big Bang, but is due to gravity; they suggest the earth is in the center of the universe, but may revolve around the sun. They also suggest that the sun's 'missing' neutrinos indicate it is powered by gravitational collapse rather than nuclear fusion, and is quite young. Evolutionists search for extraterrestrial life to prove evolution: "Man seems determined to prove that he is the result of blind chance rather than special creation!"—but absence of ET life is evidence of creation. The notion of beings on other planets makes a mockery of Christ as Saviour. The heavens are inhabited by angels, though. Stars were created as signs for man. Williams criticizes Big Bang model with thermodynamics arguments. Wilt discusses nucleosynthesis, and suggests areas for creationist criticisms of the Big Bang theory and stellar evolution. Steidl claims that space probes have produced evidence contrary to evolutionist theories about planets and asteroids, and reviews creationist arguments about comets. Hinderliter, in two articles reprinted from *CRSQ,* argues for a young sun based on observed shrinkage and other solar phenomena. Whitcomb's chapter "The Bible and Astronomy" is straightforward Bible-science; it is also pub. separately (Whitcomb 1984). The authors reject the Big Bang scenario, claiming instead that God created the planets and stars *in situ.* Biog. notes. Tables, diagrams, charts.

1163 Munk, Eli. *The Seven Days of Creation.* 1974. Jerusalem/New York: Feldheim Pub. Jewish. Addressed to student confused by the apparent contradiction between the Torah and modern views. The analysis is mostly linguistic.

1164 Murray, John. *Skepticism in Geology.* 1877. London: Murray. Published under pseudonym "Verifier." Murray was Lyell's and Darwin's publisher, but he remained opposed to evolution. This book attacked Lyell's uniformitarian geology in order to uphold Flood Geology.

1165 Myers, E.C. *Constructing a Creationist Geology.* 1984. Unpub.(?) thesis: Dallas Theological Seminary. M.A. thesis (Dallas Theol. Sem); presumably strict young-earth creationism. Includes discussion and criticism of radiometric dating. (Cited by S.R. Schrader, in **Youngblood** 1986.)

1166 Myers, Ellen. *Presenting the Biblical Creation Position on the Secular Campus* (pamphlet). 1979. Wichita, Kansas: Creation Social Science and Humanities Society. *Creation Social Science and Humanities Quarterly* Special Reprint Series: No. 1. Expanded version of article pub. in 1979 *CSSHQ.* Written "in private consultation with" Gerald Paske (philosophy prof. at Wichita State). *Myers:* born in Germany; came to U.S. after WWII; has written novels about Nazi Germany and Christianity; foreign language teacher; student at Wichita State Univ.; co-founder of Creation Social Science and Humanities Soc.; frequent contributor to *CSSHQ.* Philosophical arguments for presenting biblical creationism to college students. Defends strict young-earth creationism and Flood Geology. Among the many topics Myers has written about for *CSSHQ* are the creationist movement in Europe, **C.S. Lewis**'s creationism, evolutionist influence on Nazism, and the population non-problem.

1167 ________. *The New Age Movement: An Emerging Threat to Biblical Christianity* (report). 1984. Minneapolis: Bible-Science Association. From address given 1984. The New Age Movement—occultism, mysticism, paranormal forces, pantheism—arose from evolutionism, and is directly opposed to Christianity. Cites New Age advocates of "emergent evolution" and vitalism; says she was a vitalist before conversion in 1960. Myers considers it highly ironic that many New Agers are anti-Darwinian: "One can get fine critiques of Darwinian evolution models from 'New Agers'"—cites **J. Rifkin** and others. New Age "centering" involves demons. Discusses gnosticism and Rosicrucianism; denounces **Teilhard** as a gnostic-mystic pantheist. T. Roszak was influenced by Lamarck, **Bergson, Blavatsky.** Accuses Assoc. Humanistic Psychology of advocating yoga, meditation and ESP. Criticizes the "myth" of overpopulation. Follows views of **Cumbey, D. Hunt,** and **C.S. Lewis.** Like most fundamentalist creationists, Myers sees the New Age Movement as a logical development from, and ally of, evolutionary philosophy, intended to undermine biblical Christianity.

1168 Nachtwey, Robert. *Der Irrweg des Darwinismus.* 1959. Berlin: Morus-Verlag. In German. Discusses Darwin, Spengler, Nietzsche, **Hitler.** Nachtwey wrote other works about instinct, and about micro-organisms.

1169 Nada-Yolanda. *Evolution of Man: 206,000,000 Years on Earth.* c1971. Miami, Fla.: Mark-Age MetaCenter. "Channeled by the Spiritual Hierarchy through Nada-Yolanda." Uses Theosophical schemes and jargon. The Bible and other creation myths are true, but symbolic. Fourth-dimensional spirits became trapped in earthly three-dimensional bodies. Adamic group of Elders descends in half-physical form to help. Some of the etheric beings wish to enslave human sub-race, others want to assist them regain fourth-dimensional form. These two groups—the Cains and Abels—fought a psychic battle at Mt. Shasta, Calif. Cains form caste system; etheric elders depart from earth. Civilizations of Mu and Atlantis misuse psychic/spiritual powers; land mass is broken up. Land submerges in period of the Noahs and the Flood. Sananda, leader of Abels, returns as Jesus; will return again in end days: the Mark Age. There will be destruction, purification, and Judgment in the Mark Age, then the Aquarian Age, and reentry into fourth-dimensional planetary society. "Man Not From the Lower Kingdoms": we have a race memory of telepathic communion with lower physical elements, and some men psychically incarnated themselves in lower forms as experiments, but mankind did not arise out of these. Man came from etheric light, and that is where he must evolve to again.

1170 Nafziger, Edgar. *The Grand Canyon and Creation* (tract). Undated. Minneapolis: Bible-Science Association. Hypothesizes about Flood origins of major fetures of the Canyon. The Precambrian Vishnu Schist was formed when the dry land was separated from the water, or when the fountains of the deep opened during the Flood. Nafziger has led many BSA Grand Canyon tours.

1171 Navarra, Fernand. *L'Expedition au Mont Ararat.* 1953. Bordeaux, France: Biere. *Navarra:* French industrialist (demolitions). Navarra climbed Ararat searching for Noah's Ark several times and has made the most sensational Ark-eology claims. This is an account of his first expedition, in which he saw the Ark under glacial ice at an altitude of 13,800 feet. "The shape was unmistakably that of a ship's hull: on either side the edges of the patch curved like the gunwales of a great boat. As for the central part, it merged into a black mass the details of which were not discernible. Conviction burned in our eyes: no more than a few yards of ice separated us from the extraordinary discovery which the world no longer believed possible. *We had just found the Ark.*" (Quoted in **J.W. Montgomery** 1974; also in several other Ark-eology books.) Photos.

1172 ________. *The Forbidden Mountain.* 1956. London: Macdonald.

1173 ________. *J'ai Trouvé l'Arche de Noë.* 1957 [c1956]. Paris: Editions France-Empire. ("I found Noah's Ark.") An account of Navarra's 1955 expedition on Ararat, in which he and his 11-year old son descended a crevasse in the glacier where he had located the Ark and hacked off a five-foot piece of wood from a large hand-worked beam buried in the ice. Navarra split this piece into smaller pieces to smuggle past the Turkish soldiers at the bottom of the mountain, and later had the pieces analyzed by several labs. Book includes reprint of report from Forestry Institute of Madrid, estimating the age of a fragment at 5,000 years; also a report from the Dept. of Anthropology and Prehistoric Studies of Univ. Bordeaux, France, which described it as dating from "remote antiquity." An official of the Cairo Museum said he thought the fragments were 4,000 to 6,000 years old. (However, three American C-14 labs later dated the wood to early medieval times.) Navarra and his son took many photos of the discovery.

1174 ________. *The Noah's Ark Expedition.* 1974. London: Coverdale House. Edited with **Dave Balsiger.** Photos, maps. (British edition of 1974, below?)

1175 ________. *Noah's Ark: I Touched It.* 1974. Plainfield, N.J.: Logos International. "Edited with" **Dave Balsiger.** Balsiger also listed as U.S. agent for book. **R.H. Utt** acknowledged for help in translation. Includes summary of standard Ark-eology tales, and chapters on Navarra's several Ararat expeditions. Reviews his 1952, 1953, and 1955 climbs. Also includes account in diary format of the 1969 SEARCH expedition. Navarra guided the SEARCH team to the Ark site, and the team recovered several small pieces of wood from a glacial lake. Appendices include Flood tales from around the world; ancient Ark-eology reports and descriptions of the Ark; and reprints of letters regarding lab tests on the Ark wood samples. Photos, drawings. Balsiger endorses Navarra's claims wholeheartedly, but many other Ark-eologists are skeptical. **Cummings** (1982) and **Noorbergen** (1974) both describe in detail Navarra's suspicious relations with the SEARCH team (an outgrowth of **Vandeman**'s A.R.F. Ark-eology organization), and accuse him of deliberately leading them away from his Ark site on several SEARCH climbs prior to 1969. They suspect that Navarra had earlier planted the wood he "discovered" in the 1969 expedition. **J. Morris** (in **LaHaye** and Morris) is also highly suspicious of Navarra's claims.

1176 Nee, Watchman. *The Mystery of Creation.* c1981. New York: Christian Fellowship Publishers. Orig. Chinese; published as a series, "Meditations on Genesis," in the magazine *Christian,* 1925–27. Mostly theological, but also discusses relationship of Genesis to science, especially chapter "Genesis and Geology." "Genesis is God's revelation, but geology is man's invention. Since God knows all the facts, His revelation can never be in error. ... If Genesis and geology differ, the error must be on the side of geology, for the authority of the Bible is beyond questioning." Strong presentation of Gap Theory creationism; cites **Chalmers,** and follows **Pember** closely. Includes a chapter on "The Original World"–the pre-Adamic world before the six-day re-creation. Pre-Adamites came under Satan's rule after he fell into evil, and exist now as demon spirits. God laid waste to this world, then re-created it in six 24-hour days. On the fourth day He repaired the sun, causing it to shine again.

1177 Nelson, Byron C. *After Its Kind.* c1967 [1927]. Minneapolis: Bethany Fellowship. Orig. c1927. Rev. ed. 1952. Foreword by **J.C. Whitcomb.** *Nelson:* studied science and philosophy at Rutgers and Wisconsin; Th.M.–Princeton Sem.; Lutheran pastor (Norwegian Synod) in Wisconsin. A classic creation-science work. Demonstrates the impossibility of evolution from scientific evidence. Strongly biblical; insists on strict creationism. Includes standard creation-science arguments. Fossil forms identical with modern organisms. Species can show great variation. Nelson argues that biblical 'kind' equals "species"–the only real, clearly demarcated category–not some higher (arbitrary) taxonomical level. Morphological similarity shows common design, not evolution. Convergence allows evolutionists to accommodate contradictory evidence. "Evolutionary" stages in embryology are illusory. "Vestigial" organs are not really useless. New and complex organs could not have evolved: they would have been selected against in their early, nonfunctional stages. "Mendel's discovery has done great damage to the theory of evolution. Mendelism says: *After its kind.*" Discusses alleged "missing links" of human evolution; shows they are imaginative reconstructions or remains of degraded true humans. Suggests that "dislike of the idea of creation is in fact the underlying reason for belief in evolution by many leading evolutionists." Many scientific references. Photos, diagrams; index.

1178 ________. *The Deluge Story in Stone: A History of the Flood Theory of Geology.* c1968 [1931]. Minneapolis: Bethany Fellowship. Orig. 1931 (Minneapolis: Augsburg). Foreword by **H. Morris.** Nelson's previous book *(After Its Kind)* focused on biology; this concerns geology. Nelson, who was influenced by **G.M. Price,** calls for a return to Flood Geology and strict young-earth creationism. The early church fathers believed in the Flood. Early geologists correctly believed that fossils and strata resulted from the actions of the Flood. **Cuvier**'s theory of multiple catastrophes started the trend away from Flood Geology, followed by **Buckland, Penn,** and other compromisers. With the rise of uniformitarian geology and evolution, it was completely eclipsed–though various scholars have continued to uphold Flood Geology against thc dominant theory. This book is a good source of information on ancient commentators, classic Flood Geologists, and more recent critics of evolutionary geology. Nelson discusses and quotes **Steno, Kircher, Burnet, Ray, Woodward, Whiston, P. Cockburn, Hutchinson, John Williams, Fairholme, Twemlow, Bosizio, G. Wright, Howorth, Price,** and others. Standard creation-science arguments; emphasizes out-of-order strata sequences. "The Flood theory has not been abandoned because it does not satisfy actual geological conditions. There is nothing known about the earth's geological state today which makes the Deluge theory any less satisfactory an explanation of the fossiliferous strata than in the days when the leading scholars of the world accepted it. . . . It is a *disregard for God and the sacred record of his acts,* and nothing else, which has caused the discard of the Flood theory to take place." Discusses construction of the Ark (quoting **Totten**), and worldwide Flood traditions. Photos, diagrams; index.

1179 ________. *Before Abraham: Prehistoric Man in Biblical Light.* 1948. Minneapolis: Augsburg. This book focuses on anthropology: fossil evidence of the alleged descent of man from pre-human creatures. Strict creationism. The Flood destroyed all human remains: we have no fossils of antediluvian man. The glacial epoch occurred after the Flood, and there are many human fossils from this period. Neanderthal and Cro-Magnon fossils are of mankind after the dispersion from Babel. Nelson argues that Genesis genealogies contain gaps; thus mankind could have been created thousands of years before **Ussher**'s date–the Ice Age dates back at least 20,000 years. Ice Age man in Europe was highly intelligent, and there were civilized areas in North Africa and the Near East during this time. Compares Neanderthals with modern human individuals to show they were not a primitive, less evolved type. Ice Age man used stone tools because that was the best material in those circumstances. "The theory of man's evolution is dangerous. It is the philosophy and tool of militant atheism. Few people realize that **Hitler,** in bringing about the war, merely put into practice what he believed about human evolution. . . . The worst feature of the theory from a doctrinal point of view is that it denies the Fall and thus undermines all Christian theology which makes Christ a genuine redeemer." True science supports the Bible and creationism. Photos, drawings, diagrams; bibliog., index. (Nelson's grandson Paul Nelson writes well-informed reviews and articles for the **Students for Origins Res.** journal, under the name "Peter Gordon.")

1180 Nelson, Ethel R., and Richard E. Broadberry. *Mysteries Confucius Couldn't Solve: Analysis of Ancient Facts Shared with Hebrew Scripture.* 1986. South Lancaster, Mass.: Read Books. "Analysis of Ancient Characters Reveal Intriguing Facts Shared with Hebrew Scriptures." *Nelson:* medical pathologist at New England Mem. Hosp. (Stoneham, Mass.) when book was written; formerly lived in Thailand; now lives in Tenn. *Broadberry:* medical lab specialist in Taipei; fluent in Chinese. This book is a sequel to **Kang** and Nelson's 1979 *Discovery of Genesis.* Analysis of Chinese characters proves that the 4,000-year-old Border Sacrifice ritual of the Chinese emperors is a version of Genesis. ShangTi, worshiped in this ritual, is the Hebrew Creator God. Confucius knew this ritual was very important, but did not know its origin. The whole story of Genesis and of Christ can be discovered in various Chinese characters. Many characters illustrated. Authors select characters from ancient bronzeware and oracle bone scripts as well as later hieroglyphic scripts. Photos; index.

1181 Nesbitt, Jacques. *Creation et Evolution: Problemes d'Origines.* c1976. La Colinne, France: Editions MEAF. (Listed in Ehlert bibliog.)

1182 Nevins, Stuart E. *The Origin of Coal* (pamphlet). 1976. El Cajon, Calif.: Institute for Creation Research. ICR "Impact" series #41. "Nevins" is pseudonym of **Steven Austin.** Austin used pseudonym for various creationist publications before he received his Ph.D. (on coal formation). At ICR, he wrote other "Impact" articles—*Evolution: The Oceans Say No!* (#8), *Planet Earth: Plan or Accident* (#14), and others under the name Nevins, plus later ones under his real name.

1183 Newell, Philip R. *Light Out of Darkness: The Six Days of Creation and the Problem of Evolution.* Undated. Chicago: Moody Press.

1184 Newman, Robert C., and Herman J. Eckelmann, Jr. *Genesis One and the Origin of the Earth.* 1977. Grand Rapids, Mich.: Baker Book House. *Newman:* astrophysics Ph.D.—Cornell; M.Div.—Faith Theol. Sem.; New Test. prof. at Biblical Theol. Sem.; co-author (with **Stoner**) of *Science Speaks;* currently with Interdisciplinary Biblical Res. Inst. (Hatfield, Penn.). *Eckelmann:* M. Div.—Faith Theol. Sem.; pastor; researcher at Cornell Radiophysics and Space Center. Dedicated to F. Drake, T. Gold, C. Sagan, et al. Well-written; scientifically knowledgeable. Progressive creationism: intermittent day theory. The six days of Genesis were separated by long creative ages; the seventh is yet to come. First section of book consists of scientific proof of age of the earth and universe—strong refutation of young-earth creationism. Second section presents theological arguments for old-earth creationism; correlates Genesis with scientific theories of earth's origin. Universe started with Big Bang. Day 1 intervenes after planets are formed from nebular clouds. Day 2: follows out-gassing of ocean and atmosphere from hot primitive earth. Day 3: after continents are formed and land vegetation appears. Day 4: after atmosphere is altered and cleared by photosynthetic organisms. Includes sections by **D. Wonderly** on non-radiometric dating; R. John Snow (math M.A.; M.Div.; Grace Brethren minister; former H.S. math teacher) on the length of the sixth day; and reprint of William Henry Green's 'Primeval Chronology" (1890), which argues for genealogical gaps in Genesis and against precise biblical dating of Creation or the Flood. Index.

1185 Newton, Benjamin Willis. *Re-*

marks on "Mosaic Cosmogony." 1882 [1864]. London: Houlston. 3rd ed., rev. Orig. 1864 (London). Apparently a reply to *Mosaic Cosmogony* by C.W. Goodwin. Concerns Creation and cosmology. K.A. Smith, a flat-earther, calls Newton a "faithful writer" and quotes his blasts at those who question the Mosaic cosmogony. (Newton himself does not seem to have been a flat-earther.) Newton wrote many books on biblical themes: the End Times, Antichrist, etc., some of which were reprinted by Seventh-day Adventists (Smith was an Adventist). (From Schadewald files.)

1186 Newton, Sir Isaac. *Chronology of Ancient Kingdoms Amended.* 1728. London. *Newton:* considered by many the greatest scientist who ever lived; best-known for development of the calculus, research in optics, and discovery of laws of motion and universal gravitation. Creationists point out he spent more time, and wrote more, on theology than on science (he wrote a great deal on alchemy and other mystical subjects). He calculated earth was 50,000 years old from rate of cooling, but assumed this was in error, since his biblical studies convinced him Creation was about 3500 B.C. Newton was greatly impressed with **Whiston**'s 1696 book, and made him his successor at Cambridge. But he later considered Whiston's scheme of catastrophic cometary intrusions too radical, and retreated to a more fundamentalist position. Main purpose of this book, which occupied much of Newton's final years, was to refute Whiston's theory. Newton's book is largely concerned with determining length of year in ancient calendar systems by astronomical dating methods. He rejected the notion that the present state of our solar system is the result of purely naturalistic interaction with comets. Comets, with their highly eccentric orbits, could not account for the present order—they are too unstable and "chaotic." Insisted that the highly ordered planetary orbits do not result from purely natural forces, but were formed by a single, direct act of Creation. Conceded his own gravitational laws would eventually cause orbits to alter, so he insisted that God must directly intervene from time to time to correct orbits and straighten things out. Though Newton's "mechanical universe" inspired deistic (and atheistic) interpretations, he advocated theism (continuous and direct supervision—intervention—of God). He strongly affirmed this theistic view in his renowned *Philosophiae Naturalis Principia Mathematica* (1687). In his *Opticks,* he opined that "if traditional views of cosmic order were abandoned, the foundations of morality would be undermined" (as described by Stecchini). Newton, like Whiston, believed in the Unitarian Arian "heresy" (Christ not co-equal with God; denial of the Trinity).

1187 Nichol, Francis David. *God and Evolution* (booklet). 1965. Washington, D.C.: Review and Herald. (Listed in SOR CREVO/IMS.)

1188 _______ (ed.). *Genesis and Geology: Two Chapters Reprinted from Seventh-day Adventist Bible Commentary: Vol. I.* 1978 [1953]. Washington, D.C.: Review and Herald. *Nichol:* co-author of **A. Baker** and Nichol 1926. Standard creation-science. Many well-known creationists listed on editorial board of Seventh-day Adventist Bible Commentary.

1189 Nichols, Herbert L., Jr. *Science Blundering: An Outsider's View.* 1984. Greenwich, Conn.: North Castle Books. Includes section on evolution. "Natural selection is not—cannot be—the dominant force in Evolution." Nichols has also written books about cooking and excavation equipment.

1190 Nicholls, J. *Bacterium E. Coli: Facts Support Creation* (booklet). Undated. England: Evolution Protest Movement. EPM pamphlet #190.

1191 Niessen, Richard. *Starlight and the Age of the Universe* (pamphlet). 1983. El Cajon, Calif.: Institute for Creation Research. ICR "Impact" series #121. *Niessen:* M.A.—Trinity Evangel. Div. Sch.; Ph.D. cand. at Aquinas Inst. Theol.; apologetics and Bible prof. at Christian Heritage Coll. (ICR affil.). Proposes four young-earth creationist solutions to problem of light from stars more than 6–10,000 light-years away: Distances may be result of false measurements; light may travel in curved "Riemannian" space; speed of light may have slowed drastically since Creation; and/or light may have been created en route (created with appearance of age).

1192 ________. *Introduction to Bible-Science: Apologetics 102 Notebook—CHC* (notebook). Undated [1985]. El Cajon, Calif.: Christian Heritage College. Published course material for CHC course Apologetics 102. Consists partly of original lecture notes and outlines, and partly of reprints of articles by Niessen (from *Bible-Science Newsletter,* ICR "Impact" series, *CRSQ,* etc.). Includes refutations of old-earth theories; discussion of canopy theory (vapor or other models); starlight and age of universe; dinosaurs in the Bible; radiometric dating; cave men; Noah's Ark (rediscovery would parallel Methusaleh's death—signal the imminent Judgment); rotating earth. Creation: 4174 B.C. Flood: 2518 B.C. Niessen, though a theologian rather than a scientist, is quite familiar with all the standard ICR creation-science arguments, and this course notebook is a good summary of ICR's scientific-biblical creationism.

1193 Nilsson, Nils Heribert. *Synthetische Artbildung: Grundlinien einer exacten Biologie.* 1953. Lund, Sweden: C.W.K. Gleerup. "Synthetic Speciation." 2 vols. In German; includes a 105-page English summary. *Nilsson:* Ph.D.; geneticist; director of Botanical Inst. (Lund, Sweden). Strong attack on evolutionary theory; refutes gradual Darwinian evolution by mutations. "The idea of an evolution rests on pure belief." Evolutionary theory is not "an innocuous natural philosophy," but a "serious obstruction" to biological research because "everything must ultimately be forced to fit this theory"—thus precluding an "exact biology." "It may, therefore, be firmly maintained that it is not even possible to make a caricature of an evolution out of paleobiological facts. The fossil material is now so complete that . . . the lack of transitional series cannot be explained as due to the scarcity of material. The deficiencies are real, and they will never be filled." His botanical research refutes as "absolutely impossible" the idea that organisms undergo spontaneous mutations leading to new forms when acted on by natural selection; shows instead genetic stability. Mutation theory of evolution has degenerated into same old Lamarckism. Denies principle of gene linearity; argues that enzymes are also genes. He concluded there has been no evolution of forms at all, but stability of types, and occasional periods of catastrophic destruction followed by sudden appearance of completely new types. Advocated Hörbiger's theory (see **Bellamy**) of immense tidal waves caused by descent of successive moons. As the moon approached the earth, it became locked in stationary position, pulling air and water into huge piles, then collapsing into earth. Declaring that "Biblical deluges will occur," Nilsson resurrected **Cuvier**'s theory of successive life-destroying flood catastrophes, using Hörbiger's theory as mechanism for his neo-Cuvierian interpretation. During these catastrophic periods, new organisms are created suddenly by "emication"—a drastic alteration or production of gametes. A few survive as totally new forms. Inspired by Oparin's theory of spontaneous origin of life, Nilsson argues that these gametes, of entirely new organisms, could form spontaneously and polyphyletically, out of the mix of biocatalytic substances engendered during the catastrophic episodes. "During paleobiological times whole new worlds of biota have been repeatedly synthesized." Nilsson declares that organisms such as orchids and elephants were "instantly created out of non-living material." Argues that German coal beds refute evolutionist theory of *in situ* formation—says coal formed from catastrophic flooding of materials "from all over the earth." Refutes evolutionary horse series; says the early forms have "nothing to do with the horse." Bibliog. Very widely quoted by creationists as scientifically devastating refutation of evolution.

1194 ________. *Synthetische Artbildung.* 1973. Victoria, B.C., Canada: Evolution Protest Movement. Reprint of English summary from Nilsson's 1953 book. Foreword by **G.F. Howe** and W.D. Burrowes (head of British Columbia Evolution Protest Movement—now N. Amer. Creation Mv't.). They emphasize that Nilsson scientifically demonstrated that gradual evolution has not and cannot occur. Because he was unable to conceive of special creation (or was afraid to advocate it) he devised his bizarre emication theory.

1195 Noble, Barclay. *The Bible and Science in the High School Curriculum.* 1962. Unpub. thesis: Fuller Theological Seminary (Pasadena, Calif.). Fuller Theol. Sem. thesis: M.Relig.Educ. Focuses on evolution. Evolution is still not proved (though Noble does not claim that it cannot be in the future). Largely relies on **Ramm** and **ASA** views. Discusses different approaches to science education and evolution by Christians.

1196 Nolan, Frederick. *The Analogy of Revelation and Science Established in a Series of Lectures Delivered Before the Univ. of Oxford....* 1833. Oxford, England: J.H. Parker. 1833 Bampton Lecture. Millhauser says this work was among those which expressed a general resentment against all geology for being infidel. Nolan wrote other books on Chaldean grammar, problems of historical chronology, Greek music, and the date of the Millennium.

1197 Noone, Richard W. *Ice: The Ultimate Disaster.* c1982. Dunwoody, Ga.: Genesis Publishers. Every 6000 years there is a sudden shift of the earth's axis, related to precession of the equinox, which is caused by accumulation of polar ice. These shifts result in drastic climate changes. Between 4500 and 4000 B.C. such a shift occurred. It was accurately predicted by survivors of a previous shift, and it ushered in a new zodiacal "Age." The gnostic wisdom enabling ancient man to predict these shifts was encoded in Great Pyramid, which is a model of the earth. Noone presents much Pyramidology, relying on P. Tompkins, P. Flanagan, **W. Fix, L. Dolphin, D. Davidson,** others. Also includes lost continents of Mu and Atlantis, psychic energy, **Velikovsky**'s cosmic cataclysms, **Goodman**'s psychic earthquakes, reincarnation, **Noorbergen, Chatelain, von Däniken,** Tom Valentine, C. Hapgood, **Ouspensky**, Manley Hall, **B. Steiger,** significance of **Ussher**'s calculations, the apocryphal Book of Enoch, the Kabbalah, Tibetan and Egyptian Book of the Dead, sacred Ark of the Covenant, Shroud of Turin, CIA reports, etc. Stresses significance and secret knowledge of Knights Templars and Masonic symbolism and ritual (names many astronauts as Masons). Describes Paluxy manprints; suggests they are genuine. Quotes **Mulfinger** at length, also **B. Nelson,** to demonstrate that Darwin's theory of evolution is wrong. Proposes third alternative to Darwinism and biblical creationism—but stresses that Bible contains much prophetic wisdom and esoteric knowledge about these events, describes Jesus as mystical initiate, etc. Discusses community in Stelle, Ill. which is preparing for coming pole-shift disaster. Many photos, drawings, diagrams; bibliog.

1198 Noorbergen, Rene. *The Ark File.* 1974. Mountain View, Calif./Omaha, Neb.: Pacific Press. *Noorbergen:* born in Holland; attended La Sierra Coll. (Calif.; Seventh-day Adventist); M.A.—Univ. Groningen (Netherlands); worked for Ford Motor Co. publications; writer and journalist, also T.V. documentaries; public relations director for SEARCH Ark-eology organization; Seventh-day Adventist. "Long-time friend" of E. Cummings (they met at La Sierra Coll.) —though **V. Cummings** 1982 complains that this book is too "negative." Noorbergen assisted Aaron J. Smith's 1949 Ararat Ark search; Smith (pres. of People's Bible Coll., N.C.) then bequeathed his Ark files to Noorbergen. "Many times I have imagined listening to the shouts of ridicule, the echoing chorus of sneering laughter, and the barrage of discouraging remarks aimed at Noah and his sons while they labored on the ark for over one hundred years." Noorbergen accepts the literal truth of Genesis and is convinced Noah's Ark is atop Ararat, and here investigates several of the most exciting Ark-eology claims. He reluctantly concludes they are hoaxes. Discusses memory of Flood in Chinese language and other cultures; tales that Adam's body was embalmed and taken aboard Ark; G.F. Dodwell's theory that tilt of earth's axis caused Flood; Ben Uri's Ark model. Thorough account and refutation of the **Roskovitsky** Ark story (discovery by Russian airmen in WWI and subsequent Czarist expeditions), quoting several versions, including Col. Koor's derivative story. Similar debunking of Archbishop Nouri's story. Noorbergen was active member of 1960 A.R.F. expedition examining boat-shaped object seen in aerial photo, led by **Vandeman.** Recounts initial plans for financial backing by high-powered investors. Reviews **Navarra**'s claims; concludes they are highly suspicious. A.R.F. offered Navarra large sums to lead them to his site on several later expeditions. Finally, Navarra found more wood when guiding a related team, the 1969 SEARCH climb. Tells of being in **Jeane Dixon**'s office when billionaire H.L. Hunt called asking her whether to fund SEARCH. (Some SEARCH members later consulted her – though others strongly objected to her ties with the occult. Noorbergen supported her.) Less skeptical of G. Hagopian's claim to have seen Ark; recounts interview. Examines "Daryl Davis" claim that remains of the Ark were hidden in the Smithsonian following a joint expedition with the National Geographic Society. After initial enthusiasm, Noorbergen analyzed his interview tapes of Davis with lie detector (PSE – vocal analysis), and concluded the tale was phony. Photos, maps.

1199 ________. *The Soul Hustlers.* c1976. Grand Rapids, Mich.: Zondervan. Psychic phenomena are Satanic deceptions to lure us away from Christ. Book based on interviews with psychics, witches, faith healers, astrologers, other practitioners of the paranormal and the occult. Noorbergen then administered PSE (Psychological Stress Evaluator) voice analysis tests of the interview tapes. These showed that Satan was controlling their answers. Book includes graphs, survey and computer results. Satan is undermining belief in the Bible. Noorbergen exposes **Jeane Dixon** as a liar and a fraud – she does possess some psychic power, but it is from Satan, not God. (But cf. **Dixon** 1969, written with Noorbergen.) Psychics promote (mental) evolution rather than God's creation.

1200 ________. *Secrets of the Lost Races: New Discoveries of Advanced Technology in Ancient Civilization.* c1977. Indianapolis/New York: Bobbs-Merrill. "Researched by **Joey R. Jochmans.**" Illustrations; bibliog., index. Noorbergen has also written *You Are Psychic* (1971) and a biography of Seventh-day Adventist prophet Ellen G. White.

1201 Nordenskiöld, Erik. *The History of Biology: A Survey.* c1928 [1920–24]. New York: Tudor. Swedish; orig. pub. in 3 vols., 1920–24, by Bjorck & Borjesson (Stockholm). New English ed. 1935. Also pub. by Tudor (New York; 1928) and Kegan Paul (London; 1929). Based on lectures given at Univ. Helsingfors (Helsinki), Finland, in 1916–17. *Nordenskiöld:* well-known Finnish-Swedish zoologist and historian of biology; zoology and anatomy prof. at Univ. Helsinki; then zoologist at Univ. Stockholm; became Swedish citizen; member Finnish Acad. Science and Letters. Comprehensive survey of biological thought since ancient times; includes useful accounts of many pre- and anti-Darwinists (**Linnaeus, Cuvier, Steno, Ray, Bonnet, Owen,** many others), stressing relation of their biological theories to cultural, philosophical and historical backgrounds. Strongly critical of Darwinism; considers Darwinian natural selection totally inadequate as an explanation of evolution, and believes it is an obsolete and discredited theory. Criticizes Darwin for admitting to no law-bound limits to selection, and for relying on chance variations in any and all directions. Says Darwin is often "a speculative natural philosopher, not a natural scientist." Claims it is "irrational" to consider natural selection a scientific law on a par with gravity. ". . . Darwin's theory of the orign of species was long ago abandoned. Other facts established by Darwin are all of second-rate value." But praises him for establishing evolution and "development" as a guiding idea in all fields of knowledge.

1202 Norman, Eric. *Gods and Devils from Outer Space.* c1973. New York: Lancer. "Is earth the vast project of cosmic beings? A new accounting of history. Spaceships and atomic energy cited in Genesis . . ." Flying saucers. Relies heavily on **Trench**'s UFOlogical interpretation of Genesis. Biblical creation story refers to creation of different races of mankind by space visitors. Describes Bob Geyer's Church of Jesus the Saucerian – Second Coming as flying saucer landing. Satan is also a saucerian. Also cites G. Adamski, **L. Dolphin, B. Steiger,** E. Cayce, George King's Aetherius Society, assorted psychics, UFO prophets and contactees. Visitations from UFOnauts to Jefferson, Malcolm X. Fantastic scientific knowledge of ancient cultures transmitted by spacemen. Quotes **Blavatsky** at length on survival of giants ("nephilim") described in Genesis 6. Says this is predicted by Darwin's theory, which is included in Genesis. Describes theory of Roger Westcott (anthropology prof. at Drew Univ., N.J.) that UFOs came to earth 10,000 years ago to teach mankind, but withdrew to underwater bases when man got unruly.

1203 Norman, Ernest L. *The Infinite Concept of Cosmic Creation.* El Cajon, Calif.: Unarius. *Norman:* psychic; co-founder, with wife **Ruth Norman,** of Unarius UFO cult; d. 1971. Was Jesus in an earlier reincarnation.

1204 Norman, Ruth. *History of the Universe (And You—A Star Traveler.* c1981, 1983. El Cajon, Calif.: Unarius Educational Foundation. 3 vols. "Inspired by Uriel" (Norman's extraterrestrial spirit manifestation)—but actually written by Charles von Spaegel, the "sub-channel" used by "Uriel." Norman has written 85 books (all pub. by Unarius) and recalls 55 past lives, including reincarnations as Mary Magdelene, Socrates, Buddha, Charlemagne, and King Arthur. She founded the Unarius UFO cult, located in El Cajon (quite close to ICR)—now called the Unarius Acad. of Science—with her husband **E. Norman.** In 2001, flying saucers from 33 planets will stack up on earth to usher in the age of Unarius and welcome us into the Intergalactic Confederation. Norman's books are uninspired, repetitious ramblings about reincarnations, psychic channelling, energy forces, higher vibrational planes, different dimensions, Atlantis and Lemuria, etc. Exhorts followers to achieve higher consciousness to fulfill "progressive evolution" towards psychic and spiritual power. Biblical themes mixed in indiscriminately; Unarius Age is Christ's Second Coming, Satan is an evil force seeking to prevent our psychic evolution who gained control over the demonic sub-astral planets of Orion. "Aryans" from Aries have achieved the highest consciousness.

1205 Norris, George L. *Creation--Cataclysm—Consummation.* c1973. Fort Worth, Texas: Marno. *Norris:* son of famed fundamentalist preacher J. Frank Norris (founder of World Baptist Fellowship, to which Falwell belongs; J. Norris's influence diminished after he shot and killed a man); pastor of Gideon Baptist Church in Fort Worth; theology prof. at Arlington Baptist Schools; wrote this book when he learned he was dying of cancer. Strong refutation of the insidious theory of evolution; also refutes Gap Theory creationism and Flood Geology. There is evidence from the Bible and from secular sources of several cataclysms which have catastrophically affected earth. The first cataclysm was Noah's Flood (there were earlier ones unrecorded by man). The second involved the formation of the continents (biblical "dividing of the earth"), the events of Exodus, the sun standing still for Joshua and moving backwards for Hezekiah. Following the Tribulation there will be another earth-destroying cataclysm. Denounces evolution as unscientific as well as evil; praises anti-evolutionism of young-earth creationists but says that Flood Geology compromises biblical creationism. Advocates **Velikovsky**'s scheme of cataclysms resulting from ejection of Venus from Jupiter, followed by near-collisions with Mars. Rejects recent creation; suggests that there were long ages between creation days. Argues against "concordist" theories harmonizing Bible with current scientific theories. Presents some of the standard creation-science arguments. Endorses Paluxy and similar evidence that extinct species co-existed with man. Strongly pre-millennial. Photos, diagrams.

1206 North, Gary. *Unholy Spirits: Occultism and New Age Humanism.* c1986 [1976]. Ft. Worth, Texas: Dominion Press. Rev.; orig. 1976 (Arlington House), then titled *None Dare Call It Witchcraft. North:* history Ph.D.—Univ. Calif. Riverside; attended Westminster Sem.; with Found. Econ. Educ.; president of Inst. Christian Economics (Tyler, Texas); editor of *J. Christian Reconstruction;* author of 17 books; a leading theoretician of the Christian Reconstruction movement; has been featured on **Pat Robertson**'s and **D.J. Kennedy**'s T.V. shows. Occultism is based on Satan's original lie that man can become God, and on the denial of the Creator/created distinction – man as his own savior. Occultists, mystics, magicians, Gnostics, Eastern religions, **Teilhard** – all aspire to evolve into gods. North insists that humanism (including evolution) is fundamentally the same as Satanism – occultism. Presents massively documented chapters asserting the reality of paranormal phenomena: witchcraft, ESP, telepathy, spontaneous human combustion, Carlos Castaneda, UFOs, psychic healing, and other psychic powers. Argues (following **Van Til**) that humanism has, since the Renaissance, been based on irrationalism as much as rationalism. This society underwent a major shift to the occult and New Age philosophy in 1965, prepared by Darwinism and the new physics. Such shifts signal the end of civilizations; the West, having embraced atheism, will not be able to survive unless it returns to Christianity. "Thus, as the West has become increasingly atheistic and Darwinian, it has become vulnerable to anti-rational social philosophies and practices." Praises **C.S. Lewis** and **Rushdoony** (his father-in-law). New Age humanism says man arose by chance from slime, yet seeks to control man's destiny and all of nature; gnostic goal of power and salvation through special knowledge. North believes the most sensational reports of the paranormal and occult, attributing them to the powers of Satan and his demons. Concludes with persuasive argument for post-millennial Christian reconstructionist program. Criticizes **Hunt** and McMahon for equating emerging post-millennialist view with New Age humanist expectation of an earthly Millennium. Index.

1207 _______. *The Dominion Covenant: Genesis.* 1982. Ft. Worth, Texas: Dominion Press. First vol. of Bible commentary series. This book is "specifically an *economic* commentary," presenting the biblical approach to economic theory and business. "It offers the basis of a total reconstruction of economic theory and business. "It offers the basis of a total reconstruction of economic theory and practice. It specifically abandons the universal presupposition of all modern schools of economics: Darwinian evolution. *Economics must begin with the doctrine of creation. The Dominion Covenant: Genesis* represents a self-conscious effort to rethink the oldest and most rigorous social science in terms of the doctrine of creation. Every social science requires such a reconstruction. . . . The message of this groundbreaking work is that God has given us clear, infallible standards of righteousness – practical guidelines by which Christians are to reconstruct civilization." North had intended to write a book specifically on evolution, but did not complete it. One chapter, however, appears as an appendix in this book: "Cosmologies in Conflict: Creation vs. Evolution." Bibliog., index.

1208 _______. *Conspiracy: A Biblical View.* c1986. Ft. Worth, Texas: Dominion Press. Conspiracy view of history: a political and economic conspiracy, past and present. The conspirators use false religious principles to control us, such as the unity of mankind, the ability of man to control his own destiny, and evolution and the ability of man to further evolve. Determinist views of history seek to discredit conspiracy theories, which would place blame on those in power. The Bible reveals that all conspiracies are part of the larger Conspiracy – Satan's attempt to thwart God, and to convince man that he can be as God. "The chief premise of the modern conspirator is this: *Man, the savior of Man.*" Other than this unity against God, however, the various conspiracies are not united and they do not necessarily cooperate. (Thus the contradictions between various conspiracy theories, and the conflict between different groups of conspiracy advocates.) Discusses Illuminati, Trilateral Commission, Council on Foreign Relations, and other targets of conspiracy advocates. Cites and endorses **Rushdoony, C.S. Lewis,** Anthony Sutton, Otto Scott, C. Quigley, J. Billington, and G. Allen and L. Abraham's 1971 *None Dare Call It Conspiracy.* Refutes world government and brotherhood-of-man arguments by declaring there is a continuing war between good and evil men; also between angels and devils. Revolutionaries and anti-Christian activists have always been heavily involved in secret societies and the occult. Lists many conspiracy-theory and Christian reconstructionist organizations. Index.

1209 _______ **(ed.).** *Symposium on Creation.* 1974. Vallecito, Calif.: Journal of Christian Reconstruction. Premier issue of the *Journal of Christian Reconstruction* (vol. 1, no. 1). (All issues are in book format, and are devoted to single topics.) Authors: North; **S. Nevins [Austin]** on earth history (reprint of ICR "Impact" #12); **C. Clough** on biblical presuppositions and geology; **Lammerts** on a personal history of the creationist movement; **Davidheiser** on 1911 Antarctic penguin expedition attempting to prove evolution of birds from reptiles (excerpt from his 1969 book); **C. Van Til** on doctrine of creation and Christian apologetics; Greg Bahnsen (Th.M.—Westminster; philosophy Ph.D.—USC) on Darwinism as worship of "creatures" rather than Creator; **Poythress** on creation and math; D.E. Johnson (M.Div.—Westminster Sem.) on evolution and literature; **Rushdoony.** "This journal is dedicated to the fulfillment of the cultural mandate of Genesis 1:28 and 9:1—to subdue the earth to the glory of God." Published by the Chalcedon Foundation, Rushdoony's Christian Reconstructionist organization. The Reconstructionists are post-millennialists; they preach "dominion theology"—that Christians must entirely restructure all aspects and institutions of society according to biblical (Old Testament) law. They chastize pre-millennialist dispensationalists for their "Armageddon theology" and for fatalistically awaiting the worldly triumph of Satan. Reconstructionists insist that Christians will triumph, and will create the Millennium here on earth (unlike the pre-mills, who believe Christ will come to rule over a fallen, evil world which has rejected His word.) Not all authors in this volume are post-millennialists, however. Bahnsen, North, and Rushdoony are leading theoreticians of the Reconstructionist movement. Bahnsen and Clough here emphasize (following Van Til) the "presuppositional" nature of all knowledge; our view of the world derives either from humanist or biblical assumptions, which cannot be tested. They insist that science (as well as all other aspects of society) must be totally reconstructed according to biblical truth, and praise **Whitcomb** and **Morris**'s *Genesis Flood* for recognizing this. They reject the alternatives of reinterpreting Genesis to harmonize with science, or of assuming that Genesis has no relevance for science (the view that science is a separate domain). Science must conform to the Bible, not vice-versa. Also includes reviews of **Gish** 1972b, **Slusher** 1973 and **T. Barnes** 1973, by **G.F. Howe**; and of Gish 1972a, by Rushdoony.

1210 Northrup, Bernard. *Light on the Ice Age* (booklet). Undated [1975?]. Bible-Science Association. 1975 lecture at Colonial Hills Baptist Church, E. Point, Ga. *Northrup:* Old Testament and Semitic Lang. Th.D.—Dallas Theol. Sem.; pastor in Redding Calif.; geologist (has led several Bible-Science Assoc. Grand Canyon tours). "Genesis and geology, I have discovered, have exactly the same testimony." Book of Job was written during the Ice Age; Job lived shortly after the Flood. Geological evidence shows that the continents have divided. This occurred 550 years after the Flood. Atmospheric material lofted up by vulcanism following the continental split caused the Ice Age. Northrup insists that the geological record can be aligned with the biblical account (not vice-versa); that natural revelation is a corollary (but only a corollary) to the Bible. He first proposed his own harmonization of science and the Bible in 1968, and has been developing it ever since. He is sharply critical of Flood Geology, which he accuses of being unscientific, "uniformitarian" creationism. There were many catastrophes in earth history, not just the Flood, and there was considerable post-Flood fossilization. Paleozoic strata are mostly the result of the Flood, which took perhaps a thousand years to subside completely. Mesozoic strata result from this period of the retreat of the Flood. The Cenozoic lasted much longer. Dinosaurs died during the Mesozoic, but lived in the Paleozoic and earlier (they survived 150 days of the Flood, then came ashore). The wind which dried up the Flood waters was a jet stream lowered by God. Unlike most creationists, Northrup believes that the pre-Flood Water Canopy produced a hostile (rather than an Edenic) environment on earth, lowering the reproductive rate by a factor of 15. Conifers and reptiles predominated under the harsh Canopy conditions. Northrup suggests human remains can be found in Paleozoic strata—but probably only in Mesopotamia.

1211 The Northwestern Lutheran. *Is Evolution the Answer?* 1967. Milwaukee: Northwestern. Orig. appeared serially in *The Northwestern Lutheran,* official publication of the Wisconsin Evangelical Lutheran Synod, 1965–66. Werner H. Franzmann: managing editor. Authors (mostly teachers at Lutheran high schools, seminaries and colleges): Franzmann, Ulrik Larsen, Carl Lawrenz, Paul Eickmann, Luther Spaude, Siegbert Becker, Gerald Mallmann, Heinrich Vodel, Walter Sebald, Robert Adickes, John Denninger, Eugene Kirst, Armin Schuetze. Advocates recent six-day creationism, but does not specify date. Not all geological strata are a result of the Flood; some were formed at Creation, at the Fall, and as a result of post-diluvial activity.

1212 Northwestern Publishing House. *How Can I Answer the Evolutionist?* (tract). Undated. Milwaukee: Northwestern. Tract no. 6 N 21. Science becomes scientism if it tries to give answers about origins. No scientific fact contradicts the Bible, but intepretations and theories may.

1213 Nothdurft, Milton H. *Between Two Worlds!* c1985. Prescott, Ariz.: Mountain Valley Press. Foreword by **B. Steiger.** B. and F. Steiger listed as "editorial consultants." *Nothdurft:* Methodist minister from fundamentalist background; M.Th.—Boston Univ.; attended Harvard and Oxford. The two worlds are Christian orthodoxy and UFOlogy, which Northdurft embraces enthusiastically. Also reincarnation (which Jesus taught), trance channeling, Ascended Masters and psychic phenomena. Claims that Americans are the Lost Tribes of Israel. Discusses **B. Keith**'s *Scopes II* book; agrees that teaching of origins should not be limited to one theory. Both creation-science and evolution theories miss the truth. Space Masters explained to him that warlike consciousness units were exiled to Earth, which accounts for man's belligerency. But he expresses hope we can join the Galactic Pact soon. Space aliens planted good seed as well as bad on earth. Endorses all New Age and Aquarian beliefs; knows most prominent UFOlogists personally.

1214 Nott, Josiah C. *Biblical and Physical History of Man.* 1849. New York: Bartlett and Welford. Also 1969 reprint ed. pub. by Negro Universities Press. An apologia for slavery. Argues that there were many creations of man, not a unity of the species. Progressive, old-earth creationism. Appeals to German higher criticism of the Bible in defense of his position. Nott wrote other books explicitly defending slavery; he compared distribution of the inferior races with that of apes, described the "instinctive" opposition of Negroes to agricultural labor, and edited and added to a book by Joseph Gobineau.

1215 ________. *Types of Mankind; Or, Ethnological Researches Based upon the Ancient Monuments, Paintings, Sculptures and Crania of Races and upon Their Natural, Geographical, Philological and Biblical History.* 1871 [1854]. Philadelphia: Lippincott. 10th ed. Orig. 1854 (Lippincott, Grambo). Includes selections written by Samuel George Morton, **L. Agassiz,** and two others.

1216 Nursi, Bediuzzam S. *The Supreme Sign.* 1979. Risale Nur Institute. Presumably Islamic.

1217 Nutting, David Irwin. *Origin of Bedded Salt Deposits: A Critique of Evaporative Models.* 1984. Unpub. M.S. thesis: Institute for Creation Research (El Cajon, Calif.). ICR geology thesis. Salt deposition explained as due to the Flood, rather than from evaporation of seas over long periods. After getting their advanced degrees from ICR, Nutting and wife **M.J. Nutting** founded Alpha Omega Institute in Grand Junction, Col. They put out a creation-science periodical aimed at schoolchildren, *Think and Believe,* which repeats the standard ICR creation-science arguments. The Nuttings also give creationist presentations in public and private schools.

1218 Nutting, Mary Jo. *A Rationale for the Christian College Biology Curriculum: A Case Study at Christian Heritage College.* 1983. Unpub. M.S. thesis:

Institute for Creation Research (El Cajon, Calif.). ICR biology M.S. thesis. *Nutting:* wife of **D. Nutting.**

1219 Oakwood, Jay S. *And There Shall Be Light.* 1980. Vantage.

1220 Oberst, Fred W. *The Almighty God Declares Evolution False* (booklet). 1976. Dublin, Ga.: Rev. Tondee, First Independent Methodist Church. *Oberst:* organic chem. Ph.D.—Iowa State Univ.; research scientist for 40 years; member of Evangelical Methodist Church, Dublin, Md. Includes several standard creation-science arguments. Discusses mystery of strong nuclear force, the fact that ice is lighter than water, and examples of design in nature. "There are many other oddities in nature that cannot be explained, except that God made them that way. They just do not fit into the evolutionary ladder."

1221 O'Connell, Patrick J. *Science of Today and the Problems of Genesis: The Six Days of Creation, the Origin of Man, the Deluge and the Antiquity of Man.* 1969 [1959]. Christian Book Club of America. (The Christian Book Club of America is now listed as an imprint of Noontide Press, the parent group of the Institute of Historical Review—see **Hoggan.**) "A Vindication of the Papal Encyclicals and Rulings of the Church on These Questions." 2nd ed. Orig. 1959 (St. Paul, Minn.: Radio Replies Press Soc.). Also French and Italian eds. Book I: The Six Days of Creation and the Origin of Man. Book II: The Deluge and the Antiquity of Man. Rev. O'Connell: former missionary in China; then living in Oregon; Catholic (Jesuit). Book has Nihil Obstat and Imprimatur, plus assurance from Los Angeles Archdiocese regarding orthodoxy of 2nd ed. Quotes review in Vatican journal which says book is in "perfect agreement" with *Humani Generis.* "The object of this book is to give the scientific conclusions about the Six Days of Creation and the origin of man arrived at during the past few years, . . . and to show that these conclusions are in agreement with the Mosaic account of Creation, and are a vindication of the Papal Encyclicals issued for its interpretation." Rejects evolution of major types and of man, but admits of formation of new species. 1893 Encyclical *Providentissimus Deus* of Leo XIII declares that divine truth of inerrant Bible applies to physical and historical matters as well as faith and morals; reaffirmed by Pius XII in *Divino Afflante Spiritu.* 1950 *Humani Generis* of Pius XII explains that evolution of man has not been proved. Catholics are *allowed* to investigate evolution of man but *forbidden* to teach it as definite. Evolution "has been a potent factor in promoting atheism and communism." Atheistic evolution is official creed of Freemasons (cites Churchward's *Signs and Symbols of Primordial Man*). Darwin became an atheist after his *Origin,* the "bible of atheists'—thus "his whole system, which is a tissue of absurdities, is based on a supreme absurdity." Genesis account of the order of Creation is scientifically accurate, but creation 'days' are long ages (Day-Age creationism): "'Let light be made' refers to the creation of the fiery nebulae, the source of light; the earth was once flat, without mountains, and covered with water; the earth was formed before the sun; the rest of the Mosaic order of creation is confirmed by what geology has disclosed; the argument laboriously built up from palaeontology against the special creation of man has collapsed." Asserts science proves earth never had molten core; earth formed from nebulae but Laplace's theory is wrong. Water of early earth existed first as "mighty pall of vapour" before separation into ocean and clouds. The addition of Mendelism to Darwinism proves it is "devoid of scientific foundation." The Church opposed Galileo's denial of Joshua's Long Day, not his heliocentrism. The church was right, since the sun is not immovable, as Galileo claimed. Joshua's miracle involved additional movement of the sun—perhaps similar to Fatima miracle. Discusses and criticizes many Catholic works relating to creation/evolution. Refutes **Teilhard,** Dorlodot, Nogar, many others; cites several Catholic anti-evolution books (esp. **Ruffini**). Quotes at length from *Sacrae Theologiae Summa* written by Spanish Jesuits (1952; Madrid), especially section in Vol. II on origin of man by Fr. Sagues, which rejects evolution. Long quotes from **Dewar** and Shelton 1947; also quotes **G.F. Wright, Vialleton, L.M.**

Davies, other creationists. Refutes all alleged ape-men. Science shows there was definite "hiatus"—the Flood—at close of Ice Age, about 7,000 B.C., between Mousterian and Aurignacian periods. Date confirmed by evidence from Jericho and many other sites; Scandinavian glacial varves suggest date of 6839 B.C. Probably only one Ice Age; melting accelerated by Flood, which explains freezing of Siberian mammoths. Flood was not worldwide, but included all areas inhabited by man (covered most of Europe, N. Africa, much of Asia and N. America). All mankind perished (except Noah's family), but not all animals. Describes impossibilities of global Flood—lack of space in Ark for all animal species, etc. Neanderthals were degenerate pre-Flood race (hunters—descendants of Cain). Study of Bible chronologies, archeology and other scientific evidence shows creation of man occurred between 5,000 (or even less) and 15,000 years before the Flood. O'Connell had "private revelation" that Adam and Eve moved to Palestine and were buried at Cavalry. Photos; bibliog.

1222 ________. *The Origin and Early History of Man.* 1968 [1962]. Houston, Texas: Lumen Christi Press. O'Connell was strongly opposed to **Teilhard,** suggesting it was he who perpetrated the Piltdown hoax, and also accusing him of falsifying evidence for Peking Man.

1223 ________. *Original Sin in the Light of Present-Day Science.* Contains summary of *Science of Today and the Problems of Genesis* dealing with origin and early history of man, refutation of new proposed "missing links," proof that biblical story of Adam and Even, Cain and Abel, and the Flood are true history. Science shows there is no genetic link between man and beast.

1224 Odeneal, W. Clyde. *Segregation: Sin or Sensible?* (pamphlet). c1958. Merrimac, Mass.: Destiny Publishers. Reprinted from *Destiny* magazine. Defends and praises racial segregation. God created races after their own kind; nature attests to this "Divine law." "Segregation is an Anglo-Saxon principle because more than all the other races combined, the Anglo-Saxon, Celtic and related races are predominantly the Bible-reading, Bible-disseminating peoples of the world." Declares that "miscegenation was the principle [sic] sin which brought on the great flood of antiquity." God set bounds for the various sons of Adam, which they are not to transgress. Mankind is not "of one blood"—this is not in the original Bible. Galatians 3:28 ("ye are all one in Christ") refers not to race or sex, but to souls. Racial equality is a communist notion; notes that leading anti-segregationists are non- or anti-Christian. Cites anatomical differences between whites and Negroes.

1225 Old Paths Tract Society. *A Scientist Discovers God: The Testimony of Dr. Jerome Stowell* (tract). Undated. Shoals, Ind.: Old Paths Tract Society. Reprinted from the *Midnight Cry.* Stowell used to be an atheist, and experimented with "bad" devices. When he attempted to "find the wavelength of the brain," he discovered that each was different: God actually keeps records of our thoughts, like fingerprints. He tested to see what occurs at death, using a device to measure positive and negative charge units, with nine points being the equivalent of a 50-kw radio station. A dying patient praying to God registered over 500 on this scale. Another patient with venereal brain damage registered over 500 on the negative scale when cursing God.

1226 Oller, John W. *Not According to Hoyle* (pamphlet). 1984. El Cajon, Calif.: Institute for Creation Research. ICR "Impact" series #138. *Oller:* Ph.D.; linguistics prof. at Univ. New Mexico; member of ICR Technical Advisory Board. **Fred Hoyle**'s probability arguments against evolution and the origin of life. Oller also wrote "Impact" #112, on learning and evolution.

1227 Olmstead, James M. *Noah and His Times.* 1854. Boston: Gould and Lincoln. Cited in **Whitcomb** and **Morris** 1961. Olmstead was apparently one of those who, according to Whitcomb and Morris, appealed to "the views of prominent geologists" as the "criteria for exegeting the early chapters of Genesis," with the result that "the great Flood began a slow but steady retreat from its recorded position as the greatest catastrophe of geologic history." Olmstead cites Lyell and **Buckland.**

1228 Oltrogge, V.C. *The Bible and*

Science (booklet?). Undated. Los Angeles: American Prophetic League. (Listed in Ehlert files.)

1229 O'Neill, T. Warren. *The Refutation of Darwinism.* 1880. Philadelphia: J.B. Lippincott. Subtitle: "And the converse theory of development; based exclusively on Darwin's facts, and compromising qualitative and quantitative analyses of the phenomena of variation; of reversion; of correlation; of crossing; of close-interbreeding; of the reproduction of lost members; of the repair of injuries; of the reintegration of tissues; and of sexual and asexual generation."

1230 Osborne, Chris Dexter. *A Reevaluation of the English Peppered Moth's Use as an Example of Evolution in Progress.* 1985. Unpub. M.S. thesis: Institute for Creation Research (El Cajon, Calif.). ICR biology thesis. The peppered moth of the industrial Midlands of England is a classic example of adaptation and micro-evolution: of a "shift in gene frequencies" (a standard textbook definition of evolution). Creationists argue that such micro-evolution cannot be extrapolated beyond certain immutable limits; that the shifting ratios of the light and dark varieties of peppered moth—a response to changing environmental conditions (the changing color of the tree bark they hide on)—does not involve creation of any new forms. The moths remain moths of the same species.

1231 Ostermann, Eduard. *Das Glaubensbekenntnis der Evolution.* c1978. Neuhausen-Stuttgart, Germany: Telos-Haensler. In German. "Die Evolutionslehre ist eine Religion ohne Gott, eine Religion ohne Christus und daher die Religion des Antichristus." Also by Ostermann: *Unsere Erde—Eine Junger Planet; Das Ende einer Legende.* 1978. Neuhausen-Stuttgart: Telos-Haensler.

1232 Otey, W.W. *Creation or Evolution.* 1930. Quoted in **Graebner** 1932. If all upward improvement is the result of summing up of very small improvements, as evolutionists affirm, then what caused the great gaps separating all major groups? There should be countless transitional forms. "Why did they not leave fossils? Had they done so, the fossil record would be so gradual in improvement that the difference between the higher and next lower group would be divided into four million parts. . . . These are real difficulties that must be got rid of at least by some plausible theory."

1233 O'Toole, George Barry. *The Case Against Evolution.* 1925. New York: Macmillan. *O'Toole:* Ph.D.; theology and philosophy prof. at St. Vincent Archabbey, Penn.; biology prof. at Seton Hill Coll. Nihil Obstat & Imprimatur. Literate attack on evolution, full of scientific quotes and references. No biblical or religious references. Book demonstrates "that Evolution has long since degenerated into a dogma, which is believed in spite of the facts, and not on account of them." Unlike the theory of heliocentrism, which was quickly confirmed, "the whole trend of scientific discovery has been to destroy, rather than to confirm, all definite formulations of the evolutional theory, in spite of the enormous erudition expended in revising them." Origins can't be scientific. Theory of natural selection is dead, because variations are not hereditary; "no modern biologist attaches very much importance to natural selection..." Lamarckian vs Darwinian Transformism: each proves the other wrong. Darwin was a pretender—Mendel was the true genius. Mendelian genetics forbids natural selection, which was the only original aspect of Darwin's theory—the only difference between it and Lamarck's. Criticizes fellow Catholic Dorlodot for uncritical praise of Darwinism; also Zahm. Criticizes anti-evolutionist **McCann** 1922 for sophistry, extreme bias, and inaccurate arguments. Admires **G.M. Price**—quotes him at length; also **Fabre, Dwight,** others. Quotes much of Bateson's 1921 Toronto address. Exposes circular reasoning in dating of rocks and fossils: "Our present classification of rocks according to their fossil contents is purely arbitrary and artificial, being tantamount to nothing more than a mere taxonomical classification of the forms of ancient life on our globe, irrespective of their comparative antiquity." "The paleontological argument is simply a theoretical construction which presupposes evolution instead of proving it." Second half of book deals with origin of life, of the soul and the body. Discusses spontaneous generation and abiogenesis; explains that "a chaos of unassorted elements and undirected forces" cannot produce life. If the soul is spiritual, it "cannot be a product of organic evolution," for this would imply it is inherent in matter, and that the difference between man and beast is not "essential" but only a matter of degree. Asserts that the soul must be directly created by God—rejects materialistic monism. Refutes standard arguments and evidences for evolution: homology, vestigial organs, embryology, resemblance of man to apes, the various alleged fossil ape-men. "Had evolutionary enthusiasts adhered more strictly to the facts, had they proceeded in the spirit of scientific caution; had they shown, even so much as a common regard for the simple truth, the 'progress of science' would not have been achieved at the expense of morals and religion. As it is, this so-called progress has left behind a wake of destruction in the shape of undermined convictions, blasted lives, crimes, misery, despair, and suicide. It has, in short, contributed largely to the present sinister and undeserved triumph of Materialism, Agnosticism, and Pessimism ... fittingly characterized as the three D's of dirt, doubt, and despair." If man is but an animal, the product of chemical reactions, if immortality and free will are illusions, "then morality ceases to have meaning, right and wrong lose their significance, virtue and vice are the same." Index.

1234 Otten, Herman J. *Baal or God.* Undated [1965]. New Haven, Missouri: Leader. Undated, but Introduction dated 1965. *Otten:* Lutheran pastor (New Haven, MO); graduate of Concordia Sem.; editor of *Lutheran News.* Protests and refutes spread of liberalism and modernism within Christianity; defends fundamentalism. The difference between these is "the difference between God and Baal." Presents the various fundamentalist doctrines. Chapter "Creation" describes capitulations to evolutionism by most denominations. Otten, however, insists on literal truth of Genesis account—special creation. "Christianity rejects the theory of evolution because it is diametrically opposed to the biblical account of creation." If evolution is true, there was no Fall, and Christ's atonement is "utterly meaningless," "a complete hoax." Quotes the Creation Research Society's affirmation of strict creationism, and quotes several CRS scientists. Also presents many standard anti-evolution quotes by various scientists. Index.

1235 Ouspensky, Peter D. *The Psychology of Man's Possible Evolution.* 1974 [c1950]. New York: Vintage-Random House. 2nd ed. Orig. 1950. *Ouspensky:* born in Russia; influenced by Theosophy, then became a disciple of Gurdjieff; a "philosopher-mathematician." The teachings of Gurdjieff: mysticism and Sufism. Based on lectures delivered to study groups in England (most originally priv. pub. in 1940). Ouspensky later moved to New York. "As regards ordinary modern views on the origin of man and his previous evolution I must say at once that they cannot be accepted. We must realize that we know nothing about the origin of man..." Fossil remains of so-called human ancestors may belong to "some being quite different from man" and presumably unrelated to us. On the other hand, ancient monuments show that ten or fifteen thousand years ago there existed a "higher type of man" than today. Ouspensky is concerned with the evolution of consciousness. "Our fundamental idea shall be that man as we know him *is not a completed being;* that nature develops him only up to a certain point and then leaves him, to develop further, *by his own* efforts and devices, or to live and die such as he was born, or to degenerate and lose capacity for development. ... We must start with the idea that without efforts evolution is impossible; without help, it is also impossible."

1236 Ouweneel, Willem J. *What Is the Truth: Creation or Evolution?* (booklet). 1974. Believers Bookshelf. Orig. in Dutch. 100,000 copies in several languages by 1978. *Ouweneel:* Ph.D.—Univ. Utrecht (Netherlands); geneticist and embryologist at Royal Dutch Acad. Sciences 1971–76; co-founder of Evangelical College in Amersfoort, where he teaches philosophy and psychology; now pursuing second doctorate in philosophical anthropology; editor of *Bijbel en Wetenschap* ("Bible and science"); featured geneticist in Films for Christ *Origins* film series (co-produced by the Dutch Evangelical Broadcasting Co. and based on their Dutch T.V. series); co-author with **Glashouwer** of 1980 book version of film series (from which **McDowell** and Stewart 1984 is adapted); active creation-science lecturer in Holland and U.S. This booklet is aimed at young people.

1237 ________. *Notes on Genesis One.* 1974. The Netherlands. Orig. Dutch; also translated into German and English. *Gedanken zum Schöpfungsbericht* (1975; Neustadt, Germany: Paulus), cited in **Schirrmacher** 1985, is presumably German version.

1238 ________. *Operation Superman.* 1975. The Netherlands. In Dutch. Ouweneel (1978) says this is his "major book"; also that it "was quickly sold out and of which one of the best theologians in Holland said that it broke the power and monopoly of the evolution doctrine in this country [The Netherlands]."

1239 ________. *The Ark in Agitation.* 1976. Amsterdam: Guigten & Schipperheijn. (Cited in Ouweneel 1978.)

1240 ________. *Youth in a Dying Age.* 1977. The Netherlands. In Dutch. On the "philosophical and moral consequences of the evolution doctrine."

1241 ________. *Creationism in the Netherlands* (pamphlet). 1978. El Cajon, Calif.: Institute for Creation Research. ICR "Impact" series #56. The rise of creation-science in The Netherlands since 1974, as told by one of the most prominent and active Dutch creationists. Cites many Dutch publications, organizations, and individuals. Useful background information.

1242 Overn, William M. *A Trap for God's People* (pamphlet). 1969. Evangelical Lutheran Synod. Orig. presented at Idaho Creation Seminar, 1969. *Overn:* senior staff scientist at Sperry Univac—electromagnetics, computer and space technology; president of Industrial Enterprises (energy management and consulting); field director of Bible-Science Assoc. Satan's modern trap is the widespread belief that the achievements of science show that the Bible is a myth. "It is the purpose of this essay to show God's people, especially the young people, that there is good reason, scientific reason in fact, not to accept the popular belief in evolution." Overn discusses his pioneering scientific work, and his role as judge at science fairs. Science "is a study of facts." We all approach science with one of only two assumptions: either there is a Creator (creationism), or there isn't (evolution). God made all the laws of nature at creation, and He can miraculously suspend them. Presents some standard creation-science arguments.

1243 ________. *The Creator's Signature* (pamphlet). 1984 [1982]. Minneapolis: Bible-Science Association. BSA Reprint Series. Orig. 1982 in *Bible-Science Newsletter.* About **R. Gentry**'s radiohalos. "It is as if the Creator left His signature disbursed among the rocks, trademarks as it were, to proclaim the sudden creation process. No other rational explanation is available in terms of all present scientific knowledge."

1244 Overn, William M. (ed.). *Proceedings of the Third Creation Science Conference.* 1976. Minneapolis: Bible-Science Association.

1245 Overton, Basil. *Evolution or Creation.* 1973. Nashville, Tenn.: Gospel Advocate Co. (Listed in Ehlert files.)

1246 ________. *Evolution in the Light of Scripture, Science, and Sense.* 1981. Winona, Miss./Singapore/New Delhi: J.C. Choate. *Overton:* D.Hum.—Morehead State Univ.; prof. at and vice-pres. of International Bible College (Florence, Ala.); founder and editor of *World Evangelist.* Foreword by A. Doran (emer. president—Morehead State). A fairly comprehensive presentation of the usual creation-science arguments. Strict young-earth creationism and Flood Geology. Argues that atheism is illogical; that "Evolutionists Are Dogmatic." Discusses many wonders of nature which show God's glory. Declares that "the evidence is overwhelming that the general structure of the fossil record must have been laid down by the sudden and catastrophic action of a universal flood." Quotes **B. Nelson** at length; also many other creation-scientists. Endorses Paluxy manprints and other standard creationist evidence refuting evolutionary geology. Darwin turned away from the Bible, but not because he found any scientific evidence that contradicted it. "He formulated his views regarding evolution without scientific evidence. His doctrine was a sort of combination of some of the myths of the ancient Greeks and others regarding origins." Predicts that man won't be able to create life, but even if "life is actually created in the test tube, such a feat will only serve to prove that intelligence and intelligent work are required for the creation of life." Emphasizes saving grace of Christ and evil results of evolution. "Jesus Christ was not an evolutionist."

1247 Oviatt, Patricia C. *Genesis in the Science Lab.* 1980 [c1971]. Denver: Accent-B/P. Accent on Life Bible Curriculum. Openly Christian; intended for students. Full of biblical references. Strict creationism; standard arguments. References mostly religious; some scientific. Includes very little discussion of science. Mostly affirms truth of Bible and presents parables of school life: each chapter opens with a little drama of Christian students who are confronted by secular and anti-biblical teachings in school, which they overcome by relying on their faith in the absolute truth of the Bible and by discovering creation-science. Endorses pre-Flood Canopy Theory. Discusses Ben-Uri's theory of Noah's Ark. "Just think what [the discovery of the Ark on Ararat] will do to all those critics who say it was just a fairy story!"

1248 Owen, Richard. *Palaeontology, or a Systematic Summary of Extinct Mammals and Their Geological Relations.* 1861 [1860]. Edinburgh: Block. 2nd ed. Orig. 1860. *Owen:* renowned comparative anatomist; first head of Natural History Dept. of British Museum. Owen, the "British **Cuvier,**" held that the progression of fossil forms illustrated a gradual progressive transformation away from original Platonic archetypes. This structuralist evolution—morphological development of homologous structures—was similar to the German Naturphilosophie advocated by Goethe and continental scientists such as Oken and Saint-Hilaire. Owen saw adaptive changes as secondary "masks" modifying, but not obscuring, the essential archetype. He remained implacably opposed to the functional explanation of Darwin, his contemporary (and former friend), who rejected the essentialist concept and viewed adaptation as the driving force of evolution. Owen, though not a special creationist, is generally believed to have coached Wilberforce in the famous 1860 debate against Huxley, who defended Darwin's theory. Owen did not believe in the immutability of species, and resented Darwin for stating that he (Owen) did. He was reticent about disclosing his actual beliefs on species development, however. (Before Darwin's *Origin,* even scientists who were skeptical of species immutability did not publicly dispute it.)

1249 ________. *On the Anatomy of Vertebrates.* 1866–1868. London: Longmans, Green. 3 vols. Owen, who studied under **Cuvier,** discerned a common morphological plan uniting all organisms. All vertebrate modifications—their different skulls, limbs, etc.—resulted from transformations, governed by laws of form, of the archetypal vertebrae. He proposed this earlier in works such as *On the Archetype and Homologies of the Vertebrate Skeleton* (1848; London: Van Voorst) and *On the Nature of Limbs* (1849; London: Van Voorst). Owen rejected "the principle of direct or miraculous creation," but also rejected accidental or chance natural selection, maintaining that there was design and purpose in nature. An aim of this book was to apprehend "the unity which underlies the diversity of animal structures: to show in these structures the evidence of a predetermining Will, producing them in reference to a final purpose; and to indicate the direction and degrees in which organisation, in subserving such Will, rises from the general to the particular" (quoted in Hull). The horse, e.g., was "predestined and prepared for Man."

1250 Ozanne, C.G. *The First 7000 Years: A Study in Bible Chronology.* c1970. New York: Exposition Press. *Ozanne:* Ph.D.; studied theology, Hebrew and Aramaic. Bible numerology; presents various numerical schemes concerning Bible chronology. Seven thousand-year ages. Creation: 4004 B.C. Flood: 2348 B.C. Millenium: will begin A.D. 1996 (i.e., after six thousand-year ages). Derives chronology from internal evidence of Bible—not external studies of the world. No creation-science arguments.

1251 Paine, Martyn. *Review of Theoretical Geology.* 1856. Cited by Millhauser as a response to and refutation of **H. Miller**'s Day-Age reconciliation of Genesis and geology. Attacks modern geology; insists on literalism and religious orthodoxy.

1252 Paley, William. *Natural Theology; or, Evidence of the Existence and Attributes of the Deity Collected from the Appearances of Nature.* 1802 [1801]. London: R. Faulder. 2nd ed. pub. 1828 by J. Vincent (Oxford). Also pub. in modern reprint edition by St. Thomas Press (Houston), and in abridged version by Bobbs-Merrill (c 1963; Indianapolis). An undated edition by American Tract Society (New York) was widely used in American colleges and seminaries in the second quarter of the 19th century. *Paley:* Anglican; Archdeacon of Carlisle; fellow of Cambridge Univ.; supporter of liberal causes. The classic formulation of the venerable Argument from Design. Paley did not originate this argument but here gives it its definitive form. Paley begins with the famous example of coming across a watch on the ground. It would be illogical to suppose that it had originated there, like a stone. Our reason would of course force us to admit that a watchmaker had constructed it rather than that it had formed by itself: "design must have a designer." So much more so for the infinitely greater complexity and design of living creatures. Paley describes the intricate design of the human eye and concludes that just as a telescope must have had a designer, so must the eye. This example alone, declares Paley, is "sufficient to support the conclusion... as to the necessity of an intelligent Creator." The unity of nature, including the resemblance between different species, proves the "unity of the Diety." The generally beneficial effects of "contrivances" (adaptations) in nature and the existence of sensations of pleasure by animals is proof of "divine goodness." "Natural theology" is the doctrine, greatly respected in England at this time, that God's truth can be discovered in the "book of nature" as well as in Scripture, and that these two books complement each other. Paley's *Natural Theology* was universally praised as a devastatingly convincing argument for the existence of God (at least until Darwin provided a plausible alternative to the argument from design), and even modern critics concede that Paley constructed a brilliantly persuasive argument. Darwin himself professed to have admired Paley's book immensely. **Kerkut** notes that until 1927 all Cambridge undergraduates were required to demonstrate knowledge of another of Paley's books, *Evidences of Christianity* (1794).

1253 Panin, Ivan. *Bible Chronology.* Undated (1950?) [1923]. Ft. Langley, B.C., Canada: Association of the Covenant People. Editor's Preface dated 1950. Orig. pub. 1923 by Armach Press (Toronto). Bible numerology—sevens and thirteens are prominent. Adam was created 4003 B.C. Flood: 2347 B.C. Panin, who died in 1942, presented these ideas in his journal *Bible Numerics* in 1916. According to **K. Brooks,** he was a "converted Russian Nihilist—a Harvard scholar and a mathematician."

1254 Parker, Gary E. *From Evolution to Creation: A Personal Testimony* (booklet). 1977. San Diego: Creation-Life. Transcripts of four radio talks from the ICR series "Science, Scripture and Salvation." *Parker:* M.S. and Ed.D.—Ball State Univ.; biology prof. at Institute for Creation Research; prominent ICR speaker and debater until "promoted" in 1987 to head Christian Heritage Coll. (ICR affil.) biology dept. Account of Parker's conversion to creationism after teaching evolution in college. Widely cited as an example of a scientist trained in evolution who later realized the superiority of creation-science. Parker says evolution was a "very emotional experience for him"; it was a "faith and heart commitment, a complete world-and-life view; in other words, a religion." Evolution makes assertions about God, sin, and salvation. Parker converted to creationism while teaching at a Christian college near Philadelphia (which did not advocate strict creationism).

1255 ________. *Biological Diversity in Christian Perspective* (booklet?). 1977. San Diego: Christian Heritage College. (Listed in Ehlert bibliog.)

1256 ________. *Fossil Record in Christian Perspective* (booklet?). 1977. San Diego: Christian Heritage College. (Listed in Ehlert bibliog.)

1257 _______. *Ecology in Christian Perspective* (booklet?). 1977. San Diego: Christian Heritage College. (Listed in Ehlert bibliog.)

1258 _______. *Dry Bones ... and Other Fossils.* 1979. San Diego: CLP (Creation-Life). Illustrated (drawings) by Jonathan Chong. Children's book. Written as first-person account by Parker of fossil-hunting trips with his family. (Parker's wife Mary is curator of the ICR Museum of Creation.) Parker's four children ask him questions, and he replies by explaining how fossils were formed by the Flood. Explicitly biblical; many references to Christ. "'Why do you believe in evolution, Dad?' 'I thought I believed it because of all the evidence. I really believed it because I didn't believe in God.'" States that creationism cannot be believed in without belief in a Creator. Presents young-earth creationism and Flood Geology. This book was quoted by R. Doolittle in his televised debate with **Gish** (hosted by Jerry Falwell), because it was one of the creationist books used in a Livermore, Calif. public elementary school.

1259 _______. *Creation: The Facts of Life.* 1980. San Diego: CLP. Book's biog. note says Parker has a "doctorate in biology, with a cognate in geology (paleontology)" [cf. above]. This book is same as first half of *What Is Creation Science?* by **H. Morris** and Parker. A clear and easy exposition of the standard creation-science biology and paleontology arguments, written in a breezy, humorous style. Last chapter explicitly Christian. Makes much of rhetorical question posed by G. Hardin: "Was **Paley** right?"—demonstrates that he was. DNA shows creative design and organization. Mutations are deleterious, not beneficial. Includes arguments against molecular phylogenies; claims the results are inconsistent and therefore no proof of evolution. Evolutionists just choose the results that fit their scheme. Emphasizes lack of transitional fossils, and scientific challenges to neo-Darwinism. Endorses bombardier beetle, Paluxy manprints; presents other standard creation-science examples. Photos; many drawings. Bibliog.; list of creationist resources.

1260 Parkinson, James. *Organic Remains of a Former World—An Examination of Mineralized Remains of the Vegetables and Animals of the Antediluvian World: Generally Termed Extraneous Fossils.* 1833 [1804]. London: M.A. Nattali. 2nd ed. Orig. 1804–08 (two vols.). Described in **D. Young** 1982 as a "synthesis of Cuvierianism and Moses in which the days of Genesis were treated as vast periods of time" (Day-Age creationism). Also discussed in **B. Nelson** 1931. Includes long section on formation of coal and petroleum from the Flood. Branches and trunks are only visible in upper coal layers where it is covered with earth; the lower layers were fermented under pressure. Marine fossils are found in these deposits.

1261 Parrot, Andre. *The Flood and Noah's Ark.* 1955 [1953]. New York: Philosophical Library. Also pub. by SCM (London). *Parrot:* archeologist; curator-in-chief of French Nat'l. Museum; prof. at the Louvre. Critical of Ark-eology: not to be confused with **J.J.F. Parrot.** Includes section on "Invention of the Ark." Traces origin of tale of WWI Russian discovery of the Ark: started in Holland in 1923 as a joke. Also includes sources of many other Ark-eology tales. Discusses the two different Flood stories in the Bible (J and P sources). Notes that Woolley's Ur flood deposits are found only in Mesopotamia. Discusses other Mesopotamian versions of Flood story. Photos, maps, charts; index. Parrot also wrote *Tower of Babel* (1956).

1262 Parrot, J.J. Friedrich. *Journey to Ararat.* 1859 [1845]. New York: Harper & Bros. English translation. 1845 ed. pub. by Longman, Brown, Green, and Longmans (London). Several eds. pub. by Harper up to 1859. Orig. French. Parrot's 1829 climb, the "first modern ascent" of Mt. Ararat, is cited by all Ark-eologists. Parrot believed that Noah's Ark could be resting on Ararat.

1263 Patten, Donald Wesley. *The Biblical Flood and the Ice Epoch.* 1966. Seattle: Pacific Meridian. Slightly rev. ed. 1967. *Patten:* geography M.A.—Univ. Washington; owns microfilm business in Seattle; Baptist. Cosmic catastrophe theory to explain Bible events, strongly influenced by **Velikovsky.** "The central proposition of this book is to demonstrate the superiority of the theory of astral catastrophism over and against the uniformitarian view of earth history. . . . Among the cataclysmic forces which engaged our fragile sphere were both gravitational and magnetic forces of planetary magnitude. The results included tidal waves of subcontinental dimensions. We do not maintain that the period of crisis referred to in Genesis as the Flood was the first conflict nor the last; we only maintain that it was the worst." The Ice Age (singular) and the Flood were caused by fly-by of comet which dumped vast amounts of ice on earth in 2800 B.C. Ice was pulled in at earth's magnetic pole—Ice Age deposits are centered there. The global Flood was tidal, caused by the comet's gravitational pull. The comet altered earth's orbit, tilted its axis and caused condensation and collapse of earth's Vapor Canopy (decribed by Patten). Discusses orogenesis: geographical pattern of mountain ranges supports sudden origin from astral catastrophe. Ararat is geographic center of Eurasian-African land-mass. Frozen mammoths prove suddenness of Ice Age. Relates worldwide traditions of Flood and astral catastrophes; also previous catastrophe theories (discusses **G.M. Price, B. Nelson, Rehwinkel, H. Morris,** Hapgood, Ivan Sanderson, Velikovsky, D. Hooker, **Whiston;** mentions **Vail**). The Flood, and other, lesser astral catastrophes involved "counter-dominating gravitational forces and magnetic fields, resulting in (1) much tidal upheaval within our oceans; (2) surging spasms or tides or lava (fluid magma) from within the Earth's thin crust; and (3) further discharges of an electrical nature." References mostly scientific, some biblical. Many diagrams, charts, maps.

1264 _______, Ronald R. Hatch, and Loren C. Steinhauer. *The Long Day of Joshua and Six Other Catastrophes.* 1973. Seattle: Pacific Meridian. "A Unified Theory of Catastrophism." *Hatch:* engineering and computer work. *Steinhauer:* aeronautics and astronautics Ph.D.—Univ. Washington. Forewords by William Thompson (astrogeophysics M.S.; U.S. Dept. Transportation) and C. McDowell (Semitic studies Ph.D.; chair of geography, history and political sci. dept. at Cuyahoga Comm. Coll., Ohio). Includes biog. notes. Noah's Flood—2500 B.C.; Babel dispersion—1930 B.C.; Sodom and Gomorrah destroyed—1877 B.C.; Exodus—1447 B.C.; Joshua's Long Day—1404 B.C. (etc.). All these events except the Flood were caused by a fly-by of Mars. The Flood was caused by fly-by of Mercury and a comet. There were pre-Flood and pre-Adamic catastrophes also. Mars had a 2:1 resonant orbit which crossed earth's. In 701 B.C.—the time of the Isaiahic catastrophe—Mars, perturbed by earth, settled into its present non-resonant orbit, and the astral catastrophes ceased. Authors discover cyclic pattern: two major catastrophes in a century every five to six centuries, plus lesser catastrophes 108 years after these major ones. Much discussion of biblical events concerning these catastrophes; also other astral myths and cosmologies. Charts, photos, diagrams, maps.

1265 _______ (ed.). *A Symposium on Creation II.* 1970. Grand Rapids, Mich.: Baker Book House. (First volume of *Symposium on Creation* series listed under **H. Morris,** ed.; vols. II–V under Patten.) Authors: Patten, **Chittick, Nevins [S. Austin], McCone, C.E.A. Turner, W.H. Tier.** Includes biog. notes. Standard creation-science topics.

1266 _______. *A Symposium on Creation III.* 1971. Grand Rapids, Mich.: Baker Book House. Authors: **Tier, Nevins,** D. Tilney (botany and zoology B.Sc.—Univ. London; lecturer at Spencer-Churchill Coll., Oxford; member of EPM), **F. Cousins, Davidheiser,** R. Daniel Shaw. Includes biog. notes. Standard creation-science chapters. Davidheiser's chapter corrects many "myths" about the Scopes Trial (though he ignores or accepts others).

1267 _______. *A Symposium on Creation IV.* 1972. Grand Rapids, Mich.: Baker Book House. Authors: W. Dennis Burrowes (editor of the Canadian *North American Creation Science Newsletter;* Patten is member of its Council), **Custance, C.E.A. Turner, G.F. Howe, E. Shute, R.E.D. Clark, McCone,** L. Steinhauer (co-author, with Patten, of *The Long Day of Joshua*), **T.W. Carron.**

Includes biog. notes.

1268 ________. *A Symposium on Creation V.* 1975. Grand Rapids, Mich.: Baker Book House. Authors: Gordon Homes Fraser, **R. Whitelaw,** T.H. Leith, **Nevins,** Steinhauer, **Davidheiser,** Burrowes. Includes biog. notes.

1269 Patterson, Alexander. *The Other Side of Evolution: Its Effects and Fallacy.* 1912 [c1903]. Chicago: Moody Bible Institute Colportage Association. 3rd ed. Orig. pub. 1903 by Christian Alliance (New York) and Winona (Winona Lake, Ind.). Introduction by **G.F. Wright,** who assisted with the book. Scientists admit that evolution is not proven, and many reject it. Evolution cannot be verified; its arguments are false and unscientific, and are "violently opposed" to the Bible and to Christian faith. Evolution "originated in heathenism and ends in atheism." It is the doctrine of Chance. Many scientific quotes. Index.

1270 Patterson, Mabel E. *In the Shadow of Darwin.* 1982. Priv. pub. Darwin's contemporaries and the impact that his theory of evolution had upon them. In ICR Library.

1271 Patterson, Robert. *Fables of Infidelity and Facts of Faith* (tract series). 1875 [1858]. Cincinnati: American Reform Tract & Book. "A series of tracts on the absurdity of atheism, pantheism, and rationalism."

1272 ________. *The Errors of Evolution.* 1893 [1885]. Boston: H.L. Hastings. 3rd ed. "An examination of the nebular theory, geological evolution, the origin of life, and Darwinism" (title page).

1273 Pattison, S.R., and Friedrich Pfaff. *The Age and Origin of Man Geologically Considered.* 1883. London: Religious Tract Society. Also pub. by American Bible Society (New York; undated). Pattison and Pfaff also wrote *The Earth and the Word; or, Geology for Bible Students* (1858; Philadelphia: Lindsay & Blackiston). Pattison also wrote *Gospel Ethnology* (1887 [1860]; London: Religious Tract Society).

1274 Pavlu, Ricki D. *Evolution: When Fact Became Fiction.* c1986. Hazelwood, Mo.: Word Aflame Press. *Pavlu:* botany B.S. and plant systematics M.S.—Louisiana Tech; minister of First Pentecostal Church (Church Point, La.). Wholly unoriginal rehash of the standard creation-science arguments. Same examples, same quotes, same references. Strict young-earth creationism, Flood Geology, vapor canopy. Quotes H.H. Newman (1932) a lot; also R. Dickerson's 1978 *Sci. Amer.* article. "If a person rejects the Old Testament account of creation and classifies Genesis 1 and 2 as mere fables, then he must also reject the New Testament, for the New Testament accepts as valid the creation, Adam and Eve, and the curse. If any portion of God's Word is in error, then we must reject the whole, including the message of salvation." This book was featured and distributed by the Creation-Science Legal Defense Fund. Drawings; bibliog.

1275 [Payne, B.H.]. *The Negro: What Is His Ethnological Status?* 1867. Written under pseudonym "Ariel." Cited in **Carroll** 1900, who says Rev. Payne was the originator of the theory that Genesis describes a pre-Adamic creation of Negroes. Negroes are the most cunning of the "beasts," and were created to be servants of the Adamic (white) race.

1276 Pearce, E.K. Victor. *Who Was Adam?* c1969. Exeter, England: Paternoster. *Pearce:* anthropology degree from Oxford Univ. Book incorporates much anthropological material and information; reconciles it with Genesis. Proposes that there were two Adams. The Adam of Genesis 1 lived in the Old Stone Age; the Adam of Gen. 2 in the New Stone Age. The second Adam, who lived 10–12,000 years ago, was the first of the New Stone Age farmers. The rest of the Bible is about the Fall and Salvation of this second Adam. Includes a chapter advocating Day-Age creationism. Explains that Adam's 'rib' was an X chromosome, which God removed (then doubled) to create Eve. Jesus was created when God miraculously inserted a Y chromosome into Mary. The Word was DNA.

1277 ________. *The Origin of Man* (booklet). 1967. London: Crusade Magazine. Discussed and criticized in **D. Watson** 1976. Identifies prehistoric Danubian culture as man's age of innocence after leaving Eden.

1278 Pearcey, Nanci [Nancy]. *Teaching Creationism* (pamphlet). Undated (1980s?). Bible-Science Association.

Pearcey: M.S.—Covenant Sem. (St. Louis); philosophy grad. study at Inst. Christian Studies (Toronto); former children's writer—elementary school Readers grades 3-6; contributing editor of *Bible-Science Newsletter;* niece of **W. Overn;** now runs a writing and editing service for creationist authors with her husband in Toronto. "Children need the tools of thinking, of learning, of arguing, of clarifying, of logical inference"—not just mechanical learning. Endorses **Thurman**'s book for similarly encouraging teaching children how to think. Strict creationism. Pearcey now writes the "World View" articles for *Bible-Science Newsletter.* These articles, which affirm creationism, are philosophical in approach, and often concern scientific theories and paradigms. In general they are diligently researched.

1279 Pearman, Robert, Meryl Fergus, and Pat Alexander (eds.). *Wonders of Creation.* 1975. Glendale, Calif.: Gospel Light Regal Books. "Twenty articles cover different aspects of science such as atoms, balance of nature, astronomy, and ocean life." For children. Presumably anti-evolution. Illustrated.

1280 Pember, G.H. *Earth's Earliest Ages.* 1975 [1876]. Grand Rapids, Mich.: Kregel. New ed., edited and with additions by G.H. Lang. Orig. pub. 1876 by Fleming H. Revell (New York) and by Hodder and Stoughton (London). A presentation of the Gap Theory of creationism, first proposed by **Chalmers** and later popularized by **Scofield.** Pember here develops the theological aspects of the Gap Theory; this book greatly increased the influence of the Gap Theory. Between Genesis 1:1 and 1:2 there was a 'gap' which includes all of geological time. The Hebrew 'asah' should be translated "make over"—not "create." After the original creation, ages passed; then God destroyed the earth—"the earth became desolate and void"—and recreated it (the familiar Genesis creation account). Satan, after he fell, ruled over the earlier pre-Adamic creation. All fossils date from this sinful pre-Adamic world. There were pre-Adamic people too, since the Bible speaks of "ruined cities." "It is thus clear that the second verse describes the earth as a ruin; but here is no hint of the time which elapsed between creation and this ruin. Age after age may have rolled away, and it was probably during their course that the strata of the earth's crust were gradually developed. Hence we see that geological attacks upon the Scriptures are altogether wide of the mark, are a mere beating of the air. There is room for any length of time between the first and second verses of the Bible." Pember finds various Bible passages which he interprets as references to Satan's pre-Adamic rule; most later Gap supporters cite these same passages. Demons are the spirits of the pre-Adamic creatures. Their bones lie at the bottom of the sea (the Abyss). The second half of the book is primarily an argument against "spiritualism": reincarnation, witchcraft, fortune-telling, seances, clairvoyance, trance-mediums, Theosophy, occultism, Buddhism, and other heathen beliefs and practices denounced in the Bible. Spiritualism is the work of Satan and his demons. Index. Lang, a preacher, has also edited several other of Pember's books.

1281 Penn, Granville. *A Comparative Estimate of the Mineral and Mosaical Geologies.* 1844 [1822]. London: James Duncan. 3rd ed. Orig. 1822. Penn was one of the four great Scriptural Geologists (Millhauser). This book refutes reconciliationist theories (Gap and Day-Age creationism), presenting an uncompromising and strongly literalist creationism. Penn realized the challenge of the new geology to literalism, and sought to resist rather than accommodate it. All rock strata are the result of the creation—the separation of the land from the water—and the Flood. **Nelson** 1931, who praises Penn as a Flood Geologist, says he emphasized that "vast accumulated masses" of floating vegetation—mats or rafts—were swept away by the Flood, and later turned into coal. (Cf. **Nevins [Austin]** 1976.) Penn apparently also originated **Gosse**'s Omphalos creationism theory.

1282 ________. *Conversations on Geology; Comprising a Familiar Explanation of the Huttonian and Wernerian Systems; the Mosaic Geology as Explained by Mr. Granville Penn.* 1840 [1828]. London: J.W. Southgate.

1283 Perry, Charles. *Science and the Bible* (booklet). 1869. London: S.P.C.K. *Perry:* Bishop of Melbourne, Australia. A White 1896 calls this "a most bitter book"; Perry declared that the obvious object of Chambers, Darwin, and Huxley "is to produce in their readers a disbelief in the Bible."

1284 _______. *Creation vs. Development; a Review of the Lecture by the Rev. J.E. Bromby* (booklet). 1870. Melbourne, Australia.

1285 Petersen, Dennis R. *Unlocking the Mysteries of Creation: Vol. One.* 1986. South Lake Tahoe, Calif.: Christian Equippers International. Based on Petersen's "multi-media seminar" of the same title, available in several sessions on different creationist topics. (The seminar consists of slides, plus either live or taped narration.) *Petersen:* M.A. in historical museum admin.; prof. of Bible. Sections: "Unlocking the Mysteries of the Early Earth," of "Evolution," of "Original Man and the Missing Links," of "Ancient Civilizations." Petersen has declared that there are "large audiences craving to hear that the God Who saves them is also the God Who created them." Book has drawings on every page. (These are attractive, but lack the considerable impact of the colorful flood of slides in his seminar presentation.) Includes standard creation-science arguments. Petersen relies heavily on **Velikovsky**'s catastrophism.

1286 Peterson, Everett H. *The First 2000 Years.* Undated. Unpub. MS. Unpub. book MS in ICR Library. Strict creationism. The Flood, standard creation-science arguments, Bible references.

1287 Phelps, Guy Fitch. *The Absurdities of Evolution.* c1924. Chicago: Bible Institute Colportage Association (Moody).

1288 Philips, John. *Exploring Genesis.* 1980. Chicago: Moody Press. Admits earth is old, but skeptical of Darwin and of scientific theories—they are all eventually overthrown.

1289 Phillips, Max G. *He Made Planet Earth.* c1978. Mountain View, Calif.: Pacific Press. Presumably Seventh-day Adventist.

1290 Phillips, Walter. *The Defense of Christian Beliefs in Australia, 1875–1914: The Response to Evolution and Higher Criticism.* (Listed in Ehlert files.)

1291 Phin, John. *The Chemical History of the Six Days of Creation.* 1872 [1870]. New York: Handicraft. Pub. 1870 by American News (New York).

1292 Piaget, Jean. *Le Comportement, Moteur de l'Évolution.* 1976. France: Gallimard. ("Behavior, the Driving Power Behind Evolution.") *Piaget:* Swiss psychologist famous for his studies of the mental development of children. Based on his 1929 observations of a freshwater mollusk species, Piaget asserted the inheritance of acquired characteristics in a 1964 book. Here, according to **Blocher,** he "builds a broad and vigorous case in favour of the same thesis 'upheld by certain free minds, but contrary to current opinion', *i.e.* neo-Darwinism. . . ." In Piaget's view, neo-Darwinism has shown itself incapable of explaining macro-evolution, in particular the unbelievably precise adaptations of instinctive behavior (such as 'the paralysing stings which the Ammophila gives to caterpillars, thus immobilizing their nervous systems without killing this prey intended for its larvae'), and the construction of new genes during the course of evolution." Piaget argues that the genome is not "a mass of independent particles," but "a system of interactions" integrated into the functining of higher systems. Behavior influences the modifications of the genotype.

1293 Pierson, Arthur T. *"Many Infallible Proofs:" The Evidences of Christianity; or, The Written and Living Word of God.* 1886. New York: Fleming H. Revell. *Pierson:* D.D.; Bethany Presbyterian Church, Philadelphia. Pierson, who used to be a doubter himself, here presents definitive proof that Christianity is the "one and only Divine Religion." "God . . . asks us of no blind faith. . . . Nothing is to be accepted unless based on good evidence. . ." Asserts that science should be unbiased, and not based on preconceived theories which hinder impartial investigation. Warns against relying on appeals to feeling—conviction rather than logical persuasion. Advocates common-sense approach to science and truth; triumph of rational investigation and logic; perspicuity of argument and evidence. Discusses fulfilled prophecy of

Bible; rationality of and evidence for miracles. Refutes Ingersoll's *[Some] Mistakes of Moses.* Affirms scientific accuracy of Bible: "not one scientific error, blunder or absurdity has ever been found there!" Explains why Genesis doesn't employ scientific language. "When the modern science of Geology began to unwrap the earth's coverings and read the records of the rocks, timid faith grew pale and trembled for the Word of God." Some responded by denouncing science, saying God created fossils *in situ,* or by advocating Flood Geology. But the truth is that the six days of creation are really "six periods of vast length"—Day-Age creationism. "The correspondence between the Mosaic account of creation and the most advanced discoveries of science proves that only He who built the world built the Book." The order of creation is exactly the same. The Flood was not worldwide, though it covered all areas inhabited by man. Presents theory of H.W. Guion that earth's pre-Flood landmass consisted of a single continent: an elevated "dome" with uniform climate and terrain, and vast, luxuriant vegetation. The Flood was caused by collapse of this dome into the Pacific ocean. Presents various scientific discoveries predicted or described in the Bible, including the nature of light, air pressure, insect behavior, blood circulation, unity of mankind, etc. "Again we say, show us one undoubted fact, revealed by scientific studies for two thousand years, that cannot be harmonized with the word of God!" Also discusses moral and theological issues. Index.

1294 Pilkey, John. *Origin of the Nations. Pilkey:* Dallas Theol. Sem.; now at a Baptist college in Los Angeles; dispensationalist. Discusses Noah's Flood; cites **H. Morris.** Enoch or Melchizedech built the Pyramids. Discusses the Gundestrap Cauldron. (Referred to by televangelist Gene Scott. May advocate version of British-Israelism, like Scott.)

1295 Pimenta, Leander R. *Fountains of the Great Deep.* 1984. Chichester, England: New Wine Press. Listed in 1986 *Acts & Facts* as a recommended creationist book.

1296 Pinkston, William S., Jr. *Biology for Christian Schools.* 1980. Greenville, S.C.: Bob Jones University Press. Bob Jones Univ. "Science for Christian Schools Series": this book for grade 10. Also teacher's edition and lab manual. Openly creationist; stresses infallibility of Bible on scientific matters. "All distortions of evolution and humanism have been eliminated, making this series fully consistent with God's Word while accurately presenting what man has learned about God's creation." Discusses evolution and various concordistic theories, but advocates recent creation and Flood Geology. Says genetic engineering is not inherently "unnatural"—it can be used in biblical or evil ways. Illustrated; includes creationist bibliog.

1297 ________. *Life Science for Christian Schools.* 1984. Greenville, S.C.: Bob Jones University Press. Bob Jones Univ. "Science for Christian Schools Series": this book for grade 7. Also teacher's edition and lab manual. Illustrated.

1298 Pitman, Michael. *Adam and Evolution.* 1984. London: Rider (Hutchinson). *Pitman:* science B.A.—Open Univ. (England); classics M.A.—Oxford Univ.; biology teacher in Cambridge, England; runs the University Origins Discussion Group (in which professional scientists discuss "alternatives and modifications to neo-Darwinism"). Introduction by Bernard Stonehouse (former prof. at Oxford, Yale, etc.; director of Scott Polar Res. Inst. at Cambridge). Pitman says he "started as devil's advocate for the creationist view and came, in principle, though not according to any religious creed, to prefer it." Book is strongly antimaterialist. Pitman is knowledgeable about biology, and writes well. Typological approach: argues that life's diversity results from the mixing of different genetically programmed "sub-routines"—coded design. Archetypal forms—molecular, cellular, organic, species-level—are varied by permutations of subroutines by the designing intelligence. Discusses Second Law and requirement of energy conversion mechanism. Repeats many creation-science examples and arguments (and errors). "The assumption of evolution is the basis upon which index fossils are used to date the rocks; and these same fossils are supposed to provide the main evidence for evolution" (evolution based on circular reasoning). Endorses or quotes approvingly **Gentry, F. Marsh, Shute, Utt, Watts, Bowden, H. Morris, Dewar** and other creationists. Acknowledges help of **R.E.D. Clark** with book. "The idea of creation is primary because it states that functional design . . . was designed. The idea of evolution is secondary because it states that such design was accidental but, to explain such a self-contradictory notion, cannot render a state-by-stage account of any but the most superficial accidents. In short, it offers an explanation without sense or substance." Question of age is not important: "The idea of *biotechnological design,* in modern terms, is a rational explanation of the scientifically-discovered facts." Age is crucial for evolution, but deliberate design—not age—is the critical issue for the creationist. "A creationist is not obliged to embrace Old Testament fundamentalism; indeed, many believe the creation of life may have happened in stages, each generating life-forms more complex than the one before." Stonehouse says he did not find all Pitman's arguments convincing, but feels that Darwinism is accepted too uncritically and welcomes Pitman's exposures of its weaknesses. Index.

1299 The Plain Truth. *The "Missing Link" . . . Found!* (pamphlet). c1973 [1970]. Pasadena, Calif.: Ambassador College. No author listed. Credited to **H.W. Armstrong**'s *Plain Truth* magazine (presumably reprint). Mankind is only about 6,000 years old. Evolution is not seen in the fossil record—there is only confusion. Refutes claims of fossils of early man. The real missing link is our Creator God. Includes anthropological references.

1300 Pohle, J. *God: The Author of Nature and the Supernatural.* 1942 [1934]. St. Louis: Herder Book Co. Orig. 1934. Modified Day-Age creationism. Pohle is Catholic. The main purpose of the Bible is theological and religious—not scientific. The days/ages may not be exactly chronological—may overlap. **Ramm** 1954 says Pohle presents a "pictorial-revelatory" view also: Creation was revealed—not originally executed—in six days.

1301 Polk, Keen. *"Everything After Its Kind"* (pamphlet). Undated [1930s?]. Bellflower, Calif.: Dr. Roy Gillaspie (priv. pub.?). "A new and critical study of the origin of the Negro, according to the Holy Scriptures." Distributed by Assoc. of the Covenant People. *Polk:* U.S. Signal Corps photographer with AEF in WWI. Negroes were created before Adam: they are the "beasts of the field" of Genesis. The Bible speaks of three animal types: creeping, four-legged 'cattle,' and the two-legged beasts of the field. The 'serpent' in Eden was a Negro (snakes don't eat dust). Negroes were intended to help Adam to care for the Garden. The Adamic line interbred with the Negro (sons of God and daughters of men of Gen. 6); Noah, however, was of "pure blooded stock." Leviticus specifies the death penalty for sexual relations with 'beasts'—i.e. with Negroes. Lists scientific differences between Negroes and whites. "The Negro has been put here upon earth by God for some purpose. We may rest assured that purpose was not racial or social equality." Insists that the biblical principle of "everything after its kind" justifies racism.

1302 Pollock, Algernon James. *Evolution: Unscientific and Unscriptural.* Undated. London: Central Bible Truth Depot. (Listed in Ehlert files.) *Pollock:* died 1957; also wrote other religious books.

1303 Popoff, Peter. *Calamities, Catastrophies [sic], and Chaos.* c1980. Upland, Calif.: Faith Messenger Publications. "What the Bible Says About End-Time Disasters . . ." A crescendo of natural and man-made disasters; also UFOlogy, mass murders, the occult—"FEAR!!!" Are these events "part of a sinister diabolical plot to destroy humanity?" *Popoff:* faith-healer and televangelist who boasts of personally smuggling Bibles into Russia and floating Bible-carrying balloons into other Communist nations. In 1986 his faith-healing was exposed as a hoax (he relied on secret radio messages and other tricks to deceive his audience). This book warns that present catastrophes signal the imminent Great Tribulation. Popoff does not comment on evolution directly, but likens the present wicked and disaster-filled age to the days before Noah's Flood. Popoff insists that God personally showed him, in seven hours, the history of humanity and scenes of the imminent Tribulation.

1304 Porter, John W. *Evolution—A Menace.* 1922. Nashville: Sunday School Board of the Southern Bapt. Conv. Also pub. by Baptist Book Concern (Louisville, Ken.). *Porter:* editor of the (Baptist) *Western Recorder;* leader in the effort to ban evolution from Kentucky schools; became president in 1922 of the Anti-Evolution League of America. Creationism is "flatly contradicted" by evolution. Christianity is based on supernaturalism, which evolution denies. Evolution denies moral responsibility, and destroys belief in the Bible and in Christianity. Evolution "logically and inevitably leads to war." (Excerpted in Gatewood.)

1305 Powell, E.C. *The Challenge to Uniformitarian Geologists* (pamphlet). Undated. England: Evolution Protest Movement. EPM pamphlet #196.

1306 Poythress, Vern S. *Philosophy, Science and the Sovereignty of God.* 1976. Nutley, N.J.: Presbyterian and Reformed. Cited in **Whitcomb** and **DeYoung** as among books by Christian theologians and philosophers which have "contributed significantly to the current renaissance of Biblical creationism..." *Poythress:* A.B.—Harvard; Th.D.—Westminster Sem.

1307 Pratney, Winkie. *Creation or Evolution?* (tracts). c1984 [1980]. Lindale, Texas: Last Days Ministries. 3-part tract series. Orig. appeared in *Last Days Newsletter. Pratney:* "gave up a promising career in organic chemistry to begin a full-time, globe-trotting ministry..."; lives in New Zealand; also based in Lindale, Texas, where he works with Last Days Ministries (headed by **M. Green**); proselytizes extensively. Standard creation-science. Includes scientific references. Part I includes replies to criticisms of creation-science. Part II—"The Historical Record," lists great creationist scientists of the past. Part III—"The Fossil Record," includes a creation-science book list, and a Paluxy photo [of a fake]. These tracts have been widely distributed.

1308 Prentiss, George. *The Ages of Ice and Creation; A Lifelong Study of Natural Laws, Which Are God's Laws, in Creating the World and Transforming It from a Burning World—A Sun Blazing in the Heavens—Up to Its Present State of Development, Teeming with Plant and Animal Life.* c1915. Chicago: Common Good Company. **Vail**'s Canopy Theory, with various special and continuing creation features worked in.

1309 Price, George McCready. *Outlines of Modern Science and Modern Christianity.* 1902. Los Angeles: Modern Heretic Co. Price attended Battle Creek Coll. (Seventh-day Adventist; Mich.), then taught school in his native Canada. He took seriously the works of Adventist prophetess **E.G. White,** who stressed strict young-earth creationism and claimed that the fossil record was entirely a result of Noah's Flood. Price taught himself geology in order to prove literal creation and refute evolution, and became a science professor at Adventist colleges in California, Nebraska and Washington. (He does not cite White in his books, however.) Fundamentalist creation-science owes an enormous debt to Price, who developed (or re-invented) many of its principle anti-evolution arguments. His Flood Geology is the basis of young-earth creationism. At the Scopes Trial, Price was the only living creationist scientist **Bryan** could name (Darrow retorted that Price was a charlatan). (Price, then teaching at Stanborough Missionary Coll. in England, could not attend the trial.) Modern Heretic Co. is Price's home address. This book covers: "The Evolution Theory in its whole range, from the Nebulous Cloud, the Cooling Earth, and the Origin of Life, through Geology and Biology up to the Moral Nature of Man, Carefully discussed in a Popular Style. No one, after reading it, could for a moment suppose that the Evolution Theory had been proved by sound scientific arguments, while the moral and religious tendencies of the doctrine are shown to be anti-Christian to the last degree." Calls for a return to "primitive" Christianity. "No believer in the Sabbath as the divine memorial of creation's week will hesitate to give as the distinct, positive teaching of Genesis that life has been on our globe only some six or seven thousand years; and that the earth as we know it, with its teeming animal and vegetable life, was brought into existence in six literal days." Argued for God's direct intervention in creation and all aspects of nature. Geological record does not prove a succession of ages: it shows a taxonomic series representing different—but contemporaneous—zones of antediluvian life.

1310 ______. *Illogical Geology: The Weakest Point in the Evolution Theory.* 1906. Los Angeles: Modern Heretic Co. Later expanded into *Fundamentals of Geology* (1913). Price offers $1000 for proof of difference in age of fossils. Refutes the "onion-skin" hypothesis of rock strata: the assumption of orderly worldwide superposition of strata; argues instead that fossil types represent zoological zones. Strata can occur in any order. Discusses "fossil graveyards" as evidence of worldwide catastrophe. Also sudden climate change—the Ice Age. Fossils show general degeneration rather than evolution. "Inductive geology can never prove creation." But, it "removes forever the succession-of-life idea," thus clearing the way for creation by demonstrating the falsity of evolution. Darwinism requires geology's theory of succession of ages and of life. If that theory is refuted, evolution is also: it "has no more scientific value than the vagaries of the old Greeks—in short, from the standpoint of true inductive science it is a most gigantic hoax, historically scarce second to the Ptolemaic astronomy. . . . With the myth of a life succession dissipated once and forever, the world stands face to face with creation as the direct act of the Infinite God." "Illogical Geology" was the title of an 1890 essay by evolutionist H. Spencer. Spencer, like Price, rejected the onion-coat theory. The theory was devised by Werner, who proposed that each layer was a different rock type. Geologists have rejected Werner's Neptunian version, says Price (and Spencer), only to substitute for it a modern version in which worldwide successive layers are assumed on the basis of index fossils.

1311 ______. *God's Two Books: Or Plain Facts About Evolution, Geology, and the Bible.* 1911. Washington, D.C.: Review and Herald. Price was then a geology and physics prof. at College of Medical Evangelists (now Loma Linda Univ.) in Calif. The Med. Coll. awarded him a B.A. in 1912. Religions are concerned with origins, natural phenomena, and world-views. Evolution is a religion, but it is much older than Darwinism. Emphasis on "Moral and Social Aspects" of evolution. "Surely the Evolution theory converts into a gigantic farce the old, old story of the cross, and makes the whole Scripture a jargon of unmeaning folly." Were evolution true, the creator is a "tyrant and a fiend." Acceptance of evolution will lead to a religio-political despotism which will crush individual freedom. Quotes **Donnelly** and **Howorth** favorably. Shows that civilizations cannot have arisen independently; they are all derived from the full-blown civilizations of Egypt and the Near East. Argues that evolution, a "flimsy tissue of guesses," became accepted because of the rise of geological uniformitarianism. Refutes the "onion-coat" hypothesis of geology, and argues that the principle of superposition does not prove long ages or a uniform worldwide succession of strata. States that all extinct forms existed contemporaneously with man, and all fossils resulted from burial by the Genesis Flood. Stresses need for geology to return to methods of true Baconian inductive science; praises Bacon and **Newton.** God, in these Last Days, is allowing science to conclusively prove the truth of creationism, demonstrating the "absolute harmony between the book of nature and God's written word." Photos, drawings; index.

1312 ______. *The Fundamentals of Geology: And Their Bearing on the Doctrine of a Literal Creation.* c1913. Mountain View, Calif.: Pacific Press. Longer version of *Illogical Geology.* Dedicated to Francis Bacon and **Isaac Newton,** "who realized most clearly the true objects of *natural science,* the methods by which it should be pursued, [and] its limitations . . ." Cites and praises geologist E. Suess; says he refutes uniformitarianism by demonstrating the difference between ancient strata and modern deposits and proving the fixity of continents. Photos, drawings, charts; index.

1313 ______. *Back to the Bible: Or the New Protestantism.* 1931 [1916]. Washington, D.C.: Review and Herald. 3rd ed. Orig. pub. 1916 as pamphlet. 2nd expanded ed. c1920. Written when Price was at Pacific Union Coll. (Calif.). Calls for a second Protestant Reformation: "the problem for the Reformers of the twentieth century is to vindicate a despised and discarded Bible against so-called science already grown arrogant and dogmatic... Why not take the Mosaic account of Creation and the Deluge at their face value, and examine the claims of the popular Evolution Philosophy in the light of primitive Christian principles, without any compromise whatever?" Great Britain and America are the mightiest and most civilized nations on earth because the Anglo-Saxon races are Bible-believers. Were it not for the beclouding influence of sin, God's book of nature could be read as clearly as His later revelation in the Bible. Price rejoices that the truths of science are now proving the Bible true, and demonstrating the harmony between God's two books. Strongly and repeatedly urges a return to the true inductive science of Bacon and **Newton,** and away from the subjective philosophical speculations of evolutionary geology and biology. Warns against socialism, Bolshevism, anarchism; the erosion of individual rights and the aggrandizement of the State: the "coming slavery" of the World Federation. Emphasizes that the "scoffers" prophesied in II Peter, who doubt the reality of Noah's Flood and the Creation, are uniformitarian evolutionists. Summarizes his standard Flood Geology arguments. Suggests that the Flood was caused by sudden change of earth's axis from perpendicular to present tilt, resulting in enormous twice-daily tidal waves. These successive waves account for layered geological strata. Stresses that science cannot tell us anything about origins, though uniformitarian evolution assumes that it can. Matter and life "must have come into existence at the beginning *by laws and methods not now operating anywhere on earth.* Not Uniformitarianism and Evolution, but Creation, is what modern science is teaching us by facts so large and fixed that there is no prospect of their ever being overthrown by any fresh discoveries." We "must simply follow the record of Genesis regarding [creation], which must ever lie beyond the reach alike of man's guessing and of his research." Discusses laws of thermodynamics and biogenesis as proof of creation. Criticizes American eugenics and German evolutionist racism, but cites Galton's claim of mental inferiority of Negroes in support of his claim that mankind has degenerated.

1314 ________. *Q.E.D.: Or New Light on the Doctrine of Creation.* c1917. New York: Fleming H. Revell. Price was then geology prof. at Union Coll. (Nebraska). Stresses that there are only two theories of origins—evolution and creation—and that these are mutually exclusive. Complains that science has become a collection of specialties, each ruled by authority; he (Price), however, can stand back and survey the whole domain of science. Protoplasm is the very basis of life, marvels Price; it cannot be analyzed scientifically, because life is thereby destroyed. Reviews biogenesis argument and the wonders of the cell. States that the naturally delimited (immutable) groups in biology are not our modern species, but "kinds" at the genus level or even larger. Identifies these with **Linnaeus**'s broader species concept of created types. Quotes many advocates of rival interpretations of evolution: "If each of these opposed schools of scientists are right in *what they deny,* the whole theoretical foundation for the origin of new kinds of animals and plants is swept away—absolutely gone. For if an individual really cannot transmit what he has acquired in his lifetime, how can he transmit what he has not got himself, and what none of his ancestors ever had? And if natural selection cannot start a single organ of a single type, what is the use of discussing its supposed ability to improve them after the machinery is all built?" Includes usual Flood Geology arguments. Though science can tell us nothing about the supernatural processes used during Creation, modern scientific discoveries have established the doctrine of Creation like "a mathematical Q.E.D." Jesus said you must believe Moses to believe me, and belief in Creation and the Flood implies belief in the coming End of the world by fire.

1315 ________. *A Textbook of General Science.* 1917. Mountain View, Calif.: Pacific Press.

1316 _______. *Poisoning Democracy.* 1921. New York: Fleming H. Revell. Quoted in **H.W. Clark** 1966. Astronomy, anthropology and biology all teach theories which contradict the Bible account of creation. The teachings of modern geology, especially, "flatly contradict the Bible." "The alleged fact that the world is old beyond computation, that life has existed through successive ages covering millions of years—who does not recognize in this a direct contradiction of the most obvious teachings of the Bible?"

1317 _______. *The New Geology.* 1926 [1923]. Mountain View, Calif.: Pacific Press. 2nd ed. Orig. 1923. A 726-page "Textbook for Colleges." Same Flood Geology arguments as Price's other books, but presented here in their fullest form. Bristling with scientific references; appears quite scholarly. Accuses evolutionists of circular reasoning: that the fossils are assumed to exhibit evolutionary progression, and are thus used to date the rocks they are found in; then the rocks are used to date the fossils. Usual emphasis on out-of-order strata. Price insists the geological column only applies to the corner of Europe in which it was developed, and that strata can and do appear in any order. Explanations by geologists of out-of-order sequences by thrust faulting and folding of layers are merely attempts at avoiding the obvious. Fossiliferous layers can appear in any order, and no fossil can be shown to be any older than another. This is Price's "*great law of conformable stratigraphic sequences* . . . by all odds the most important law ever formulated with references to the order in which the strata occur." Price continually insists on the difference between facts and theories. Proper science should be built inductively on facts—not on *a priori* theories such as evolutionism. Skeptical of evidence of Ice Ages; attributes this to Flood. Explains coal formation as due to Flood also. Final chapter asserts creation of man; refutes animal ancestry. Photos, charts, diagrams, maps. This book established Price as the principle creationist scientific authority. Price was then convinced the world would soon abandon evolution and convert to creationism.

1318 _______. *Science and Religion in a Nutshell* (booklet). 1923. Washington, D.C.: Review and Herald.

1319 _______. *The Phantom of Organic Evolution.* 1924. New York: Fleming H. Revell. Arguments against biological evolution. Proposes that the family is the "primary organic unit": the immutable "kind." Different species and even genera can be produced by variation according to Mendelian principles. Kinds, however, cannot evolve into other kinds. Discusses implications of embryology for creation.

1320 _______. *The Predicament of Evolution.* c1925. Nashville: Southern Publishing Association.

1321 _______. *Evolutionary Geology and the New Catastrophism.* c1926. Mountain View, Calif.: Pacific Press. Also pub. in photocopy ed. by **Corliss**'s Sourcebook Project (Glen Arm, Md.). Rewritten version of *Fundamentals of Geology* (1913). Price's standard Flood Geology arguments. Most of the material is contained in Price's other books (in fact most of Price's books consist largely of re-cycled material). Full of scientific references and technical arguments; quotes a few of the classical Scriptural Geologists. No biblical references. Begins by demolishing the "onion-coat" theory of uniform geological strata and fossil succession. Presents example after example of wrong-order strata and fossils. "Every scrap of physical evidence tends to show that these rocks were actually deposited in the order in which we find them." It is only the evolutionary theory which causes geologists to ignore these plain facts. Reliance on such a speculative hypothesis is a "mere travesty on the methods of Bacon and **Newton**. . ." Alleged chronological relationships between fossils are entirely dependent on geologists' assumption of evolution, Price declares. Evolutionary geology assumes the various fossil assemblages were distributed worldwide in invariable order (the modern version of the onion-coat theory). It further assumes these layers represent long ages, for which there is no evidence at all, and it denies the possibility of different zoological zones. Price insists that all fossils existed simultaneously, in different zones, and were contemporary with man. Price acknowledges that his law of conformable stratigraphic sequences was recognized by **Fairholme.** Presents young-earth argument: antecedent rivers, old elevated shore lines, deep ocean fossils, the fact that mountains are composed of "young" layers. Points out that most modern forms have fossil representatives in the "oldest" rocks—and often skip all intermediate layers. Emphasizes fossil graveyards, which must have been caused by a sudden catastrophe (cites **Howorth** frequently). Endorses 1923 book by J.M. McFarlane *Fishes the Sources of Petroleum* (Macmillan). Argues that evidence of previous global warm climate is evidence for the Flood. Reviews alleged evidence of "Fossil Man," making many of the same criticisms used by modern creationists. Present various evidences that man existed in the Tertiary and Paleozoic, and was contemporaneous with dinosaurs. Ridicules evolutionary exhibits of the horse series as well as of fossil hominids as purely arbitrary arrangements. "By these methods of strict inductive science, we shall not be able to avoid the conclusion that our world has witnessed an awful aqueous catastrophe, and that back of this lies a direct and real creation as the only possible origin of the great families of plants and animals. In short, a strictly inductive and mature study of the facts of geology as known to modern science confirms in a very marvelous way the literal interpretation of the first chapters of Genesis, which a pseudo-criticism and the infant lispings of science supposed they had consigned to the realm of fable and myth." Pictures, photos, diagrams; index.

1322 ________. *A History of Some Scientific Blunders*. 1930. New York: Fleming H. Revell. **H.W. Clark** 1966 says Price wrote this in response to the Scopes Trial.

1323 ________. *The Geological-Ages Hoax: A Plea for Logic in Theoretical Geology*. c1931. Chicago: Fleming H. Revell. Insists that proper logic demands geological investigation "begin at the surface of the earth and work downward, instead of beginning at a supposed bottom of the fossiliferous strata and working upward." "The correct scientific method would be to begin with the present world, with all that we know about our modern earth and its living inhabitants and the forces now operating over the earth's surface; then by working backwards into the past" we can explain the past. Geologists, however, begin at the bottom and speculate recklessly up to the present. Previous geologists were "wild dreamers and speculators; for the relics of ancient plants and animals were used by them as mental spring-boards from which to launch away on the wings of airy fancy about how the world was made and what innumerable vicissitudes it had experienced in remote ages." The proper approach leads to conclusion of a single catastrophic Flood. Calls for "ecdysis" of geology—that it shed its old rigid shell of accumulated facts and obsolete theories. Price no longer expects mass conversion to Flood Geology: "I am not so optimistic as to think that all of my readers will agree with my attempt at a constructive programme. . ." But his criticisms decisively refute evolutionary geology, even if the Flood is considered outside the realm of science. "It is no answer to my criticism of the grotesque logic offered for the geological ages to say that my hypothesis of a great world-catastrophe as the cause of much (if not most) of the fossiliferous deposits is incredible and absurd. Perhaps it is." "The time values of the fossils must be forever discredited in the minds of those who value facts alone as the solid ground on which any science can be built that is to endure." "The evolutionary geologists have already made the world believe in their mystical ability to separate the fossils of one 'age' from those of all others; but there is absolutely no scientific evidence on which to base this ability of discrimination." Gives a popular presentation of his claim of circular reasoning in geology. Emphasizes out-of-order strata and "missing ages" as proof that geological column is arbitrary and false. The geological-ages theory "seems to paralyze the reasoning faculties, and often hypnotizes its followers into doing almost any strange logical antic." "What scientific justification can we plead for clinging to a metaphysical theory in the face of objective evidence which so squarely contradicts it?" Discusses recent land submergence. Says all mountain-building occurred simultaneously, after other geological processes ceased. Claims

Cambrian rocks are the world's oldest. Refutes the "Ice Age nonsense"—the evidences claimed for it are really after-effects of the Flood. Lists and answers common objections to his Flood Geology (use of miracles in science, source of Flood water, speed of deposition, existence of other methods to date rocks besides fossils, biogeography, etc.). Replies defensively to accusation he is not a working scientist: "one does not have to spend his life in pounding out specimens ... to be able to judge of the logic of a scientific statement." Points out the failure of **W.J. Bryan** and others who tried to reconcile Genesis with long geological ages. Concludes with "charitable view" that the geological-ages theory is a "hoax"—a bluff, a joke—rather than a deliberate fraud.

1324 _______. *Modern Discoveries Which Help Us to Believe.* 1934. New York: Fleming H. Revell. "The doctrine of Creation stands vindicated on all four of its most important points: (1) Modern physics shows that matter must have been created, (2) spontaneous generation has been discredited, (3) variations in plants and animals have all been within the original 'kinds,' and (4) it is impossible to discover among the fossils any real historical sequence for the world as a whole." (Quoted in **H.W. Clark** 1966.)

1325 _______. *The Modern Flood Theory of Geology.* c1935. New York: Fleming H. Revell. "It is pure unscientific dogma to affirm that the trilobites, the dinosaurs, and the mammals ... must represent successive ages of the earth's history merely because they are found here in beds which were deposited in some sort of chronological sequence." The physical evidence cannot tell us whether the strata were deposited millions of years or mere minutes apart. "In ... geology, there would seem to be only two alternatives possible: ... either uniformity, or the Flood theory, the theory of a great world-catastrophe. The latter is just as truly scientific as is the former; and in every geological study the two hypotheses should always be considered." "All the great major facts of the rocks are much more easily explained on the basis of a universal Flood than on the current theory of many long ages of slow changes like those said to be taking place in our modern world." (Quoted in **H.W. Clark** 1966.)

1326 _______. *Some Scientific Stories and Allegories.* c1936. Grand Rapids, Mich.: Zondervan.

1327 _______. *Genesis Vindicated.* 1941. Takoma Park, Md.: Review and Herald. Price's last major work. **H.W. Clark** 1966 says Price attempted to review the latest discoveries in all fields of science, relating these to his own geological "discoveries." Argues for recent six-day creation, and refutes old-earth creationist theories. Declares we cannot know age of universe. Skeptical of new view of interchangeability of matter and energy. Reaffirms belief in validity of **Linnaeus**'s concept of species as originally created forms. What taxonomists now call 'species' are really varieties. Mendelism supports this concept. Cites Goldschmidt's new book as a denial of Darwinism. Concedes greater role to glaciation than in previous books, but still skeptical of Ice Ages. Largely abandons former optimism regarding imminent triumph of creationism.

1328 _______. *If You Were the Creator.* Mountain View, Calif.: Pacific Press. "If you could plan everything from the beginning, you would have to make the *matter* of the universe; then the chemical elements would have no properties you did not put into them. You would merely have to speak them into existence. *There is no other way.* All things, inorganic and organic, came into existence at His word. Philosophically, this is the only rational view of direct creation." Science proves the universe could not have always existed, nor could its present forms have arisen by chance. The theory of evolution "makes the cruelty and treachery of the surviving animals the very ladder by which the human race came into existence, and the same bloody ladder the means by which the modern races have risen to their present superiority above 'primitive' man. Thus, according to this theory, God must have put a high premium on all those revolting and blameworthy characters which we now have to condemn and punish in human society. ... Obviously there must have been something radically wrong with such an unethical method by which the human race arose, if a kindly and wise Creator had anything to do with man's origin." (Quoted in **H.W. Clark** 1966.)

1329 _______. *Were the Fossils Contemporary?* (pamphlet). 1949 [1941]. Malverne, N.Y.: Christian Evidence League.

1330 _______. *How Did the World Begin?* 1942. New York: Fleming H. Revell. Refutes hypotheses of continuous creation and of stellar evolution; cites **Jeans.** Energy conservation law proves creation. Einstein's relativity is only a mathematical model and is not the only way to mathematically describe the universe. It is "a sort of weird magic," and not necessarily true. Scientists invent many such speculative theories, often "inspired by a secret wish that evolution might be true, but without any basis in scientific fact." Refutes deism (the notion of God as "absentee landlord") as libelous to God: says gravity and other natural phenomena are manifestations of His direct fiat control. Reviews biogenesis argument and embryology. Microbes have existed unchanged throughout the fossil record, and thus "refute evolution." Chapter "The Stones Cry Out" is summary of Price's Flood Geology. Notes that applied geology does not depend on the "successive-age theory" for its success. "Evolution Foretold": the prophecy of II Peter. "Cycle of Life" chapter presents **Gosse**'s Omphalos argument about life-cycles of organisms. Using Gosse's diagrams, Price maintains that these prove that the "beginning" must have been an act of creation breaking into these continuous cycles. Admits that species and even genera are not immutable, but asserts immutability of family level: "nobody can imagine how any of these families could possibly become changed over into another." It is clear that God intended for *some* taxonomic level to consist of members fertile with each other but cross-sterile with other groups. "Otherwise, untold confusion and utter chaos would result." Higher level taxa such as phyla, though, "*are not entities,* they are mere idealized collections or abstractions."

1331 _______. *Common-Sense Geology.* 1946. Mountain View, Calif.: Pacific Press. "A simplified study for the general reader." Review of Price's geology and paleontology arguments. There are but two possible hypotheses for geology: uniformity or the Flood theory. Price notes that as he was "the first in modern times to revive this ancient and honorable Flood theory, it is a satisfaction to be able to say that this view has been adopted by a large number of modern Christian apologists..." Many photos, also drawings, diagrams, maps.

1332 _______. *Feet of Clay: The Unscientific Nonsense of Historical Geology* (pamphlet). 1949. Malverne, N.Y.: Christian Evidence League.

1333 _______. *The Story of the Fossils* (booklet). 1954. Mountain View, Calif.: Pacific Press.

1334 _______. *Problems and Methods in Geology* (pamphlet). 1956. Malverne, N.Y.: Christian Evidence League.

1335 _______. *The Time of the End.* c1967. Nashville: Southern Publishing. Eschatology. Bible prophecy: Daniel and Revelation; Armageddon.

1336 _______. *Report on Evolution.* 1971. Malverne, N.Y.: Christian Evidence League. Reprints of Price articles and other items.

1337 Pride, Mary. *The Big Book of Home Learning.* 1986. Westchester, Ill.: Crossway Books (Good News). *Pride:* formerly a radical feminist; now (since becoming a Christian) a radical anti-feminist and home-schooling advocate. (Her book *The Way Home* explains the different roles God gave men and women. Wives should be submissive, and should stay at home to teach their children. A 1987 review in *Apologia* describes Pride as "utterly opposed to birth control, day-care, both private and public schools and what she calls careerism.") This is a reference book and manual for Christian home teaching, listing recommended materials and publications for home-schooling so that mothers do not have to send their children out to school. The "Science" chapter recommends Master Books (ICR), Christian Light Publications, the Bob Jones Univ. Press science textbook series, the CSRC "Science and Creation" series, plus non-religious suppliers and publishers. Also includes a creation-science bibliography listing books, films and periodicals. Pride contrasts theistic ("biblical") science with the humanist approach. Includes section on how to start your own private school (which Pride says is acceptable for older children); cites **Blumenfeld.** Index of suppliers. Pride also wrote *The Child Abuse Industry* (1986; Crossway).

1338 Pro-Family Forum. *Are Evolutionists Like Three Blind Mice?* (pamphlet). Fort Worth, Texas. Quoted at length in **B. Keith** 1982. Evolutionists are like three blind mice because: "They can't see the past! They can't see the present! They can't see the evidences!" "If evolution really happened in the past, it should be a continuing process, observable today—or at least during recorded history." But no evolution has ever been observed in 6,000 years of recorded history. Asks if we see "apes moving into houses," "fish sprouting legs" or "snakes growing wings." Affirms that all animals reproduce only their own kind. Accuses evolutionists of circular reasoning: "The rocks are dated by evolution, then used to prove evolution!" Evolutionists "tell us to believe that all forms of life—unbelievably complex—just happened by accident!" Protests that for several decades creationist evidence "has not been given to students in our public schools." Pro-Family Forum has been an active supporter of the Louisiana creation-science bill. It presented testimony to the Louisiana legislature, plus a written "Summary of Scientific Evidence for Creation," which defined creation-science as including belief in a young earth, in Flood Geology, and in fixity of plant and animal "kinds." It also presented to the court a list of creation-science books appropriate for public school use.

1339 Professional Educators' Workshop. *Alternatives in the Classroom: Manuscript from the Professional Educators' Workshop on Implementation of the Creation/Evolution Equal Presentation Resolution.* Undated [1975?]. Unpub.(?) report (Lubbock, Texas). Meeting held 1975 in Lubbock, Texas. The sponsoring teachers here "protest the presence of State-sponsored humanism in the public schools." Contributors include **Clough,** Bean Boulter (Amarillo lawyer) on legal issues, **T. Barnes,** and a long, annotated creation-science bibliography by Mrs. Harvey Olney.

1340 Prophet, Elizabeth Clare. *Forbidden Mysteries of Enoch: The Untold Story of Men and Angels.* 1983 [c1977]. Malibu, Calif.: Summit University Press. Prophet is the leader of the Church Universal and Triumphant (founded by her deceased husband), which recently moved its headquarters from Malibu to Montana. Prophet's teachings are borrowings from Theosophy and other occultic and gnostic doctrines. In a previous reincarnation she was told by Jesus that she would teach the truth of the Ascended Masters from the Great White Brotherhood (a community of reincarnated spirits possessing ancient and secret knowledge) at the dawning of the Aquarian Age. This book is based on the apocryphal Book of Enoch. Angels are material beings. Some lusted after human women and descended to earth. These became the 200 "Watchers," who taught mankind sorcery and other evil arts. Evil giants were offspring of the mating of angels (Nephilim) and humans. God wiped out these offspring in the Flood, but they returned after Atlantis sank. The Watchers remain, and are the source of all the sinister conspiracies in the world. Prophet has also written books on Jesus's "lost years" and "lost teachings": he studied with Ascended Masters in the Himalayas. In addition to its eschatological emphasis, the Book of Enoch is more cosmological than the canonical books of the Bible: it describes explictly the structure of the heavens and the (flat) earth. Allusions to Enoch are incorporated into the canonical books of the Old Testament.

1341 Prout, William. *Chemistry, Meteorology, and the Function of Digestion.* 1834. London: Pickering. One of the Bridgewater Treatises.

1342 Pun, Pattle P.T. *Evolution: Nature and Scripture in Conflict?* c1982. Grand Rapids, Mich.: Zondervan. *Pun:* biology Ph.D.—SUNY Buffalo; assoc. prof. of biology at Wheaton College; does research in biochemistry and microbiology. Foreword by **R. Mixter.** Appendix by **J.O. Buswell II** on length of creation days. Book is expanded version of 1977 paper in *J. of the* **American Scientific Affiliation.** Part I of book describes history and evidences of evolutionary theory and presents a critical evaluation. Part II is "A Christian View of the Origin of Life." Part III surveys the relationship of naturalism and Darwinism, and Part IV presents "A Christian Attitude on Evolution." Presentation of evolution is knowledgeable and factually accurate and gives a fairly thorough summary of evolutionary theory. Accepts limited evolution, and argues strongly for an ancient earth, but stresses that evidence for macro-evolution is inconclusive. Pun favors progressive creationism: long overlapping creation "days," or an intermittent-day model —long overlapping ages in between the creation days. Defends *ex nihilo* creation; argues that chance evolution is inadequate. Advocates regional (not worldwide) Flood. Describes the various exegetical approaches and different evangelical interpretations of Genesis. Discusses the negative effects of Darwinism in many fields. Intelligent and well-written. Many good charts and diagrams; index.

1343 Raaflaub, Vernon (ed.). *The Creation Alternative.* 1970. Minneapolis: Bible-Science Association. Essays selected from *Bible-Science Newsletter. Rev. Raaflaub:* acad. dean and teacher at Canadian Lutheran Bible Inst., Alberta.

1344 Radcliffe-Smith, Alan. *Botany and Evolution* (pamphlet). Undated. England: Evolution Protest Movement. EPM pamphlet #206. *Radcliffe-Smith:* London research botanist; Secretary of EPM—1976–81.

1345 Radl, Emanuel. *Geschichte der Biologischen Theorien der Neuzeit.* 1913 [1907]. Leipzig. *Radl:* physiology prof. in Prague. Discussed in **Nordenskiöld.** Radl was a vitalist who rejected Darwinistic morphology as a "soulless description and specification of different developmental forms one after another..." He favored an "ideal" morphology; "its aim is stated to be to discover the ideas according to which the forms of living organisms are constructed." "Many ideas compete at the root of organic life for precedence, and an ideal structure forms the basis of every organism." Argues for a return to the idealistic morphology of the early nineteenth century.

1346 Rahmeh, Farid Abu. *Creation or Evolution: Does Science Have the Answer?* (booklet). Undated [1981?]. Priv. pub. (England?). Text of lecture given at Bakewell, Derbyshire, England. Undated; autographed copy inscribed to **H. Morris,** 1981, in ICR Library. *Rahmeh:* civil engineering M.Sc.—American Univ. Beirut; currently in Ph.D. program at Univ. Sheffield (England). Usual creation-science arguments. Relies heavily on ICR material. Rahmeh does not state whether he is Moslem or Christian.

1347 Ramm, Bernard. *The Christian View of Science and Scripture.* 1954. Grand Rapids, Mich.: William B. Eerdmans. *Ramm:* philosophy Ph.D.—USC; also studied under Barth; prof. of systematic theology and apologetics at American Baptist Sem. of the West. Dedicated to **F.A. Everest.** This is a scholarly book and an excellent reference source for various interpretations and writings regarding the relation of science to the Bible. Ramm urges a return to the tradition of the late nineteenth-century conservative evangelical scholars who diligently and carefully tried to harmonize science and Scripture (he praises **J.W. Dawson, Pye Smith, H. Miller,** Asa Gray, **Dana, Rendle-Short, A. Fleming, Bettex,** e.g.). He laments the abandonment of science to materialists who ignore the Bible, and also criticizes "hyperorthodox" interpretations as naive, unscientific, and self-defeating. The Bible is neither full of scientific error nor filled with modern scientific predictions and theories. This book contains lengthy chapters on Bible-science interpretations of astronomy, geology, biology, and anthropology. Ramm (after this book) led the ASA resistance to the strict young-earth creationism and Flood Geology of **Morris** et al.; here, he sharply criticizes their Flood Geology predecessors **(Price, Rehwinkel, H. Clark, B. Nelson).** He argues that the language of the Bible is phenomenal: it uses popular (not technical) terminology, expressed in terms of the cultures of the time, and deals with appearances of things and events rather than with any scientific theorizing. The creation days were "pictorial-revelatory," not literal: they were revealed to Moses in six visions or six days. Ramm advocates "progressive creationism": God created major types by direct fiat, but accomplished this over long ages (similar to Day-Age creationism, but creative acts not limited to six). He rejects dogmatic atheistic, materialistic evolution, but realizes there is irrefutable evidence of development of life. Flood was regional, not global. Thoroughly researched, well-written, and persuasively argued. Good classified bibliog.; index.

1348 Rand, Howard B. *The Stars Declare God's Handiwork* (booklet). c1944. Merrimac, Mass.: Destiny Pub. Reprinted from *Destiny* magazine. The message of the Gospel writ in the stars: the constellations depict Bible themes. Based on **Bullinger.** The Great Pyramid also incorporates testimony of the stars. God declared the stars to be "signs" testifying to the eternal truth of Scripture. A response to anti-Christian astrology. *Destiny* advocates Anglo-Saxon-Celtic race as modern and rightful descendants of Israel – God's Chosen People.

1349 ________. *The World That Then Was* (pamphlet). 1954. Merrimac, Mass.: Destiny Pub. Reprinted from 1954 *Destiny* magazine. Concerns the highly advanced antediluvian civilization. Pre-Flood earth was covered by a Water Canopy, which created an ideal climate worldwide; the earth was watered by a mist from the ground (from Genesis). The sea level was much lower: only one-seventh of earth's surface was covered by ocean. Presents geological evidence of the original extent of land. Eden is now submerged. Rand also wrote booklet on *Flying Saucers* (1975) in which he explains that UFOs are caused by disembodied spirits trapped in the Abyss: the offspring of the Nephilim – the giants produced by the mating of the Sons of God (fallen angels) with humans.

1350 Ranganathan, Babu G. *Evolution: Science or Fiction?* (pamphlet). c1981. Vernon, Conn.: Better Life Ministries. Standard creation-science.

1351 Raphael. *The Starseed Transmissions: An Extraterrestrial Report.* 1983 [c1982]. Kansas City, Mo./Walpole, N.H.: Uni*Sun/Stillpoint. Extraterrestrial wisdom channeled by "Raphael" through Kenneth X. Carey (who wrote the book). When God created, He preordained a rhythmic cycle of expansion and contraction. The Second Coming of Christ is really the midpoint of the cycle, and will occur in A.D. 2011. The Creator is ALL; the Source. Our consciousness became the Son – Christ; they (we) have existed for billions of years seeking a suitable planet. The Fall is the gradual entrapment into physical existence, caused by lack of faith, which began in the Garden of Eden. This process will be reversed; earth-life will depart from this planet in A.D. 3011.

1352 Rapkin, Geo. *Genesis in Harmony with Itself and Science.* 1899. London: Christian Commonwealth. Advocates Gap Theory creationism, but is otherwise a strict creationist. Accepts **Ussher**'s chronology for the re-creation, and emphasizes effects of the Flood. Discusses existence of Pre-Adamites at length: ". . . we know races existed prior to Adam." The Nephilim – giants – were an aboriginal pre-Adamite race.

1353 Rasmussen, B. Gorm. *Ancient Thought Models* (booklet). 1973. Hokksund, Norway: Eiker Trykkeri & Papirhandel. "Paper for research of Biology, Geology, Nuclear Natural Science, and Cosmic Energy in regard to space, time, and life in the history of the earth." Scripture as science, including – and especially – creationism.

1354 ________. *Bibelen og Evolutionslaeren.* Norway. (Mentioned in Rasmussen 1973.)

1355 Rasmussen, Stanley A. *A Determination of the Time of the Flood from the Geologic Ages of River Deltas.* 1984. M.S. thesis: Institute for Creation Research (El Cajon, Calif.). ICR geology M.S. thesis. Formation and age of river deltas. Multiplying the rate of sedimentary depositions times the total deposit gives age of delta. No deltas are older than several thousand years. Rasmussen disputes uniformitarian suggestion that total volume is actually greater – that it is pushed down into the crust (subsidence); demonstrates that this is untenable. Discusses isostasy. Deltas have same superstructure as if there was added weight, because of deeper subsidence of lower levels. Rasmussen joined ICR staff after completion of this thesis.

1356 ________. *Geologic Age of River*

Deltas. 1987. El Paso, Texas: Geo/Space Research Foundation. Published version of Rasmussen's 1984 thesis.

1357 Ray, John. *The Wisdom of God Manifested in the Works of Creation*. 1691. London: Samuel Smith. Three editions enlarged by Ray were pub. in 1692, 1701 and 1704. At least 12 editions were published within the next century. Ray taught at Cambridge Univ.; later began to catalogue all plant and animal species, which work earned him election to the Royal Society. His pioneering taxonomical works were influenced by the Cambridge Platonists. Ray stressed that all science served a religious end, arguing that study of nature proves God's power, wisdom and goodness. Ray was very influential in promoting the tradition and genre of "natural theology." In this work, Ray discusses the solar system, the theory of matter, geology, plants and animals, and human anatomy and physiology, demonstrating the impressive range of his scientific knowledge. "The concept of nature set forth in Ray's pages was to dominate the study of natural history for nearly two hundred years to come. Profoundly nonevolutionary in character, it was to constitute the chief obstacle to the rise of evolutionary views" (J.C. Greene 1959). A static view of the earth.

1358 ________. *Three Physico-Theological Discourses*. 1732 [1693]. London: W. Innes. "Concerning I. The primitive chaos and creation of the world. II. The general deluge, its causes and effects. III. The Dissolution of the world and future conflagration." Orig. pub. 1693 by S. Smith (London). A revised version of Ray's 1692 book *Miscellaneous Discourses Concerning the Dissolution and Changes of the World*. "Consideration of the Deluge and Creation, in an attempt to arrive at scientific theories reconcilable with the Scriptures, was the dominant feature of the *Discourses" (Dict. Sci. Biog.)*. Ray argued—contrary to many authorities of his time—that fossils ("formed stones") were organic (petrified) remains: the result of the Flood. But he also realized many fossil forms were not identical to modern organisms. This violated his static conception of earth history; he could not accept the idea of extinction, and suggested representatives of fossil forms were still living in the ocean depths. He was also troubled by the distribution of fossils, apparently sensing that the Flood explanation was inadequate. **D. Young** 1982 says Ray suggested earthquakes may have elevated sea floors to the required elevations, but this suggested the notion, unpalatable to Ray, that the earth might be older than six thousand years.

1359 Read, John G. *Fossils, Strata and Evolution: A Test of the Credibility of the Evolution Theory* (booklet). c1979. Culver City, Calif.: Scientific-Technical Presentations. Written "with assistance from" **C. Burdick.** *Read:* engineering B.S.; grad. of Northwestern Univ.'s Technological Inst.; senior engineer with Hughes Aircraft (Los Angeles). Read's Scientific-Technical Presentations produces creation-science filmstrips, other AV materials and literature. This booklet presents standard geological creation-science arguments. Emphasizes overthrusts, wrong-order strata, and catastrophic origin of fossils. Asserts that geological evidence damaging to evolutionary concepts is ignored. Advocates Flood Geology. The Flood is proven by geology as well as the Bible: "the Great Flood is one of the best attested events in history." Scientific references. Mostly photos; also charts, diagrams.

1360 Ream, Robert J. *A Christian Approach to Science and Science Teaching*. 1972. Nutley, N.J.: Presbyterian and Reformed. *Ream:* B.S.—Elizabethtown Coll.; B.D.—Faith Theol. Sem.; Ph.D.—Dropsie Univ.; teacher at Philadelphia Montgomery Christian Academy. Argues against Thomistic attempt to harmonize rationalism and revelation; says that the Bible must be given primacy. Scripture is "foundational" for science; reason must become "servant" of revelation. There is "no truth that is not Christian." Purpose of science is to discover God's plan of creation. "The necessary implication of the Bible's conception of reality is the complete absence of what is called 'chance.'" Praises Kuhn for demonstrating role of paradigms in science, citing science's rejection of Paluxy manprints, but laments his relativistic conception of truth. "The evolutionary paradigm is a prize example of a model spawned in a framework which rejects a personal Creator and sets up a blind universe operated by chance." Adam's Fall, through which sin and death entered the world, has resulted in deterioration of nature. This biblical view flatly contradicts evolution.

1361 Reed, Lucas A. *Astronomy and the Bible: The Empire of Creation Seen in the Dual Light of Science and the Word.* c1919. Mountain View, Calif.: Pacific Press. The stars and the wonders of the heavens declare the magnificence of the Creator. The Bible speaks "with authority in the realm of science." "Truth is conformity to fact... The Bible is 'the truth,' and nature consists of facts; and so there is perfect conformity between the Bible and nature." The Bible explains **Newton**'s universal gravitation, which is God's sustaining Word. Describes other scientific discoveries found in the Bible (atmospheric pressure, etc.). True science rejects evolution. Does not discuss age of universe, but implies recent creation. Describes immense distance light can travel – has traveled – in 6,000 years, but suggests universe is larger even than this. Refutes notion that the stars must inevitably die. In chapters "God's Dwelling Place, Where?" and "The Open Space in Orion," Reed defends the old Seventh-day Adventist doctrine that the Throne of God is literally inside the Orion Nebula, which is the center of the universe, quoting **E.G. White** and astronomers Sir John Herschel and E.L. Larkin. Photos, drawings.

1362 Rehwinkel, Alfred M. *The Flood: In the Light of the Bible, Geology, and Archeology.* 1951. St. Louis, Mo.: Concordia. *Rehwinkel:* (geology?) M.A. – Univ. Alberta; B.Div.; theology prof. at Concordia Theol. Sem. (St. Louis); Missouri Synod Lutheran. Flood Geology and strict young-earth creationism. Largely inspired by **G.M. Price.** The worldwide Flood is the greatest single event in earth history since Creation. "Nothing comparable with it has happened since nor will happen until the final destruction of this universe in the fire of Judgment Day. The Flood marks the end of a world of transcendent beauty, created by God as a perfect abode for man, and the beginning of a new world, a mere shadowy replica of its original glory. In all recorded history there is no other event except the Fall which has had such a revolutionary effect upon the topography and conditions of this earth... No geologist, biologist, or student of history can afford to ignore this great catastrophe." To deny it "means to question the infallibility of the Bible and that of Christ Himself." The antediluvian earth was uniformly mild, pleasant and luxuriant; the oceans were much reduced. The human population was large, widely distributed, and highly advanced. Rehwinkel, praising **Kircher,** discusses the size, construction and cargo of Noah's Ark. Relates many worldwide Flood traditions. Quotes in full the 1942 **Roskovitsky** account of the Ark's discovery by Russians. Asserts that geological deposits and fossils are proof of a worldwide Flood. Includes a chapter based on **Howorth**'s *Mammoth and the Flood.* Argues that the glacial theory doesn't make sense unless linked with the Flood. Presents the Flood as the "Prototype of the Final Judgment," and concludes by warning that the world today is as evil as in the days preceding the Flood and that the Final Judgment is imminent. Its cause was "not geological or cosmic, but ethical and moral. ... A study of the Flood would therefore be incomplete without a reference to the moral depravity of that generation which was responsible for the destruction of the earth and without some application to the world of today." References mostly scientific; biblical references largely restricted to chapter endings. Seventeen printings through 1978. Photos, drawings; index.

1363 ________. *The Age of the Earth and Chronology of the Bible.* 1967 [1966]. Adelaide, Australia: Lutheran Publishing House. 2nd ed. Orig. 1966. Written while guest lecturer at Concordia Sem. in Adelaide, Australia. Presents chronology based solely on the Bible. Shows that the three greatest events in history – man's creation, fall, and redemption – are intimately connected and occurred within a relatively short span. Relies on chronology of the Septuagint rather than **Ussher**'s. Creation probably 5556 B.C.; Flood 3324 B.C.

1364 ________. *The Wonders of Crea-*

tion: An Exploration of the Origins & Splendors of the Universe. c1974. Minneapolis: Bethany Fellowship. "Genesis is the only possible source for knowledge concerning the origin of the universe; science is limited to the here and now . . ." An unbiased study exposes evolution as "absurd, impossible, and unscientific." Standard strict creation-science arguments; also includes several poems. The enemies of Genesis are: rationalism, higher biblical criticism, Protestant neo-orthodoxy (e.g. Barth), and evolution. Includes biog. note. Index.

1365 Reid, James. *God, the Atom, and the Universe.* c1968. Grand Rapids, Mich.: Zondervan. *Reid:* systems engineer. Bible-science: scientific discoveries found in the Bible—e.g. '$E = mc^2$' in Hebrews 11:3. Old-earth, but solidly creationist. The Big Bang was Creation. "Without form and void" (Gen. 1:2) refers to the atoms at creation. The separation of waters was the formation of the galaxies. The sun was "dark" before it was 'created' on the fourth day; the plants in existence before then were different. Quotes **von Braun** a lot. Bible catastrophes: Joshua's Long Day and the Ice Ages were caused when the earth's crust shifted. Darwinism is only a hypothesis. Presents mathematical proofs that life did not arise by chance. God must have helped, though He may have used some evolution. Discusses experiments at Duke Univ. in the 1950s which proved the power of prayer. Index.

1366 _______. *Does Science Contradict the Bible?* 1971. Grand Rapids, Mich.: Zondervan. (Listed in SOR CREVO/IMS.)

1367 Rendle-Short, Arthur. *Modern Discovery and the Bible.* 1942. London: Inter-Varsity Fellowship of Evangelical Unions. *Rendle-Short:* prof. of surgery at Univ. Bristol (England). Cites scientists who professed their Christian faith. Evolution has been seized upon to attack Christianity, but Darwin believed in God. Earliest religions were monotheistic. "Purpose and plan in nature": astronomy, physics, and chemistry demonstrate that the universe must have been created. Pasteur disproved spontaneous generation, and thus evolution also. "We must either accept the Bible doctrine that God created life, or go on making improbable speculations." Natural selection is theoretically possible, but evidence for common ancestry of all life is "totally insufficient." New species and genera may have evolved, but higher taxa seem to have appeared suddenly. Cites many anti-evolutionist and anti-Darwinian scientists, especially **L. Berg.** Genesis account refers to stages of earth's evolution from cloud-covered body with universal ocean. Interprets creation of life as harmonizing with order of fossil record: Day-Age creationism. Fossil men may be pre-humans (pre-Adamites), but Adam, with a human soul, was a *de novo* creation. Also describes archeological confirmation of the Bible, and medical knowledge contained in the Bible. Index.

1368 _______. *Wonderfully Made.* 1951. London: Paternoster Press.

1369 Rendle-Short, John. *Man—Ape or Image: The Christian's Dilemma.* c1984 [1981]. San Diego: Master Books. 2nd ed. Orig. 1981. *Rendle-Short:* son of **A. Rendle-Short;** chairman of child health at Queensland Univ. (Australia). Rendle-Short was a theistic evolutionist until 1976; now he is a strict creationist (unlike his father). This book is largely a defense of young-earth creationism. The arguments, which are fairly thoughtful, are mostly biblical rather than scientific. Some creationists attack evolution by demonstrating its scientific fallacies. This approach is valid, but "to try to counter evolution by scientific argument *alone* is, I submit, futile." Presents the "Scriptural answer to evolution" here. Feels obliged to deny evolution in order to preserve literalist interpretation of the Bible. The Bible "flatly contradicts" the theory of evolution. "Therefore, if evolution is true, the Genesis narrative must be largely myth or poetry, with a spiritual, non-historic basis." Evolution thus undermines the authority of the Bible. "If evolution is true, there was no historical Fall. If man did not Fall, there is no need for a Saviour." The Bible says there was no death before the Fall. Also: "Evolution gives the scientific justification for such creeds as Marxism, Nazism, racism, and many other 'isms which exploit humanity." Index.

1370 Reno, Cora A. *Evolution: Fact or*

Theory? 1953. Chicago: Moody Press. *Reno:* B.S.—Wheaton College; M.A.—Univ. Michigan; working on Ph.D. at Berkeley. Addressed to Christian students. "In your classroom you are in an entirely foreign atmosphere, faced with the theory of evolution. You don't know which to believe—the Biblical account or the evolutionary theory." True science cannot conflict with the Bible. Shows that evolution is "a mere theory which is unsustained by scientific proof and that the facts of science do give support to the doctrine of creation." Presents, and refutes, standard evidences for evolution given in textbooks. Covers morphological similarity, fossil distribution, taxonomy, biogeography, embryology, horse series, vestigial organs, alleged evolution of man, origin of life, proposed evolutionary mechanisms. Quotes from many school texts. "In Genesis we are told that God created living things after their kind. Exactly what this means we may never know until we get to Heaven!" Noah's Ark did not contain all varieties of animals—only larger categories. Denies that belief in ancient earth is inconsistent with view of inerrancy in Genesis. "The proofs that our earth is very old are irrefutable." Declares complete absence of fossils in pre-Cambrian strata; also lack of transitional forms between major groups. Continued existence of simple life-forms contradicts evolution. Asserts that no fossil hominids are "missing links." "We can be very emphatic about the fact that there is no graded series of connecting links between man and the lower animals." "Are you not glad that you were created in the image of God instead of being some higher form of a beastlike creature?" Includes analysis of evolution coverage in eleven textbooks.

1371 ________. *Evolution on Trial.* c1970. Chicago: Moody Press. Favors Day-Age creationism, but mentions possibility of six 24-hour creation days separated by long ages (intermittent day theory). Presents standard creation-science arguments.

1372 ________. *Evolution and the Bible.* c1972. Chicago: Moody Press. A condensed version of *Evolution on Trial.* Foreword by **K. Taylor.**

1373 Reymond, Robert L. *A Christian View of Modern Science.* 1974. Philadelphia: Presbyterian and Reformed. *Reymond:* Old Testament prof. at Bob Jones Univ.; editor at Craig Press. Mostly philosophy of science and theology. The "Biblical cosmogony demands that the scientist reject all evolutionary speculations." Christians must resist those who deny the supernatural. Approves of **Dooyeweerd** and **Van Til** in this regard.

1374 Rice, John R. *Evolution or the Bible: Which?* (booklet). c1963. Murfreesboro, Tenn.: Sword of the Lord. *Rice:* evangelist; founder and editor of *Sword of the Lord.* Supports "true science," but denounces "infidel guesses in the name of science." "I am against a so-called science which is based, not on fact, but on the opinions of wicked men. I am for the Bible doctrine of creation as against the theory of men about evolution without God and contrary to the Bible." Hammers home theme that evolution and the Bible are utterly opposed. Uses standard creation-science arguments. Cites **B. Nelson, Rehwinkel, G.M. Price** approvingly. II Peter foretells of evolutionists (uniformitarians) who are "willingly ignorant" of creationism. These wicked people do not heed the message of the Flood. Also argues that the musical scale, math equations, and gravity all constitute proof of the Creator.

1375 ________. *"In the Beginning . . ."*. c1975. Murfreesboro, Tenn.: Sword of the Lord. Commentary on Genesis "With Detailed Studies on Creation Vs. Evolution, the Flood, etc." 559 pp.; much of it on evolution and creationism. Recommends and quotes **Whitcomb** and **H. Morris, B. Nelson, G.M. Price, Rehwinkel, Rimmer, Bible-Science Association.** Includes chapter "The Evolution Fantasy." Strict creationism. **Ussher** was almost correct. Standard creation-science arguments, Flood Geology, Water Canopy. "Much Evidence That the Flood in Noah's Days Accounts for All Fossils, Coal Beds, and Other Changes in Earth's Surface, for Which Scientists Would Like to Have Millions of Years." "The reason people believe in evolution is not that it is reasonable but that it is an excuse for not believing in God and direct creation." Creation was a miracle; Christianity is "a religion of miracles." If we acknowledge Jesus as Creator we can expect His return to rule on earth—a new Eden. Long refutation of the "unscriptural" Gap Theory. Insists on recent creation in six literal days. Presents scientific evidence of impossibility of evolution, the "unprovable, illogical folly and faith of infidels" (quotes from **Meldau** 1974, others). Denounces the "wicked, dishonest fakes" of ape-men presented in textbooks and museums. **Velikovsky** (though unsaved) and **D. Courville** were right about Egyptian chronology. Evolutionists scoff at the Flood, just as the wicked did prior to the Flood as described in II Peter; likewise the world will again be destroyed. Earth was created perfect; animals were all vegetarian. Discusses pre-Flood water canopy. Pre-Flood earth had advanced civilizations of many millions. Sethites intermarried with wicked Cainites (Gen. 6). Includes tale of **Newton**'s scale model of the cosmos ("Who made this?" the atheist asked; Newton replied this query demonstrates that the real cosmos must also have been created).

1376 Richards, Larry. *Science and the Bible: Can We Believe Both?* (booklet). c1973. Wheaton, Ill.: Victor Books. Opens with anecdote of a professor publicly attacking a college student's belief in the Bible. Discusses T. Kuhn; philosophy of science. Includes sections on creation/evolution, the Flood, and miracles (affirms they are real).

1377 Richards, Lawrence O. *It Couldn't Just Happen: Faith-Building Evidences for Young People.* c1987. Ft. Worth, Texas: Sweet Pub. "Consulting editors": **J.N. Moore** and **J.N. Clayton.** *Richards:* soc. psych. Ph.D.—Northwestern Univ.; has written over 100 Christian books, mostly for children. "The Lord has given us literally thousands of evidences to prove that He created and sustains the universe. . . . It's not safe to assume your kids will reject the scientific Theory of Evolution. Give them real answers. . ." Richly and profusely illustrated: color photos and drawings. Praises design of the universe, of earth, of life. Evolutionists cannot explain mysteries of earth and of life. Suggests worldwide Genesis Flood is necessary to explain geology and paleontology. Presents standard objections to evolutionist dating methods, but allows for either young- or old-earth creationism and presents various creationist theories. Rejects **Ussher**'s chronology; states that the Bible does not tell us when God created. Accuses evolutionists of wishful thinking and of unscientific reasoning. Creationism is far better science than evolution. "But in the end we do not believe in creation just because it is more likely scientifically. We believe in creation because we believe in God and because we trust what God has revealed to us in the Bible." Criticizes origin-of-life theories; presents probability arguments and many examples of design in nature. Refutes proposed "apemen." "Where did humans really come from? The answer is found in the Bible." There are only two possible approaches to origins: evolution is directly opposed to the Bible. Describes supernatural origin and accuracy of the Bible; biblical miracles and prophecy (cites **Stoner**). Recommends many creation-science books. Index.

1378 Ridderbos, Nicholas H. *Is There a Conflict Between Genesis I and Natural Science?* 1957. Grand Rapids, Mich.: William B. Eerdmans. *Ridderbos:* Old Testament prof. at Univ. Amsterdam (The Netherlands). Theological arguments rather than creation-science. Does discuss relationship of Genesis to natural science, though, and affirms its historicity. God created; materialism should be rejected. Suggests Framework Hypothesis interpretation of creationism: the days of creation were topical or symbolic rather than sequential and chronological. Allows for old earth. Stresses urgent need for Calvinists to study historical anthropology and related fields in order to decide upon such matters.

1379 Riddle, Lawson. *What in Creation?* c1979. Encino, Calif.: Enigma. Account of a fictional trial; written in dialogue form. The teacher before the court has been presenting strict creationism in class. The lawyers discover that there is a huge and compelling body of creation-science evidence. Creation-science wins! Presents a naive view of legal proceedings.

1380 Ridenour, Fritz (ed.). *Who Says God Created?* c1967. Glendale, Calif.:

G/L Regal Books. "Has Science proved the Bible obsolete?" *Ridenour:* youth editor—Gospel Light (G/L) Publications. "Writing and research consultants": David Harvey (authorship of Bible), Georgiana Walker (archeology and Bible prophecy), Jack Durkee (evolution and science), David F. Siemens (evolution and origin of man). Siemens: teacher at Los Angeles Pierce Coll. Real issue is not science vs. Bible, but naturalism and rationalism vs. supernaturalism and faith in the Christian God. Affirmation of supernatural origin of Bible. Bible contains solutions for all life's problems. Evolutionary ideas predated Darwin, but people embraced Darwin's theory because of increasing anti-Bible sentiment. Reviews standard creation-science objections to evolution. Ridenour says the Bible does not state when creation occurred, but he seems to favor recent creation. "It all comes down to this: The Bible plainly teaches that God created the heaven and the earth, but it gives no scientific specifics. It cannot be proved or disproved according to current scientific evidence." Cites **A. Standen** and **M. Adler** approvingly; recommends creation-science works. Ridenour has also written other religious books for Gospel Light.

1381 Riegle, David D. *Creation or Evolution?* c1971 [1962]. Grand Rapids, Mich.: Zondervan. 1971 ed. completely revised. Orig. 1962. *Riegle:* teacher in Danville, Ill.; Church of the Nazarene. Foreword by **J.N. Moore.** "My main criticism of [teaching of evolution] is that pupils do not get an opportunity to read materials presenting the Bible story of Creation." They lack information with which to oppose evolution, and their faith in the Bible is shaken. "The main purpose of this book is to present ... some of the arguments for anti-Biblical evolution [i.e. macro-evolution], followed by the Bible version of the same material." "When men reject the Bible story of creation, they are faced with the near-impossible task of explaining the existence of everything known to man." Design of the universe and of earth demands creation. Evolution cannot account for origin of life, gaps between major kinds, and development of complex organs. Riegle concedes some arguments can be made for the animal ancestry of man, though the idea is "distasteful" to many. But the differences between man and ape are too great for evolution to be possible. On the other hand: "One does not need much imagination to grasp the story of the Creation as related by Moses." Descriptions of cave-men in books are primarily attempts to promote evolution and deny biblical creation. Riegle considers arguments for Day-Age creationism, but favors Gap Theory creationism. He suggests that the re-creation following the gap was a literal six-day creation. The earth may perhaps be billions of years old, though evolutionists base their dating methods on shaky assumptions which may be wrong. But creation was instantaneous. "Let us accept the Bible story of Creation, in its entirety, and have faith enough in the God of Creation to believe there is purpose in the things which we do not understand." Drawings; bibliog.

1382 Riem, Johannes. *Die Sintflut in Sage und Wissenschaft.* 1925. Hamburg, Ger.: Das Rauhe Haus. In German. ("The Genesis Flood in Legend and Science.") "Among all traditions there is none so general, so widespread on earth, and so apt to show what may develop from the same material according to the varying spiritual character of a people as the flood tradition. Lengthy and thorough conversations with Dr. Kunike have convinced me of the evident correctness of his position that the fact of the great Deluge is granted because at the basis of all myths, particularly nature myths, there is a real fact, but during a subsequent period the material was given its present mythical character and form." Discusses Flood and rainbow traditions around the world. (Quoted in **Rehwinkel** 1951.)

1383 Riemensnyder, J.B. *The Six Days of Creation; The Fall; The Deluge.* 1886. Philadelphia: Lutheran Publication Society. (Cited in **Ramm** 1954.)

1384 Riffert, George R. *Great Pyramid Proof of God.* 1932. Destiny Publishers. 10 eds. by 1952. A popular version of **Davidson**'s pyramid theory and Anglo-Israelism. "The popularity of evolution as an anti-Christian explanation for the universe *is not due to anything that science has yet discovered in its favor,* but rather to the decline of Christianity in the life of the people, the weakening of confidence in the Bible as God's Book of Truth . . ." (quoted in **Destiny Pub.** 1967). M. Gardner says the first ed. predicted Armageddon and the Second Coming would begin on Sept. 16, 1936; the world would seek to destroy the divinely-protected Anglo-Saxon race (the "true" Israel). The 10th ed. says "By 1953 the present Babylon-Beast-Gentile type of Civilization, the Capitalistic System of Money profits by exploitation and usury, the Armageddon Conflict, the Resurrection and Translation of God's spiritual Israel preparatory to their administrative service in the New Social or Economic Order, the overthrow of dictatorships, the regeneration and transformation of the Anglo-Saxon Nations into the world-wide Kingdom of God, and literal return of Jesus Christ as King of Kings to prepare and perpetuate the Millennial Age, will have come to pass."

1385 Rifkin, Jeremy. *The Emerging Order: God in the Age of Scarcity.* c1979. New York: G. Putnam's Sons. Written by Ted Howard. *Rifkin:* influential Washington lobbyist; heads Foundation on Economic Trends. Here, Rifkin foresees —and applauds—a "second Protestant Reformation" in the U.S. led by fundamentalists, evangelicals and charismatics. They are creating a new theological as well as economic order. The old order is based on maximization of profits and exploitation of resources (the Genesis "dominion mandate"); the new order is based on "stewardship" (also a biblical term) of resources, and will result in a "steady-state" society. Rifkin shows that there is a massive religious storm brewing; he describes the vast communications network of the evangelicals and fundamentalists, and the impressive infrastructures already in place. Rifkin sees religious views (and also scientific theories) as propaganda—as means of justifying economic paradigms, which are the fundamental realities. He supports this new Protestant Reformation as a means of implementing his "steady-state" society. (Fundamentalists, however, advocate growth and reject Rifkin's "steady-state" notions.)

1386 ________. *Algeny.* 1983. New York: Viking Press. "In collaboration with" Nicanor Perlas. A warning against bioengineering and genetic manipulation. We are shedding the mechanistic Darwinist paradigm used to justify the Industrial Age in favor of a new cybernetic concept of nature in order to justify "algeny"—manipulative bio-technology. Rifkin argues that scientific theories are reflections of and rationalizations for economic worldviews and for the dominant, underlying attitudes concerning man's relationship to nature. They are socially determined ideological constructs used by the ruling elites. Darwinism has outlived its usefulness, and scientists are abandoning it in favor of the emerging cybernetic paradigm. The theory of evolution "provides a convenient rationale for avoiding responsibility for our activities." "What Darwin discovered was not so much the truths of nature as the operating assumptions of the industrial order, which he then proceeded to project onto nature." Rifkin quotes **Grassé, Himmelfarb,** Colin Patterson, **Kerkut, Macbeth.** Non-creationists are often quoted misleadingly or misrepresented. Also quotes **Wysong, Gish, Wilder-Smith** and **G. Parker** as scientific authorities critical of Darwinism—without mentioning that they are creationists. Urges a return to non-manipulative coexistence with all of nature. Persuasively written, but sensationalized. Index.

1387 Riley, William Bell. *The Finality of the Higher Criticism; or, the Theory of Evolution and False Theology.* c1909. Priv. pub. (Minneapolis?).

1388 ________. *The Menace of Modernism.* c1917. New York: Christian Alliance. Includes denunciations of evolution. "Darwinism is a theory, not a demonstration; a speculation, not a science; and yet it is confessedly the foundation of so-called modern history, philosophy and science . . ." It leads students away from the Bible. Universities are saturated with destructive German theories and philosophies. "I believe in Science, but not in an anti-scriptural one! I believe in the college, but not if it deny my Christ. . . . When Christ is no longer worshipped, men will sink back into cannibanlism . . ." Objects to mandatory vaccination, but asserts anti-biblical teaching is far worse.

1389 ________. *Inspiration or Evolution*. c1926. Cleveland, Ohio: Union Gospel Press. 2nd ed. *Riley:* pastor of First Baptist Church in Minneapolis; founded World's Christian Fundamentals Association in 1919; founder and president of Northwestern Bible College. Vigorous anti-evolution argument by one of the most prominent fundamentalist leaders of the 1920s. A "call to arms" to protect American democracy and Christian faith against the destructive doctrines of modernism and evolution. Riley emphatically declares that evolution promotes atheism and anarchy. Evolution is not science, and it destroys Christian beliefs of students. "Science is now the subtle word of Satanic employment." "The Product of Evolution Theory is Bestial Bolshevism!" The Soviets actively seek to control U.S. education by planting the evil seed of evolution. Darwin's book is based on speculative theories rather than facts (he wrote "we may suppose" over 800 times). Quotes academics and scientists who oppose evolution or Darwinism. Quotes historian Carlyle's denunciation of Darwinism as the "Gospel of Dirt." Notion of survival of the fittest leads to militarism, the bloodshed brought on by Nietzsche and Treitschke's promotion of race-superiority. Armageddon and the Great Tribulation will result if Darwinism triumphs. Assails modernism and liberal theology. Strongly premillennialist. Eloquent defense of divine origin of Bible—vehemently denies it "evolved." Study questions (like a catechism) for each chapter.

1390 ________. *Darwinism; or, Is Man a Developed Monkey?* (pamphlet). 1929(?). Minneapolis(?).

1391 ________. *Daniel versus Darwinism*. N.d. Minneapolis: H.M. Hall.

1392 ________**, and Harry Rimmer.** *A Debate: Resolved, That the Creation Days in Genesis Were Aeons, Not Solar Days*. 1929. Minneapolis. "W.B. Riley for the Affirmative; Harry Rimmer for the Negative." Debate held at Riley's Northwestern Bible College in 1929. Though Riley and Rimmer were both vehemently opposed to evolution, Riley advocated Day-Age creationism while Rimmer supported the Gap Theory (with a re-creation in six literal days). Here, they engage in a friendly but serious debate.

1393 ________. *The Creation Days of Genesis: A Profound Debate Between Dr. W.B. Riley and Dr. Harry B. Rimmer*. c1974. Colton, Calif.: World Prophetic Ministry. (Listed in Ehlert files. May be new edition or version of Riley and Rimmer 1929.)

1394 ________**, and C. Smith.** *Should Evolution Be Taught in Tax Supported Schools?* 1928. Minneapolis. A debate.

1395 Rimmer, Harry. *Modern Science, Noah's Ark and the Deluge* (pamphlet). 1925. Los Angeles: Research Science Bureau. Later incorporated as chapters in Rimmer 1936. *Rimmer:* Presbyterian minister; attended Whittier College and Bible Inst. Los Angeles; set up one-man Research Science Bureau in L.A. to promote Bible-science (he fancied himself a research scientist); later became field sec'y of **Riley**'s World's Christian Fundamentals Assoc. A delightfully flamboyant lecturer and debater as well as popular writer; debating evolutionists was child's play for him.

1396 ________. *Monkeyshines: Fakes, Fables, Facts Concerning Evolution* (pamphlet). c1926. Los Angeles: Research Science Bureau.

1397 ________. *Theories of Evolution and the Facts of Paleontology* (pamphlet). 1926. Duluth, Minn.: Research Science Bureau. Also the title of a chapter in Rimmer's 1966 [1935] book *The Theory of Evolution and the Facts of Science*. This pamphlet is presumably the same as the book chapter. Apparently the book chapters were issued originally as pamphlets.

1398 ________. *The Facts of Biology and the Theories of Evolution* (pamphlet). Undated [1929]. Los Angeles: Research Science Bureau. Presumably same as chapter of the same title in Rimmer 1966 [1935].

1399 ________. *Modern Science and the First Day of Genesis* (pamphlet). Undated [1929]. Los Angeles: Research Science Bureau. Possibly the same as the chapter "Modern Science and the First Day of Creation" in Rimmer's 1941 [1937] book *Modern Science and the Genesis Record*.

1400 _______. *The Theories of Evolution and the Facts of Human Antiquity* (pamphlet). Undated [1929]. Los Angeles: Research Science Bureau. Presumably the same as the chapter of the same title in Rimmer 1966 [1935].

1401 _______. *Embryology and the Recapitulation Theory* (pamphlet). c1935. Duluth, Minn.: Research Science Bureau. Presumably the same as the chapter of the same title in Rimmer 1966 [1935].

1402 _______. *The Theory of Evolution and the Facts of Science.* 1966 [c1935]. Duluth, Minn.: Research Science Bureau/Berne, Ind.: Berne Witness (co-pub.). 10th ed. Also pub. 1951 and 1966 by William B. Eerdmans (Grand Rapids, Mich.). Orig. 1935 (Research Science Bureau). The topic of origins is philosophy, not science. "The last twenty-four months in scientific research have not only failed to demonstrate the fact of organic evolution, but have actually produced new facts of science that render the theory wholly untenable." Rimmer's confident, breezy style was very effective and popular. He appeals to scientific evidence and arguments, throwing in a lot of impressive scientific terms and references, managing to sound quite authoritative and knowledgeable, but his prose is simple and easy to understand, and is sprinkled with anecdotes and humor. One of the four chapters is a refutation of the "recapitulation theory"; Rimmer accuses scientists of deliberately misleading students into believing that "ontogeny recapitulates phylogeny" in order to promote evolution. Chapter on paleontology emphasizes the "absolute fixity" of many species throughout the fossil record, asserts that evidence shows degeneration rather than development, and ridicules the horse series as evidence of evolution. The fourth and final chapter refutes proposed human ancestors. Rimmer declares [erroneously] that Prof. Newman confronted **Bryan** at the Scopes Trial with the evidence of Nebraska Man (later admitted to be a pig's tooth). Includes correspondence to Rimmer from Dubois' son concerning Java Man.

1403 _______. *The Harmony of Science and Scripture.* c1936. Grand Rapids, Mich.: William B. Eerdmans. Eleven eds. up to 1945. Orig. pub. as pamphlet series 1925–34 by Glendale Pub. (Glendale, Calif.). Lively writing style; uses a lot of folksy dialogue, homespun analogies and jokes. Boasts that he learned "double-jointed, twelve cylinder, knee-action" scientific vocabulary at medical school (one term at a homeopathic college) and that he can out-argue any scientist. Argues aggressively for the scientific inerrancy of the Bible and for the superiority of Bible-science. Uses scientific arguments but no specific references. "If the words of the Great Book are not in full accord with all known *fact,* then we have been mistaken in calling it the Word of God. We use the word 'fact' in its accepted meaning, as distinct from theory and unproved hypotheses. Our main objection to the pseudo-scientific philosophy of this present generation is that it manifests an amazing willingness to surrender the eternal verity of God's revelation for the unfounded theories propounded by men who are utterly without ability to prove their wild imaginings. And science, we must repeat, is a correlated body of absolute knowledge." The Bible does not use scientific language, but it contains no scientific error, and in scores of cases it has "anticipated the discoveries of modern science." Has a chapter "Modern Science in an Ancient Book." Other chapters: "Modern Science, Jonah, and the Whale" (the sea monster was supernatural; anyway, some fish can still swallow humans); "Modern Science and the Deluge" (Rimmer stresses the literal truth of the Flood and Noah's Ark, pointing to worldwide Flood traditions and Woolley's excavations as proof); and the "Long Day of Joshua" (citing **Totten,** he declares science has proved a missing day, but he also uses **Maunder**'s argument that Joshua's command was for the sun to "be silent"—he wanted relief from the sun's heat). Cites Caltech Nobelist R. Millikan frequently as a believing scientist.

1404 _______. *Modern Science and the Genesis Record.* 1941 [c1937]. Berne, Indiana: Berne Witness Co. 11th ed. Orig. 1937. Also published by Eerdmans (Grand Rapids, Mich.). Some scientists quoted, but mostly a vigorous defense of Bible-science, concentrating on the argument from design. Presented as commentaries on each of the creation days of Genesis; advocates Bible-science interpretations of meteorology (Day Two), oceanography, geology and botany (Day Three), astronomy (Day Four), zoology (Day Five), and human physiology and anthropology. Chapter on Day One largely concerned with light: Rimmer explains how light could exist before the sun was created. In the first chapter ("The Prologue to Genesis"), Rimmer defends Gap Theory creationism: "It is apparent . . . that the first verse of Genesis refers to a work of origination which antedates the Creative Week with which Moses deals." Rimmer declares: "even if we view the first chapter of Genesis as a theory only, it is at least a reasonable theory and may be scientifically adopted as a working hypothesis. On the other hand, the alternate theory, that of evolution, is utterly discredited scientifically. We do not advance the first chapter of Genesis as a theory, however, but boldly contend that it is a scientific record of absolute facts." (This book and Rimmer 1936 are part of a six-part series by Rimmer on apologetics and Christian evidences—the John Laurence Frost Memorial Library. Other volumes deal with biblical archeology, internal evidence of the Bible, and the deity of Christ.)

1405 ______. *That Lawsuit Against the Bible.* 1956 [1940]. Grand Rapids, Mich.: William B. Eerdmans. 9th ed. pub. 1945. Orig. 1940. In 1939 Rimmer offered $1000 for proof of any scientific error in the Bible. William Floyd sued to collect the money, but Judge B. Shalleck of New York ruled in favor of Rimmer. This book is "the record of the trial which ended in legally establishing the position of all who hold that the Word of God is inerrant." **Bennet** was Rimmer's lawyer. One of plaintiff's claims was that the Bible declares creation occurred in six days, whereas the universe has gradually evolved over millions of years. Rimmer refuted this in part by arguing that the original creation may have occurred ages before the six-day (re-)creation (the Gap Theory). He also attacks geological dating methods. Rimmer refuted the claim of a contradiction in Noah's being told to bring two, then seven, pairs of each clean animal by showing that the clean animals were used for sacrifices on board the Ark.

1406 ______. *Lot's Wife and the Science of Physics.* 1947. Grand Rapids, Mich.: William B. Eerdmans. Title essay suggests Lot's wife became petrified from volcanic emissions. Most of the rest of the book is a creationist interpretation of biology and geology.

1407 Ritchie, Archibald Tucker. *The Dynamical Theory of the Formation of the Earth.* 1855. 3 eds. Orig. 1855. Described by Millhauser as an elaborate version of a strange variant of Day-Age creationism.

1408 Ritland, Richard M. *A Search for Meaning in Nature: A New Look at Creation and Evolution.* 1970 [1966]. Mountain View, Calif.: Pacific Press. Orig. 1966; titled *Meaning in Nature* (Dept. of Educ., General Conference of Seventh-day Adventists). *Ritland:* Seventh-day Adventist; director of Geoscience Research Inst., the creation-science group now at Loma Linda Univ., then at Andrews Univ. (Berrien Springs, Mich.). Careful, reserved exposition of creation-science arguments. The Geoscience Research Inst. is strongly creationist but is more skeptical and scientifically rigorous than groups such as ICR. GRI exists primarily to provide Seventh-day Adventist teachers with expert scientific advice rather than to evangelize to the public. It criticizes the poor science and careless methods of other creation-science groups freely. This book is designed for school use. Discusses presuppositional nature of both evolutionism and creationism, design in nature, improbability of chance origin of life, the uncertain nature of the fossil record (especially hominids), the evidence for catastrophism, and the speculative nature of various "proofs" of evolution. States that not all fossils resulted from the Flood; doubts authenticity of Paluxy manprints. Many scientific references; also biblical quotes. **E.G. White** cited frequently; also BSCS textbooks. Photos.

1409 Robbins, Dorothy E. Kreiss. *Can the Redwoods Date the Flood?* (pamphlet). 1984. El Cajon, Calif.: Institute for Creation Research. ICR "Impact" series #134. **Robbins:** writer; lives in Redding, Calif. The oldest redwood trees are the same age as the Flood; ca. 2000 B.C. Q.E.D.

1410 Roberts, Jane. *Dreams, "Evolution," and Value Fulfillment (A Seth Book).* 1986. New York: Prentice Hall. 2 vols. *Roberts:* Elmira, N.Y. writer (d. 1984). Robert F. Butts, Roberts' husband, transcribed Roberts' Seth sessions, and wrote the introductory essays and notes. Seth is an "energy personality no longer focused in physical form" who speaks through Roberts in regular trance sessions. Seth presents knowledge of the universe from his past lives. This book (one of many Seth books) includes discussions of two creation-science books Roberts had read. Seth says both evolution and creation-science are too limited, and offers an amalgam. Roberts/Seth was obviously swayed by many of the creationist arguments, such as the sudden appearance of 'kinds' and the Flood, but she doesn't accept recent creation or (of course) fundamentalist creationism's denial of reincarnation.

1411 Robertson, Pat. *Answers to 200 of Life's Most Probing Questions.* c1984. Nashville/New York: Thomas Nelson. *Robertson:* M.Div.—N.Y. Theol. Sem.; J.D.—Yale Law; president of Christian Broadcasting Network (Virginia Beach) and host of its *700 Club* T.V. show; founder and chancellor of CBN Univ.; Southern Baptist minister; 1988 presidential candidate. Two hundred questions from a CBN-commissioned Gallup Poll. Includes "Does the Bible teach evolution?" and related questions. The Bible teaches creation, answers Robertson, not random evolution. Evolution "eliminates belief in God..." Robertson asserts the evidence shows "Creation in ascending order." The origin of life cannot be known apart from the Bible or other revelation (Robertson, a charismatic, often receives "words of wisdom" directly from the Holy Spirit). There is no case of any crossover between class or phylum, he points out—even the mule is sterile. On the question of the origin of man, he states that Adam and Eve are real. So is the Flood. The races of man are derived from Noah's sons. Index. Robertson has hosted many creationists (both old- and young-earth) on his *700 Club,* and is a strong supporter of creation-science.

1412 Robinson, Jim and (Rowena) Darlene Robinson. *Children's Travel Guide & Activities Book.* c1981. San Diego: Creation-Life. The Robinsons (husband and wife) are former editors and writers of the BSA *Children's Science Readers* series (see **Bartz**) from 1977–1979. (Kevin and Laurie, their children, are also listed on the title page.) They now run the Creation Center of Colorado (Arvada) and publish *Creation Readers* for grades 1–8 on biblical and scientific creationism. Coloring-book format. Covers Colorado, Utah, New Mexico, presenting "the creation view of the Bible as various national parks, monuments, and special places are featured."

1413 Rodd, Thomas. *A Defence of the Veracity of Moses in His Records of the Creation and General Deluge.* 1820. London: T. Rodd. Published under pseudonym "Philobiblos." Millhauser says this work was among those expressing general resentment against all geology for being infidel.

1414 Roget, Peter. *Animal and Vegetable Physiology.* 1834. London: Pickering. One of the Bridgewater Treatises.

1415 Rosenberg, Alfred. *Myth of the Twentieth Century.* 1982 [1930]. Torrance, Calif.: Noontide Press. Orig. 1930; German. *Rosenberg:* official Nazi philosopher; editor of Nazi paper *Der Völkische Beobachter.* Inspired by Meister Eckhart, Houston Stewart Chamberlain's racist theories; also by Wagner's romanticism and Aryanism. "The Old Testament of the Nazi religion is [**Hitler**'s] *Mein Kampf,*" says P. Viereck; its New Testament is Rosenberg's *Myth.* Outlined in 1917, completed 1925, first pub. 1930. 142 eds. by 1938. By "myth" Rosenberg means transcendent reality, necessary faith. "This is the mission of our century: out of the new Life-myth to create a new human type." A racial interpretation of history. God created man as separate races—not as individuals or mankind as a whole. Only the race has soul, and no two have the same. The higher races must rule over the lower, not interbreed with them. Cross-breeding destroys the literally divine combination of physical heredity and spirit. Organic metaphor of the race and the State: Nazis must repurify the race soul by ruthlessly eliminating alien non-Aryan elements as cancer is cut from diseased body. Lost continent of Atlantis was origin of the brilliant prehistoric Nordic culture. Migrating Atlanteans gave world its Aryan ruling castes. Rosenberg proves Egyptian, Libyan and Indian rulers were pure Nordics. Atlantean master-race spread in four waves: over North Africa; the Indo-Aryan migration to Persia and India, and the Doric Greeks and Latins; the Teutonic migration; and the recent colonizing by Germanic western Europe. Jesus was also a pure Nordic, untainted by Jewish blood. Rosenberg advocated a "positive" Christianity (inspired by pantheistic pagan nature-worship and Teutonic gods) but despised the Church's "negative" Christianity, asserting it was corrupted by Paul, the Etruscans and the Jews. In his mystic conception of race the Germans are the Chosen People and Hitler the Messiah sent by God. All the past empires were weakened by race-mixing. Rosenberg calls for new Teutonic Order, modeled after the Catholic Jesuit hierarchy. Contrasts noble German idealism with Jewish materialism. Urges that German women become state-subsidized breeding-machines. Rosenberg, with Hess, was a member of the mystic Thule Society (predecessor of his Nordic Societies) before meeting Hitler; Thule is the legendary Nordic island. Rosenberg was also an extremely influential pamphleteer who spread the *Protocols of the Elders of Zion* and other anti-Jewish propaganda; his influence on Hitler's *Mein Kampf* was enormous.

1416 Rosevear, D[avid] T. *Genealogies and Early Man* (booklet). Undated. England: Evolution Protest Movement. EPM pamphlet #219. *Rosevear:* organometallic chemistry Ph.D.—Univ. Bristol; lecturer at Portsmouth Polytechnic; chairman of Creation Science Movement (formerly Evolution Protest Movement); Baptist. Rosevear also wrote EPM pamphlet #214: *Enzymes.*

1417 Rosin, Jacob. *The Predestined World.* 1976. New York/Hollywood: Vantage Press. *Rosin:* director of Montrose Chemical Co. 1946–73; now lives in Israel; presumably Jewish. Part I: exposes the contradictions and paradoxes of evolutionary theory. Part II: "Vitachemistry." The genetic pool is a big "vitachemical" molecule. Viruses and amebas "commit suicide" in order to obey the biblical command to multiply. Argues that the synchronized features of a bee colony had to have appeared simultaneously—couldn't have evolved gradually or piece-meal. Regarding the alleged transformation of one species into another, Rosin asks: Why would a species commit "treason" to its purpose? "Programmed World" chapter: evolution is unavoidable; it is a progressive chemical process which is identical throughout the universe. But it does not have a single goal, as in orthogenesis; it is a programmed *process.* Dinosaurs may be in existence now on other planets, propelled by the same vitachemistry process. Eventually we will evolve into *Homo sempervirens,* who may also already exist on other planets. Rosin first proposed *Homo sempervirens* in a previous work, *In God's Image.*

1418 Roskovitsky, Vladimir. *Noah's Ark Found?* (tract). Undated? (North Syracuse, N.Y.): Book Fellowship. (Listed in SOR CREVO/IMS.) (Same tract as **Darby?**) Roskovitsky is the Russian flyer who saw Noah's Ark from his plane during World War One. Czarist troops were then dispatched to examine the Ark, the story goes, but this irrefutable proof of the Bible was deliberately destroyed when the communists seized power. Roskovitsky, it is claimed, later sold Bibles in the U.S. His account appeared in *New Eden* magazine ca. 1939; **Cummings** 1982 claims there was a 1920 version also. **Rehwinkel** 1951 reprints a 1942 version from the Reformed Church *Banner.* Many creation-scientists dispute the whole story as a wishful fabrication. (In fact the author of the *New Eden* account later admitted it was almost entirely fiction.)

1419 Rostand, Jean. *Evolution.* 1962 [1960]. London: Prentice-Hall International. Orig. 1960; French. Solidly evolutionist, but cited by creationists for various evolution-as-hypothesis statements. Includes useful section on French ideas and opinions regarding "transformism"; discusses several French anti-Darwinists. Expresses dissatisfaction with neo-Darwinian explanations, which Rostand feels are inadequate. He does not think that mutations, even in cooperation with natural selection, can account for evolution. Mentions the eye and the brain as objections—impossible for them to have evolved by neo-Darwinian means. Rejects Lamarckism—a "fairy tale for grown-ups"—as a solution. Suggests collective germinal variations, and the genesis of types before species.

1420 Rowe, Ed. *New Age Globalism: Humanist Agenda for Building a New World Without God.* 1985. Herndon, Va.: Growth Publishing. Foreword by **T. LaHaye.** *Rowe:* president of Christian Mandate for America, which is affiliated with **D.J. Kennedy**'s Coral Ridge Presbyterian Church (Ft. Lauderdale, Fla.). Includes a chapter "New Age Globalism and Evolution," in which the only "evolutionist" views acknowledged are (nonscientific) Utopian New Age notions of the transformation of human development and morality. Rest of the book deals with the ACLU, the UN, education, values, the family, etc. Reprints the *Humanist Manifestoes.*

1421 Rue, Hazel May. *Bomby, the Bombardier Beetle.* 1984. El Cajon, Calif.: Institute for Creation Research. **Richard Bliss:** Writing Project Director (headed the team of writers producing various children's creationist books). Sandy Thornton: illustrator. Illustrated children's book about the bombardier beetle: a favorite creation-science example. Explains that its chemical defense system is too complex to have arisen by chance: no part of the integrated system could have evolved without the prior existence of all the other parts. (The bombardier beetle sprays its enemies with hot bursts which are propelled by mixing two chemicals in a special combustion chamber.)

1422 Ruffini, Ernesto. *The Theory of Evolution Judged by Reason and by Faith.* 1959. New York. Orig. pub. in Rome; in Italian. **Cardinal Ruffini:** former prof. of Scripture in Propaganda University, Rome; later member of Pontifical Biblical Commission. Quoted and discussed in **O'Connell** 1969, who says Ruffini "gives evidence from the Old and New Testament, from the writings of the Fathers, the Schoolmen and the theologians of the Church and from the various Papal Encyclicals on the subject to show that the theory that man's body was evolved from a lower animal is incompatible with the teaching of the Catholic Church properly understood." Examines and refutes all the arguments for evolution made by Darwin and his followers. Shows there is no proof of common descent, nor even of evolution of any species from another. Alleged examples (horse series, *Archaeopteryx*) do not demonstrate genuine transitions. Paleontology furnishes no transitional forms. It shows, rather, that the invertebrates appeared suddenly, all together. Vertebrates appeared successively, but also suddenly, in sharply defined groups. If it is admitted God created man's soul, why not allow that He created man's body directly from the dust, instead of bestowing a soul on evolved beasts?

1423 Rundus, Onesimus J. *Created or Evolved?* (pamphlet). Undated. Rosemead, Calif.: Narramore Christian Foundation. ". . . I sincerely believe evolution is a disgrace to science." Darwin uses the phrase "we may suppose" over 800 times: "This is hardly scientific." "If we think of ourselves as nothing more than the chance product of some blind, unintelligent force, if we trace our pedigree to a purely material origin, we shall never rise very high in moral character." Agrees with H.G. Wells (in his strongly pro-evolution *Outline of History*) that if evolution is true then there was no Adam and Eve, no Fall, etc.—and the whole basis of Christian morality collapses.

1424 Rusch, Wilbert H. *The Argument: Creationism vs. Evolutionism.* 1984. Norcross, Ga.: Creation Research Society Books. CRS Monograph No. 3. *Rusch:* biology M.S.—Univ. Mich.; emer. prof. of biology, geology, science and math at Concordia College (Mich.); Missouri Synod Lutheran; president of Creation Research Society. Standard creation-science arguments. Discusses history of evolutionary thought, fossil evidence, young-earth evidences. Stresses that limited variation has occurred within created 'kinds.' Endorses **Gosse**'s 'omphalos' argument: "The writer, as a creationist, would postulate that the result of a creation was a 'breaking into the time cycle,' with everything suddenly a going concern. This would imply that the earth looked as if it had been around a long time..."—though in fact it was young. Adam and Eve were a genetic mixture of all human types now in existence. Argues against exclusive teaching of evolution, especially in public schools, where attendance is compulsory: "this whole matter of origins, including which theory the student is to accept, should be a matter of choice on the part of the students and teachers, rather than a one-sided brainwashing." Includes list of critical quotes by evolutionists; directory of creationist organizations; bibliog. (The copy referred to here is a priv. printed edition, c1983.)

1425 Rushdoony, Rousas J. *The Messianic Character of American Education: Studies in the History of the Philosophy of Education.* 1963. Phillipsburg, N.J.: Presbyterian and Reformed. *Rushdoony:* M.A.—Univ. Calif.; Orthodox Presbyterian minister; missionary among Paiute and Shoshone; leading theoretician of the Christian Reconstruction movement and president of Chalcedon Foundation, the leading Reconstructionist center. Stresses that education is inevitably religious and inevitably indoctrination. The Fascists used it to indoctrinate; so do Marxists and democratic statist governments. Rushdoony and other "Christian Reconstructionists" emphasize the "presuppositional" nature of all knowledge. Since knowledge, and all human institutions, are built from such presuppositions, they should be based on the absolute revealed supernatural authority of God in the Bible rather than on sources which presuppose the autonomy of man. Rushdoony is implacably opposed to the anti-biblical heresy of democracy. He advocates a strict theocracy based explicitly on biblical (Old Testament) law. Rushdoony was instrumental in getting **Whitcomb** and **Morris**'s *Genesis Flood* published by C. Craig's strongly Calvinist press, Presbyterian and Reformed.

1426 ________. *The Mythology of Science.* 1967. Nutley, N.J.: Craig Press. Criticizes the arrogance of science; of man playing God. Some scientific references; largely philosophical in approach. "Evolution requires chance, whereas science rests on absolutely determined factors and causality. The doctrine of evolution is thus basically hostile to science. Again, evolution is a theory which is radically hostile to biblical religion. The Bible clearly asserts that God created heaven and eaarth ... in six days. If this statement be allegorized or interpreted away, no meaning stands in Scripture." Evolution is a "cultural myth," seized upon in order to avoid God and because it promises the benefits of religion with none of its responsibilities. "The convincing thing about evolution is not that it proves man's origins or even gives anything resembling a possible theory but that it dispenses with God." Rushdoony insists that "education apart from God is enslavement"; evolutionists strive to remake man according to their utopian notions, whereas Christians acknowledge human limitations and defer to the absolute standards of God. Evolutionists seek total control over man: it is an "inescapable fact that evolutionary thinking *requires* totalitarianism. If the education of a people is dedicated to teaching evolution, it will also teach socialism or communism." Criticizes attempts to control and create life. Relies heavily on **Van Til.** Evolutionists are "parasites" who are "living off the unearned capital of Christian civilization, on the impetus, law, and order of centuries of Christianity." They are destroying their host, Christianity, as well as themselves. "If they were faithful to their philosophy, these scientists could have no science, because they would have to say that the world is a world of brute factuality, without meaning, purpose, causality, or law." Includes separate, critical reviews of **Lever** 1958 and **Mixter.** Index.

1427 ________. *The Necessity for Creationism* (pamphlet). 1980 [1967]. Minneapolis: Bible-Science Association. Reprint of a chapter from Rushdoony's 1967 *Mythology of Science.* Strict creationism. Widely reprinted and distributed. Appeared in *Creation Soc. Sci. & Hum. Q.* 1980. **J.G. Vos** cites 1966 "script, tape and disc" of same title by Rushdoony, pub. by Jotham Publications (Pasadena, Calif.).

1428 ________. *The Myth of Over-Population* (booklet). 1969. Phillipsburg, N.J.: Craig Press. (Cited approvingly in **H. Morris** 1137.)

1429 Rusk, Roger. *The Flood* (cassette). 1971. Unpub.(?). Audiotape cassette cited in **Dillow** 1981. *Rusk:* physics prof. at Univ. Tennessee; brother of former Sec'y. of State Dean Rusk. Rusk has appeared in creation-science films supporting the Flood model. He suggests the earth's axis was tilted, resulting in a massive tidal Flood. Stresses that this supports the biblical account. (Rusk strongly influenced David McQueen, now with ICR, when McQueen was a Univ. Tenn. student.) SOR CREVO/IMS lists a book by Roger Rusk: *Atoms, Men and the Stars* (1937; Alfred A. Knopf).

1430 Rutherford, Joseph Franklin. *Creation: The Scriptural Proof of the Creation of Things Seen and Unseen, Showing the Unfolding of Divine Plan from the Logos to the Completion of the Royal Family of Heaven and the Restoration of Man.* c1927. Brooklyn, N.Y.: International Bible Students Association—Watchtower Bible and Tract Society. Over two million copies. *Judge Rutherford:* succeeded founder Charles Taze Russell as leader of the Jehovah's Witnesses. Rutherford rejected Pyramidology, which the Witnesses had previously emphasized, saying the Great Pyramid was built by Satan to mislead the faithful. Predicted 1925 as great jubilee year.

1431 Rybka, Theodore W. *Consequences of Time Dependent Nuclear Decay Indices on Half Lives* (pamphlet). 1982. El Cajon, Calif.: Institute for Creation Research. ICR "Impact" series #106. *Rybka:* Ph.D.—Univ. Oklahoma; part-time physics prof. at ICR; also works for General Dynamics (San Diego).

1432 Ryrie, Charles C. *We Believe in Creation* (pamphlet). 1967. Dallas: Dallas Theological Seminary. Dallas Theol. Sem. Office of Publicity publication. *Ryrie:* Th.D.—Dallas Theol. Sem.; Ph.D.—Univ. Edinburgh; former president of Philadelphia College of the Bible; theology prof. at Dallas Theol. Sem.; was on ICR Technical Advisory Board. This pamphlet presents official position of Dallas Theol. Sem. on evolution. Man is a recent creation, but the creation "days" may be long (Day-Age creationism); also allows for possibility of Gap Theory creationism. Old-earth, but resolutely anti-evolution.

1433 ________. *You Mean the Bible Teaches That?* c1974. Chicago: Moody Press. Topics covered include abortion, women's rights, activity of demons, etc. Chapter on race explains that Noah's curse on Ham's sons was against Canaanites, not Negroes. Final chapter, "Evolution," orig. appeared in *Bibliotheca Sacra* (the Dallas Theol. Sem. journal) in 1967; this is a revised version. "Creationism offers, even today, a reasonable and accurate account of the origin of man and the world." Evolution discredits the Bible and Christ; therefore it is false. Includes many scientific quotes and references (mostly old). Stresses circularity of evolutionists' arguments. Serology proves creationism by contradicting evolution (uses Nuttal's 1904 data). Endorses creation-science probability arguments. Rejects **Ussher**'s chronology; suggests Gap Theory and/or Day-Age Theory creationism instead. Man is recent—the final creation. Discusses evidence of the Flood, and notes that earth and life must have been created with appearance of age.

1434 Saarnivaara, Uuras. *Can the Bible Be Trusted?* 1983. Minneapolis: Osterhus Publishing House. Massive volume; intended as seminary reference book. Answers "Bible difficulties." Combines biblical literalism and creation-science with Velikovskian catastrophism. The section on Genesis includes Flood Geology and standard creation-science arguments; cites **Barnes, Whitelaw, Gentry.** Favors date of 3400 B.C. for Flood. Accepts **Velikovsky**'s idea of the recent origin of Venus from Jupiter. As a 'comet,' Venus caused worldwide catastrophes in 1400 B.C. (Joshua's Long Day; the end of the Post-Flood Ice Ages), in 750 B.C. (Uzziah's earthquake; the Siberian mammoth freezings), and 701 B.C. (Hezekiah's sundial). Many scientific references.

1435 Sachs, Elmer B. *The Creation Science Dictionary* (booklet). c1980. Bakersfield, Calif.: Sky Pilot Press. Funk & Wagnalls definitions versus Sachs' creation-science definitions.

1436 ________. *Who Fathered "Mother" Nature?* (booklet). Bakersfield, Calif.: Sky Pilot Press. "A Creation Science Textbook by a Professional Police Detective for God, and HIS ownership of all Creation. A former school Assembly speaker. Telling it like it is. The Sky Pilot..." Evolution equals atheism and communism. Clarence Darrow said "It is bigotry for American schools to be permitted to teach a 'One-sided' theory on the origins of life and species, to the utter exclusion of another theory." Evolution is a Trojan Horse: don't let it in the schools. Sachs calls Mother Nature "A Myth But Never a 'Mythess.'"

1437 Sage, Bengt. *Noah and Human Etymology* (pamphlet). 1980. El Cajon, Calif.: Institute for Creation Research. ICR "Impact" series #83. *Sage:* Australian businessman; born in Sweden. Words related to Noah and the Flood are similar in languages all around the world, and can be traced to the post-Ararat migrations following the Flood.

1438 Saidookhail, Mohammed Ayub Khan. *The Missing Link: An Antithesis* (booklet). 1971(?). Sialkot, Pakistan: Saidookhail Traders. *Saidookhail:* presumably Islamic. Quotes from the Koran, also from Genesis. "Since long, my mind has been agitating to write something relating to 'The Missing Link of Man' because I could not relish the idea that my ancestors were apes." "The man [sic] got on earth by creation and not by evolution. The monkey or baboon never evolved into a man, but man was made as such from the very beginning."

1439 Saint, Phil. *Fossils That Speak Out.* 1980s. *Saint:* brother of Nate Saint, who was murdered by Auca Indians in the 1950s; the Indians were later converted by Saint's sister. Book includes 60 illustrations by Saint. Advertised by Creation-Science Legal Defense Fund.

1440 Sala, Harold J. *Science and God in the 80s.* 1980. Irvine, Calif.: Harvest House. Creationism and Bible-science. Copy in ICR Library (inscribed "Thanks for your help").

1441 Salet, Georges. *Hasard et Certitude: Le Transformisme devant la Biologie Actuelle.* c1972. Paris: Editions Scientifique Saint-Edme. Critical of mechanistic evolution—of Monod's "chance and necessity."

1442 Salisbury, Frank B. *The Creation.* 1976. Salt Lake City: Deseret Book Co. *Salisbury:* Caltech Ph.D.; prof. of plant physiology at Utah State Univ.; Mormon. Salisbury is favorably impressed by ICR books and arguments, and also other creation-science works, many of which he cites here. Concludes with exposition of several scenarios of origins. Doesn't make definite choice but seems to favor the last scenario presented as being closest to Mormon doctrine. Notes that Duane Jeffrey (an active anti-creationist) shows that the Mormon Church hasn't yet taken any official position on the creation/evolution issue. Likes **Patten**'s ice comet theory, but favors the idea that fossils are left over from earlier creations.

1443 ________. *The Utah UFO Display.* 1974. Greenwich, Conn.: Devin-Adair. According to **Lang**'s 1975 UFO booklet, which is primarily a discussion of this, Salisbury became interested in UFOs while writing a 1962 *Science* article on Martian biology, and worked closely with UFOlogist Hynek. Salisbury also wrote a favorable introduction to a 1967 UFOlogy book by Lorenzen. This book reports on 80 sightings. Salisbury takes UFOs very seriously, but apparently does not claim to be sure what they are.

1444 Sanden, Oscar E. *Does Science Support the Scriptures?* c1951. Grand Rapids, Mich.: Zondervan. *Sanden:* attended Columbia Theol. Sem. and Austin Theol. Sem., Bible Inst. Los Angeles, and Yale School of Applied Psychology; dean of Northwestern Schools (Minneapolis); Presbyterian minister. Foreword by Billy Graham (then president of Northwestern). [H. Morris says **W.B. Riley** offered him the presidency of Northwestern Bible College before Graham, but Morris declined.] Bible-science. "Why should not the great Bible schools and seminaries, the great Gospel centers, the Bible-preaching pulpits of the land, be known as the custodians and users of true science, for is this not the great book of God in nature whereby He confirms the Book of His inspired Word, the Bible?" Sanden praises the wide acceptance of **Velikovsky,** and shows that science is proving the Bible correct in every field. Argues that the six geological eras equal the six days of Genesis (Day-Age creationism), and that the sequence of life on earth shown by science is "virtually identical" to the Mosaic account. Gen. 1:1–2 refers to time, space, matter, energy and motion in that order, as Herbert Spencer earlier observed. Woolley proved the Flood at Ur. On Galileo: Joshua's Long Day proves that the Bible teaches heliocentricity, because the moon also stopped. Includes an autobiographical section. Bibliog.

1445 _______. *Twelve Bridges No Evolutionist Has Ever Crossed* (pamphlet). c1961. Lincoln, Neb.: Back to the Bible Broadcast. The bridges—obstacles to evolution—include: man's mind, morality, creativity, morphology, etc.

1446 Sargent, Melvin P. *Photo Drama of Creation* (multi-media). 1914. Recorded talks, slides and motion picture show which "gave quotations from **Vail** that gave physical support to the Bible history of Creation." Sargent, a follower of Vail, now living in San Jacinto, Calif., helped produce this show, and presented it over the next twenty years to 50,600 people. "The account of the days of creation given in Genesis relates not to the construction of our globe but to the ordering of it for human habitation. There are various theories regarding its formation. We follow the one most closely harmonizing with the Bible. It is called the Vailian. It assumes that Saturn's rings and Jupiter's belts illustrate the earth's development as a planet." Earth history divided into seven 'days' or periods, each 7,000 years long. First 'day': earth was molten; water and minerals were lofted above surface as great rings, kept aloft by centrifugal force. Second 'day': rings start to descend, spreading out into canopies, then falling to the surface at the poles, forming the various rock strata. Third 'day': accumulated depositions from fallen rings caused earth to buckle, forming mountains and ocean beds. Fourth 'day': sun becomes visible as remaining vapor canopy disperses. Fifth 'day': modified atmosphere permits air-breathing animals and birds. Sixth 'day': continued modification of atmosphere, new forms of animals, and man. Man created 6,000 years ago, about 42,000 years after creation of earth, in God's image, at close of day 6. Noah's Flood caused by collapse of last canopy, composed solely of water. Seventh 'day' will end in 1,000 years. The deserving of mankind will then inhabit a restored Edenic earth. Described and quoted by Sargent in 1982 *Stonehenge Viewpoint*.

1447 Sauer, Barbara. *Walk the Dinosaur Trail.* 1981. San Diego, Calif.: CLP. **Richard Bliss:** Writing Project Director of ICR book series of which this is a part. "Designed for use by children in elementary grades in public schools, these two-model books contain creation/evolution discussions on an introductory scientific level" (ICR catalog). This book for grades 3–4. Accompanying Teachers Guide.

1448 Schaars, Herman W. *Nature and Nature's God.* c1970. Dept. Christian Educ., Lutheran Church—Missouri Synod (S. Wisc. District). Produced under the direction and auspices of The Department of Christian Education: The Lutheran Church—Missouri Synod; S. Wisconsin District. Editorial work by Fritz A. Callies. *Schaar:* emer. teacher at Immanuel Lutheran Church (Milwaukee). Schaar wrote a column "Nature and Nature's God" for the *Badger Lutheran* from 1965–70. Intended "primarily for pupil use." Discusses various plants, animals, and natural phenomena, arranged by their appearance in the six days of creation.

1449 Schaeffer, Francis A. *Genesis in Space and Time.* c1972. Downers Grove, Ill.: InterVarsity Press. Also included in Schaeffer's *Complete Works: Vol. 2* (1982). *Schaeffer:* fundamentalist theologian and philosopher; founder and sponsor (with his wife) of L'Abri Fellowship retreats in Switzerland. Historicism of Genesis: the Bible consists of propositional truths rooted in actual history, and the New Testament depends on this historicity of Genesis. Regarding the two allegedly contradictory creation accounts in Genesis, Schaeffer argues that Gen. 1 presents the cosmic setting—creation as told from God's perspective; Gen. 2 tells the story of creation told from the perspective of man. Stresses the parallel between Adam and Christ, and Noah and Christ. Advocates a pre-millennial Second Coming. The creation "days" may not be literal 24-hour days. The Flood probably occurred over 20,000 years ago. Schaeffer does not deal with evolution directly, and does not employ creation-"science" arguments.

1450 ________. *How Should We Then Live? The Rise and Decline of Western Thought and Culture.* 1983 [c1976]. Westchester, Ill.: Crossway Books (Good News). 2nd ed. Orig. a film and T.V. series with accompanying book (Gospel Films of Muskegon, Mich.); the concept, research, and film were all by Schaeffer's son Franky Schaeffer V, a painter and filmmaker. Our life and culture has degenerated to the extent that we have forsaken Christian ethics, morality and belief. Includes a chapter on the rise of science, which was due to Christianity, and on "The Breakdown in Philosophy and Science" caused by evolutionism and the destructive philosophies it has spawned. Darwin's idea "opened the door for racism" and reached its "logical conclusion" in Nazism. Full-page photos of Darwin, then of **Hitler.** Also discusses influence of Christianity and humanism/evolutionism on philosophy, theology, music and art.

1451 Scherer, Siegfried. *Wort und Wissen: Fachberichte 1.* 1983. Neuhausen, Germany: Hänssler-Verlag. Cited in **Schirrmacher** 1985. "Wort und Wissen" is a German creation-science organization, and Fachberichte Wort und Wissen is its monograph series. Citation may refer to volume in this series [but cf. **Scheven**]. *Scherer:* biochemist at Univ. Konstanz (Germany); doctorate for biochem. and physiol. of blue-green algae; wrote article criticizing Woese's origin-of-life model in *J. Mol. Evol.* (1985); member of Evangelische Stadtmission Konstanz.

1452 Scheuchzer, Johann Jacob. *Homo Diluvii Testis.* 1726. In a 1708 book *The Fishes' Complaint and Vindication,* Scheuchzer, a Swiss, has fossil fish argue against the then-common inorganic interpretation of fossils. They claim to be once-living fish who perished in the Flood along with man. Later, Scheuchzer came across a fossil which he thought was the skeleton of a human drowned in the Flood. He named his specimen *Homo diluvii testis* ("human witness of the Deluge"), and published a description of it (apparently with the same title) in 1726. Scheuchzer's "Deluge Man" turned out to be an extinct giant salamander. The fossil, about three feet long, showed a head and backbone, but otherwise did not look especially human. Scheuchzer was convinced, however, that his fossil was "the bony skeleton of one of those infamous men whose sins brought upon the world the dire misfortune of the deluge," and began an enthusiastic study of the Flood (Cohen 1983). He dated this book "In the Year (4032) after the Great Flood." "Mixing together various texts of Scripture with notions derived from the philosophy of Descartes and the speculations of **Whiston,** he developed the theory that 'the fountains of the deep' were broken up by the direct physical action of the hand of God, which, being literally applied to the axis of the earth, suddenly stopped the earth's rotation, broke up 'the fountains of the great deep,' spilled the water therein contained, and produced the Deluge" (White 1896). Scheuchzer later prepared an edition of the Bible which included many engravings illustrating his theory of the Flood.

1453 Scheven, Joachim. *Data Pertaining to the Teaching of Evolution in Biology Instruction.* 1980(?). Stuttgart-Neuhausen: Hänssler-Verlag. Volume 1 of Wort und Wissen series. Series editors: **H. Beck, H. Schneider** and Theodor Ellinger. An "excellent and profusely illustrated work by one of Germany's most active and knowledgeable creationists" **(E. Myers,** in 1980 *CSSHQ*). High school and beginning college level. A "thoroughgoing refutation of evolutionism from the basis of actual facts of geology..." Scheven has a fossil collection—a creation-science museum—in Enneptal.

1454 Schiffner, Alexander. *The Origin of the Races; and Pre-Adamic Man.* 1968 [1961]. Spokane, Wash.: Prophetic Herald. Vol. 29, No. 9 of *Prophetic Herald* (1968). Contents also incorporated in Schiffner's book *Seven Plateaus to Glory.* "Pre-Adamic Man" section c1961. **Schiffner:** editor of *Prophetic Herald.* The Negro is the "beast" of Genesis; the Bible says the beast—the Negro—was created before Adam, as was the yellow race ("living creature" of Genesis). The biblical "beast" has hands, can talk, etc. Schiffner allows for Day-Age or Gap Theory creationism; Negroes were created in the six-day creation. Adam—

white man—was created on Day Eight, but fell, and now resembles the "beasts." The sons of God—the white men—mated with Negroes. God punished them for this sin with the worldwide Flood. Negroes were taken on board the Ark along with the other animals.

1455 Schirrmacher, Thomas. *The German Creationist Movement* (pamphlet). 1985. El Cajon, Calif.: Institute for Creation Research. ICR "Impact" series #145. *Schirrmacher:* director of German Center for World Mission (Erfstadt, Germany). The recent upsurge of interest in creation-science in Germany. Provides many names and references.

1456 Schmeling, William A. *Creation versus Evolution? NOT REALLY!* 1976. St. Louis, Mo.: Clayton Publishing House. The Bible contains theological truth—not cosmological or geographical. Scripture is scientifically primitive.

1457 Schnabel, A.O. *Has God Spoken?* 1974. Priv. pub. (Tampa, Fla.). *Schnabel:* Boeing engineer; Church of Christ. "This work has been written to demonstrate that between the observations of science and a simple, direct interpretation of the Bible narrative there exists a harmony such as would be expected of a book having the same Author as the physical world." Quotes and references as to accurate biblical scientific facts and predictions in astronomy, geology, oceanography, meteorology, physics, biology, and archeology. Accepts ocean floor spreading, and stellar distances of millions of light-years. Final chapter is on "Scientific Difficulties of the Theory of Evolution"; many quotes, mostly old. Mentions most standard creation-science arguments. Also published in undated (presumably earlier) booklet version, minus last three chapters (archeology, Bible, and evolution).

1458 Schoepffer, Carl. *Die Bibel Lügt Nicht! Erklärung der Mosaischen Schöpfung-Urkunde, Oder Beweis [etc.]* (booklet). 1854. Nordhausen, Germany: A. Buechtung. Listed in Schadewald files. ("The Bible Doesn't Lie! Exposition of the Mosaic Witness to Creation, or Proof...")

1459 Schonberg, David J. *On Purpose or by Chance? (Or Does It Matter?)* (booklet). c1974. Priv. pub. (Holmes City, Minn.). Includes scientific quotes on the problems of evolution (**Kerkut**, G. Wald, Blum, Eiseley, **Moorhead** and Kaplan, etc.). Cites creation-scientists also. In his introduction, Schonberg reveals that he wrote this book on evolution because evolution dismisses the idea of a Creator and of moral standards.

1460 Schreur, Clarence. *Genesis & Common Sense: The Reality: Creation vs. Evolution.* 1983. Prescott, Ariz.: Ralph Tanner Associates. *Schreur:* chemist with High Energy Astrophysics Lab at NASA's Goddard Space Flight Center. Wants to mediate the creation/evolution controversy and reconcile science with the Bible. Argues against both Gap Theory creationism and Flood Geology. Suggests a non-literal interpretation. Genesis describes the Big Bang cosmogony. Presents virtually no references. Schreur mentions that his interest was sparked by an astronomer on a Noah's Ark documentary on T.V. Discusses persistence of species. The Bible gives the explanation of the strong nuclear force. Endorses the Pre-Flood Vapor Canopy. An ice comet over Turkey caused the Ice Age. The continental shelves prove that the sea level was previously much lower.

1461 Schroeder, J. *Man: A Unity of Matter and Spirit.* 1984. Priv. pub. (Waterloo, Canada). 3rd rev. ed. *Schroeder:* prof. of civil engineering at Univ. Waterloo (Canada). Discusses nuclear physics, the Noachian Flood, and creationism. Refutes materialistic view of man. Inspired by Duhem, Schroeder says science represents rather than explains reality, and questions whether it can ever lead to truth. "Faith in Christ depends on concepts revealed by God to man, whereas science depends on concepts imagined by man." We must choose between accepting Christ by faith, or man-made ideas. Evolution cannot be science because it cannot be observed or experimentally validated. It has survived mainly "because it evolved into a man-made secular religion as contemporary evolutionists conjecture it to be true without any possibility of ever proving it."

1462 Schroeder, John R. *Answers from Genesis* (booklet). c1973. Pasadena, Calif.: Ambassador College. *Schroeder:* writer for *Plain Truth* (H.W. Armstrong's Worldwide Church of God). Question and answer format; questions from readers. Includes arguments for Gap Theory creationism.

1463 Schroeder, Werner. *Man—His Origin, History, and Destiny.* c1984. Mt. Shasta, Calif.: Ascended Master Teaching Foundation. *Schroeder:* civil engineer in Germany; now a U.S. citizen. Uses terminology and mythical schemes of Theosophy. Occult evolution: hidden mystical knowledge of the reincarnated Ascended Masters of the Great White Brotherhood, the Akashic Records, Atlantis, the seven Root-Races of mankind, and cycles of everything.

1464 Schultz, Alfred P.K.E. *The End of Darwinism: Not Change But Persistence Is Characteristic of Life* (pamphlet). c1911. Monticello, N.Y.: A.P. Schultz. "Every change is essentially a persistence; only what persists can change." Schultz also wrote *Race or Mongrel:* "A brief history of the rise and fall of the ancient races of earth: a theory that the fall of nations is due to intermarriage with alien stocks: a demonstration that a nation's strength is due to racial purity: a prophecy that America will sink to early decay unless immigration is rigorously restricted" (1908; Boston: L.C. Page).

1465 Schwarze, Carl Theodore. *The Harmony of Science and the Bible.* c1942. Grand Rapids, Mich.: Zondervan. *Schwarze:* prof. of civil engineering at NYU; AAAS Fellow; member of Plymouth Brethren. Much of book orig. broadcast on Erling Olsen's New York radio program 1935–39. Presented in dialogue format. Primarily an exposition of the Pre-Flood Canopy Theory. Earth is older than 6,000 years but younger than evolutionist estimates. Gap Theory creationism: most life was destroyed in the cataclysm of Gen. 1:2. Plants and bacteria survived. God exploded the earth's surface; water was blasted up first and farthest, beyond the stratosphere, and became an ice layer "miles in thickness, forming an oblate spherical canopy around the earth." Dirt and dust settled back to earth, but the ice-lens remained suspended in the sky. The Canopy resulted in a worldwide greenhouse effect, which explains the pre-Flood conditions described in Genesis: the universal mild climate, lack of any weather, extreme longevity, etc. The Flood was caused by the collapse of the Canopy. Schwarze, whose Canopy Theory is an adaptation of **Vail**'s, was influential in popularizing it amongst creationists. "No sane person has any good reason for *not* believing the statement of Scripture. Other statements, at variance with it, are all pure and unadulterated figments of man's invention." People are so opposed to God they accept any silly theory that "gives the lie to Scripture." Evolution has been taught since before Moses. Describes pre-Adamic demon followers of Satan, who are now disembodied spirits. Out-of-control atom-smashing experiments by vain scientists may trigger the end-of-the-world nuclear catastrophe prophesied in the Bible. Book also discusses other Bible-science topics, especially civil and sanitary engineering, and Cain's wife (his sister).

1466 ______. *Evolution* (booklet?). 1957. Toronto: ICC (International Christian Crusade). Quoted in **J.G. Williams** 1970. "The doctrine of evolution would be an insult to anyone's intelligence." Evolution is "accepted by scholars and scientists because it categorically denies the word of God, which they hate."

1467 ______. *The Marvel of Earth's Canopies.* c1957. Westchester, Ill.: Good News Publishers. The Pre-Flood Ice Canopy. The canopy collapsed when the moon was ejected from the Pacific basin of the earth (which also caused the Mid-Atlantic Rift—Atlantis—to sink). All creatures were vegetarian before the Flood; the Canopy conditions provided luxuriant vegetation. Fermentation was impossible under Canopy conditions; Noah's episode of drunkenness after the Flood was due to ignorance of this hitherto unknown process. The Canopy will be restored at the Millennium. We will all return to vegetarianism then. The destruction referred to in Gen. 1:2 (Gap Theory creationism) resulted from a great atomic explosion caused by Satan. The future destruction described in II Peter 3:10 is also an atomic explosion.

1468 Scofield, Cyrus I. (ed.). *The Scofield Reference Bible.* 1917 [1909]. Oxford: Oxford University Press. Rev. ed. Orig. 1909. Scofield: lawyer; later a Congregational minister. Scofield's annotated

Reference Bible was extremely influential in shaping the doctrines and beliefs of modern fundamentalism. Its notes popularized dispensationalism (historical periods defined by differing 'dispensations' or covenantal relationships with God) and the pre-millennialist interpretation of the Second Coming (Christ will return to earth to rule at the Millennium after the world sinks into evil, following Armageddon and the Rapture of the faithful). Most fundamentalists have been dispensational pre-millennialists, largely due to Scofield's influence (though post-millennialism is now starting to make a comeback). Scofield also included notes endorsing Gap Theory creationism, legitimizing it and increasing its popularity enormously. All the geological ages occurred in the 'gap' between Gen. 1:1 ("In the beginning God created heaven and earth") and 1:2 ("And the earth was [became] without form and void"). "Relegate fossils to the primitive creation," comments Scofield, "and no conflict of science with Genesis cosmogony remains." (It is ironic, given Scofield's influence on fundamentalism, that the strictest, most literal fundamentalists today tend to reject the Gap Theory as an evil compromise with evolutionism, insisting instead on a young earth.) The Gap Theory was downgraded in the rev. 1967 edition of the *Scofield Reference Bible* (Scofield d. 1921)—relegated to notes outside of Genesis and mentioned only as a possibility.

1469 Scott, John D. *The Four Most Glorious Events in Human History: Or the Refutation of Evolution.* Undated [1970s?]. Priv. pub. Main section of book is not directly concerned with evolution or creationism; it is a theological discussion of the four events, from Christ's birth through His Ascension. The introductory section, written by S.G. Posey, however, does focus on evolution. (*Posey:* First Baptist Church of Austin, Texas; prof. at Golden Gate Sem.) Posey is angry that atheist evolutionism is featured on television, but insists that it will never disprove God's Revelation.

1470 Scripture Press. *Evolution: Fact or Fiction?* (pamphlet). c1979. Wheaton, Ill.: Scripture Press. Listed in Ehlert files.

1471 Sears, Jack Wood. *Conflict and Harmony in Science and the Bible.* c1969. Grand Rapids, Mich.: Baker Book House. Based on lecture presented to Univ. Christian Student Center at Univ. Miss. *Sears:* Ph.D.—Univ. Texas; Tour Lecturer for Amer. Chem. Soc.; now head of biology dept. at Harding College (Searcey, Ark.). Concerns the theme "Science, the Bible, and Evolution." Presented in a calm, reasoning tone. Many scientific references and quotes. Evolutionist explanations are fraught with speculation. Warns against the dangers of scientism; urges consideration of the alternative explanation presented in the Bible. Science changes, but the ultimate truth of the Bible is immutable. The subject of origins is not science but history. Notes problems with radiometric dating, but is willing to provisionally assume that dates, and geological ages, are correct. Relies heavily on **Kerkut.** Discusses the truth of Bible prophecies. Includes good charts, tables and drawings.

1472 Sedgwick, Adam. *Discourse on the Studies at the University of Cambridge.* 1850 [1833]. London: J.W. Parker. *Sedgwick:* ordained as Anglican minister and appointed geology professor at Cambridge the same year; Darwin's geology teacher at Cambridge. Sedgwick remained opposed to evolution. This book was orig. written in 1833 as a defense of the new academic discipline of natural science at the University. The 5th ed. (1850) included a huge new Preface, five times the size of the original text, refuting Chambers' *Vestiges of Creation* (the quasi-scientific work promoting evolutionism which pre-dated Darwin's *Origin.*) Sedgwick attacks *Vestiges* for being bad science but also for its irreverence. Elsewhere, he expressed vehement contempt for *Vestiges* with its notions of monkeys breeding men and its reflections of continental infidels. If evolution were true, then "religion is a lie, human law is a mass of folly..., morality is moonshine..." (Quoted in Millhauser.)

1473 Segraves, Kelly L. *Jesus Christ Creator.* 1973. San Diego: Creation-Science Research Center. *Segraves:* son of **Nell Segraves;** a leader of the Creation-Science Research Center (San Diego); initiated the famous 1981 court case *Segraves v. California,* in which his son was plaintiff, which sought to balance evolution in public school textbooks with presentation of creationism. Book opens with reprint of **von Braun**'s 1972 letter to the Calif. Board of Education affirming created design. Presents the basic creation-science position and the usual creationist examples; the references are mostly biblical rather than scientific. Eve was created from Adam: the male Y chromosome was canceled out, and the X chromosome doubled. Noah's sons were ancestors of the races of mankind. Ham's descendants are materialistic inventors; Shem's are spiritual; Japheth's are intellectuals who improve the contributions of the others. Includes a bibliog. of creationist sources "for further study." Simply written. Segraves has written over 30 books—most are largely reworkings of the same material.

1474 _______. *The Great Dinosaur Mistake.* 1975. San Diego: Beta Books. Dinosaurs did not die out millions of years before man. Man and dinosaurs coexisted, and dinosaurs are mentioned in the Bible.

1475 _______. *The Great Flying Saucer Myth.* c1975. San Diego: Beta Books. Rev. and expanded ed. The new religion of UFOlogy is "really Satan's last attempt to falsely fulfill Biblical prophecy in an effort to deceive mankind in the end times. Satan and his followers are preparing now to establish the kingdom of the beast, the antichrist, and advocates of UFOlogy are playing directly into his hands." Extraterrestrial flying saucer inhabitants are really fallen angels—sons of God (Gen. 6) who mated with human women, producing an unbelieving race. Discusses and refutes UFO advocates **von Däniken,** Dione, **Norman;** relates his adventures at a UFO convention.

1476 _______. *Search for Noah's Ark.* c1975. San Diego: Beta Books. Adapted from popular 1973 CSRC filmstrip. Kent Kraber: illustrator. Slender book: mostly photos with captions. Standard Ark-eology tales of Ararat climbs and sightings of Noah's Ark, including **E. Cummings'** 1969 expedition. States that the pre-Flood population of earth was three billion.

1477 _______. *The Way It Was.* c1976. San Diego: Beta Books. Standard creation-science, presenting simplified versions of the usual arguments for young readers. Biblical and scientific references. Cites **Moorhead** and Kaplan, **Kerkut, M. Cook,** Pasteur, **G. Wald, Lammerts,** etc. Cites Levi-Strauss as proof that so-called primitives are really as intelligent as us. "If man is going to learn anything about his origin, he is not going to find out much by looking downward and trying to determine where he came from." Instead, we must believe in God and Creation: "We then have an adequate explanation of the beginning." Last chapter about Satan. Pictures, photos.

1478 _______. *Everything You Always Wanted to Know About Dinosaurs, Flying Saucers, the Beginning and the End.* 1980. San Diego: Beta Books.

1479 _______ **(ed.).** *And GOD Created: Volume 1.* c1973. San Diego: Creation-Science Research Center. Standard strict creation-science topics and arguments. Authors: **G.F. Howe** (on the Creation Research Society, and Pasteur), H.D. Dean (Pepperdine Univ. biology prof., on origin of life and of man), **J.D. Bales** (on Genesis), and **Davidheiser** (on history of evolution).

1480 _______ **(ed.)** *And GOD Created: Volume 2.* c1973. San Diego: Creation-Science Research Center. Standard strict creation-science topics and arguments. Authors: **Bales** (on evolutionism), **Whitcomb** (on the origin of life, planets, the creation days, the Gap Theory), and **Moore** (on the influence of evolution, and fossil gaps and frauds).

1481 _______ **(ed.)** *And GOD Created: Volume 3.* c1973. San Diego: Creation-Science Research Center. Authors: **J.C. Whitcomb** (on Creation, Genesis, and the Curse), **Davidheiser** (theistic evolution), **T. Barnes** (thermodynamics and origin of life), and **H. Morris** (Day-Age and Gap theories of creationism).

1482 _______ **(ed.)** *And GOD Created: Volume 4.* c1973. San Diego: Creation-Science Research Center. Authors: **H. Morris** (the Flood, the Ark, and fossils), **J.N. Moore** (archeology and creationist apologetics), **Burdick** (geology of Ararat), and **E. Cummings** (the search for Noah's Ark).

1483 Segraves, Nell J., and Jean E. Sumrall. *A Legal Premise for Moral and Spiritual Guidelines for California Public Schools* (booklet). Undated. San Diego:

Creation-Science Research Center. (Listed in Ehlert files.) *Segraves:* mother of **Kelly Segraves.** Segraves and Sumrall, angered by the 1961 Supreme Court decision protecting atheist students against required religion in public schools, responded by devising a retaliatory argument, which would insure that Christian, creationist students would not be subject to teachings offensive to *their* religious beliefs. They persuaded the California Board of Education in 1969 to include creationism in textbooks in order to balance the dogmatic presentation of evolution. Unlike groups such as ICR which argue that creationism is an equally valid *scientific* alternative to evolution, and ought therefore to be presented in science classes, the Creation-Science Research Center, led by Segraves and Sumrall (and K. Segraves and **Kofahl**), takes the position that exclusive teaching of evolution is a violation of the *religious* rights of Christian students. They fight for direct legal enforcement of these religious rights.

1484 Seibel, Alexander. *Relativitätstheorie und Bibel* (booklet). 1974. Augsburg, Germany: Verlag Lebendiges Wort. **Schirrmacher** 1974 says this is a booklet by a German scientist which rephrases published creationist material.

1485 Seiss, Joseph A. *The Gospel in the Stars.* c1972 [c1882]. Grand Rapids, Mich.: Kregel. Orig. pub. c1882 as *The Gospel in the Stars: or, Primeval Astronomy* (Philadelphia: E Claxton). *Seiss:* "one of the most effective and popular Lutheran preachers of the 19th century." The Good News of salvation by Christ is revealed in the heavens. Seiss realized the constellations had a divine, prophetic source after studying the "marvellous wisdom embodied in the Great Pyramid of Gizeh." The signs of the Zodiac are interpreted as illustrating themes such as the Suffering Redeemer, Christ as the Lion, etc. Based on literal interpretation of Genesis and recent creation of Adam. "Was primeval man a gorilla, a troglodyte, a brutish savage, a wild man without knowledge?" No: "the doctrine that modern man is a mere evolution from savageism . . . *is a lie.*" Scholarly style; cites diverse ancient sources. Frances Rolleston's 1863 *Mazzaroth; or, the Constellations* (Keswick, England), which Seiss cites, was the first major presentation of this view. Drawings; index. Seiss also wrote other works such as *The Apocalypse* and *Miracle in Stone* (the Great Pyramid).

1486 Senter, Mark H. *Evolution: Why a Bible Believer Cannot Accept It* (tract?). Christian Missions Press. (Listed in SOR CREVO/IMS.)

1487 Servants of Christ. *After His Kind: Creation vs. Evolution* (tract). Phoenix, Ariz.: Servants of Christ. Argues for "persistency of species" and geological evidence of the Flood. "We, as Christians, . . . believe that the days of creation were literal 24 hour days, and that fossils are the remains of life that perished in the flood of Noah's time." Reproduces chart from **Rehwinkel.**

1488 Setterfield, Barry. *The Velocity of Light and the Age of the Universe.* 1983. Creation Science Association (Australia). *Setterfield:* described as an "astronomer and physicist." When Setterfield was agonizing over the dilemma that the vast distances of the stars posed to creationist theory, God dramatically revealed to him that the speed of light has not been constant. This book, a result of Setterfield's subsequent study, is a confirmation of that direct revelation. Light has slowed down exponentially since creation. This reconciles star distances of millions of light-years from earth with a recent creation. This book is buttressed by impressively complex computations and technical data. Computer analysis of speed of light measurements since 1675 shows slowdown matching log sine curve, with creation occurring at 4040 B.C. The speed of light has become constant, however, since 1960. Radioactive decay rates and 17 other physical constants have similarly declined. This would cause rapid heating of rocks during creation week, accounting for the scenario described in Genesis: formation of oceans and atmosphere, movement of land mass, etc. The Flood released molten rock of varying isotope ratios, with rocks which appear 'older' deposited first (lowest). God created two sets of stars, on Day One and on Day Four. The first set underwent fantastically intense nuclear reactions, burning most of its fuel in a few hours. The light from these stars travelled across the galaxy to earth in 3.2 seconds; the survivors appear vastly older than the Day Four stars, though they only predate them by a few days. The universe is contracting, not expanding; the expected Doppler blue shift of light is compensated for by an even greater red shift resulting from the slowing speed of light. "Setterfield had done other work which indicates that the Flood of Noah was probably caused by a huge meteor which plowed into the earth cracking the crust and releasing huge amounts of molten rock from deep below the surface." Setterfield's work "should lay the groundwork for a complete reappraisal of modern astronomy, geology and physics. The implications are so staggering that virtually every aspect of science will have to be checked for its consistency with his new findings." (Quotations from 1985 interview in *Bible-Science Newsletter*). The BSA has been enthusiastic about Setterfield's theory, though many other creation-science groups refuse to endorse it. Setterfield is also editing a book on Australian astronomer George Dodwell's theory on the tilting of the earth's axis and the Flood.

1489 _______, and Trevor G. Norman. *The Atomic Constants, Light, and Time.* c1987. Menlo Park, Calif.: Stanford Research Institute International. "Invited Research Report." "Prepared for **Lambert T. Dolphin**," senior research physicist (geoscience and engineering) at Stanford Research Center. A reworking of Setterfield's *Velocity of Light and the Age of the Universe. Norman:* mathematics instructor at Flinders Univ. of S. Australia. SRI is well known for laboratory testing of paranormal phenomena; it has tested and endorsed as genuine the claims of psychic abilities by Uri Geller and many other psychics. Technical report: many graphs, tables, equations. Includes "non-technical summary." Experiments show there is a discrepancy between "atomic clocks" and "dynamical time" (time based on orbital periods of earth or other bodies). The authors claim the atomic clock is slowing down relative to dynamical time. Thus, the speed of light is decaying, and other atomic constants are varying proportionally. Comments by technical referees endorsing this research are included: Ker Thompson (geophysics prof.—Baylor Univ.), **Walter Brown,** Dean Kenyon, D. Russell Humphreys, and Dolphin. It is not stated that these referees are creationists, though; nor do the authors anywhere state the creationist implications of their conclusions.

1490 Sewell, Curtis, Jr. *Evolution? No! Creation? Yes! A Scientific Alternative.* 1974. Unpub. report. *Sewell:* electronics engineer; worked on Manhattan Project (WWII); at Los Alamos til 1957; chief engineer at Isotopes, Inc.; at Lawrence Livermore Labs since 1962. Long unpublished report; standard creation-science, Flood. Includes appendix on carbon-14 dating methods and statistics. In ICR Library.

1491 Shadduck, B.H. *Rastus Augustus Explains Evolution* (pamphlet). 1928. Rogers, Ohio: Homo Pub. Also pub. by Higley Press (Butler, Ind.). Part of a six-part pamphlet series against evolution.

1492 _______. *Puddle to Paradise* (pamphlet). [1930s?]. **H. Morris** 1984 says this was one of a "widely used series of humorous cartoon booklets against evolution." Presumably part of Shadduck's six-part anti-evolution pamphlet series.

1493 _______. *The Toadstool Among the Tombs* (pamphlet). [1930s?].

1494 Shaw, George Bernard. *Back to Methusaleh: A Metabiological Pentateuch.* 1947 [1921]. New York: Oxford University Press. Orig. 1921. Shaw, the English playwright, was an outspoken advocate of evolution by inheritance of acquired characteristics ("Lamarckism"); the huge prefaces to several of his plays are essays promoting this view. Shaw proclaims it the philosophical salvation of the evolutionist movement; it is unclear whether his support for Lamarckism—which by that time was discredited among biologists—was due to ignorance of biology or wishful defiance (Bowler). Here, Shaw denies selection via accident and chance, and evolution as a senseless struggle. Shaw praises **S. Butler** and denounces materialism. "We are intellectually intoxicated with the idea that the world could make itself without design, purpose, or intelligence; in short, without life." We "revel in demonstrating to the Vitalists and Bible worshippers" that random forces operating over unlimited amounts of time can produce the perfect adaptations which **Paley** saw as proof of the Creator. "We took a perverse pleasure in arguing, without the least suspicion that we were reducing ourselves to absurdity, that all the books in the British Museum library might have been written word for word as they stand on the shelves if no human being had ever been conscious." He urges greater effort to better ourselves and evolve to higher forms. The play itself consists of five separate plays, the first set in Eden, the last in A.D. 31,920. Adam and Cain debate the proper lifestyle. The serpent, meanwhile, teaches Eve about reproduction and creative evolution – controlling the body with the mind. Through successive ages, humans begin to understand and use creative evolution. Future humans hatch at age 17, then, after a short active life, become eternal, disembodied ancients. The conclusion is spoken by Lilith, who preceded Adam and Eve (creating them by dividing herself into halves). Shaw's play *Man and Superman* (1901) also advocates this non-Darwinian evolution.

1495 Sheldrake, Rupert. *A New Science of Life: The Hypothesis of Formative Causation.* c1981. Los Angeles: J.P. Tarcher (Houghton Mifflin). *Sheldrake:* formerly at Cambridge Univ., Harvard Univ.; Res. Fellow of Royal Soc.; studies physiology of tropical crops in India. Blurbs by **Koestler,** Gribbin, *Brain/Mind Bull.* ("new age"). Criticizes mechanistic explanations in biology, and especially for behavior; denies that living organisms can be fully explained by physics and chemistry. Mechanistic view is used to justify a conservative approach to biological problems. Emphasizes "epigenetic" nature of biological development: growth is more than mere unfolding of the egg; there appears to be some external factor – cites **Driesch.** Discusses vitalist alternative to mechanistic view, but favors a more radical alternative: organismic or holistic philosophy, which conceives of universe in terms of hierarchies of complexity. Suggests that repetition of form and patterns of organization result from "morphogenetic fields," which are probability distributions actually caused by the existence of previous forms. Past forms influence and encourage development of similar forms by "morphic resonance"; development of the forms becomes increasingly canalized. Biological growth is not merely the result of genetic informaation, but also the causative influence of all past forms of that morphogenetic field. Stresses unproved and speculative nature of mechanistic neo-Darwinian theory; notes disagreement whether micro-evolution can be extrapolated into macro-evolution. Unconstrained by detailed fact or experimental test, evolutionary theory consists largely of elaboration of initial assumptions (mechanistic or vitalistic). Argues that Lamarckian theory provides plausible explanation for many hereditary adaptations, but lacks a mechanism. Cites several experiments which confirm inheritance of acquired traits (Waddington, **McDougall**). Mental phenomena follow laws of their own, not physical laws. Suggests that parapsychological phenomena may be due to formative causation (psychokinesis, telepathy as morphic resonance). Hypothesis of formative causation predicts that all learned activities should become progressively easier to acquire. Sheldrake proposes several possible metaphysical interpretations of formative causation: as a modification, but not a denial, of materialism; conscious self as causal agent; "creative universe" – causative force immanent in life or universe as a whole (or composed of hierarchies); or of a conscious agent which gives purpose to the universe but which itself transcends it. Cites Eccles, Hardy, **Smuts,** Heisenberg, A.N. Whitehead, R. Thom's catastrophe theory, Galton's family composites, **D'Arcy Thompson.** Drawings, diagrams; bibliog., index. The Tarrytown Group (New York) offered $10,000 for best test of Sheldrake's theory.

1496 Shelley, J.E. *The Evolution Question.* 1969. Brisbane, Australia: Evangelical Press.

1497 ________. *The Flood Vindicated* (pamphlet). Undated. England: Evolution Protest Movement. EPM pamphlet #160.

1498 Shelly, Rubel. *Simple Studies in Christian Evidences.* 1970. Birmingham, Ala.: Bible & School Supply. (Listed in **B. Thompson** C/E bibliog.; presumably strict creationism.)

1499 Shipley, Maynard, Francis Nichol, and Alonzo Baker. *The San Francisco Debates on Evolution.* c1925. Mountain View, Calif.: Pacific Press. Transcript of creation/evolution debate series held in 1925. Shipley, who argued for evolution, was president of the Science League of America; wrote anti-fundamentalist book *The War on Modern Science* (1927). **Nichol** and **Baker** are Seventh-day Adventists. Part I debates the proposition that earth and life are the result of evolution. Part II debates whether the teaching of evolution should be forbidden in public schools. Nichol and Baker argue that evolution as well as Genesis should be kept out of public schools, and make an eloquent appeal to democracy. Shipley notes the value of the just-discovered Taung (australopithecine) fossil.

1500 Shockey, Don. *Agri-Dagh—Mount Ararat: The Painful Mountain.* c1986. Fresno, Calif.: Pioneer Publishing. "With a presentation of evidence for a scientific case for creation, entitled 'A Look at Genesis and Science,' by Dr. **Walter Brown.**" *Shockey:* anthropology B.A.; former science teacher; later anthropology teacher at Univ. New Mexico—Taos; founder and president of Foundation for Int'l. Biblical Exploration and Research (FIBER) of Albuquerque; member of 1984 Ararat Ark-eology expedition; currently an optometrist. Book is primarily the story of Ed Davis, who claims that Kurds showed him the remains of Noah's Ark in 1943 (he was stationed in Iran during the war). Davis kept a record of his Ararat climb in his Bible, and only went public with his story at the 1986 Ark Conference, shortly after talking to Shockey and relocating his lost Bible. At the Conference he was thoroughly and enthusiastically interrogated by various Ark-eologists. Book also includes account of Dave GuMaer, who claims that A. Arutunoff showed him Ark photos taken by Russians during WWI (and explains that Bolshevik Revolution was a conspiracy financed by international banking groups). Shockey suggests that God will "use the re-discovery, the physical proof of the Ark's reality as the last sign to unbelievers of the need to get on board before God again shuts the door" as He did in Noah's time. Brown's section of the book presents 116 evidences for creationism, supported by many scientific quotes. Covers the standard creation-science arguments. "This information is generally being withheld from students. If this evidence were not censored from the public classroom but openly presented, better science education would result." Book also includes tribute to eminent Ark researcher Dr. Howard Davis, founder and president of Artesia Christian College. Many photos; also drawings.

1501 Showalter, Lester E. *Investigating God's Orderly World: Book One.* 1970. Crockett, Ky.: Rod and Staff. A general science textbook for Christians, full of explicit Bible references. "Satan is very busy" promoting various scientific theories and superstitions. "The theory of the evolution of man . . . is wrong because it is contrary to God's Word—the Bible." "Today scientists are accumulating much data" proving the Flood—though some "ungodly men" continue to doubt it. If there is any conflict between the Bible and science "we must accept and believe the Bible and not the scientists." We "cannot explain fossils"; we "must accept Creation events by Faith." Also includes biblical health rules. "Doctors only treat disease in cooperation with God's will." Rod and Staff is a Mennonite publisher. This book, and Book Two, constitute a junior high school course (or up to grade 10). These books, according to Rod and Staff, help the teacher to: "Uphold faith in the six-day Creation as given in Genesis 1. . . Do not teach as truth that which is speculation. . . Demonstrate a personal faith in the truth of the Bible and its message . . . even in the face of unexplainable information."

1502 ________. *Discovering God's Stars* (booklet). 1968. Crockett, Ky.: Rod and Staff. Learning to identify the stars enhances understanding of God, "whose glory the stars so nobly declare."

1503 Shute, Evan. *Flaws in the Theory of Evolution.* 1962 [c1961]. Nutley, N.J.: Craig Press. Orig. c1961 by Temside Press (London, Canada). *Shute:* entered Univ. Toronto at age 14; Fellow of Royal Coll. Surgeons (Canada); editor of *Summary* (med. j.); has pub. in both *Nature* and

CRSQ. Strongly creationist, though Shute accepts old-earth chronology. Full of scientific references; no biblical references, but insists that the evidence proves a Creator. Demonstrates that the scientific evidence is against mega-evolution—"Botanists and bacteriologists must be especially aware of this." Botanists "persistently fail to find the genealogical connections between the great groupings of plants" required by evolution. Bacteria and other simple organisms, which ought to evolve the fastest, show no evolution at all—only development of different strains. Asserts that Precambrian rocks are "remarkably lifeless" (though later concedes there are "a few dubious exceptions"). Defends proposition that "The biochemical probabilities of the spontaneous origin of life are so infinitesimally small that life obviously could not have suddenly started up on its own. It must have been created." Refutes embryological evidence of evolution ("biogenetic law" of recapitulation) as "superficial and crude." Similarly refutes evidence of vestigial organs, demonstrating that alleged examples have biochemical purposes. Complex adaptive life-cycles of parasites, perfection of mimicry, interdependence of species, existence of instinct, sense of beauty, and social insects all defy evolutionary explanation. Serology (blood group) evidence does not support evolution (cites Nuttall's old data). Presents many examples of extraordinary adaptations. Discusses species problem, biogeography, lack of transitional forms, horse series. Modern man appeared suddenly in Near East 9,000 years ago. Shute suggests that Adam may have been the first of these, and that Adam's descendants intermarried with the older, more primitive hominid types. Cites **Dewar, Du Noüy, Fabre, Lunn, Acworth, Vialleton, Merson Davies,** plus other anti-Darwinian scientists. Bibliog., index.

1504 Siegler, Hilbert R. *Evolution or Degeneration: Which?* c1972. Milwaukee: Northwestern Pub. House. Strives to preserve belief in divine creation and inerrancy of Bible while admitting to some changes in organisms. Mentions **Klotz, Lammerts** and **Zimmerman** as inspirations.

1505 Sime, James. *The Mosaic Record in Harmony with the Geological.* 1854. Edinburgh: T. Constable. Millhauser says that Sime argued that the 'days' of Creation were both "literal and geologically irrelevant by reducing them to the duration of a series of six visionary trances during which Moses saw a sort of serial recapitulation of cosmic history." I.e., revelatory creationism.

1506 Simonds, Robert L. *Communicating a Christian World View in the Classroom.* c1983. Costa Mesa, Calif.: Nat. Assoc. of Christian Educators. *Simonds:* pres. of Nat'l. Assoc. of Christian Educators (NACE). "A Manual" on how to defeat secular humanism in the schools. The Christian world view includes strict creationism. Book includes "Student Questions to Use in the Classroom," provided (along with answers) by **Bliss** of ICR, which expose the weaknesses of evolution and let the student witness for creationism. How does evolution help in practical research? Give even one example of an unquestionable transitional form, and explain all the stages it went through. How often have anthropologists been wrong about human ancestors? Does evolution theory help in finding petroleum? Since life comes only from life, isn't evolution just a faith or dogma? Hasn't evolution theory resulted in harmful medical practices? Is scientific truth decided by majority vote? Did Hitler think he was practicing evolutionary theory? Also includes "Summary of Scientific Evidence for Creation" (from ICR).

1507 Simpson, David. *Evolution Cross-Examined.* 1934. Grand Rapids, Mich.: William B. Eerdmans. (Listed in SOR CREVO/IMS.)

1508 Sitchin, Zecharia. *The 12th Planet.* 1976. New York: Stein and Day. *Sitchin:* studied Hebrew and archeology at London Sch. Economics and Univ. London; journalist in Israel; now lives in U.S. "The Old Testament has filled my life since childhood." Sitchin's study of Sumerian archeology and ancient astronomical texts shows that astronauts from Marduk, the 12th Planet, settled earth and created *Homo sapiens.* Reinterprets creation mythology, including Genesis, to support theory that these extraterrestrials—the *Nefilim* of Genesis—are our ancestors. The Creation account describes formation of earth, moon and asteroids from remains of a larger planet after collision with Marduk. Earth was initially seeded with life from Marduk. *Homo erectus* evolved on earth; evolution, however, "cannot account" for the sudden appearance of modern man. The Nefilim first arrived 450,000 years ago to exploit gold and mineral resources. After a mutiny of imported workers, they created slave laborers for their mines by genetic manipulation, producing hybrids between *Homo erectus* and Nefilim. One of the Nefilim (the 'Serpent') gave these hybrids the ability to reproduce independently. By 100,000 B.P. these humans were mating with the Nefilim (Gen. 'sons of God, daughters of men'). The last Ice Age ended when a close approach of Marduk caused the Antarctic ice to fall into the ocean, creating a massive tidal wave: the Flood. The Nefilim had predicted this catastrophe and fled earth, but left mankind to be destroyed, as they were regressing to primitive types and defiling the purity of the Nefilim. Other Nefilim warned some of the humans, however. These survivors were taught agriculture and civilization during return approaches of Marduk (11,000, 7,400 and 3,800 B.C.). Sitchin's scholarly style resembles **Velikovsky**'s; his theory, a combination of Velikovsky and **von Däniken.** This 436-page vol. is Book I of "The Earth Chronicles." Book II is *The Stairway to Heaven;* Book III: *The Wars of Gods and Men.* Drawings, maps, photos; bibliog., index.

1509 Skeem, Kenneth A. *In the Beginning.... A Book About Reason, Rocks and Religion.* 1981. Oasis, Utah: Behemoth Publishing. *Skeem:* math and physics B.S.—Utah State; Mormon. Day-Age creationism. First printing published anonymously.

1510 Slusher, Harold S. *Critique of Radiometric Dating.* 1981 [c1973]. El Cajon, Calif.: Inst. for Creation Res. ICR Technical Monograph No. 2. 2nd ed. Orig. c1973. *Slusher:* geophysics M.S.—Univ. Oklahoma; D.Sc. (hon.)—Indiana Christian Univ.; Ph.D.—Columbia Pacific Univ.; ass't. prof. of physics at Univ. Texas—El Paso (since 1957); director of Kidd Mem. Seismic Observatory at UTEP; concurrently prof. and Grad. School Dean at ICR (but has since left ICR faculty); co-editor of **Moore** and Slusher textbook. States at the outset that this subject is important because evolutionists need the vast amounts of time indicated by radiometric methods; if they are wrong, evolution is impossible. "In much of the work done by geochronologists it is not always easy to separate fact from hypothesis because they fail many times to make clear where fact ends and supposition begins. It seems that scientists oftentimes try to force science to express their philosophy as to what the world was like in the past." Refutes the uniformitarian assumptions of radiometric dating methods. For radiometric 'clocks' to be valid, initial ratios of elements measured must be known, ratios or quantities of measured elements must change at a constant rate, and external processes cannot influence these amounts and ratios. Shows that all these assumptions are invalid. Scientific references; many math equations. Suggests pre-Flood Vapor Canopy minimized absorption of cosmic rays and lowered rate of C-14 formation. Vastly stronger magnetic field also affected rates. Discusses uranium-thorium-lead, carbon-14, and rubidium-strontium methods. Cites **M. Cook** frequently; also **T. Barnes.** Endorses H.C. Dudley's "neutrino sea" hypothesis of space as an explanatory framework. Diagrams, tables.

1511 ________. *The Origin of the Universe: An Examination of the Big Bang and Steady State Cosmogonies.* 1980 [1978]. El Cajon, Calif.: Inst. for Creation Res. ICR Technical Monograph No. 8. Rev. ed. Orig. 1978. There are two categories of cosmogonies which are diametrically opposed: evolutionist (naturalist) and creationist. Stresses Second Law of Thermodynamics. Big Bang scenario, says Slusher, assumes initial explosion resulted in disorder, which then evolved into order (decrease of entropy), and thus violates Second Law. "Some cosmogonists refuse to accept results when they are in conflict with their intellectually appealing views of the universe." Proposes alternative explanations of galactic red shift other than as Doppler effect (which supports vast distances): light may get "tired," increasing in wave-length; red shift may be a gravitational effect; interaction between photons. Interprets Olbers' Paradox (dark night sky despite plenitude of stars) as evidence of smaller, or young, universe. Cites Herbert Dingle's anti-relativity arguments. Cites **Hoyle**'s criticisms of Big Bang theory—then criticizes Hoyle's steady-steady model. Discusses various processes and phenomena indicating young age for universe. "Truly we have suffered too long and too disastrously under serfdom to barren and naturalistic nature-myths regarding the cosmology and the cosmogony of this actual universe. The evolutionist lives in a dream world in which any resemblance to the real world is lacking." Concludes that laws of physics prove universe was created by "the infinite, omniscient, omnipotent Creator."

1512 _______. *Age of the Cosmos.* 1980. El Cajon, Calif.: Institute for Creation Research. ICR Technical Monograph No. 9.

1513 _______. *Star Birth in the Milky Way: One Aspect.* 1986. El Paso, Texas: Geo/Space Research Foundation. Tutorial Paper No. 4. Cited by **T. Barnes** in 1987 *CRSQ*. Tutorial papers are short technical papers from Slusher's Geo/Space Res. Found. "They give strong support to equal ages for stars in the Milky Way, and young-age limits on certain astronomical bodies."

1514 _______. *The Stars—Their Birth.* 1986. El Paso, Texas: Geo/Space Research Foundation. Tutorial Paper No. 2. Cited by **T. Barnes** in 1987 *CRSQ*.

1515 _______. *The Protoplanet Hypothesis and Tidal Instability in the Solar System.* 1987. El Paso, Texas: Geo/Space Res. Found. Tutorial Paper No. 6.

1516 _______**, and Thomas Gamwell.** *Age of the Earth.* c1978. San Diego: Institute for Creation Research. ICR Technical Monograph No. 7. Consists largely of equations, which are not understandable to the non-specialist reader. Gives an impressively technical appearance, however. Presents mathematical models of length of time it would take earth to cool to present temperature, considering various initial temperatures and possibility of internal heat source (radioactivity). "It would seem that the earth is vastly younger than the 'old' earth demanded by the evolutionists." Their models show cooling times of tens of millions of years (far less than evolutionist estimates but much higher than the few thousand years allowed by the authors' own young-earth position). Graphs, tables; bibliog.

1517 _______**, and Francisco Ramirez.** *The Motion of Mercury's Perihelion: A Reevaluation of the Problem and Its Implications for Cosmology and Cosmogony.* c1984. El Cajon, Calif.: Institute for Creation Research. ICR Technical Monograph No. 11. Ramirez wrote 1982 M.S. thesis, *Secular Variations on the Orbital Motion of Mercury,* at Univ. Texas—El Paso (presumably under Slusher). Authors note that Einstein's relativity theory is "completely upsetting to our ordinary common sense way of looking at physical and astronomical phenomena." A discrepancy in the observed orbit of Mercury provided an early and crucial test of Einstein's theory. Slusher and Ramirez dispute the standard interpretation that it confirmed relativity, arguing that it can be explained by traditional Newtonian physics and is caused by the sun's oblateness. The book is a critique of relativity theory. Cites anti-relativist H. Dingle. "Modern cosmology based on relativity seems nothing but a fantasy of mathematicians who find it agreeable that the world should be made in this way. We should abandon Aristotle's approach to study of the cosmos, forsaking mathematical artifice, and develop a cosmology based on observations." Slusher thinks that modern physics is far too abstract, and ought to limit itself to real objects and direct observation, rather than indulging in speculative hypotheses.

1518 _______**, and Stephen J. Robertson.** *The Age of the Solar System: A Study of the Poynting-Robertson Effect and Extinction of Interplanetary Dust.* 1982 [1978]. El Cajon, Calif.: Institute for Creation Research. ICR Technical Monograph No. 6. Rev. ed. Orig. 1978. Stephen J. Duursma (pseud.?) listed as co-author of orig. ed. "Poynting-Robertson effect" is the theory that all interplanetary particles below a certain size (dependent on distance from sun) are slowed down by solar radiation and eventually spiral into the sun. Authors argue that according to this theory the amount of dust measured in space proves the solar system is young. "None of the mechanisms for resupplying the dust complex in the solar system seems adequate for maintaining the dust complex against the forces of extinction for any time remotely approaching evolutionary guesses on the age of the solar system." Thus the Poynting-Robertson effect puts an upper limit on the age of the solar system "vastly less than the evolutionary estimates of its age." Mostly equations; very technical. Diagrams, tables, graphs; bibliog.

1519 Smith, David W. *Problems in Evolutionary Theory* (booklet). Undated [1973?]. Inside cover says "To be used with the Scopes Simulation – Glenbrook South Social Science Department: Glenview, Ill." Latest citation 1972. Creation/evolution is not a science vs. religion issue, because both sides use science. Many people believe evolution only because they have been told it is true – not because they have analyzed it. All ancient cosmogonies are basically evolutionary – except Genesis.

1520 Smith, E. Lester. *Intelligence Came First.* 1975. Wheaton, Ill.: Theosophical Publishing House. "Edited by" Smith. Smith drafted most of text, which was revised and modified at monthly group meetings. Contributors: Dr. H. Tudor Edmonds; Prof. Arthur J. Ellison; Marion E. Ellison; Forbes G. Perry; V. Wallace Slater; Smith; Corona G. Trew; K.B. Wakelam. Theosophy. "It is rational to suppose that intelligence is primal and cosmic, the original cause of evolution and not its product." We know this intuitively. Primal intelligence is necessary for origin of life. Philosophical and scientific arguments; many references. Discusses perception, the nature of the brain, computer intelligence. Insists on mind/body duality. Stresses importance of intuition, which "represents brief communion with Cosmic Intelligence" (as does mysticism and ESP). Argues against chance formation of life. DNA could not have arisen randomly. Criticizes origin-of-life theories. Explains that "central dogma" of biology – the one-way influence of DNA on proteins – is now discredited. Discusses neo-Lamarckian theories. Optimistic about genetic engineering as means to influence heredity. Darwinism is inadequate, incomplete. By changing their habits, animals influence natural selection; Darwinian evolution then simulates Lamarckism. These changes of habit may then spread telepathically. Cites **Koestler, W.R. Thompson,** Waddington, Hardy, Bertalanffy, B. Commoner, **F. Salisbury,** other critics of Darwinian theory.

1521 [Smith, James Alexander]. *The Atheisms of Geology: Sir C. Lyell, Hugh Miller, &c. Confronted with the Rocks* (booklet). 1857. London: Piper, Stephenson, & Spense. Author listed as "J.A.S." Millhauser says this work was part of the "pamphlet war" in response to **Hugh Miller**'s attempted reconciliation of geology and Genesis. Also wrote *Errors of Modern Science and Theology* (1864; London: Murray).

1522 Smith, John Augustine. *Mutations of the Earth.* 1846. New York. Cited by Millhauser as advocate of Gap Theory creationism who maintains a "literally diluvial and Adamic theory of creation." Smith also wrote *Prelections on ... Moral and Physical Science* (1853; New York).

1523 Smith, John Pye. *On the Relation Between the Holy Scriptures and Some Parts of Geological Science.* 1854 [1839]. London: Henry G. Bohn. 5th ed. Orig. 1839 (London: Jackson and Walford). Also pub. 1852 (New York: D. Appleton). Popularly known as *Scripture and Geology. Smith:* head of Homerton Divinity Coll. Pye Smith, who was geologically knowledgeable, abandoned the idea of a worldwide Deluge in favor of a regional Flood. (Mankind's sinfulness had prevented the antediluvian population from spreading much beyond its origin.) "If so much of the earth was overflowed as was occupied by the human race, both the physical and the moral ends of that awful visitation were answered." Pursuing this reasoning, he further tried to reconcile geology with the Bible by proposing that the Creation of Genesis was local as well. Six thousand years ago, God laid waste (largely by vulcanism) and flooded a portion of the earth's surface, then restored and repopulated it as Eden to be man's abode.

The flooded region was western Asia; the Flood waters drained off into the Caspian Sea and Indian Ocean. The original creation occurred ages prior to this. Pye Smith's book was denounced as an infidel interpretation by the literalists, but was endorsed by influential figures such as **Whewell, Sedgwick,** Baden Powell and Sir John Herschel. (**Whitcomb** and **Morris,** who refer to him often, criticize his non-literal approach and "tranquil Flood" theory. The 5th ed. of his book, they complain, contains 60 pages of arguments against a global Flood.) Gillispie also cites *Lectures on the Bearing of Geological Science upon Certain Parts of Scriptural Narrative* by Pye Smith (1839; London).

1524 Smith, Ron. *Creation or Evolution?* (booklet). c1979. Bromley, Kent, England: Fishers Fellowship. Does everyone believe in evolution? If so, it is for the same reason that most Russians believe in communism: i.e., because of propaganda. Lamarckism and natural selection have been rejected as mechanisms for evolution; now mutation is accepted as mechanism. Standard creation-science.

1525 Smith, Roy. *The Folly of Evolution* (pamphlet). Undated. Bible-Science Association(?). Orig. French. *Smith:* missionary to Moslems in Marseilles. Easy, readable summary of standard creation-science arguments. Adds itemized list of main points. If evolution occurred, why are unicellular oganisms still around? Fossil hominids all either ape or man, or fraud. Says Piltdown Man is still presented in textbooks as real. Human cranial capacity hasn't changed. Compares man with animals and especially apes: animal blood can't be transfused, only man is monogamous, only man can reason and be conscious of God. Includes testimony of **du Noüy** on belief in God via rational path of biology and physics.

1526 Smith, Wilbur N. *Therefore, Stand: A Plea for a Vigorous Apologetic in the Present Crisis of Evangelical Christianity.* 1946 [1945]. Boston: W.A. Wilde. *Smith:* dean of Moody Bible Inst. The faith of many young men in the Bible and God is destroyed in college; the main purpose of this book is to defend Christianity against these attacks and ridicule. Asserts that "the facts of history, and the facts of science, are not on the side of agnosticism and atheism, but on the side of Christian truth, and that our faith is definitely not contradicted by *facts,* but is opposed only by the *theories* of men..." Refutes the propaganda of Voltaire, J. Huxley, Dewey, H.G. Wells, B. Russell, liberal theologians; points to the sad example of Russia and Germany. Praises and quotes Romanes, **Agassiz, Dana,** other believing scientists; also **Jeans, Wallace, L.T. More.** Includes long passages from many scientists and academics on the harmony of Genesis and science. Emphasizes fundamental importance of creationism in chapter "The Creation of the World by God the Apologetic for This Era of Scientific Emphasis." "Destroy faith in the Genesis account of creation, and the great structure of doctrinal truth built up through the ages, in the Word of God, is without foundation." Science cannot tell us about origins, but Divine revelation does. Affirms creation *ex nihilo* as voluntary act by eternal, transcendent God. If Genesis were scientifically inaccurate, it would not be trusted theologically, though its primary purpose is spiritual. Old-earth creationism: "we must dismiss from our mind any conception of a definite period of time, either for creation itself, or for the length of the so-called six creative days. The Bible does not tell us when the world was created. The first chapter of Genesis could take us back to periods millions of years antedating the appearance of man." Advocates Day-Age creationism; also favorable to Gap Theory and **Guyot**'s theory of original chaos. Describes how the order of creation in Genesis is "remarkably confirmed by modern geologists and the writings of contemporary scientists." Day 1: primacy of water, penetration of light. Day 2: water 'divided'—dense vapor clouds. Day 3: emergence of land, appearance of plants. Sun became visible when opaque cloud canopy was withdrawn on Day 4. Day 5: water animals—end of Paleozoic. Stresses similarity of birds to fishes. Day 6: Mesozoic/Cenozoic boundary. Genesis "certainly leads us to believe that species as such were fixed at creation." Science knows of no species changing into another. Agrees with T.H. Huxley that Darwin's *Origin* is "Anti-Genesis"—

evolution tries to challenge Genesis because it is the only religious creation account which is scientific. Index.

1527 Smuts, Jan Christiaan. *Holism and Evolution* c1961 [1926]. New York: Viking Press. Orig. 1926. Introduction by E. Sinnot. *Smuts:* Field Marshall; Prime Minister of South Africa (before this book, and again after). Smuts tries to transcend the materialist/spiritualist deadlock by urging "holism." Admires Darwin but insists evolution must be rid of reductionism. Natural selection exists, but "inner evolution" is the real shaper. The mind becomes an active part of processes of biological regulation and of evolution; its ability to learn and introduction of purpose influences development. There is "little doubt that acquired characters in the long run reach down to the hereditary germ-cell and become transmissible variations"; Lamarck was therefore right. Darwinian selection, which is mere mechanical analysis, cannot account for evolution of coordinated organisms. Smuts finds Weismann's mutation theory too crude and mechanistic to aid in understanding the gradual and delicately harmonious process of evolution. "Our crude uncritical mechanistic conceptions are the real source of the difficulty, and Holism appears to me to be the way out. The root of the error lies in our disregard of the individual organism as a living whole, and in our attempt to isolate characters from this whole and study them separately, as if they were mere mechanical components of this whole. . . . The whole is not a mechanical aggregate indifferent to and without influence on its parts. It is itself an active factor in controlling and shaping the functions of its parts." Index.

1528 Smyth, Charles Piazzi. *Our Inheritance in the Great Pyramid.* 1890 [1864]. 4th ed. Orig. 1864. *Smyth:* Astronomer-Royal of Scotland; prof. at Univ. Edinburgh. In 1859 John Taylor argued in *The Great Pyramid; Why Was It Built? And Who Built It?* that the Great Pyramid of Egypt was built by an Israelite, probably Noah. Based on the biblical cubit, it accurately expresses the earth's dimensions and all sorts of mathematical and biblical truths. Smyth enthusiastically developed these ideas, and elaborated on R. Menzies' theory that measurements of the internal passageways (one "pyramid-inch" equalling a year) encode the past history of mankind, and also the future. Proves that Creation was 4004 B.C.; also commemorates the Flood, Exodus, birth of Christ, etc. The Great Tribulation will begin 1892–1911. Smyth's book has had an enormous influence on pyramidology advocates.

1529 Snelling, Andrew, John Mackay, C. Wieland, and Ken Ham. *The Case Against Evolution: The Case for Creation* (booklet). c1984 [c1983]. Sunnybank, Australia: Creation Science Foundation. Magazine format. "Casebook I": a supplement to *Ex Nihilo* magazine, but widely distributed as separate booklet. Standard young-earth creation-science arguments, with biblical references. "Hidden beliefs" of evolution: uniformitarianism, and naturalism-atheism. Zoo animals can survive without meat: this is evidence supporting Genesis account that animals were originally vegetarian. Discusses Australian astronomer G.F. Dodwell's theory that earth's axis used to be perpendicular, but a major impact tilted it 27 degrees, then it wobbled back to its present 23 degrees. This caused the Flood, dated by Dodwell to 2345 B.C. "Dodwell's theory may also do away with the theory of continental drift." Rift systems are cracks from asteroid impact, not evidences of moving continental plates. Also discusses **Setterfield.** Photos, pictures.

1530 Society for Creation Science. *Creation-Evolution: Understanding the Issues at Hand.* c1987. E. Lansing, Mich.: Society for Creation Science. "Written and compiled" by David M. Skjaerlund. Course manual for 10-week college seminar presented by the Society for Creation Science, a ministry of Maranatha Campus Ministries (see **Broocks**). *Skjaerlund:* Ph.D. candidate in animal science at Michigan State Univ.; SCS national director. SCS goals are to "Evangelize the College Campuses" and "Influence the College Curriculum"; to "promote sound scientific investigations and to take dominion in the earth with Christians as the originators of future scientific discoveries." After registering SCS as a campus student organization, the SCS course is to be taught first to the Christian community; then to the general college community. This illustrated, loose-leaf manual quotes and cites many creationists, relying heavily on **H. Morris, K. Ham** and **Bert Thompson,** and also incorporates Christian Reconstructionist ideas. Standard creation-science arguments. Genesis is the foundation for all of Christianity, all of life's purposes. Evolution is the very antithesis of Christianity and the basis of communism, racism, abortion and humanism.

1531 SONshine Publishing. *How Old Are Those Bones—Really?* (tract). Undated. Tustin, Calif.: SONshine Publishing. Purpose of tract· "to make you (the reader) aware of the UNREASONABLE assumptions which are necessary to use the carbon-14 method of dating beyond the time of the world-wide flood." The pre–Flood Water Canopy prevented C-14 formation, thus C-14 dates from this period give ages much too old.

1532 Sooter, Wilburn B. *The Eye: A Light Receiver.* 1981. San Diego: CLP. **R. Bliss:** Project Director of children's two-model book series. "Designed for use by children in elementary grades in public schools, these two-model books contain creation/evolution discussions on an introductory scientific level" (description of series in ICR catalog). This book for grades 5–7. Discusses structure of eye as an optical instrument. Purpose of book is to enable student to choose between creationist and evolutionist interpretation of the function of the eye. Creationists argue that design of the eye shows intelligent creation; evolutionist position is that these facts support "random evolution" of the eye.

1533 Spanner, D.C. *Creation and Evolution.* 1965. London: Falcon Books. Also pub. 1965 by Zondervan (Grand Rapids, Mich.), and by Church Pastoral Aid Soc. (London). *Spanner:* reader in botany at Univ. London. Spanner wants a synthesis of creation and evolution. The Bible is true—but evolution is also probably fact. Evolution is "very extensive," though there may not have been a single origin for all of life. Urges that "six days of creation" not be taken too literally. We must be "reverent and cautious" in interpreting Scripture.

1534 Sperling, C. Nelson. *Evolution as a Passing Theory.* 1926. (Listed in SOR CREVO/IMS.)

1535 Spilsbury, Richard. *Providence Lost: A Critique of Darwinism.* 1975. Oxford, England: Oxford Univ. Press. *Spilsbury:* philosophy lecturer at Univ. Coll. of Wales; former philosophy of science prof. at Univ. Toronto; has studied genetics. Purpose of book: "to propose limits to the scientific understanding of man, with special emphasis on the orthodox Neo-Darwinian account," which has "hardened into a scientific dogma" though it has not been firmly demonstrated. Argues against Monod's "chance and necessity." The universe, and evolution, is purposive. Drafts of this book were read by **Koestler** and Waddington. Includes blurb by Koestler.

1536 Standen, Anthony. *Science Is a Sacred Cow.* 1958 [1950]. New York: E.P. Dutton. Orig. 1950. *Standen:* Oxford Univ. grad.; British chemist; taught at St. John's Coll. (Md.); Catholic convert; became U.S. citizen; an editor of *Encycl. Chem. Technol.* A clever and hard-hitting attack on scientism. Standen thinks scientists are conceited, arrogant, and have a vastly overrated conception of their importance. Smugly superior, they assume they alone possess the key to knowledge and truth. Much of book is criticism of university science education policies. Cites **W.R. Thompson, Belloc;** praises Aristotle. Says biology deals mostly with mere analogies, yet biologists are forever claiming to be testing hypotheses. (Likewise for the social sciences.) Evolution is a great generalization, but it hardly deserves the title of theory, since it "has by no means been tested by experiment." There is a vague theory of evolution and a precise theory: the first, which says that evolution occurred, is true. "The precise theory of evolution is that all forms of life on earth today came from some original form of life by a series of changes which, at every point, were natural and *explainable by science.*" Since there is no good precise theory it is wrong to treat evolution as established—though scientists have unbounded faith in the vague theory. Chides biology for rejecting teleological explanations in favor of reductionist ones. Evolutionary "trees" are full of missing links and hypothetical ancestors: no branches or trunks. "Man is not an animal." Well—maybe he is;—but the important things about man are not those things that biology (or any science) can tell us. Ethics is important, and ethics is impossible without theology. "The first purpose of science is to learn about God, and admire Him, through His handiwork."

1537 Stanton, Mary. *Can You ... Recognize Bias in History Content?* (pamphlet). 1977. El Cajon, Calif.: Institute for Creation Research. ICR "Impact" series #45. *Stanton:* Ed.D.; co-author of **Hyma** and Stanton 1976 (ICR history textbook). Protests the anti-Christian bias in textbooks. They present evolution but not Adam and Eve; they ignore Bible-believing scholars; etc.

1538 Steeg, Richard A. *Spacemen or Angels?* (pamphlet). Undated [1974?]. Bible-Science Association. Discusses 1973 "In Search of Ancient Astronauts" T.V. show based on **von Däniken.** Von Däniken was partly correct: earth was visited by outer space beings who interbred with humans; this is "actual historical fact." "The beings, however, were angels, not spacemen." The "sons of God" of Gen. 6 were angels; they mated with human women, thus committing the sin of mating with different kinds of flesh, and polluted the human seed. The forced interbreeding was evil, and brought degeneration. It was not beneficial and civilizing, as Von Däniken supposes; this is an evolutionist assumption, and evolution is false. Disputes **Scofield**'s interpretation (Sethites and Cainites).

1539 Steele, DeWitt. *Investigating God's World.* c1977. Pensacola, Fla.: A Beka Book. No. 5 (grade 5) of A Beka Book Science Series. Textbook for Christian schools; includes biblical references. A Beka Book is division of Pensacola Christian College. Part 2 (of 3), on physical science, is by Herman and Nina Schneider—reprinted from their c1973 textbook (D.C. Heath). (Part 2 does not deal with evolution or creation; **Ellsworth** and **Ellsworth** in fact criticize the Schneider's textbooks for being evolutionist propaganda.) The A Beka Science Series books are handsome, profusely illustrated, and comprehensive, covering all standard science textbook material. They are also strongly creationist, affirming special fiat creation. The sidebar sections in particular emphasize that design in nature must result from creation by God, not random evolution. Includes most standard creation-science arguments. "Fossil Record" chapter presents and endorses Flood Geology. It includes Paluxy manprints, Nebraska Man, Precambrian pollen, trilobite-in-shoeprint, etc. There is no scientific evidence at all for evolution, which is an "imaginary process." If it were true, all amoebas (e.g.) "should be extinct." Dinosaurs either died during the Flood or were killed off by man. Photos, color illustrations; index.

1540 ________. *Observing God's World.* 1978. Pensacola, Fla.: A Beka Book. No. 6 (grade 6) of A Beka Book Science Series. A textbook for Christian schools; includes biblical references. Physics, chemistry and astronomy. Presents evolution as false. Some sections explicitly creationist. Favors young-earth creationism. Photos, color illustrations; index.

1541 ________. *Science of the Physical Creation in Christian Perspective.* c1983. Pensacola, Fla.: A Beka Book. A Beka Book Science Series: secondary school level. Written with Verne L. Biddle and Lawrence S. Lynn. Biddle: analytical chem. Ph.D.—Univ. Tenn.—Knoxville; chem. teacher at Pensacola Christian Coll. Biblical references. Strongly creationist, especially biochemistry and geology sections. "Interpreting the Fossil Record" chapter is full-blown presentation of Flood Geology. "The worldwide Flood described in the Bible is an actual historical event." It provides the best explanation of geology. Cites **H. Morris,** other creation-scientists. Presents young-earth evidences. Describes and illustrates the Ark. Presents great creationist scientists of the past. Quotes evolutionists on lack of transitional fossils. Refutes proposed transitional fossils and fossil hominids. Shows fossils anomalies (trilobite-in-shoeprint, etc.). Endorses **Ussher**'s chronology. Discusses scientific explanation of Joshua's Long Day, but adds that Bible must be believed even if there is no scientific confirmation of it. Photos, color illustrations, charts; index.

1542 ________. *Science: Order & Reality*. c1980. Pensacola, Fla.: A Beka Book. A Beka Book Science Series: secondary school level. "Edited by" Laurel Hicks and Jerry Combee. Biological and physical science. Strongly creationist. Science is not absolute knowledge; the Bible is the only source of absolute knowledge. Evolutionists "want to destroy order altogether," since they believe nature evolved from disorder. "Origin of Life" chapter consists mostly of anti-evolution arguments and standard creation-science evidences. "Living things have not evolved from non-living things." "A vast body of scientific evidence opposing evolution has been slowly accumulating. The majority of scientists who accept evolution do so because they refuse to accept special creation. Perhaps the best evidence against evolution is the fact that modern science exists." Allows for possibility of Gap Theory creationism (pre-Adamic destruction of world ruled by Satan), but main emphasis is on Noah's Flood. Refutes evolutionist arguments of vestigial organs, embryology, geological column, transitional fossils, etc. Quotes scientists admitting problems with fossil record. Points out bad social consequences of evolution. Presents Bible passages predicting scientific discoveries. Criticizes radiometric dating. Photos, color illustrations, tables; index.

1543 Steele, E.J. (Ted). *Somatic Selection and Adaptive Evolution: On the Inheritance of Acquired Characteristics*. 1981 [1979]. Chicago: Univ. Chicago Press. 2nd ed. Orig. 1979 (Toronto: Williams and Wallace). *Steele:* Australian biologist. Noting that mutations occurring in body cells can be spread by viruses to other cells, Steele proposes that transmission to germ cells can result in inheritance of acquired characteristics. Steele claims his experiments support this controversial proposal: new genetic material created by the immune system (antibodies) in response to environment is transmitted to germ cells. Also suggests that other body systems pass information to germ cells. But other scientists, led by Medawar, have been unable to replicate his experiments. (Steele blames this on their lab procedure.) Book has endorsement by **Koestler** and blurb by Medawar on importance of Steele's proposal. Influenced by Koestler; also Popper, Polanyi and Kuhn. Cites **Macbeth**'s *"Darwin Retired"* [sic]. Even if true, say some evolutionists, Steele's mechanism represents an addition to—not a replacement of—Darwinian evolution.

1544 Steidl, Paul M. *The Earth, the Stars, and the Bible*. 1979. Phillipsburg, N.J.: Presbyterian and Reformed. *Steidl:* physics B.S.—Univ. Wash.; astronomy M.S. Strict creationism; presents young-earth arguments. Nicely written; gives much basic astronomical information as well as anti-evolution arguments. Cites recent creation proofs of **T. Barnes, Whitcomb** and **Morris, Patten,** etc. Many scientific references; also biblical references. The space program won't help us understand earth's origin; to "expect one planet to shed light upon another" is an evolutionist assumption. Joshua's Long Day and Hezekiah's Sign: God stopped the earth. "Objections that disastrous side effects would result from such a stoppage of the earth are of no import; if God can stop the earth, He can also stop the side effects." Joshua's event also proves Bible is not geocentric, as moon was ordered to stop also. "The earth really did reverse its spin" to move Hezekiah's shadow; this is confirmed by evidence it was observed in other places. Argues against stellar evolution. Describes the various types of stars and stellar phenomena, but says "no evolutionary relationship has been demonstrated." States that big bang cosmogony must be rejected on grounds it contradicts Genesis order of events. Steady state cosmogony is also anti-Scriptural. According to the Bible there is no extraterrestrial life (except for the heavenly beings). Photos, diagrams; index.

1545 Steiger, Brad. *Gods of Aquarius: UFOs and the Transformation of Man*. c1976. New York: Harcourt Brace Jovanovich. *Steiger:* investigator and promoter of psychic phenomena and UFOs; has written a hundred metaphysical, inspirational and UFO books. Reality of telepathy, psychic healing, pyramid power and other psychic phenomena, UFO tales, messages channeled by UFO contactees, and visions of religious archetypes prove that an Intelligence from outer space is preparing mankind for cosmic linkage. There will be a transitional period before the Aquarian Age when earth will undergo great catastrophes, both natural and man-made. The UFOs have guided our evolutionary development, and are now leading us into an evolutionary leap forward. Man no longer requires an external source of salvation. Natural evolution is finished, but man will be able to technologically manipulate and direct his future evolution: "we, too, can literally become as gods under God." "It is my conviction that the 'gods' of Aquarius, the Olympians of the New Age, will be transformed humankind." Many biblical and Eastern religious concepts. Discusses **R. Norman** and many other UFOlogists. Photos, drawings; index. Steiger also wrote *The Seed, Flying Saucers Are Hostile* (1967; with Whritenour), and many other UFO books.

1546 ______. *Worlds Before Our Own.* 1978. New York: Berkley. "Archaeologists, anthropologists, and various academicians who play the 'Origin of Man' game, reluctantly and only occasionally acknowledge instances where unique skeletal and cultural evidence from the prehistoric record suddenly appear long before they should—and in places where they should not." Presents out-of-order fossils and artifacts, including many proving that dinosaurs coexisted with man (has photos of Paluxy manprints).

1547 ______**, and Francie Steiger.** *The Star People.* 1981. New York: Berkley Books. This book is part of "Star People Series" of five books by the Steigers. Extraterrestrials are responsible for human origins. They arrived on earth and planted both "good" and "bad seed." Thousands of people are now discovering they are descendants of these ETIs, and are sent to guide humanity into the coming golden New Age. "Certain of the Star People have memories of a star ship that came to this planet thousands of years ago . . . to observe, to study, to blend with evolving *Homo sapiens.* Their own seed would enrich the developing species and accelerate the time when mankind would begin to reach for the stars—and their cosmic home." The "Star People Pattern Profile" helps identify those who are direct descendants. Francie is one of these; she does a lot of channeling. Genesis is a memory about these ETIs, who arrived 40,000 years ago. The father of Jesus was an ET. They improved human DNA by mating with the primitive earth creatures and laboratory manipulation. The new, implanted DNA is now becoming activated: this explains psychic abilities, and why people are now discovering their extraterrestrial origins. Predicts increased UFO activity, pole shift in 1982–84, Armageddon in 1989–90.

1548 Steiner, Rudolf. *Cosmic Memory: Atlantis and Lemuria.* 1971. Blauvelt, N.Y.: Rudolf Steiner Publications. Orig. German; 1939 *(Aus der Akasha-Chronik).* Collection of essays; the first dated 1904. *Steiner:* Austrian; worked at Goethe-Schiller Archives; founder of Anthroposophy. Dense occultic philosophy; anti-materialist; derived from the Akashic Chronicles. "Man existed before there was an earth." The earth has evolved along with man's consciousness; man used to live on the moon and other planets. Occultic interpretation of Genesis Creation Days. There are seven "Root Races" of mankind, each composed of seven stages of sub-races. "Each root race has physical and mental characteristics which are quite different from those of the preceding one." Two root races are in the future. First root race is the Polarean; second is the Hyperborean. The earth, and life, was vastly different in these early stages. Humans were not divided into sexes, and had entirely different senses. They were astral or etheric beings who inhabited the atmosphere; later fire-mist beings. A cosmic catastrophe—the extrusion of the moon —caused further evolution. The Lemurians were the third root race. (Savage tribes seem to be direct, stunted descendants of these.) The fourth root race, the Atlanteans, descended from the most evolved Lemurian sub-race. The seven Atlantean sub-races are the Rmoahals, Tvalatli, Toltecs, Turanians, Primal Semites, Akkadians, and Mongols. The Aryans are the fifth root race.

1549 Steno, Nicolaus [Niels Stensen]. *Prodromus to a Dissertation Concerning*

Solids Naturally Contained Within Solids. 1669. Florence. Orig. Latin. Pub. in English in 1671 (London); 1916 (New York). *Steno:* Danish scientist—medical anatomist, then geologist; later a theologian and Catholic bishop. Steno was one of the founders of modern geology. Steno argued that fossils—"solids occurring naturally within solids"—were organic remains and were not formed by the rocks themselves. He distinguished between mineral intrusions into other rocks, and solid objects (organic remains) which subsequently became enclosed and petrified in rock. He argued that sedimentary rocks were formed from waterborn deposits, and described rock strata. Steno, who sought physical causes for the phenomena of nature, is credited with introducing the concept of chronology in a naturalistic earth history, although he assumed the validity of the Mosaic chronology. He attempted to show that fossiliferous rocks were formed by the Flood, while other rocks date from Creation. The *Prodromus* also contained valuable insights in a number of other scientific fields.

1550 Stiverson, James A. *The Harmony of God's Laws in Creation* (booklet). c1986. Priv. pub. Stiverson's wife, two weeks before her death, presented a message about Divine order and numerology (it had a lot of three's). Inspired by the Holy Spirit, Stiverson began this book a month later, and discovered that all the science books he studied confirmed his wife's message. Largely about the "electromagnetic spectrum of life": the correspondence of life-forms to the electromagnetic spectrum. Includes color theory and therapy, radionics, healing powers. (In ICR Library.)

1551 Stokes, William Lee. *The Genesis Answer: A Scientist's Testament for Divine Creation.* 1984. Englewood Cliffs, N.J.: Prentice-Hall. Modified Day-Age creationism. "This is the Genesis Code: Each creative 'day' consists of a period dominated by darkness and a period dominated by light. Earth emerged from chaos as a product of the progressive succession of six such periods. The creative days were not of equal duration and were not intended to be measures of time. They are not the periods, epochs, and eras invented by geologists. Their meaning is celestial and not terrestrial. They are God's divisions of his own creation." Explains how each Genesis period corresponds to scientific knowledge of the formation of the earth. Stokes praises Asimov's and Sagan's recent books for helping him to arrive at these conclusions, and criticizes **Jastrow** for failing to pursue the religious implications of his own book.

1552 Stoner, Peter W., and Robert C. Newman. *Science Speaks: Scientific Proof of the Accuracy of Prophecy and the Bible.* 1976 [1958]. Chicago: Moody Press. Orig. 1958; written by Stoner. 1944 version titled *From Science to Souls;* also 1952 ed. 4th rev. ed. 1976 lists Newman as co-author. Other eds. say "assisted by" Newman. *Stoner:* a founder of the **American Scientific Affiliation;** former math prof. at Pasadena City Coll. (Calif.); emer. science prof. at Westmont Coll. (Calif.). Harmony of science and scripture: modern scientific discoveries are catching up with—and confirming—the Bible. Demonstrates how scientific findings (including old-earth evidences) correspond to Newman's Day-Age creationist interpretation. "Thus we find that the thirteen things named in Genesis are in the same order that geology finds them." Computes probability of this and similar statements being true without supernatural knowledge. Section on evolution declares that God's creative acts cannot be denied. Second part of book is a probability argument based on proofs of Bible prophecies. Stoner estimates probability of each prophecy coming true, and multiplies everything together: this combined probability, which is vanishingly small, represents the chances of all coming true. Thus, the Bible is God's divine word. Continues with prophecies regarding Christ. Christ's divinity is the best-proved fact in history. The only alternative to admitting all this is to say: "I shall live a life in sin against God, and for this decision I shall spend eternity in hell with Satan."

1553 Straton, John Roach. *The Famous New York Fundamentalist-Modernist Debates: The Orthodox Side.* 1925. New York: George H. Doran. *Straton:* pastor of Calvary Baptist Church, N.Y.; celebrated opponent of evolutionist advocate

H.F. Osborn of the N.Y. Mus. of Natural History. This book presents Straton's speeches from his debate series with C.F. Potter. (Apparently he omits Potter's.)

1554 ________, and Charles Francis Potter. *Evolution versus Creation.* 1924. New York: George H. Doran. *Potter:* well-known Unitarian minister in New York. This book is record of a debate between Straton and Potter. "Second in a series of fundamentalist-modernist debates" between the two.

1555 Strickling, James E., Jr. *Origins — Today's Science, Tomorrow's Myth.* 1986(?). New York: Vantage Press. An examination of "major fallacies of Darwinism"—but also an "unprecedented critique" of creationism. Strickling describes himself as a neo-catastrophist and Velikovskian. Summarized in 1986 *Kronos* article. Folklore evidence shows that mankind's Golden Age occurred during Mindel/Riss glaciation interval. Then came the great Flood, and a sudden burst of mutations created our human races. (Previously, the only human differences were between inhabitants of East and West.) This demonstrates the sudden origin of the black race, as described in legend. Flood destroys Bering land bridge.

1556 Stringfellow, Bill. *All in the Name of the Lord.* 1980. Clermont, Fla.: Concerned Publications. Stringfellow discovered that the Lord's Day is Saturday. Sunday Sabbath is un-biblical; it is Satan's primary hoax. This counterfeit pagan practice of Sun(day) worship was initiated in Babylon by Nimrod. Mandatory Sunday worship, which is imminent, will signal Armageddon. Observing the correct Sabbath is necessary in order to affirm belief in a literal 24-hour day, six-day Creation. "It is a sign to the world that despite all the modern teaching of evolution" you believe in strict creationism. Relates story of converting a postgrad student from evolution to creationism by parable of design in a watch (**Paley**'s argument). No creation-science arguments. Does not give any denominational affiliation.

1557 Students for Origins Research. *CREVO/IMS: Creation/Evolution Information Management System* (reports). 1983–1984. Santa Barbara, Calif.: Students for Origins Research. Intended to provide college students with access to information on creation/evolution sources. Developed by Kevin Wirth (SOR director of research). CREVO/IMS is "a computerized database system which manages a wide-ranging variety of information on creation and evolution topics. Although still in its infancy, CREVO/IMS is probably the largest and most comprehensive system of its kind in the world." It consists of several sections available separately. These include: listings of 1060 magazine and journal articles arranged alphabetically by author (MAGAUT.RPT), sorted by date (MAGDAT.RPT), and by title of journal (MAGPUB.RPT); listings of 2600 books arranged by author (BOOKAUT.RPT), date (BOOKDAT. RPT), and by publisher (BOOKPUB. RPT); and 550 articles from *Creation Res. Soc. Quart.* arranged by author and date (CRSQ.RPT). These files are updated quarterly. SOR also plans to enlist a network of volunteers (SORNET) to research, edit, write, and enter data, and hopes that the files can be put online for timesharing. Customized searches are available.

1558 ________. *Branch Chapters* (booklet). Santa Barbara, Calif.: Students for Origins Research. Students for Origins Research began as a creation-science group at Univ. Calif.—Santa Barbara, and now has several branch chapters on other campuses. SOR publishes *Origins Research,* a creation-science journal distributed free to students and educators. SOR is fairly non-dogmatic and is the most open-minded of the major creationist groups. It does not require members to submit to any statement of belief, and publishes anti-creationist as well as creationist opinions. This booklet (elsewhere credited to **D. Wagner**) contains information about starting and maintaining local SOR chapters. Includes sample SOR constitution.

1559 Sumner, John Bird. *A Treatise on the Records of Creation, and on the Moral Attributes of the Creator.* 1816. London: J. Hatchard. Six eds. up to 1850. *Sumner:* Archbishop of Canterbury. A 'liberal,' Sumner urged reconciliation of geology with biblical revelation. Argues that Moses, because he was speaking to a pre-scientific audience, simplified the creation account accordingly, and spoke of only the last of a whole series of creations. The six-day creation of Genesis was thus a rearrangement of the wreckage of previous worlds (a modified Gap Theory view). Lyell cited Sumner for the effectiveness of his plea for reconciliation. A. White says that Sumner elsewhere voiced concern over "higher criticism" of the Bible, but felt it could not easily be refuted.

1560 Sunderland, Luther D. *Prominent British Scientists Abandon Evolution* (pamphlet). 1982. Richfield, Minn.: Onesimus. Orig. 1982 article in *Contrast* (*Bible-Science Newsletter* suppl.); reprinted as pamphlet. *Sunderland:* B.S.—Penn State Univ.; aerospace engineer with General Electric specializing in automatic flight control systems; developed his own experimental plane assembled from kit; very active creation-science promoter and lecturer; lobbied for Louisiana creation-science bill; lived in Apalachin, N.Y. (d. 1987). **Hoyle** and Wickramasinghe's denial of random evolution, and Colin Patterson's iconoclasm. Largely about 1981 talk by Patterson, a paleontologist at Brit. Mus. Nat. Hist., to an Amer. Mus. Nat. Hist. discussion group (Sunderland obtained, then publicized—"unethically" according to Patterson—a tape of that talk.) Patterson complains that evolution is believed and taught dogmatically, and that smug acceptance inhibits productive research. (A cladist, Patterson feels that classification based on evolutionist assumptions interferes with objective taxonomy.) He declares that nothing certain is known about evolution and that it conveys "anti-knowledge." Sunderland and Patterson rely on N. Gillespie's 1979 book arguing that Darwin consciously attempted to replace the creationist paradigm with a positivist one. Patterson's talk is also subject of 1982 ICR "Impact" series pamphlet #108 by Sunderland and **Parker:** *Evolution? Prominent Scientist Reconsiders.*

1561 ________. *Darwin's Enigma: Fossils and Other Problems.* 1984. San Diego: Master Books. Preface by Genevieve Klein: N.Y.. State Board of Regents (emer.). Includes blurbs by T. Black (Board of Regents Emer. Chancellor), R. Frantz (mineral engin. prof.—Penn State), R. Jenkins (biol. prof.—Ithaca Coll.), **Austin, Fix,** and T. Bethell (*Harper's* contr. ed. who has written several anti-evolution articles and had planned a book.) Based primarily on interviews with five prominent paleontologists: Colin Patterson (Brit. Mus. Nat. Hist.), Niles Eldredge (Amer. Mus. Nat. Hist. N.Y.), David Raup (Field Mus. Chicago), David Pilbeam (Yale Peabody Mus., now at Harvard), and Donald Fisher (state paleontologist of New York). Sunderland conducted these interviews for a 1978 study by N.Y. State Educ. Dept., commissioned by Board of Regents, on how to present origins theories. Complete transcripts available from ERIC doc. Repr. Serv.: microfiche ED 228 056 *(Darwin's Enigma: The Fossil Record).* Includes Patterson's 4/10/79 letter to Sunderland (widely quoted by creationists) admitting absence of transitional forms. Quotes **Macbeth, Hitching,** Fix, **Hoyle, Koestler** and other anti-Darwinians. Macbeth told Sunderland that Eldredge and Gould invented punctuated equilibria theory of evolution "partly in response to" Macbeth's 1971 book. Belittles Darwin's talents and originality, citing **Himmelfarb** and "historian" **B. Thompson.** Praises Baconian scientific methodology, and Popper's criterion of falsifiability. Accuses Eldredge of publicly claiming horse series was good evidence of evolution after privately conceding it was "lamentable." Discusses Marxist implications of punctuated equilibria theory. Says "noncreationist" **Velikovsky**'s catastrophism is now being proven true. Quotes G. Bateson and M. Sahlins as saying Darwinian natural selection is bunk. Presents lengthy summary of **Moorhead** and Kaplan symposium; also of Macbeth's 1983 Harvard debate. Sunderland asserts paleontologists realize that evolution has been falsified, but that educators prevent this knowledge from reaching students. He argues for inclusion of evidence for "abrupt appearance" and of worldwide catastrophism. This book is very highly praised as a non-religious refutation of evolution. Photo, charts, diagrams; bibliog., index.

1562 ________. *A Critique on Creation Models* (pamphlet). Undated. Bible-

Science Association. Primarily a critique on Flood Geology model of creationism, which Sunderland says doesn't hold water. Sunderland's model is based on **Northrup**'s "Catastrophe Series Theory." Scripture tells of five catastrophes. The first occurred during the literal six-day Creation (universal flood of Gen. 1:2). Second: the orogeny of Gen. 1:9 when God lifted land out of sea. Third: Noah's Flood—worldwide tidal waves which deposited Proterozoic and Paleozoic strata. Fourth: retreat of Flood waters, lasting a thousand years. Mesozoic strata formed by jet stream used by God to dry up Flood; dinosaurs—swamp and marine creatures, not land animals—died in the new post-Flood environment. Fifth: Cenozoic phase of intense vulcanism, and bursting free of Flood waters trapped inland.

1563 _______. *Evolution of Evolution, and the Creation of New Tactics* (pamphlet). Undated [1980s]. Minneapolis: Bible-Science Association. The new punctuated equilibria theory of evolution is merely a restatement of Goldschmidt's "hopeful monster" theory, which refutes evolution. Advocates such as S.J. Gould and Eldredge know that the fossil record refutes gradualism, yet they now defend Darwinian evolution in public statements in order to combat creationism. Also recommends Steven Stanley's 1979 book on macroevolution as demonstrating the poverty of punctuated equilibria theory. Many arguments same as later presented in 1984 book.

1564 _______. *Hopeful Monsters—Evolutionists' Last Chance* (tract). Undated. Priv. pub.? (Bible-Science Assoc.?). Sunderland points out that he, and **Gish,** had earlier used Goldschmidt's "hopeful monster" theory as an example of how desperate evolutionists had become. Evolutionists responded that this was unfair—but then revived the hopeful monster theory themselves. Sunderland praises **Himmelfarb;** also **Kerkut, N. Macbeth.** S.J. Gould, he says, rejects natural selection at the macro level. Argues that evidence for the hopeful monster theory (systematic macro-mutations resulting in new forms in single jumps) is identical to evidence for creation. Hopeful monster theory violates all laws of science, but creation-science agrees with all laws.

1565 Swaggart, Jimmy. *The Pre-Adamic Creation and Evolution* (audiocassettes). c1984. Baton Rouge, La.: Jimmy Swaggart Ministries. *Swaggart:* Pentecostal televangelist (Assemblies of God) based in Baton Rouge; has second-largest U.S. religious T.V. audience (after **Robertson**'s *700 Club*); claims to reach 500 million viewers in 140 nations; editor of *The Evangelist;* founder and chancellor of Jimmy Swaggart Bible Coll.; claims to have sold more gospel music albums than any other musician; first cousin of rock'n'roll pioneer Jerry Lee Lewis; militant opponent of secular humanism. Set of three audiocassettes. First half is exposition of Gap Theory creationism. The pre-Adamic creation, ruled by Satan, is of unknown age, but could be millions of years old. God destroyed it prior to the six-day re-creation of Genesis. Second half is vitriolic attack on evolution. Denounces arrogance of intellectuals, academics and scientists. Ridicules evolution by relating various old and silly anti-evolution tales and arguments, including several used by **W.J. Bryan.** Emphasizes dire consequences of communist and atheist educational indoctrination which promotes evolution. Attempts to cite some scientific evidence against evolution. Swaggart frequently vilifies evolution on his telecasts.

1566 _______. *Questions & Answers: Bible-Based Answers to Your Questions About Life.* c1985. Baton Rouge, La.: Jimmy Swaggart Ministries. Includes section on evolution. One cannot believe in both creation and evolution. Denounces and ridicules evolution, which "degrades God's image to nothing more than a mere beast." Most fossils of prehistoric man are fakes. "The evolutionists teach that *hair* is but elongated scales of prehistoric animals. They teach that *legs* of all animals developed from warts on aboriginal amphibians. They teach that *eyes* are but an accidental development of freckles or blind amphibians that responded to the sun. They also teach that *ears* came about by the airwaves calling to spots on early reptiles. They teach that *man* came from monkeys. They teach that the vast *universes* came from a few molecules.

They actually teach that *nothing* working on *nothing* by *nothing* through *nothing* for *nothing* begat *everything!*" Strongly pre-millennialist. Denounces **H.W. Armstrong.** Advocates speaking in tongues and other Holy Spirit manifestations. Stresses reality of Hell and Satan.

1567 _______. *Rape of a Nation.* c1985. Baton Rouge, La.: Jimmy Swaggart Ministries. Foreword by **LaHaye.** Material from Swaggart's crusade sermons assembled in book form. Sounds the alarm to arouse Christianity into action against the sinister forces of humanism which are destroying America. Chapter "They Became Fools," an attack on public education, focuses on evolution. Says teaching of evolution in public schools began in 1925. "Evolution is a bankrupt, speculative theory (a guess)—not a scientific fact. And only a spiritually bankrupt society would stubbornly persist in such foolishness."

1568 _______. *The Creation of Man and Fallen Angels* (audiocassettes). [1980s?]. Baton Rouge, La.: Jimmy Swaggart Ministries.

1569 Swaggart, Jimmy (ed.). *Issues of the Eighties* (booklet). 1985. Baton Rouge, La.: Jimmy Swaggart Ministries. Magazine format; special issue consisting of articles from Swaggart's magazine *The Evangelist.* Distributed at Swaggart crusade services. Swaggart's son Donnie also listed as editor. Articles on rock music, incest, pornography, alcohol, abortion, communism, humanism, the occult, behavioral sciences. Includes anti-evolution statements. Recommends ICR and BSA in section on how to combat secular humanism ("the new religion: false teaching"), but lists obsolete addresses. *The Evangelist* has frequently attacked evolution.

1570 Swift, Wesley A. *In the Beginning God.* Listed in Identity book adv., in Christian Defense League Report. Swift also wrote *Were All the People of the Earth Drowned in the Flood?* and other works. The Identity movement claims the Anglo-Celtic race is the true descendant of Israel, God's favored people, and that inferior races, which were created prior to Adam, survived the Flood. Swift, of Lancaster, Calif., is a prominent advocate of Anglo-Israelism.

1571 Talbot, Louis T. *God's Plan of the Ages.* 1946 [1936]. Grand Rapids, Mich.: William B. Eerdmans. 1983 reprint of 3rd ed. Orig. 1936. *Talbot:* chancellor of Biola Coll. (Calif.) and Talbot Theol. Sem. "A Comprehensive View of God's Great Plan from Eternity to Eternity, Illustrated with Chart." Dispensational pre-millennialism. First chapter "Earth's Earliest Ages" is presentation of Gap Theory creationism. Original creation was corrupted by the fallen Lucifer (Satan). This may have occurred millions of years ago. "Many people imagine the Bible teaches that the earth was *created* in six solar days, but nowhere does the Bible say so." "Very clearly the Scriptures teach that God renovated a chaotic earth; and having brought order out of chaos, He created a new being—man—and gave him dominion over the renovated earth." Largely concerned with Bible prophecy and the coming Millennium.

1572 _______**, and Harry Rimmer.** *The Inquiring Student and the Honest Professor* (pamphlet). c1940. Glendale, Calif.: Church Press. "A Radio Skit Presented over KMTR" by Talbot and **Rimmer.** (Listed in Ehlert files.)

1573 Tanner, Jerald. *Views on Creation, Evolution, and Fossil Man.* 1975. Salt Lake City: Modern Microfilm Co.

1574 Tatford, Frederick A. *Is Evolution True?* (pamphlet). Undated [1969]. Evolution Protest Movement. EPM pamphlet #174. SOR CREVO/IMS lists it pub. 1969 by Prophetic Witness Pub. House.

1575 Taton, René (ed.). *A General History of the Sciences: Vol. 4.* 1966 [1964]. London: Thames & Hudson. Orig. pub. 1964 by Presses Universitaires de France (Paris); French *(La Science Contemporaine II).* Discussed and quoted in **Hayward** 1985. Hayward says that Andrée Tétry, who wrote the section on evolution in vol. 4, is one of France's greatest biologists and evolutionists, but that she rejects Darwinism, Lamarckism and other theories of evolution as inadequate. Especially critical of neo-Darwinism—says evolution by chance mutations is impossible. She implies that neo-Darwinism is not much accepted in France, though highly regarded in England and the U.S. Tétry believes evolution occurred, but doesn't know how.

1576 [Tayler, William Elfe]. *Voices from the Rocks; or Proofs of the Existence of Man During the Paleozoic or Most Ancient Period of the Earth.* 1857. London: Judd & Glass. Published anonymously in response to **H. Miller**'s attempted reconciliation of Genesis and geology, but credited to Tayler by Millhauser. Quoted and endorsed by **G.M. Price** 1926 (and third-hand by **R.M. Daly** 1972). Price: "This book was carefully and candidly written in the light of the best knowledge then obtainable, for it was during the very heat of the discussion as to whether or not man is ever found in the fossil state. Some of the examples there recorded have since been 'explained' in one way or another, while others have been ignored entirely." Describes human fossils which refute evolutionary sequence. Tayler writes: "These discoveries are so clear and incontrovertible that impartial inquirers after the truth are amazed at the obstinacy with which geologists persist in shutting their eyes to the real facts in the case. The world affords no parallel to such conduct, unless, perhaps, that of the Church of Rome in reference to the discoveries of Galileo." Also quoted and endorsed by **B. Nelson** 1931, who says the author described several geological phenomena indicating rapid successive deposition of strata. Describes polystrate fossils: "Amongst the various phenomena which the indefatigable researches of geologists have lately brought to light, few have awakened more attention than the discovery of a vast number of fossil trees and plants, standing in some cases in an erect, and in others an oblique position, and *piercing through successive beds of stone.*" "A few passing waves, or at least a few days of the agitated and turbid waters of the Deluge, must have been sufficient for the whole formation of the bed" in which such trees are found. Also cites perfect fossilization of delicate plants, and describes fossil sites containing thousands of tightly packed fish. "All these fishes must have died suddenly on the fatal spot, and have been *speedily buried* in the sediment then in the course of deposition. From the fact that certain individuals have even preserved *traces of color* upon their skin, we are certain that they were entombed before decomposition of their soft parts took place."

1577 ________. *Geology, Its Facts and Its Fictions; or, The Modern Theories of Geologists Contrasted with the Ancient Records of the Creation and the Deluge.* 1855. London: Houlston & Stoneman. Tayler also wrote *Popery: Its Characters and Its Crimes.*

1578 Taylor, Carol. *Fossil Footprints Found* (booklet). c1973. Caldwell, Idaho: Bible-Science Association. *Taylor:* elem. school teacher (Tacoma, Wash.). Coloring-book format. The Paluxy River fossil footprints, which many creationists say include human as well as dinosaur tracks. "The proof is here before us/ Man lived with Allosaurus/ (Or one like him.)"

1579 Taylor, Charles V. *The Oldest Science Book in the World.* 1984. Priv. pub. (Assembly Press: Slacks Creek, Queensland, Australia). *Taylor:* language, music, and theology B.A.s; applied linguistic M.A.; Ph.D. in Central African languages; has written books on language; taught at Univ. Sydney and Univ. Mich. Bible-science. Preface by **D. Watson.** Strict six-day recent creationism; worldwide Flood, Vapor Canopy. Argues that Genesis is collection of nine distinct sources, edited by Moses, and linked by colophons (closing title phrases). Opening verses written by God; the next few by Adam. Genesis is straight history, not religion, and only history—not science—can tell us about the past. But the Bible harmonizes perfectly with science. Evolution and Big Bang cosmogony, unlike Genesis, are similar to folklore. Emphasizes contradictions between "science fiction" of evolution and the Genesis account. Refutes notion that Genesis cosmology is a pre-scientific flat-earth view. 'Firmament' is atmosphere. Admits Bible does not directly state *ex nihilo* creation. Endorses **Setterfield**'s ideas; also creation with appearance of age. Discusses snake's locomotion prior to Fall. Presents (oddly) 1771 *Encycl. Brit.* passage endorsing "plurality of worlds" view as example of science's former acknowledgment of God. Says that early theologians erroneously interpreted Genesis as referring to primordial (pre-existent) chaos—an interpretation contradicted by modern science. Haydn's oratorio *Creation* follows this erroneous view. Argues for pre-Flood population of millions or billions. "Miracles aren't anti-scientific." "Once we reject the Bible at one point we become judges of the Bible and can pick and choose which points we accept or reject. . . . If man arose from the animals, then it's hard to see how sin could be the result of the Fall and hard to appreciate the value of Jesus' death." Drawings, diagrams; index.

1580 Taylor, George. *The Indications of the Creator; or, The Natural Evidences of the Final Cause.* Undated [1850s?]. Glasgow: William Collins. Reprinted and bound with **Croften** 1853. (Listed in Schadewald files.)

1581 Taylor, Gordon Rattray. *The Great Evolution Mystery.* c1983. New York: Harper & Row. *Taylor:* grad. of Cambridge Univ.; reporter; BBC chief science advisor; science writer (*Biological Time Bomb; Natural History of the Mind;* etc.); d. 1981. A literate critique of Darwinian theory. Accepts evolution but denies it occurred only by natural selection working on random variations. Presents usual arguments regarding its inadequacies. Taylor suggests "third alternative"—neither creationist nor Darwinian—to resolve chance vs. purpose dichotomy. Argues that Darwinism cannot account for intricately coordinated change. Eloquent plea for reconsideration of Lamarckism. Posits an inherent drive within evolution, involving indirect feedback from the environment to genetic inheritance. Variation not entirely random. Seeks laws of form. Asks why primitive forms are still in existence. Presents examples of alleged orthogenesis: argues that these "evolutionary trends sometimes acquire a momentum which carry them forward long past the point at which natural selection should have eliminated them." Denies universal competition in nature. Darwinian explanation of giraffe's neck is "nonsense" because female is shorter. Appearance of structures before need—before acted on by selection (e.g. *Archaeopteryx* feathers evolved before flight). Flightless birds, extreme perfection of the eye, transformation of the ear, puzzle of instinct: all impossible to explain by current theory. Implies that orthodox Darwinians can't see, or refuse to face, contradictions. Cites **Koestler, E. Steele, Moorhead** and Kaplan, **Piaget, Willis,** plus many other critics of Darwinism. Many scientific references; good descriptions of the various scientific arguments and sources. Photos, drawings; bibliog., index. Book-of-the-Month Club alt. selec. Much cited by creationists.

1582 Taylor, Hebden. *Evolution and the Reformation of Biology.* 1967. Nutley, N.J.: Craig Press. "A Study of the Biological Thought" of **H. Dooyeweerd** and **J.J.D. de Wit.** *Taylor:* born in Congo (medical missionary parents); missionary in Alaska; Cambridge Univ. grad.; now pastor in England; author of books on Christian law and politics. Follows Dooyeweerd's approach; endorses Dooyeweerd's Reformed anthropology for a philosophy of man's nature. Urges biblical creationism rather than the "apostate humanist theory" of origins by chance. "The Reformed scientific approach to modern biology is the only one which can effectively answer the modern apostate evolution. The facts of science can be interpreted in either of two frames of reference: (1) evolutionary naturalism, or (2) the Biblical account of creation. As a result the Christian believes that the universe derives its existence from Almighty God who created it for His own glory out of nothing. It follows that scientific thought and research are fundamentally a religious activity." Advocates Flood Geology: "the Great Deluge alone offers a plausible solution to the enigma of the fossil record." Genetics does not allow for "continuous progressive evolution by means of mutations"; nor does natural selection. Science shows only variation within types. "Only by accepting God's Word as the ordering principle in scientific study can we make sense of the data of science." Includes standard creation-science arguments and quotes; also much philosophy and theology. Rejects E. Mayr's rejection of typological thinking in biology. Says pre-human remains are of "devoluted human tribes."

1583 Taylor, Ian T. *In the Minds of Men: Darwin and the New World Order.*

1984. Toronto: T.F.E. *Taylor:* research metallurgist; lives in Toronto. Taylor was researcher and writer for "Crossroads" creation-science T.V. series (on televangelist D. Mainse's Toronto-based *100 Huntley Street*). This book grew out of that research, and incorporates much material from the T.V. series—though Taylor alludes only once to a "documentary film series." "By keeping a low profile, he hopes to have provided a book attractive to the secular book trade in addition to the Christian market" *(Bible-Science Newsl.)*. Book binding reinforced to meet NASTA specs. for U.S. school textbooks. Impressive, handsome, scholarly volume: "easily the most ambitious creationist book" in some time. Comprehensive presentation of usual creation-science evidences and arguments; avoids many of the commonest creationist errors. Traces rise of humanism from Greeks to Darwin, who gave it scientific justification, and to the modern humanist attempts to found an elitist one-world government. Praises Plato's theism; condemns Aristotle's naturalism. Emphasizes that socialism depends on validity of evolution, which is criticized in detail. In contrast to American fundamentalists, Taylor argues that the American as well as the French revolution was based on socialist rejection of the Bible and theism. English humanists (especially Lyell), unable to get rid of the Bible directly, instead subtly created doubt about biblical doctrines of Creation and the Flood. Exposes **Newton**'s anti-Trinity and anti-supernatural beliefs. "An actual conspiracy is not being suggested, however, but rather a deeper motivation that lies hidden in the recesses of the human mind and one to which kindred spirits gravitate." "There is evidence to indicate that a cadre of keen and influential minds had been skillfully prepared to be apostles of the new faith for some time before the publication of Darwin's *Origin*." Explains fallacy of Malthus' argument. Shows that Lyell's geology was founded on assumptions and circular reasonings. Evidence points to catastrophism. Criticizes **Whitcomb** and **Morris**'s Flood Geology; prefers **G. Morton**'s model. Describes Darwin's Unitarian ties. Darwin married Wedgwood cousin (Taylor says she was his aunt) to acquire mate from "closely related 'superior' stock." Debunks Lady Hope tale of Darwin's deathbed confession. Mendel's genetics, which challenged Darwinism, was deliberately discouraged. Accuses conspiratorial evolutionist 'X' Club of cover-up of truth about Haeckel's primordial sea-bed Monera. Claims that textbooks deliberately suppress all evidence which does not support evolution. Fossils are tested by radiometric methods until a date is obtained which fits preconceived notions; all other dates are then discarded. Endorses **C. Baugh**'s claims, **Kang** and **Nelson, Setterfield,** creation with appearance of age. "Indebted" to **Macbeth.** Exposes fallacies of old-earth dating methods; endorses many methods showing recent creation. Says Marxists realize they can infiltrate Catholicism through **Teilhard**'s theistic evolution. Darwin's evolution is the foundation for Nazism. Masterminded by J. Huxley, the U.N.'s "new world order" is explicitly based on evolution. Photos, drawings, charts; bibliog., index.

1584 Taylor, Kenneth N. *Creation and the High School Student*. c1969. Wheaton, Ill.: Tyndale House. "Compiled and edited" by Taylor. *Taylor:* president of Tyndale House Publishers; former director of Moody Press. Small book; mostly photos and diagrams. Appeals primarily to argument from design. Presents examples of complex insect societies; awesome design of universe; wonders of the cell. If God exists we must worship Him. "But if the universe originated through chance and there is no Creator, then nothing in this universe or in our lives has purpose."

1585 _______. *What High School Students Should Know About Evolution*. c1983 [1969]. Wheaton, Ill.: Tyndale House. Orig. c1969; titled *Evolution and the High School Student* ("compiled and edited" by Taylor). Purpose of book: to discuss evolution from strict creationist viewpoint; to show why evolution is "merely a theory, rather than being a fact on which a case against the Bible can be built." Stresses that "presuppositions" govern our opinions on creation vs. evolution. Evolution has never been observed. Fossil record is no proof—lack of transitional forms. Cave men merely human varieties. Chance origin of life impossible. Mentions usual creation-science arguments. "To me it appears that God's special creative acts occurred many times during six long geological periods, capped by the creation of Adam and Eve perhaps a million or more years ago." Photos, drawings.

1586 Taylor, Paul S. *The Bible and "The Great Dinosaur Mystery"* (pamphlet). c1981. Elmwood, Ill.: Films for Christ Association. "A Young People's Guide," to accompany Taylor's film *The Great Dinosaur Mystery* (Films for Christ). *Taylor:* prod. director of Films for Christ; son of Films for Christ founder Stan Taylor who made *Footprints in Stone,* the celebrated 1972 movie about the Paluxy manprints. "People and dinosaurs *did* live at the same time long ago. Animals and people did *not* evolve." "Evidence from the Bible and from scientists" shows that dinosaurs—and the earth—are only a few thousand years old. Shows that biblical Behemoth and Leviathan are dinosaurs. Refers to proof in "wonderful film" *Footprints in Stone* that dinosaurs and man co-existed. Dinosaur fossils are result of Noah's Flood. Noah "must have" taken dinosaurs aboard Ark, though post-Flood conditions (collapse of Water Canopy) and human hunters finished them off. Lots of drawings. Aggressive appeals to turn to Christ.

1587 ———. *The Great Dinosaur Mystery and the Bible.* 1987. El Cajon, Calif.: Master Books. Illustrators: Jonathan Chong, Todd Tenant, Gary Webb and Paul Taylor. Lavish color illustrations on every page. "Special note" (intro.) by **H. Morris.** Based on Taylor's Films for Christ movie *The Great Dinosaur Mystery.* Index.

1588 Teachout, Richard A. *Noah's Flood—3398 BC: A New Case for Biblical Chronology* (booklet). 1971. Caldwell, Idaho: Bible-Science Association. *Teachout:* theologian and missionary. Bible genealogies and Bible chronology. Some creation-science references. Creation: probably 5654 B.C.

1589 Teilhard de Chardin, Pierre. *The Phenomenon of Man.* 1965 [1955]. New York: Harper Torchbooks/Harper & Row. Orig. 1955 (Paris: Editions du Seuil); French. Laudatory introduction by Julian Huxley (grandson of T.H. Huxley, Darwin's champion, and a founder of neo-Darwinian evolutionary theory), who, like Teilhard, finds cosmic significance in evolution. *Teilhard:* Jesuit priest and paleontologist: geology prof. at Catholic Inst. Paris; Dir. of Lab. Advanced Studies in Geology and Paleontology (Paris); involved in Piltdown and Peking Man digs; spent final years with Wenner-Gren Found. (New York). Forbidden by Catholic Church to publish this book during his lifetime. Mystical view of evolution, presented in obscure poetic style. Influenced by **Bergson** and German Naturphilosophie. Declares that this book is not to be read as metaphysics or rekligion, but "purely and simply as a scientific treatise." Evolution is the fundamental process of the universe. Posits immanent progressive drive embracing entire universe. Through love, people grow closer, eventually evolving into a single superorganism. This is the "Omega Point," the ultimate destiny of evolution. People retain their own souls, but also merge into the world soul, which Teilhard equates with Christ. Man becomes deified; transmuted with the divine cosmic center. Key concepts in Teilhard's evolution are "noogenesis" (evolution of mind), "cosmogenesis," and "complexification." "Man discovers that he is nothing else than evolution become conscious of itself"; man's brain "proves that evolution has a direction." This book remains enormously popular, although strongly condemned by creationists, who consider Teilhard the chief inspiration and patron saint of theistic evolution, and denounced as well by many secular evolutionists.

1590 Thaxton, Charles B., Walter L. Bradley, and Roger L. Olsen. *The Mystery of Life's Origin: Reassessing Current Theories.* 1984. New York: Philosophical Library. Foreword by Dean Kenyon (San Francisco State Univ. biology prof.; author of *Biochemical Predestination;* later convert to creation-science; supported Louisiana creation-science bill with affidavit). *Thaxton:* chemistry Ph.D. —Iowa State Univ.; director of curriculum research at Foundation for Thought and Ethics (Richardson, Texas), an organization which promotes presentation of a theistic worldview in science teaching and textbooks. *Bradley:* materials science Ph.D. — Univ. Texas; prof. of mechanical engineering at Texas A&M Univ. *Olsen:* geochemistry Ph.D. — Colo. Sch. Mines; project supervisor with D'Apolonia Waste Management (Colo.). A knowledgeable critique of origin-of-life theories and experiments. Praised by many non-creationists as a well-researched and valuable contribution, but dismissed by others as unoriginal criticism masking anti-evolutionist motivation. Authors do not question ancient earth and succession of life-forms over geological ages, but assert that origin of life by random, purely naturalistic processes is fundamentally implausible and "probably wrong." Book emphasizes thermodynamic argument against chemical evolution and abiogenesis: reactions destructive to necessary organic compounds far outweigh reactions which form them. Origin-of-life scenarios make unwarranted assumptions about conditions of early earth and atmosphere in order to fit hypotheses. Refutes "myth of the prebiotic soup." In "Epilogue," authors urge consideration of the supernatural in scientific explanation. To break out of origin-of-life impasse they consider alternatives of panspermia, "special creation by a Creator within the cosmos" (**Hoyle** and Wickramasinghe), and — the position they favor — "special creation by a Creator beyond the cosmos." Argue for sharp distinction between "operation science" (testing of normal, recurrent phenomena) and "origin science," which deals with singular and non-repeatable events. The "God Hypothesis" is illegitimate in operation science but perhaps necessary for origin science. Many scientific references. Drawings, tables, charts; index.

1591 Thieme, R.B. *Creation, Chaos, and Restoration.* 1973. Houston, Texas: Berachah Church. Cited by R. Price in 1982 *Creation/Evolution.* Gap Theory creationism.

1592 Thomas, Cal. *Book Burning.* 1983. Westchester, Ill.: Crossway Books/Good News. *Thomas:* writer and broadcast journalist (formerly with NBC); syndicated columnist; vice-pres. of Moral Majority. Dedicated to Jerry Falwell and **F. Schaeffer.** Asserts that secular humanists are the real censors in our society. They suppress Christian books and harass those espousing Christian opinions. Fundamentalists are discriminated against by the non-religious media, schools, libraries, the ACLU and the courts. Says Christian books and television are relegated to "Negro league" ghetto — Thomas wants them to become mainstream force. Secular humanist dogma is spread by thought-control and violates Constitutional rights. Includes charges of censorship against creationist books, and imposition of evolutionary viewpoint. Discusses censorship of creationist opinions in reporting of 1981 Arkansas trial. "Plan for Action" appendix lists books on creation-science plus other Christian Right issues. Urges Christians to make libraries, schools, newspapers and politicians take notice of fundamentalist opiion. Cites the **Gablers**, Solzhenitsyn, **Geisler.** "It is time to move out of the ghetto and begin to reclaim the land for Christ and his kingdom."

1593 Thomas, J.D. *The Doctrine of Evolution and the Antiquity of Man* (booklet). 1961. Abilene, Texas: Biblical Research Press. *Thomas:* Bible prof. at Abilene Christian Coll. (Texas). "Evolution is just a faith!" Evolution can't explain mind, spirit or values. Includes standard creation-science arguments. Gap Theory creationism, Day-Age theory, strict creationism and other theories are considered as possible interpretations. **Ussher**'s chronology is inaccurate, but man seems recent. Recent origin of man is not a necessary biblical interpretation, however.

1594 ________. *Facts and Faith. Vol. I: Reason, Science and Faith.* Abilene, Texas: Biblical Research Press.

1595 Thomas, Louise. *The Myth of Evolution.* 1985. Pompano Beach, Fla.: Exposition Press of Florida. Naive attack focused on human evolution. (Listed in Schadewald files.)

1596 Thomasson, William. *The Glacial Period and the Deluge: Announcement of Epoch-Making Discoveries, with a New and Startling Interpretation of Revelation, Daniel and Isaiah* (booklet). 1910. Chicago: Aragain Publishing. Maps, illustrations.

1597 Thompson, Adell. *Biology, Zoology, and Genetics: Evolution Model vs. Creation.* 1983. Washington, D.C.: University Press of America. *Thompson:* Univ. of Missouri–Kansas City; state director of Missouri Acad. Science. Presents scriptural basis for creationism, but states "there is no scientific data that supports creationism..."

1598 Thompson, Bert. *Theistic Evolution.* c1977. Shreveport, La.: Lambert Book House. *Thompson:* food microbiology Ph.D.–Texas A&M Univ.; former ass't. prof. of veterinary public health at Texas A&M; now prof. of Bible and Science at Alabama Christian Sch. of Religion; Church of Christ; co-leader, with **W. Jackson,** of Apologetics Press (Montgomery, Ala.), and co-editor of its journal *Reason & Revelation.* Strong defense of strict young-earth creationism against the heretical compromise of theistic evolution.

1599 _______. *Can American Survive the Fruits of Atheistic Evolution?* (tract). c1981. Fort Worth, Texas: Pro-Family Forum. Orig. pub. by Apologetics Press. Widely distributed as **Pro-Family Forum** tract. Mostly quotes, by scientists and creationists, about the influence of evolutionist philosophy. Declares that the fruits of evolution are Nazism and communism. "Evolution is built upon the assumption that man is only an animal; consequently, morals and values are useless and not to be incorporated into the system." "Evolution, though its advocates do not like to admit it, leads to out-and-out racism." "America's ruin will be hot on the heels of the popular acceptance of evolution."

1600 _______. *The History of Evolutionary Thought.* c1981. Fort Worth, Texas: Star Bible & Tract Corp. Dedicated to Russell Artist (emer. biology prof. at David Lipscomb Coll., Tenn.). Foreword by **W. Jackson.** Declares that this is the only complete and well-documented book on this topic. Pretentious and repetitious little book; consists largely of quoted passages. Compiled from creation-science and standard sources. Surveys evolution from ancient Greeks to current controversies. Birdsell's *Human Evolution* cited a lot; also relies on **R. Clark** and **Bales, Rusch.** Follows **Davidheiser** in Scopes Trial chapter. Quotes **Barzun** passage four times: "Darwin was not a thinker, and he did not originate the ideas he used." **Paley**'s argument from design is "incontrovertible" Includes many references (from Boller's 1969 book) to nineteenth-century opponents of Darwinism. Index.

1601 _______. *The Scientific Case for Creation* (booklet). c1985. Montgomery, Ala.: Apologetics Press. Stresses that evolution and creation are mutually exclusive and diametrically opposed views. Each is also a scientific "model." Presents standard creation-science arguments, quoting many creationists. Following **Thaxton** et al., argues for distinction between "operation science" and "origin science." Quotes **Jastrow** at length as proving universe was created. Since the scientific evidence fits the creation model so much better than the evolution model, Thompson says that "The one-sided indoctrination of students in this materialistic philosophy in the tax-supported public schools in our pluralistic, democratic society is a violation of academic and religious freedoms. Furthermore, it is *poor* science and *poor* education. To remedy this intolerable situation, creation scientists are asking that, excluding the use of the Bible or any other religious literature, only the scientific evidences that can be adduced in favor of creation and evolution be presented thoroughly and fairly in public schools." Let the students decide! Let evolution and creation "compete freely in the marketplace of ideas!"

1602 _______. *The Global, Universal, Worldwide Flood of Noah (Genesis 6–9)* (booklet). 1986. Montgomery, Ala.: Apologetics Press. "That You May Believe" booklet series on Christian apologetics and evidences. Observes that Noah's Ark and the Flood is the most ridiculed story in the Bible, and deplores the fact that even many so-called Christians now doubt it. Insists on literal interpretation; the Bible demands a worldwide Flood. "We firmly contend that a true exegesis of Scripture yields absolute Truth which is both *ascertainable* and *knowable.*" Praises and quotes global Flood supporters (**Rehwinkel, H.W. Clark, J.W. Dawson,**

Kearley, Patten, Morris, Whitcomb and others). Castigates compromisers and local-Flood advocates such as **J. Clayton, Ramm, J.W. Montgomery, Custance.** Does not necessarily accept claim that Ark is on Mt. Ararat. Mentions many Flood Geology arguments and cites worldwide Flood myths, but relies primarily on biblical evidence. Flood had to be universal because of man's sin and total rebellion against God.

1603 _______. *Is Genesis Myth? The Shocking Story of the Teaching of Evolution at Abilene Christian University.* 1986. Montgomery, Ala.: Apologetics Press. Introduction by **W. Jackson.** In 1985, Thompson, an alumnus of Abilene Christian Univ., heard complaints from students that two ACU professors were teaching "raw evolution"—even using anti-creationist books—and were treating Genesis as a myth. This is an account, in excrutiating, repetitious detail, of Thompson's grim crusade to stamp out creeping evolutionism. With intensely righteous wrath, he embarked on a personal investigation of this insidious heresy. Reproduces many of the materials used in the ACU classes, Thompson's correspondence and interviews, and his inquisitorial, heresy-exposing questionnaires. Thompson presents himself as acting with scrupulous fairness, bending over backwards to allow promoters of heresy to repent and return to true Christian teachings.

1604 _______. *American Family Crisis: The Attack of Evolution* (pamphlet). Undated. [Montgomery, Ala.:] Apologetics Press.

1605 _______. *The Doctrine of Special Creation* (booklet). Undated [1970s?]. [Montgomery, Ala.: Apologetics Press. Undated; references up to 1977. Apologetics Press monograph series; spiral-bound. Strict creationism; usual creation-science arguments.

1606 _______. *An Extended Bibliography of Works Dealing with Special Creation and Organic Evolution* (pamphlet). Undated. [Montgomery, Ala.:] Apologetics Press. Undated; citations up to 1979. 218 works listed, both creationist and evolutionist. Not annotated.

1607 Thompson, D.A. *A Major Challenge to Evolution: Made by Prof.* **W.R. Thompson** (pamphlet). Undated [England:] Evolution Protest Movement. EPM pamphlet #143.

1608 Thompson, D'Arcy Wentworth. *On Growth and Form.* 1942 [1917]. Cambridge: Cambridge Univ. Press. Orig. 1917. Expanded ed. 1942. Also abridged ed. c1961, with Foreword by J.T. Bonner. *D'Arcy Thompson:* anatomist, physiologist, mathematician; also a classicist; natural history prof. at St. Andrews Univ.; F.R.S.; pres.—Royal Soc. Edinburgh. This celebrated work, which has had a profound if indirect effect on many areas of research, is non- rather than anti-evolutionist. D'Arcy Thompson ignores (and seemingly denies) evolutionary causation, concentrating on proximate causation. He does, however, acknowledge evolutionary history, though his approach is thoroughly a-historical. His approach is a mathematical investigation of functional and "dynamical" morphology. This, he argues, can be more productive than the often sterile speculations regarding descent relations and adaptive value of traits. "So long and so far as 'fortuitous variation' and the 'survival of the fittest' remain engrained as fundamental and satisfactory hypotheses in the philosophy of biology, so long will these 'satisfactory and specious causes' tend to stay 'severe and diligent enquiry [quoting Bacon] . . . to the great arrest and prejudice of future discovery'." He does not deny natural selection, but does suggest it only eliminates the unfit and is not progressive. "My sole purpose is to correlate with mathematical statements and physical law certain of the simpler outward phenomena of organic growth and structure or form. . . . We want to see how, in some cases at least, the forms of living things, and of the parts of living things, can be explained by physical considerations, and to realise that in general no organic forms exist save such as are in conformity with physical and mathematical laws." Unlike the physical scientist, the biologist always explains in terms of evolutionary concepts, but, says D'Arcy Thompson, the physical and mathematical causes he is concerned with operate now on organisms just as they did in past geological ages; consequently, it is "by no means certain that the biologist's usual mode of reasoning is appropriate to the case, or that the concept of continuous historical evolution must necessarily, or may safely and legitimately, be employed." D'Arcy Thompson assumes that genetic make-up of organisms can alter to adapt to functional problems—"direct adaptations" which are impressed upon the organisms by immediate physical, environmental forces, rather than by accumulation of hereditary adaptations. Discusses the relation of size to form (akin to allometry), geometry of forms, growth of forms as mathematical functions, and structural efficiency. Stresses discontinuities between major forms. The most famous chapter, "On the Theory of Transformations, or the Comparison of Related Forms," demonstrates lawful deformations of closely related organisms by morphological mappings on Cartesian coordinates. Concludes with praise for **Fabre.** Drawings; index. Medawar called this book "the finest work of literature [in English] in all the annals of science."

1609 Thompson, W.R. *Science and Common Sense.* 1965 [1937]. Albany, N.Y.: Magi Books. *Thompson:* F.R.S.; director of Commonwealth Inst. of Biological Control (Ottawa, Can.). Orig. 1937 (Longmans, Green). Argues that "the concept of organic evolution was an object of genuinely religious devotion"; "this probably is the reason why the severe methodological criticism employed in other departments of biology has not been brought to bear against evolutionary speculation." Thompson later claimed that if he were re-writing this work now he would have made a much stronger case against evolution. At the time he had to get it passed by a referee with strong evolutionist views, therefore he was "forced to modify his own position to get the book published." (Quoted in **Zimmerman** 1959; discussed in **Zimmerman** 1972.)

1610 ________. *New Challenging 'Introduction' to the Origin of Species* (pamphlet). 1967 [1956]. [England:] Creation Science Movement. Thompson's "Introduction" to 1956 J.M. Dent edition (also pub. by E.P. Dutton: New York) of Darwin's *Origin* (Everyman Library No. 811), reprinted here separately. Footnotes added by **F. Cousins** of Evolution Protest Movement. Highly critical introduction. Darwin, says Thompson, didn't prove that evolution by natural selection had occurred—only that it might have. He "fell back on speculative arguments" because he had no experimental evidence. Explanation in terms of evolutionary survival value (pan-selectionism) is vague, slippery, and untestable. Darwin ignored **Cuvier**'s adaptive correlation of traits, treating them as independently variable. Argues that Mendelian inheritance has rendered Darwinian evolution less rather than more plausible. So-called vestigial organs have present causes and functions. Denies that classification 'naturally' tends to conform to genealogical relations. Because Darwinian evolution relies on chance variations, "the last thing we should expect on Darwinian principles is the persistence of a few common fundamental structural plans. Yet this is what we find." These major types should not, according to Darwinism, persist. Paleontology continues to demonstrate lack of transitional forms. Rocks are dated according to the fossils they contain, and many strata do not appear in the accepted order. Evolution stimulated biological research initially, but now tends to hinder it: much research is now "directed into unprofitable channels or devoted to the pursuit of will-o'-the-wisps. . . . Much time was wasted in the production of unverifiable family trees." "A long-enduring and regrettable effect of the success of the *Origin* was the addiction of biologists to unverifiable speculation"—speculations on the evolutionary origins and adaptive values of structures and traits. "Thus are engendered those fragile towers of hypotheses based upon hypotheses, where fact and fiction intermingle in an inextricable confusion." "The success of Darwinism was accompanied by a decline in scientific integrity." Cites the reckless, devious and fraudulent manipulations of Haeckel, T.H. Huxley, and Dubois. "This situation, where scientific men rally to the defense of a doctrine they are unable to define scientifically, much less demonstrate with scientific rigour, attempting to maintain its credit with the public by the suppression of criticism and the elimination of difficulties, is abnormal and undesirable." Accuses Darwin and his followers of being "strongly anti-religious." They attempted to deny finality and design of biological adaptations, insisting instead on chance and undirected processes. They strove to deny that God providentially controls evolution. Their failure to address these religious issues "indicates a regrettable obtuseness and lack of responsibility." Thompson argues that the ability of many organisms to feel, and the ability of man to reason, constitute transitions which Darwinism cannot explain. Cites **D'Arcy Thompson, Guyenot.**

1611 Thomson, J. Arthur. *The Bible of Nature.* 1922. New York: Charles Scribner's Sons. 1907 Bross Lectures (established to illustrate from science the divine origin and authority of Scripture, prove that science and revelation coincide, and prove the existence of the Christian God). *Thomson:* natural history prof. at Univ. Aberdeen (Scotland). "Wherever you tap organic nature it seems to flow with purpose." "The old special arguments from design are replaceable by a wider teleology based upon the fundamental proposition of evolution." Accepts evolution but warns it doesn't really explain: we must add Logos. Suggests that man is a "great exception" to naturalistic evolution; probably a huge mutation.

1612 Thomson, William (ed.). *Aids to Faith: A Series of Theological Essays.* 1861. England. Quoted and discussed in **Wonderly.** Series of nine essays by Church of England clergymen, defending the Bible against liberal assaults. *Thomson:* Lord Bishop of Gloucester and Bristol. Rev. **A. McCaul** (Hebrew and Old Testament prof. at King's Coll., London, wrote the essay "The Mosaic Record of Creation." Genesis is historically and scientifically accurate. Exegesis shows that Genesis allows an ancient earth, as proven by geology. Advoctes Day-Age creationism. Marvels at anticipation of scientific discoveries by Moses, and observes he got the order of creation exactly right.

1613 Thurman, L. Duane. *How to Think About Evolution: & Other Bible-Science Controversies.* c1978 [c1977]. Downers Grove, Ill.: InterVarsity Press. Orig. c1977; titled *Creation and Evolution: The Renewed Controversy* (Burgess Pub.). 2nd ed. c1978. *Thurman:* botany Ph.D.—Berkeley; biology prof. at Oral Roberts Univ. Fair-minded and reasoned tone. Emphasizes proper scientific method, interpretation of evidence, facts and theories; how to avoid fallacious arguments. Chides both creationist and evolutionist extremists for unfair arguments. Deplores appeals to authority in science and dogmatic attitudes. Stresses need for proper definitions of creationism and evolution: complains that "evolutionists use the most extreme, least-known version of creation as representative of creationists in general." They also assume macro-evolution true because micro-evolution occurs. Argues that creation-evolution controversy is healthy because it exposes religious nature of many evolutionist arguments, allows presentation of alternative scientific theories, and challenges scientists to produce evidence to support their claims. Quotes and analyzes creation-evolution correspondence in *BioScience* and *Amer. Biol. Tchr.* Section on 1969 Calif. educ. ruling—says creationists wanted to expand, not limit, options in textbooks. Quotes **Kerkut** and other scientists critical of dogmatic presentation of evolutionary scenarios and phylogenies. Presents pros and cons of various creationist theories; says he has "no firm choice," but seems to betray discreet preference for Gap Theory creationism. Urges students to examine and weigh all theories. Says young-earth Flood Geology advocates are the "most vocal and dogmatic." "The Bible is quite accurate and specific about some scientific matters," but does not deal with the "how" of creation. Discusses proper methods of biblical interpretation, urging caution. Charts, diagrams.

1614 Tier, W.H. *Creation or Evolution: Nature's Wonders* (pamphlet). Undated. [England:] Evolution Protest Movement. EPM pamphlet #89. Tier also wrote *Miracles of Evolution* (pamphlet #104) and *Instincts Contradict the Theory of Evolution* (#145).

1615 Tilney, A.G. *Is Evolution, Or the Bible, True?* (booklet). 1957. A.E. Norris. (Listed in SOR CREVO/IMS.) *Tilney:* B.A.—Univ. London; linguist; secretary of Evolution Protest Movement (1955–76); schoolmaster and pastor in Hayling Island, Hants. (England). Tilney wrote over a hundred pamphlets and booklets for EPM.

1616 ________. *Man: Child of Ape, or Son of God?* 1960. Emsworth, England: Evolution Protest Movement.

1617 ________. *The Case Against Evolution.* 1964. Hayling Island, Hants, England: Evolution Protest Movement. Cited in **H. Morris** 1984.

1618 ________. *Without Form and Void.* 1970. The Evolution Protest Movement included old-earth as well as young-earth creationists; Tilney defended Gap Theory creationism. (His EPM pamphlets do not deal with the age of the earth; only with anti-evolution arguments.) This book cited in 1986 *Ex Nihilo;* presumably a defense of Gap Theory creationism.

1619 ________. *Evolution: The Great Delusion* (pamphlet). Undated. [England:] Evolution Protest Movement. EPM pamphlet #198. One of over a hundred EPM pamphlets by Tilney. Others were about: Darwin, Erasmus Darwin, **Dewar,** T. Huxley, J. Huxley, Haldane, Leakey, Lyell, **Teilhard, Kerkut, F. Cousins, Patten, Shute,** Freud, **Macbeth,** Adam, coelacanths, slogans, *Pithecanthropus,* anthropology, cells, disease and death, skin, population, education, speech, geological column, mind, textbooks, classification, ape-men fakes, the authority of laymen, and the EPM.

1620 Tiner, John Hudson. *When Science Fails.* 1974. Grand Rapids, Mich.: Baker Book House. *Tiner:* high school science and math teacher. This book "explodes the naive assumption that science has all the answers and challenges the reader to think independently with a Biblical faith." "The Bible is an accurate description of the universe. Science will not contradict the Bible." The Bible has been tested for 4000 years, and has never been wrong. "Is there a science book that will be completely accurate four thousand years from now?" Cases presented here are largely history of medicine (Semmelweiss, Jenner, Lister, Fleming, etc.). Also chapters on the presence of Noah's Ark on Ararat (follows and endorses **Navarra**'s account), unreliability of radiometric dating, Piltdown Man, previous disbelief in 'rocks from the sky' (meteorites), Mendel, the coelacanth (believed long extinct but found alive), discovery of Nineveh (confirming biblical history), etc. Scientists refuse to listen to amateurs who challenge entrenched dogmas, complains Tiner. This is especially true of criticisms of evolution. Cave paintings show that ancient man had great intelligence, but evolutionists denied this evidence. "For the first time in a hundred years, scientists who doubt evolution are willing to put their doubts into words. They have begun to offer solid evidence that refutes the view that man developed from lower creatures."

1621 Tinkle, William J. *Heredity: A Study in Science and the Bible.* 1970. Grand Rapids, Mich.: Zondervan. Foreword by **J.N. Moore;** Introduction by **W. Lammerts.** *Tinkle:* zoology Ph.D. – Ohio State Univ.; biology prof. at Taylor Univ. (Ind.); wrote a Christian textbook *Fundamentals of Zoology;* a founder of the Creation Research Society. Largely a presentation of Mendelian inheritance and standard genetics, but denies that genetics is a vehicle for evolution. Includes some interesting discussions of plant and animal breeding practices. Discusses eugenics; concludes it is impractical and tends to become unethical. Mendel's discoveries were ignored because of enthusiasm for Darwin. Argues that first humans possessed innate intelligence. "We believe that Adam and Eve were real persons with intelligence quotients of at least 100." Attributes success of Darwin to patronage of middle class eager to justify its new power. Refutes evolutionary recapitulation in embryology. Story of Jacob and Laban's goats involves recessive genes. Does not give age of earth; notes that "the Bible does not state the date of creation," though many churchmen have attached extra-biblical dogmas to it. Life has undergone some changes (especially parasites), but this is mere variation and degeneration caused by mutations. "Genetics itself does not teach Christianity nor any other form of religion but it allows plenty of room for Christianity and does not clamor for change. It does not supply facts to indicate a natural upward evolution of the race but indicates a horizontal tendency for the most part with loss when mutation occurs." Diagrams, drawings; bibliog., index.

1622 _______. *God's Method in Creation.* 1973. Nutley, N.J.: Craig Press. (Cited in **H. Morris** 1974.)

1623 Totten, Charles A.L. *Joshua's Long Day and the Dial of Ahaz.* c1968 [1890]. Merrimac, Mass.: Destiny Publishers. Orig. 1890. Reprinted in c1968 ed. *Totten:* prof. of military science and tactics at Yale Univ.; editor of *Our Race;* advocate of Anglo-Israelism (supremacy of English and Americans, the true descendants of Israel, who inherited God's promises to that nation). 1968 ed. also includes 1946 article by **Howard Rand** "When the Earth Turned Over" (reprint from *Destiny* magazine). Totten, eager to prove truth of the biblical story most frequently ridiculed—Joshua's command for God to make the sun stand still—here issues "A Scientific Vindication, and a Midnight Cry." "It will not do to preach Christ and deny Moses. It will not do to doubt the universality of the Flood, and ask men to accept a Saviour who alludes to it! It will not do to doubt Joshua's Long Day, with the sun and moon poised in mid-heaven while he fought, and yet stultify our hearts with hopes of a LONGER DAY when even sun and moon will not be needed!" Elaborate study of calendrical cycles, astronomical calculations and conjunctions show that 23 hrs. 20 min. were intercalated on Joshua's Day. The remaining 40 min. were added during Ahaz's Event, when God made the sun move backwards. Totten warns that the time of the End is upon us. Rand suggests a mechanism for these events: a passing comet flipped the earth end over end, but earth continued undisturbed in its orbit. The 'hail' of the Long Day was a meteoritic shower. Diagrams, tables. Totten's mathematical-astronomical proof inspired **H. Hill**'s updated version, the NASA missing day tale.

1624 _______. *The Flood the Fact of History.* 1892. New Haven, Conn.: Our Race Publishing. "A Chronological Vindication, with a Guarantee of the Second Advent." Quoted in **B. Nelson** 1931. Treatise on Noah's Ark. The Ark was 600 feet long (assuming larger cubit size), and had three decks. Carrying capacity was equal to ten thousand railway cars. To support the ability of Noah to construct such a huge vessel, specially bulkheaded to withstand the fury of the Flood, Totten "dwelt at length on the marvelous scientific and architectural skills of the ancients," such as "their unmatchable genius in the building of the pyramids."

1625 Townley, Jeffrey Kent. *A Pilot Study on the Validity of Using an Inquiry Approach in a Video Format for Origins: Two Models, Evolution—Creation in Christian Schools.* 1985. Unpub. M.A. thesis: Institute for Creation Research (El Cajon, Calif.). ICR M.A. thesis in science education. The effectiveness of teaching the "two-model" approach in a Christian high school in El Cajon, Calif. Study used **Bliss**'s 1978 *Origins: Two Models—Evolution, Creation* video and accompanying textbook. (Video was designed for high school use.) Described in 1985 *Acts & Facts.*

1626 Townsend, Joseph. *The Character of Moses Established for Veracity as a Historian, Recording Events from the Creation to the Deluge.* 1812–15. London: Longman, Hurst, Rees, Orme and Brown. Gap Theory creationism.

1627 _______. *Geological and Mineralogical Researches, During a Period of More than Fifty Years, in England, Scotland, Ireland, Switzerland, Holland, France, Flanders, and Spain; Wherein the Effects of the Deluge Are Traced, and the Veracity of the Mosaic Account Is Established.* 1824. London: Samuel Bagster. Endorses Gap Theory creationism.

1628 Townsend, Luther Tracy. *Evolution or Creation: A Critical Review of the Scientific and Scriptural Theories of Creation and Certain Related Subjects.* c1896. New York: Revell. *Townsend:* teacher at Methodist Boston Sch. of Theol.; also wrote *Bible Theology and Modern Thought* and other religious books.

1629 _______. *The Collapse of Evolution.* c1922 [1905]. Louisville, Ken.: Pentecostal Publishing. Also (orig.?) pub. 1905 (Boston: National Magazine). Also pub. c1922 by Bible Truth Depot (New York). Based on 1904 lecture to American Bible League, Boston. Townsend reprints the Lady Hope story from

Watchman Examiner in 1922 ed. (quoted in **Hardie** 1924). Cites **Bell,** James Orr. Strong denunciation of evolution.

1630 ________. *The Deluge: History or Myth?* 1907. New York: American Tract Society.

1631 Traylor, Ellen Genderson. *Noah.* 1985. Wheaton, Ill.: Tyndale House. Fictionalized account of Noah and the Flood which "accepts the Biblical account of Creation and the Flood as factual history." Reviewed in 1987 *Bible-Science Newsletter,* which says its readers "will be especially interested in Traylor's use of several Creation Science theories." Portrays collapse of Pre-Flood Water Vapor Canopy as cause of Deluge. Antediluvian and folkloric creatures explained as result of genetic tampering. "Traylor presents the 'sons of God' of Genesis 6:1-4 as fallen angels who rule a corrupt humanity by posing as 'gods'; the *nephilim* . . . are seen as their giant (and rather dim-witted) half-human offspring who provide the military muscle in a world of treachery and violence." Assumes gaps in Genesis genealogies; extends age of earth by several millenia.

1632 Trench, Brinsley Le Poer. *The Sky People.* 1960. London: Neville Spearman. *Trench:* the Earl of Clancarty; coauthor of G. Adamski's first UFO book; cousin of W. Churchill. A biblical interpretation of interaction of earthlings with the "gods from the skys"—flying saucer occupants. The Sky People have been coming to earth for millions of years, as recounted in myths and legends around the world. The 'Golden Age' of myths—when gods came down and mingled with mortals—refers to these UFOnauts. The plural *Elohim* of Genesis, which refers to a whole race of 'gods,' created two separate races: the two creation stories of Genesis. Telepathic "Galactic Man"—Adam I—was intended to populate the universe. "Chemical" or "animal man"—Adam II, *Homo sapiens*—was created ages later;—an experiment that got out of control. The Adam I race are known as the Serpent People. "The creation of . . . animal man was an illegal act," done in a special, isolated place on Mars—the Garden of Eden. Adam II females introduced Adam I males to forbidden nutrients. Humans are descendants of matings between the two races. Noah's Ark was a spaceship which rescued people from floods on Mars. God banished humans to earth, but the Serpent People visit from time to time, offering wisdom and enlightenment. Discussed in **E. Norman** 1973. In 1979 Trench addressed the House of Lords on UFOs, urging research funds, and admitting he had written about UFOs under a pen-name.

1633 ________. *Men Among Mankind.* 1962. London: Neville Spearman. Strange beings have intermittently appeared on earth, altering history and promoting human progress. A great catastrophe occurred around 4000 B.C., ushering in a new Zodiacal Age. Atlantis sank during three previous catastrophes in 25,000, 13,000 and 9500 B.C. Our Atlantean ancestors fled to Britain; the catastrophes shattered the mighty Atlantean Empire and mankind reverted to savagery. The Sky People came to earth to administer and help mankind regain civilization. This history is encoded in the megaliths and in astrological lore. The Sky People built the Great Pyramid before they departed—it encodes extraterrestrial wisdom. Modern notions of racism are atavistic memories of the catastrophic period when isolated and mutant human groups developed into distinctly different types. Discusses myths, Celtic and Arthurian legends. Prophetic nature of Shakespeare's plays, which were written by four men. The geniuses who have altered history—King Arthur, Alexander, Moses, Jesus, Leonardo, Tesla, Benjamin Franklin—have been of extraterrestrial origin. Cites **Bellamy, Davidson, Blavatsky.** Trench has also written *Operation Earth* (London: Neville Spearman; 1969) and *Mysterious Visitors: The UFO Story* (New York: Stein & Day; 1973). Photos; index.

1634 Trop, Moshe. *Creation: Origin of Life.* 1982. Israel. In Hebrew. *Trop:* chemist with Institutes of Applied Research—Ben Gurion Univ. of the Negev (Bergmann Campus: Beer-Sheva, Israel); former visiting prof. at Rutgers Univ.; Orthodox Jewish. Trop has written for *CRSQ* and other creation-science publications. He defended the claim of Lee Spetner (also a Jewish creationist) that *Archaeopteryx* is a hoax—a claim later taken up by **F. Hoyle.**

1635 Tuccimei, Guiseppe. *La Decandenza di una Teoria.* 1908. Italy. Compares evolution theory with Copernican theory. Copernican theory explains "that which is," whereas evolution tries to explain "that which was"—an insoluble problem, lacking experimental verification. The Copernican system was quickly verified; "the theory of evolution, on the other hand, is at the present day no longer able to hold its own even as an hypothesis, so numerous are its incoherencies and the objections to it raised by its own partisans." Quoted in **G.B. O'Toole** 1926. O'Toole also quotes from Tuccimei's "La teoria dell' evoluzione e le sue applicazioni," possibly a separate work. "This perverse determination to place man and brutes in the same category, interests me not so much from the scriptural standpoint as for reasons moral and social." The consequences of evolution are "socialism and anarchy" amongst the "ignorant and turbulent masses." "Socialistic doctrines are based exclusively upon our assumed kinship with the brutes . . ." If man arose from the animals, "why should we not enjoy in common with them the right to gratify every instinct? . . . why not proceed then to a general leveling of the existing social order?" If true, then free will, the soul, and life after death become myths, and morality loses its sanction: "what guilt will there be in the delinquent who lapses into the most atrocious crimes? . . . And behold the suffering, the unfortunate, and the dying deprived of their sole consolation, the last hope which faith held out to them, and society reduced to an inferno of desperadoes and suicides!"

1636 Turner, C.E.A. *A Jubilee of Witness for Creation Against Evolution: CSM/EPM 1932–1982* (booklet). England: Creation Science Movement. *Turner:* chemistry and science educ. Ph.D.; council chairman of Evolution Protest Movement since 1957. Account of fifty years of the Evolution Protest Movement (now called the Creation Science Movement) of England.

1637 ________. *Composition of Blood: Opposed to Evolution.* Undated. [England:] Evolution Protest Movement. EPM pamphlet #119. Turner also wrote EPM pamphlets *Scientific Methods and Evolution* (#58), *Horse Evolution* (#74), *Archaeopteryx: The Bird, No Link* (#76), *Unreliability of Dating Methods* (#144), *Trace Elements in Creation* (#195), *Dinosaurs: Without Descent or Descendants* (#197), and others.

1638 Turner, Sharon. *The Sacred History of the World, Attempted to Be Philosophically Considered, in a Series of Letters to a Son.* 1832. New York: William Jackson. 3rd ed. At least eight eds. up to 1859 by Harper (New York) and by Longman, Brown, Green and Longmans (London). Gap Theory creationism. Recently, says Turner, it had been "the fashion for science, and for a large part of the educated and inquisitive world, to rush into a disbelief of all written Revelation; and several geological speculations were directed against it. But I have lived to see the most hostile of these destroyed by their own as hostile successors; and to observe, that nothing, which was of this character, however plausible at the moment of its appearance, has had any duration in human estimation, not even among the sceptical." (Quoted in **Fairholme** 1833). Millhauser says Turner was a "sound secular historian," but that he was best known for this work, which "interpreted the 'interval' doctrine [the Gap Theory] to children . . ."

1639 Twemlow, George. *Facts and Fossils Adduced to Prove the Deluge of Noah, and Modify the Transmutation System of Darwin; with Some Notices Regarding Indus Flint Cores.* 1868 [1867]. London: Simpkin, Marshall. *Twemlow:* major-general in British Army. Quoted and discussed in **B. Nelson** 1931. Noting that water-borne sediments are deposited in layers according to size and specific gravity, Twemlow argues that the Flood likewise deposited all the geological strata. "Twemlow describes certain fossiliferous conditions which, he held, Darwin's uniformitarian theory could not explain." Cites deep undisturbed strata of Tibetan plain, and various fossil sites containing great masses of commingled fossils.

1640 Tyler, David J. (ed.). *Understanding Fossils and Earth History.* 1984. Glasgow, Scotland: Biblical Creation

Society. Symposium papers: special issue of *Biblical Creation* (journal of the Biblical Creation Society). Authors: **C. Bluth, J. Scheven,** Tyler, Gerald Duffett, **David Watts.** Reviewed in 1985 *CRSQ.*

1641 Udd, Stanley V. *The Early Atmosphere.* 1974. Th.M. thesis: Grace Theological Seminary (Winona Lake, Ind.). Unpub. thesis. *Udd:* Old Testament prof. at Calvary Bible Coll. (Kansas City, Mo.). A literal interpretation of Genesis. Udd analyzes the 'water heaven': the pre-Flood "waters above the firmament." Insists that a straightforward reading of Genesis requires this to be in a liquid state. Proposes it was a thin liquid plate in the plane of the equator, similar to Saturn's rings, held up by centrifugal force. May have been a spherical shell initially, later assuming ring form. "It seems quite probable, however, to suggest that a rotational motion supplied by creative fiat on the second day of creation could suspend a spherical plane of liquid water above the atmosphere. The exact mechanics of such a model cannot be supplied in this work. But it should be noted in defense of this proposal that equilibrium conditions between the forces of gravitational attraction and the inertia of matter are the mechanisms by which all celestial bodies are held in their courses." Quoted and discussed in **Dillow** 1981. Udd presents similar arguments in 1975 *CRSQ.*

1642 Ulbricht, T.L.V. *Did Life Evolve?* (pamphlet). Undated. [England:] Evolution Protest Movement. EPM pamphlet #148.

1643 Upham, F.W. *The Debate Between Church and Science.* 1860. Cited by Millhauser as advocating Day-Age theory creationism.

1644 Ure, Andrew. *New System of Geology.* 1829. "In which the Great Revolutions of the Earth and Animated Nature are reconciled at once to modern science and sacred history." Explicitly sets "the origin of the material system" at six thousand years ago. Cited by Millhauser, who says Ure was "respected by his fellow scientists as a contributor to geological method."

1645 Ussher, James. *The Annals of the World.* 1650 [1654]. London: J. Crook and G. Bedell. English ed. Orig. pub. 1650 in Latin ("Annales veteris testamenti"). Subtitle: "Deduced from the Origin of Time, and continued to the beginning of the Emperour Vespasian's Reign, and the totall Destruction and Abolition of the Temple and Commonwealth of the Jews. Containing the Historie of the Old and New Testament, [etc.]." *Ussher:* Archbishop of Armagh, Ireland; vice-chancellor of Trinity Coll. (Dublin); respected biblical scholar. Ussher calculated the date of Creation as 4004 B.C. This date was added as a marginal note to the 1701 edition of the English (King James) Bible and many subsequent editions, thereby gaining great authority. Ussher compared four sources of the Pentateuch yielding different dates, but considered the Hebrew translation the most reliable. (The Hebrew version gives the time between Creation and the Flood as 1656 years.) Ussher calculated his date not only from study of the genealogies of the Old Testament prophets, but also from analysis of astronomical and calendrical cycles. He calculated the calendrical zero-point (when the solar, lunar and Paschal cycles were all at zero) to be 4714 B.C.—i.e. 710 years before Creation. "Which beginning of time, according to our chronologie, fell upon the entrance of the night preceding the twenty third day of October in the year of the Julian calendar, 710." Thus, Creation began on Saturday evening, Oct. 22. Quoted and discussed by Brice in 1982 *J. Geol. Educ.* Brice notes that "Ussher was aware of the provisional nature of this estimate and how much it depended on various textual readings." Many fundamentalists, however, finding the date in their Bible, have assumed it to be Gospel.

1646 Utt, Richard H. (ed.). *Creation: Nature's Design and Designer.* 1971. Mountain View, Calif.: Pacific Press. "By a panel of Ten Scientists." *Utt:* Seventh-day Adventist. Preface by **Wernher von Braun,** who gives his NASA affiliation. Authors: **R.E.D. Clark, H. Coffin,** L.E. Downs, Eric A. Magnussen, Leonard Hare, **R. Ritland,** Joan Beltz Roberts, Asa C. Thoreson, James R. Van Hise, T. Joe Willey.

1647 Vail, Isaac Newton. *The Earth's Annular System: or, The Waters Above* *the Firmament.* 1912 [1874]. Pasadena, Calif.: Annular World Co. "The World Record Scientifically Explained." 4th ed. Orig. pub. 1874 as pamphlet, titled *The Waters Above the Firmament.* Rev. 400-page ed. in 1885, 1886 (priv. pub.: Barnesville, Ohio), titled *The Story of the Rocks; or, The Earth's Annular System.* 3rd ed. (rev. and enlarged) pub. 1902. *Vail:* teacher; oil and gas prospector; claims credit for discovery of Karg well in Ohio; later lived in Pasadena, Calif., involved in Los Angeles oil development; Quaker. Vail apparently originated the Canopy Theory, popular with modern creationists (e.g. **Dillow**). (But see **Webb** 1854.) Vail's model, however, is richer and more dynamic than current creation-science versions. Vail proposed that the earth originally had suspended above its skies a magnificent ring system similar to Saturn's. He argued that all planets undergo a similar evolution. In its initial condition, water and other vaporized materials were forced high above the molten surface. These were arranged in layers, with heavier elements below and lighter materials on top. Gradually, the rings lost rotational momentum, spread outward and over-canopied the entire sphere (like Jupiter's "cloud-ocean"), then gravitated to the poles and collapsed, forming the geological strata in "successive installments." The last and lightest ring, composed of water, produced the Ice Age and the biblical Flood. Genesis speaks of this great source of waters revolving above the earth: the "waters above the firmament." The sun was not directly visible under canopy conditions; thus its "creation" not until the fourth 'day.' Canopy conditions also allowed for the universal, mild tropical climate and great longevity of the pre-Flood patriarchs. Presents other confirming Flood legends; catastrophic nature of Ice Age. Geology proves ocean was greatly enlarged in recent times. Coal, oil, and other organic and mineral deposits are all fallout from canopies. Some of the rings teemed with life, seeding the surface with new forms when they collapsed. Vail claims that "the manner in which specific living organisms have succeeded each other on the earth as revealed by the geologic record demands that that [annular-canopy] system was the cradle of infant life,—the propagating beds in which the life-germs were placed by the great Gardener of nature. Men may laugh at this, but it is not half as ridiculous as to claim that all life came from monera or the rizopods in the primeval ocean on the earth. It is just as reasonable to suppose that germs took form in the waters, under the Creative Hand, before they fell to earth, as afterwards, and when we see that each and every downfall brought in new life-forms which exhibit no specific relation to previous forms, we are forced to admit that either the seed-beds of the annular system provided the undeveloped organisms or there was a special creation at each period." Each catastrophic collapse resulted in massive "life-mutations" as world conditions changed drastically. Vail repeatedly insists his conclusions follow strictly and logically from "philosophic law" (natural philosophy, i.e. science). Though he appeals to Genesis and the Flood, he refers to his theory as "evolutionary," as it assumes long ages and successive change. Drawings. Vail's contagious enthusiasm and sublime confidence, grandiloquent style, and ability to explain Everything in terms of his theory earned him a devoted following. This is pseudo-science in the grandest tradition. Vail published *Annular World Magazine;* the Annular World Society was active many decades after his death. Vail influenced **Patten, P. Johnson, Fort,** the Jehovah's Witnesses, and the Israel-Identity movement. Donald Cyr's *Stonehenge Viewpoint* journal carries on the Vail tradition, reprinting his work, comparing his canopy theory with modern creationist versions, and interpreting archeological remains (especially megaliths) and mythic symbols as literal reflections of the awesome atmospheric conditions caused by the canopies and rings. In 1987 *Stonehenge Viewpoint* reprinted Vail's orig. 1874 pamphlet: "we have the plain declaration of scripture that there were waters above and beyond the firmament; since we see waters so placed above the surface of other planets, and since such bodies of water must revolve around the central body, I claim that the earth in antediluvian times was surrounded by a huge belt of waters."

1648 _______. *Important Disclosures Connected with the Coal Problem Examined in Light of the Annular Theory* (booklet). 1886. Barnesville, Ohio: Enterprise Steam Printing. Rev. ed.

1649 _______. *Alaska: Land of the Nugget; Why?* (booklet). 1897. Pasadena, Calif.: G.A. Swerdfiger. "A Critical Examination of Geology and Other Testimony, showing how and why gold was deposited in polar lands."

1650 _______. *Ophir's Golden Wedge; Some Light on a Burning Question*. 1898. Priv. pub. (Pasadena, Calif.).

1651 _______. *The Lost Lake: A Glacial Problem* (booklet). ca 1900. [Pasadena, Calif.: Annular World Company?]

1652 _______. *The Misread Record; or, The Deluge and Its Cause*. c1905. Chicago: Suggestion Publishing. "Being an explanation of the annular theory of the formation of the earth, with special reference to the Flood and the legends and folk lore of ancient races." Reprinted in serial form in *Stonehenge Viewpoint* (1984). Reviews arguments for the annular-canopy theory. Describes ring system of Saturn and canopy of Jupiter. Argues that man observed canopy and its periodic collapses, preserving this memory in myths which have been largely misread. Explains influence of canopy on climate: created Edenic conditions; warm, rainless world of perpetual spring which allowed for antediluvian longevity.

1653 _______. *The Heavens and Earth of Prehistoric Man* (booklet). 1913. Pasadena, Calif.: Annular World Company. Reprinted in serial form in *Stonehenge Viewpoint*. Analysis of Genesis and many other myths. These stories of ancient man are "canopy fossils": memories of the canopy.

1654 _______. *Eden's Flaming Sword* (booklet). Date unknown. Title also refers to (projected?) second volume on the annular system. "I look the wide world-scene over and I see the most unmistakable proof that man certainly began his career just as all other animated beings did, in a greenhouse world under a God-made canopy of vapors. I call witnesses not alone from the wreck of ages, for, indelibly impressed on the human mind, is the human memory of world-tragedies which link the race with the reign and fall of vapor skies. My Bible tells me that the human family was cradled in an Eden clime, which is nothing less than an affirmation that God was still ruling the earth by canopy law. . . . I say we have the most conclusive proof that man came upon the earth when it was a greenhouse world, and made so by just such a vapor canopy as the wild denizens of earth beheld in the geologic past—just such a canopy as we see today around the planets Saturn and Jupiter. This is God's plan of world building—wisdom's scheme for the nursing of infant races." Noah's Flood manifests all the traits of a final canopy collapse. Eden's Tree of Life was the canopy. The flaming sword placed by God in Eden after the Fall was a manifestation of the partially unveiled sun, accompanied by atmospheric haloes, during a period of canopy collapse. Ezekiel's wheels are reflected halo phenomena.

1655 _______. *The Great Red Dragon*. Unpub.(?). Pasadena, Calif.: Annular World Company. Cited in Vail 1912 as manuscript (2nd ed.) "to be published." "The Dragon of all mythologies is here shown to have been the earth-canopy personified." Dragon was originally good; canopy produced beneficial Edenic conditions. With canopy collapse, dragon becomes evil, as reflected in all mythologies of celestial combat. Rebels against sun, which emerges as conqueror. Appearance of dragon based on serpentine coils of banded canopy beginning its collapse into polar regions.

1656 _______. *The Ring of Truth*. 1979–1981. Apparently unpub. MS, but pub. 1979–1981 in serial form in *Stonehenge Viewpoint*. The Annular-Canopy theory. Cyr, the editor, who advocates a modified Vailian canopy, adds NASA/JPL photos to Vail's drawings. "The Lyellian principle of uniformity then must apply in the development of all worlds." Earth, like all planets, began as molten mass; water and metallic elements vaporized, were lofted high above surface. Disputes **Winchell**'s claim that this water fell down again soon after; says rapid rotation of rings prevented this. Denies being influenced by nebular theory. Predicts greatest mineral deposits will be found in polar regions, where the successive canopies fell. Under canopy conditions earth enjoyed tropical greenhouse climate with evenly diffused light. Memory of this era preserved in many myths and legends, but Genesis contains an especially reliable and scientifically accurate account. Analyzes Genesis account of the Canopy. Man also witnessed periodic canopy collapses, which produced floods "vast beyond conception." Elohim of Genesis was the Canopy; Jehovah, the new god, was the interim sky or the disrupted canopy. Sky again became concealed (over-canopied), until the Flood—the final collapse of the last canopy. Ice Ages also caused by periodic canopy collapses, dumping immense amounts of ice and snow.

1657 van Delden, J.A. *Creation and Science.* 1970s. The Netherlands. *Van Delden:* former T.V. producer with Evangelische Omroep (Evang. Broadcasting), Hilversum, The Netherlands, which had a popular creation-science series; director of new Evangelical Coll. since 1977. Cited in **Ouweneel** 1978, who says it deals with "the nature and limits of science, its relation to the Bible, and the creation and evolution models."

1658 Vandeman, George E. *Tying Down the Sun* (booklet). c1978. Mountain View, Calif.: Pacific Press. *Vandeman:* T.V. evangelist *(It Is Written)* based in Thousand Oaks, Calif.; Seventh-day Adventist; leader of Archeological Research Foundation, sponsor of expeditions in the 1960s to Ararat searching for Noah's Ark. Title refers to Inca ritual. Uses argument from design to prove evolution impossible. Focuses on bird flight and bee society. "Wouldn't it be better—and easier too—to take the clear, simple, plain, understandable statement of Genesis that 'in the beginning God created the heaven and the earth'?" "Isn't that easier than believing that life, unaided by intelligence, could arise from lifelessness?" Acknowledges assistance of ornithologist Asa Thoresen, and Leonard Brand, chairman of Loma Linda Univ. biology dept. Discusses Yosemite as demonstration of global catastrophe. Advocates Flood Geology. Endorses creation with appearance of age: "If God made the earth in six days, as the book of Genesis says He did, then on Friday afternoon He looked out on a mature, grown-up creation," which included "Rocks that looked as if they had stood there from eternity." States in chapter "The Search for Certainty" that Darwin was not certain about his view (tabulates all the phrases indicating uncertainty in the *Origin*), but that the Bible is. Presents standard creation-science arguments in chapter "Darwin's Dilemma." Presents behavior of potter wasp, whose larvae slowly consume the partially paralyzed caterpillars, as proof of creationism. Jesus is either the Son of God or the "greatest imposter that ever lived." We cannot accept Him on the Cross but reject Him in Genesis. The record of creation and of the global judgment of the Flood is rejected because men seek to avoid the warning about the judgment to come. "Saddest of all is the emptiness, the hopelessness of the evolutionary theory."

1659 ________. *The Cave Dwellers* (booklet). c1984. Thousand Oaks, Calif.: It Is Written. Cavemen did exist, but are not the product of prehistoric evolutionary development. They were simply descendants of Noah who dispersed after Babel. Gives examples of biblical cave dwellers. Persecuted Christians also sought refuge in caves. Discusses Waldesian cave chapels; Dead Sea Scrolls. In the Last Days, people may again inhabit caves to flee Satan. It is much easier to believe Genesis than evolution.

1660 ________. *The Telltale Connection.* c1984. Boise, Idaho: Pacific Press. Telltale connection is the Satanic source of both evolution and of psychic and occultic phenomenon. Vandeman begins with story of Piltdown Man; accuses Conan Doyle because of his belief in spiritism. Belief in evolution led to rejection of Bible; thence to the occult, to fill the void. The speech of the serpent to Eve in Eden, caused by Satan, was the first psychic phenomenon. Satan promised Eve she would never die—that rebellion would lead to a god-like state. This has always been Satan's message. Near-death phenomena—tempting reports of a pleasant existence after death for everyone—are false promises of Satan. Satan likewise promotes belief in reincarnation (immortality; ability to atone for wrongs in future life) and in communication with the dead. Vandeman accepts the reality of all psychic phenomena; explains them all as Satan's deceptions—counterfeit miracles or demon-possessions. Discusses demon-inspired rock music, ESP, psychic surgery, Cayce. Cites **Keel**'s studies as evidence that UFOs are Satanic manifestations which will impersonate the Second Coming. "The truth is that today's liberal society is open to *any* alternate to the Genesis account of creation." Distinguishes between Satanic holistic medicine and Seventh-day Adventist version. Quotes **E.G. White** discreetly.

1661 van der Leeuw, J.J. *The Fire of Creation.* 1976. Wheaton, Ill.: Theosophical Publishing House. "The universe as a dynamic process with emphasis upon the Holy Ghost." *Van der Leeuw:* Catholic priest. Theosophy. Creation as the Eternal Mother. Creation is eternal, not a six-day affair. Brahma = God the Holy Ghost = Third Logos (Theosophy). "It has been one of the main contributions of Theosophy that it has coordinated the entire universe in one great conception of the evolution of life." Hindu as well as Christian concepts; no scientific references.

1662 Vander Lugt, Herbert. *Would a Good Rock Lie ... Or Are We Missing Its Message About Evolution and Creation?* (booklet). c1977. Grand Rapids, Mich.: Radio Bible Class. Foreword by **M.R. DeHaan II.** Influenced especially by **G.M. Price.** Presents Gap Theory, Day-Age ("progressive creationism") and strict creationism all as valid options.

1663 Van Dolson, Leo R. (ed.). *Our Real Roots: Scientific Support for Creationism.* 1979. Washington, D.C.: Review and Herald. Articles from Seventh-day Adventist magazine *Ministry,* but includes non-Adventist authors. Authors: Leonard Brand, Robert H. Brown, **H. Morris,** Ray Hefferlin, Eric Magnusson, **Overn,** Edward Lugenbeal, Asa C. Thoresen, **Javor, Sunderland,** G.E. Snow, **Nevins [Austin].** Importance of presuppositions: either biblical or evolutionist. Scientific evidence can support either, but the "Bible story of the origin of life comes from the Creator Himself."

1664 Van Haitsma, John P. *The Supplanter Undeceived.* 1941. Priv. pub. (Grand Rapids, Mich.). *Van Haitsma:* organic science prof. at Calvin Coll.; a founder of **Amer. Sci. Affil.** Cited in **Marsh** 1944, who says van Haitsma explains the story in Genesis of Laban's flocks as due to recessive genes.

1665 van Houwensvelt, S. *Darwinism Has Deceived Humanity.* 1931. London: George Routledge & Sons. Orig. Dutch. Proves that evolution, as conceived by the Darwinians, is impossible. Includes chapters on the prophecies of Nostradamus and on "iris-diagnosis" (iridology: the belief that all disease states can be diagnosed by examination of the iris): endorses both. Advanced knowledge of modern particle physics is contained in the Veddas of ancient India. Presents tale of Voltaire's death-bed turn to God. Quotes evolutionist De Vries: "Nobody was present at the time to see how things took place in the beginning... In all probability we should not understand it, even if we knew it." Says that he is now waiting for De Vries to renounce atheism and materialism.

1666 Van Til, Cornelius. *The Defense of the Faith.* 1967 [1955]. Philadelphia: Presbyterian and Reformed. 3rd ed. Orig. 1955. *Van Til:* seceded from Princeton Theol. Sem. over doctrinal issues; founded Westminster Theol. Sem. (Philadelphia), where he is prof. of apologetics. **Rushdoony,** who assisted with the manuscript, says that Van Til is a great champion of the doctrine of creation from a philosophical perspective; **Whitcomb** and **DeYoung** also praise him. Van Til developed the doctrine of presuppositional apologetics, following Kuyper (theologian and former Dutch Prime Minister). The Bible, God's Word, is the foundational premise of all thought—its truth cannot be proved, but must be presupposed, by faith. By faith we must accept the necessity and sufficiency of its divine revelation. Asserts that "science is absolutely impossible on the non-Christian principle." The non-Christian must assume that rationality and the laws of logic are products of chance. "Thus the truth of Christianity appears to be the immediately indispensable presupposition of the fruitful study of nature." In the absence of Christian theism, no facts could be distinguished from other facts; Chance would be supreme; no hypothesis could be superior to any other. Also, "it would be impossible to *exclude* one hypothesis rather than another. . . . The idea of testing hypotheses by means of 'brute facts' . . . is meaningless. Brute facts, i.e. facts not created and controlled by God, are mute facts." "The argument between Christians and non-Christians involves every fact or it does not involve any fact. If one fact can be interpreted correctly on the assumption of human autonomy then all facts can."

1667 ________. *Christian-Theistic Evidences.* 1961. Unpub. class syllabus and lectures for Van Til's apologetics course at Westminster Theol. Sem. (Philadelphia). Theology; however, the chapter "Creation and Providence" deals with creation-evolution, and contains many quotes (mostly physicists and philosophers). The premise of all human thought must be the biblical Creator God. All men are equal in their absolute total depravity.

1668 ________. *Pierre Teilhard de Chardin: Evolution and Christ* (booklet). c1966. Nutley, N.J.: Presbyterian and Reformed. Orig. pub. 1966 in *Westminster Theol. J.* Blasts **Teilhard** for assuming science is intelligible apart from biblical revelation; says Teilhard's "evolutionary pantheism" is "pure irrationalism." "The final issue then is between those who hold and those who do not hold that God has identified himself discernibly in history in Palestine as the creator and redeemer of men . . . If man's intellect is itself derived, not from the creative fiat of God but from the cauldron of Chance, then what is the difference between right and wrong and how is intellectual contradiction possible? . . . And all would end in a mystery that is meaningless unless with Luther and with Calvin we presuppose the God who has really created and who does really control all things . . ." (Quoted by Greg Bahnsen in **G. North** 1974.)

1669 Van Till, Howard J. *The Fourth Day: What the Bible and the Heavens Are Telling Us About Creation.* c1986. Grand Rapids, Mich.: William B. Eerdmans. *Van Till:* physics and astronomy prof. at Calvin Coll. (Mich.). Argues for a third option instead of the false creation/evolution debate, which is a "tragic blunder." The real debate is between atheistic naturalism and biblical theism. Objects to teaching evolution as a purely "naturalistic" process. Rejects naturalistic evolutionism "because it fails to look at the cosmos through the spectacles of Scripture." Also rejects special creationism, which fails to present a "dynamic, covenantal relationship for all time" and which describes a world whose history and behavior conflicts sharply with observation and science. Advocates "creationomic" perspective: evolution as the ongoing strategy of God's creation. "The creationomic perspective is achieved when natural science is placed in the framework of biblical theism. The foundation of this perspective on the cosmos and its history is the recognition that the entire cosmos is God's Creation." Strongly critical of creation-science, especially **ICR, H. Morris, T. Barnes, Steidl.** Opposes "two-model" education as presenting false dichotomy. Cites anti-creationist references favorably. Much of book concerns stellar evolution. Bibliog., index. Van Till, **Davis Young,** and another Calvin Coll. prof. were recently under investigation by the Christian Reformed Church on charges of heterodoxy, apparently instigated by **D. Gish.** They were fully acquitted.

1670 Vardiman, Larry. *UP, Up and Away! The Helium Escape Problem* (pamphlet). 1985. El Cajon, Calif.: Institute for Creation Research. ICR "Impact" series #143. *Vardiman:* Ph.D.—Colorado State Univ.; worked for U.S. Air Force and Dept. Interior (weather modification); now meteorology prof. at ICR. If the source of earth's atmospheric helium is radioactive decay, and it cannot escape into space, then (as argued earlier by **M. Cook**), the amount of helium in the atmosphere indicates a young earth. Vardiman's special research interest is the pre-Flood Canopy Theory, which he strongly supports.

1671 Varghese, Roy Abraham (ed.). *The Intellectuals Speak Out About God: A Handbook for the Christian Student in a Secular Society.* 1984. Chicago: Regnery Gateway. Foreword by Ronald Reagan. Includes prefatory "Message from the Vatican." Dedicated to **C.S. Lewis.** Intended as a "theistic manifesto." Some sections in form of interviews by Varghese or edited responses. Part I—

The Sciences; Introduction by **Thaxton;** chapters by **Jastrow,** Wickramasinghe (**Hoyle**'s co-author), H. Margenau (Yale physics prof.; worked with Einstein, Schrödinger and Heisenberg), Sir J. Eccles (Nobelist; neurobiologist interested in mind-body problem; co-author with Popper), **Jaki, Sheldrake,** Vitz (NYU psychologist; critic of secular psychology; author of recent study showing that discussion of religion is systematically excluded from history textbooks), D. Martin (sociol. prof. at London Sch. Econ.). Thaxton explains that evolution is philosophical rather than scientific. Jastrow claims that Big Bang cosmogony proves creation, and anthropic principle proves theism. He thinks evolutionary origin and development of life is "plausible" but not certain. Sheldrake refutes materialism. Margenau says relativity and quantum mechanics refute mechanistic physics; objects to Darwinism; agrees with **Paley** and finds anthropic principle "absolutely convincing." Eccles defends dualism and denies materialism. Wickramasinghe denounces evolutionists as "arrogant, dogmatic people" who "hold absolutely tenaciously to a point of view which has become a theological issue"; insists that a Higher Intelligence created life, and that Darwinism is fatally flawed. All insist on an ancient earth; some chide young-earth creationists for making creationism appear unscientific. Part II—Philosophy: includes **Geisler.** Part III—Apologetics and Theology: includes **W. Craig, J. McDowell.**

1672 Velikovsky, Immanuel. *Worlds in Collision.* 1968 [1950]. New York: Dell. Orig. pub. 1950 (New York: Macmillan). Many editions. *Velikovsky:* born in Russia; M.D.—Moscow Univ.; psychoanalyst; lived in Palestine, Europe and New York; Jewish. Accepts truth of biblical history, but proposes quasi-naturalistic explanations for Old Testament events and miracles. Shunned by fundamentalist creationists for denying supernaturalism of Bible, and dismissed as a pseudo-scientist by orthodox scientists, Velikovsky remains enormously popular. Creationists do, however, admire his advocacy of catastrophism in earth history, and often cite him as a victim of censorship and persecution, and of the close-mindedness of establishment science against minority opinions and competing paradigms. (Macmillan, his original publisher, transferred publishing rights to Doubleday when threatened with a textbook boycott by outraged scientists.) Venus, claims Velikovsky, was formed from a violent eruption of Jupiter. It became a comet, and its near approaches to earth in 1500 B.C. caused the various miracles described in Exodus: earthquakes, plagues, parting of the sea, etc. Venus knocked Mars into a different orbit; subsequent near-collisions of Mars with earth around 700 B.C. caused later Old Testament miracles: manna from heaven, hail of rocks, sun standing still, etc. Mankind has repressed conscious recollection of these tremendously violent cosmic cataclysms, but memories of them are preserved in myths. Much of book is examination of myth and folklore around the world for confirmation of these celestial events and catastrophes. Cites many scientific and historical sources; many early catastrophists cited (**Whiston, DeLuc, Cuvier, Donnelly,** etc.), but Velikovsky claims originality for theory. Proposes new magnetic and electrical forces in addition to gravitational effects to explain interactions of Venus, Mars and earth. Earth's axis was violently tilted; rotation was shifted; orbital period was shortened. Rejects uniformitarian explanations for evolution of life as well as earth history. Index.

1673 ________. *Earth in Upheaval.* c1955. New York: Dell. Orig. pub. by Doubleday (New York). *Worlds in Collision* dealt with the last two series of catastrophes, presenting evidence mostly from myth. This book presents geological and paleontological evidence for these as well as earlier catastrophes, including the worldwide Deluge described in Genesis and many other myths. Celestial cataclysms caused mass extinctions and upheavals. Discusses fossil graveyards and sudden onset of Ice Ages. Explains that Darwin did not invent evolution, and that he proposed natural selection as the mechanism of evolution in an attempt to refute catastrophism. Darwin viewed catastrophism as a foe of evolution, but it is actually the cause of evolution. Natural selection cannot create new species. Velikovsky advocates "cataclysmic evolution": sudden evolution of new forms caused by electrical forces, cosmic rays and radioactivity from the planetary near-collisions. Relies heavily on catastrophe arguments of **G.M. Price,** who read an early draft of several chapters. Also cites **Buckland, H. Miller, G.F. Wright,** and **Nilsson** frequently. Index.

1674 Verbrugge, Magnus. *Materialism, Animism and Evolution* (pamphlet). 1981. El Cajon, Calif.: Institute for Creation Research. ICR "Impact" series #94. *Verbrugge:* M.D.; Fellow—Royal Coll. Surgeons of Canada; son-in-law of **Dooyeweerd;** head of Dooyeweerd Foundation (La Jolla, Calif.). Materialism is a religious notion expanded into a complete philosophical system. Denial of God as Creator leads to animism. Modern materialists deny vitalism, but attribute animistic powers to other forces. "Materialists have been repeating over and over that Christians want to introduce supernatural forces into science. But it is really the materialists who want to introduce spirits and animism into science under the guise of creative forces hiding in dead molecules."

1675 ________. *Alive: An Enquiry into the Origin and Meaning of Life.* 1984. Vallecito, Calif.: Ross House Books. Foreword by **Rushdoony.** Deals with origin of life rather than evolution, but notes that these two concepts are closely linked. Those who reject God always substitute other gods—false idols. Rejection of God as Creator by modern science led to theory of abiogenesis: origin of life from inanimate matter. "Abiogenesis declares that *life is death.*" This is a return to the pagan animistic theories of the ancient Greeks, who invented various spirit forces to explain life. Since the humanistic Renaissance, scientists have continued to invent animistic spirits. Early Christian attempts to marry God's word with animistic theories have failed. Augustine argued that God created everything as a 'potential,' thus allowing for Aristotle's spontaneous generation. Both vitalism and materialism share animistic error of replacing God with animistic spirits. Animists of different persuasions refute each other's arguments. Discusses J. Monod's "demons" as animism in biochemical theory. Calls his theory of regulatory genes a "voodoo formula." Describes Oparin's attempts to fit his theory of abiogenesis to Marxist dialectics, and Lenin's political philosophy. Also criticizes Sagan. Christians must adhere to a "cosmonomic" view of creation: belief in God's maintenance of His laws in a scientific manner, whose origins humanists cannot explain. Discusses complexity of cell and genetic mechanism. Stresses Dooyeweerd's distinction between function and functor. Those who reject the Creator constantly confuse these, endowing mere *capacities* with animistic powers. God-rejecting science is a mass of circular reasoning and contradictions. Drawings; bibliog.

1676 ________. *Life: Origin and Meaning.* Undated. Typescript in ICR Library. Presumably early version of 1984 book.

1677 Vere, Francis. *The Piltdown Fantasy.* 1955. London: Cassell.

1678 ________. *Lessons of Piltdown: A Study in Scientific Enthusiasm at Piltdown, Java, and Pekin.* 1959. Stoke, HI, England: Evolution Protest Movement.

1679 Vernet, Daniel. *La Bible et la Science.* 1978. Guebwiller, France: Ligue pour la Lecture de la Bible. Cited in **Blocher,** who says Vernet does not accept evolution above the genus, or at most the family, level.

1680 Vernon Harcourt, Loveson V. *The Doctrine of the Deluge.* 1838. London: Longman, Orme, Brown, Green, and Longmans. 2 vols. Subtitle: "Vindicating the scriptural account from the doubts which have recently been cast upon it by geological speculations." *Vernon Harcourt:* civil engineer; wrote about canals, harbors, flood-control, sewage. Millhauser says he advocates Gap Theory creationism, but also insists on a primarily diluvial geology, and stoutly opposes Lyell.

1681 Vialleton, Louis. *Le Transformisme: Le Cahiers de Philosophie de la Nature.* 1927. Paris: Librairie Philosophique. Also pub. by J. Vrin (Paris). *Vialleton:* zoology prof. at Medical Faculty in Montpelier (France). Symposium by five authors who accept descent with modification but advocate finalist rather than mechanistic interpretation. Authors: Vialleton, **W. Thompson,**

Cuenot, Elie Gagnebin, Roland Dalbiez. According to A.M. Davies (*Evolution and Its Modern Critics;* 1937), Vialleton and Thompson are "hostile to evolution," while Cuenot and Gagnebin support it. Vialleton argues that diversification is different from emergence of major phylogenetic branches, which is not reducible to physico-chemical actions. **Field** 1941 says Vialleton also attacked evolution in *Morphologie Generale* (1924).

1682 ________. *La Vie et L'Evolution.* 1928. Paris: G. Beauchesne.

1683 ________. *L'Origine des Etres Vivant, L'Illusion Transformiste.* 1930 [1929]. Paris: Librarie Plon. 17 eds. in first two years. Vialleton, strongly anti-Darwinist, seems to think that sudden major changes in early embryological development account for differences between families and orders. "Clearly *Archaeopteryx* in no way enlightens us how a reptile could be converted into a bird and, above all, how it would be possible to acquire gradually" all the characteristics of bird. Fetal whale teeth are useless as proof of land animal ancestry; they are not truly vestigial, but serve as a support for growth of jaw. Asserts that man is as far separated from apes as bats and whales from other mammals. Declares it is impossible to conceive of a form intermediate to bats. Quoted extensively in **Dewar** 1957. Dewar 1947 says this book "has done much to cause many French biologists to reject what the French correctly call *Le transformisme* and what we English incorrectly call Evolutionism."

1684 von Braun, Wernher. *The Case for Design: A Letter from Wernher von Braun—NASA* (tract). Undated. Publisher unlisted. Letter dated 9/14/72 from von Braun to Calif. State Board of Educ. and read at textbook hearings by John Ford. *Von Braun:* chief German rocket engineer in WWII; key figure in American space program; NASA administrator. There must be design and purpose in the universe because there is law and order. "To be forced to believe only one conclusion—that everything in the universe happened by chance—would violate the very objectivity of science itself." If physicists accept electrons as real, why not accept the Designer as real? "Alternate theories" of origins should be presented in schools.

1685 von Däniken, Erich. *Chariots of the Gods?* 1971 [c1968]. New York: Bantam. Orig. c1968 (Econ-Verlag); in German *(Erinnerungen an die Zukunft). Von Däniken:* Swiss; former hotel manager; convicted embezzler; one of the best-selling authors of all time. Various "unsolved mysteries"—archeological monuments, artifacts, legends—prove that beings from outer space visited earth. Primitive humans could not have constructed or devised such things. These extraterrestrials were revered as 'gods.' Interprets Old Testament events and descriptions as evidence that God was an astronaut. The human race resulted from "an act of deliberate 'breeding' by unknown beings from outer space," and Noah's Flood was a "preconceived project" for "exterminating the human race" except for Noah's family. Appeals to descriptions in Genesis of 'angels,' 'sons of God and daughters of men.' "Otherwise what can be the sense of the constantly recurring fertilization of human beings by giants and sons of heaven, with the consequent extermination of unsuccessful specimens?" Photos; bibliog., index. By far the most popular of the many refutations of von Däniken is **C. Wilson**'s *Crash Go the Chariots* (Master Books; c1976).

1686 ________. *Gods from Outer Space.* 1972 [1971]. New York: Bantam. Orig. 1971; in German *(Zuruck zu den Sternen).* 1st English ed. 1971 (British ed. titled *Return to the Stars*). Genesis and other myths show that 'gods' came from outer space to earth. These extraterrestrials artificially mutated sub-human apes to produce beings in their own image. The emergence of human intelligence 40,000 years ago is totally unexplainable by other theories, though von Däniken grants the prior evolution of hominids. Creation of Adam describes "artificial mutation of primitive man's genetic code by unknown intelligences"; Eve was created by cell culture in a retort. But these new humans reverted to mating with sub-humans and animals: the Fall. A few thousand years later, the space 'gods' destroyed, in the Flood, the hybrid animal-men described in Genesis and in countless myths of half-human monsters. "'Gods' came, chose a group whom they fertilized and separated from the unclean. They imparted all kinds of modern knowledge to them and then disappeared . . ." Photos; bibliog., index.

1687 _______. *Von Däniken's Proof.* 1978 [1977]. New York: Bantam. 1977 British ed. (London: Souvenir) titled *According to the Evidence: My Proof of Man's Extraterrestrial Origins.* Includes chapter disputing Darwinian evolution and the origin of life. Discusses fossil footprints which refute evolutionist orthodoxy; endorses Paluxy manprints amongst dinosaur tracks (also included in film version of this work).

1688 von Ditfurth, Hoimar. *The Origins of Life: Evolution as Creation.* c1982 [1981]. San Francisco: Harper & Row. Orig. pub. 1981 (Hamburg: Hoffman und Campe) in German *(Wir Sind Nicht Nur von Dieser Welt: Naturwissenschaft, Religion und die Zukunft des Menschen).* German bestseller. *Von Ditfurth:* prof. of psychiatry and neurology; writes popular science books. Solidly evolutionist, but rejects purely materialistic explanations and insists on dimension of spirit. Emphasizes unity of all truth: "this book was written in the conviction that it is possible to harmonize the scientific and the religious interpretations of the world and man." Von Ditfurth quite knowledgeably discusses the evolutionary history of earth, of life, and of man, and describes various evolutionary processes. Rebukes strict creationists (especially **Wilder-Smith**) for their scientific ignorance, but also chides dogmatic positivists, and makes the case for "another world" beyond physical reality. Views evolution as the continuing process of creation. Criticizes **Teilhard**'s anthropocentric teleology. Evolution is a lawful process; God doesn't intervene miraculously. Criticizes Monod's view of pure "chance" evolution (cytochrome C example). Diagrams; index.

1689 von Fange, Erich A. *Time Upside Down* (booklet). 1981. Priv. pub. (Ann Arbor, Mich.). First work in von Fange's "Surprises in Genesis Series." *Von Fange:* psychology Ph.D. Article by von Fange of same title in 1974 *CRSQ.* Discusses nature of time and its bearing on Genesis. Discusses many fossil anomalies and artifacts which contradict evolutionary geological timetable and conventional history.

1690 _______. *Spading Up Ancient Words.* c1984. Syracuse, Ind.: Living Word Services. Words are linguistic "artifacts" which can help us reconstruct the pre-Flood world. Says "there ought to be some hint" of events such as the Flood, the repopulation of the world from Ararat, the dispersion from Babel, and the Genesis Table of Nations "half-buried in the languages which have come down to us." Von Fange argues, however, that "Pre-Flood place names have no relationship at all with the same or similar Post-Flood place names." E.g., the "Euphrates" of Genesis is not the present river of that name. The pre-Flood world consisted of a single landmass with no high mountains. The great Flood catastrophe completely altered the geography and topography of the earth and buried the old world. Cites theories of **Velikovsky** (catastrophic origin of Venus), Barry Fell (widespread transoceanic pre-Columbian colonization of New World), John Michell, and **Berlitz** favorably. Sumerian king lists are corrupted versions of pre-Flood genealogies recorded accurately in the Bible. Refutes the notion that languages have evolved; insists they have degenerated instead. Photos, drawings.

1691 Vos, Howard F. *Genesis and Archaeology.* c1985 [1963]. Grand Rapids, Mich.: Academie Books (Zondervan). Rev. and enlarged ed. Orig. 1963 (Chicago: Moody). *Vos:* Th.D. — Dallas Theol. Sem.; history Ph.D. — Northwestern Univ.; history and archeology prof. at King's Coll. Archeology proves Genesis is historically accurate. Vos does not endorse all fundamentalist claims, however. Babylonian *Enuma Elish* is older than Genesis; it shows degeneration, though, while the Genesis creation account remained pure — the "one preserved by God Himself." Location of Eden not certain because of Flood, but probably in Mesopotamia. Babylonian and Hebrew Flood accounts both derive from same original source. Local flood at Ur discovered by Woolley was not Noah's; it was later. Useful discussion and references on search for Ark on Ararat; Vos is skeptical of most Ark-eology claims. Appendix: archeological sites relevant to Genesis (includes Ebla, about which he expresses caution). Index.

1692 _______. *Genesis.* 1982. Chicago: Moody Press. Everyman's Bible Commentary Series. Critical of evolution. Rejects monophyletic evolution, but admits to variation within "kinds" and suggests occurrence of polyphyletic evolution.

1693 Vos, Johannes G. *Scriptural Revelation and the Evolutionary World View* (booklet). Undated [1966?]. Beaver Falls, Penn.: Blue Banner Faith and Life. 1966 Lecture to annual meeting of Reformed Fellowship (Grand Rapids, Mich.). Denies "double revelation" theory (that Bible and nature both proclaim God's Truth and are equally authoritative). Science falls within the Genesis cultural mandate, and is not on a par with theology. Rejects **Ussher**'s age of earth; accepts Benjamin Warfield's tolerant view (but not billions of years!). Presents three interpretations of six creation days: literal, figurative (indeterminate duration), and framework hypothesis of Augustine. Prefers literal view. Mendel was ignored because he was a theist; Darwin was applauded because he wasn't. Bibliog.

1694 _______. *Surrender to Evolution: Inevitable or Inexcusable?* (pamphlet). 1966. Grand Rapids, Mich.: Reformed Fellowship. Orig. pub. 1966 in *Torch and Trumpet.* Bibliog.

1695 Voss, John E. *The Biblical Significance of the Flood.* 1976. Unpub. thesis: Dallas Theological Seminary (Dallas, Texas). M.A. thesis. Largely theological, but discusses and endorses Flood Geology.

1696 Waggoner, E.J. *The Gospel in Creation* (booklet). 1894. Battle Creek, Mich.: Review and Herald. Reprinted 1975-1987 by Leaves-of-Autumn Books. Seventh-day Adventist. Themes from the story of Christ foretold in the Creation account. Christ as Creator. The heavens preach the Gospel. Experimental "voice-pictures," formed by agitation of powder by human voice, always produce shapes of plants and animals. Thus did God create by His Word. "It is not a trivial matter that 'the latest deductions of science' have drawn so many professed believers in the Bible to modify their views of the story of creation." Organisms possess a life-force breathed into them by God. Stresses importance of proper observance of Sabbath as memorial to creation.

1697 Wagner, A. James. *Some Geophysical Aspects of Noah's Flood* (report). 1961. Unpub.(?). *Wagner:* meteorology dept.—MIT. Cited as "private research paper" by **Dillow,** who does not state Wagner's position.

1698 Wagner, Dennis A. *Students for Origins Research* (audiocassette). Undated [1985]. Sylmar, Calif.: Bible-Science Association—San Fernando Valley Chapter. Tape of 5/18/85 talk to San Fernando Valley chapter of BSA about **Students for Origins Research.** *Wagner:* a founder of SOR; editor of SOR journal *Origins Research.* Recommends that local creation-science groups affiliate with student groups so that they can meet on campus and use campus facilities. SOR strives to promote dialogue between creationists and evolutionists, dealing only with scientific and philosophical—not religious or political—issues.

1699 Wagner, Dennis A. (ed.). *Student Essays on Science and Creation: Volume I.* c1976. Goleta, Calif.: Creation Society of Santa Barbara. Essays by Univ. Calif.—Santa Barbara students: members of Creation Soc. Santa Barbara (later SOR). Authors: CSSB/SOR founders Wagner (1976 electr. engin. B.A.; programmer), Greg Wilkerson (geology thesis on gold-silver mining) and David Johannsen (anthropology/geology), plus Stanley Rice (biology) and John Rudat (petroleum geology). A "collection of papers from science students who believe that the special creation of life is a viable scientific alternative to the chance theory of evolution. These articles were written as individual research projects, class term papers, answers to exams, critiques of lectures and almost exclusively represent thoughts that are not currently included in classroom discussion. It is our hope that the creation model of man's origin will soon be presented on an equal basis with the evolution model in the science classroom..."

1700 Wakeley, Sir Cecil Pembrey Gray. *A Layman Looks at Genetics and Natural Selection* (pamphlet?). Undated. Stokes, H.I., England: Evolution Protest Movement. *Wakeley:* Fellow—King's Coll. (London); president of Royal Coll. Surgeons of England; president of Bible League. Listed in Ehlert bibliog.

1701 Wallace, Alfred Russel. *Man's Place in the Universe: A Study of the Results of Scientific Research in Relation to the Unity or Plurality of Worlds.* 1903. New York: McClure, Phillips. Wallace was the co-discoverer, with Darwin, of the theory of evolution by natural selection. He advocated natural selection as the mechanism of evolution even more rigorously and consistently than Darwin himself—with one important exception. He came to consider that man—and especially the human brain, with its mental capacities far greater than necessary for primitive human survival—could not have evolved naturally. Man's soul must have come from God. This book argues that the design of the universe proves the earth was created especially for habitation, and is the only body capable of supporting life. "Modern skeptics, in the light of accepted astronomical theories (which regard our earth as utterly insignificant compared with the rest of the universe) have pointed out the irrationality and absurdity of supposing that the Creator of all this unimaginable vastness of suns and systems should have any special interest in so pitiful a creature as man, an imperfectly developed inhabitant of one of the smaller planets attached to a second or third rate sun, while that He should have selected this little world for a scene so tremendous and so necessarily unique as to sacrifice His own son ... is in their view a crowning absurdity, not to be believed by any rational being." But Wallace shows that, despite the vast size and age of the universe, earth and man are indeed unique and worthy of God's special attention. The universe was designed for man—the pinnacle of evolution. In Wallace's model of the universe, the sun's location within the galaxy, and the earth's location relative to the sun, demonstrate that our earth is uniquely suited for existence of life, thus proving intelligent design and disproving the notion of a plurality of inhabited worlds. Wallace says he agrees with those who "cannot believe that life, consciousness, mind are products of matter," though he concedes this is not the only possible explanation. Wallace became heavily involved in spiritualism in his later life, and also opposed vaccination.

1702 Walton, Rus. *Biblical Principles Concerning Issues of Importance to Godly Christians.* 1984. Plymouth, Mass.: Plymouth Rock Foundation. Christian Liberty Acad. Satellite Schools Special Edition. (CLASS helps organize private and home schools around the nation, and provides curriculum material.) *Walton:* journalist; syndicated columnist; editor; radio and T.V. commentator. The material in this book orig., pub. in *Letter from Plymouth Rock* and Plymouth Rock Foundation's *FAC-Sheet.* "The Plymouth Rock Foundation is an uncompromising advocate of Biblical principles of self and civil government and the Christian world and life view." Emphasizes Christian heritage of America and biblical basis of government. Endorses Christian Reconstructionist program; cites **Rushdoony, G. North, J. Whitehead.** Praises America as a Christian republic; denounces democracy as humanistic tyranny and socialistic mob rule. Chapters on how to apply biblical principles to various contemporary issues, including abortion, economics, defense, church-state relations, etc. Includes chapter "Evolution vs. Creation.' Recommends **W. Bird; H. Morris, R. Bliss, Wysong, F.N. Lee, Thaxton** et al., **Jastrow.** Advocates teaching of creation-science in public schools. Exclusive teaching of evolution violates Christian students' religious rights, unconstitutionally establishes the religion of humanism, and violates academic freedom. Mentions standard creation-science arguments; endorses Paluxy manprints. Advocates Flood Geology and young earth. The Plymouth Rock Foundation has recently organized nationwide "Committees of Correspondence" to further its program at the local level.

1703 Walworth, Ralph Franklin. *Subdue the Earth.* 1977. New York: Delta (Dell). Written with Geoffrey Walworth

Sjostrom. *Walworth:* attended Shrivenham Univ. (England) and USC; lives in St. Petersburg, Fla. An extension of **Velikovsky**'s catastrophist "revelations," which are being scientifically confirmed. Rejects uniformitarianism. Darwinian evolution is wrong; no new species have arisen gradually. Species evolve suddenly – in thousands, not millions, of years. They arise in large groups, and abruptly become extinct. Emphasizes fossil graveyards; sudden onset of Ice Ages. Chides "extremists" on both sides: young-earth creationists and uniformitarian geologists and evolutionists. 'Apparently, uniformitarians believe that if one waits long enough, the laws of physics are repealed and all manner of amazing things may occur." Discusses catastrophes in formation of plants, vulcanism, geophysical effects of earth's electromagnetic field. Explains truth of Atlantis and Flood myths. Most Atlanteans ended up in the Americas. Internal eddy currents in the earth caused periodic massive vulcanism, lofting huge amounts of dust in atmosphere. This caused the Ice Ages, which resulted in six-mile thick ice layer and consequent three mile drop in sea level. Flatly rejects continental drift theory, which "requires its adherents to have great faith, for many of the assumptions upon which that theory is based contravene observable fact." Petroleum and coal deposits formed by burial during Ice Ages. Sub-humans migrated into fertile new lands covered by volcanic ash and decomposed marine organisms which were exposed by drop in sea level. "The subhumans who left the continents underwent a metamorphosis and became modern people. The means by which this happened is mysterious, shrouded both by time and the fact that this change took place in areas that are now miles under the ocean. Perhaps the stresses of the catastrophe caused them to change form. Perhaps, as **Erich von Däniken** has suggested, visitors from another planet caused the change from subhumans to true humans. Perhaps the Biblical Genesis is correct." These were truly Edenic conditions, occurring in cycles alternating with periods of catastrophic annihilation. Gives credence to **Ussher**'s date and to Cesare Emiliani's calculation of great climatic change 11,500 years ago. We must learn to control this great cycle, to "subdue the earth" as Genesis teaches us. Index.

1704 Ward, Rita Rhodes. *The Bible Versus Evolution for Young People.* 1949. Priv. pub. (El Paso, Texas). Listed in Ehlert files.

1705 ________. *In the Beginning: A Study of Creation versus Evolution for Young People.* 1967 [c1965]. Grand Rapids, Mich.: Baker Book House. *Ward:* B.S. – Abilene Christian; M.A. – Texas Western Coll. (in science?); public school biology and physiology teacher in El Paso, Texas. A series of lessons "especially prepared to give young people a basis for resisting evolutionary philosophy and holding firmly to their faith in God and the Bible." Acknowledges assistance of **Tinkle** and Douglas Dean (Pepperdine Coll. biol. prof.). Includes scientific and biblical references. Book is mostly presentation of standard creation-science arguments, including Flood Geology. Bible doesn't state date of creation, and science cannot know either. Admits Gap Theory may be correct. "The Bible is God's revealed word and therefore is perfect." "If man evolved, there is no point at which he became in the image of God or received a soul. There is no sin, the Garden of Eden [sic], and a promise of a Saviour. There would be no sacrifice of Jesus and plan of salvation. If life evolved there would be no truth to the Bible." Nazism and communism are the results of atheist philosophy. States that vitalism is necessary to creationism. Includes study questions and activities for each chapter, plus creationist reading list. Drawings.

1706 Wardell, Don. *God Created.* 1984 [1978]. Priv. pub. (Winona Lake, Ind.). 2nd ed. Orig. 1978. Argues against young-earth creationism. Explains many geological features as evidence of ancient earth. Suggests a modified Gap Theory: earth "without form and void" refers to clouded-over, flooded condition. This stage was of unknown duration; then God dissipated the clouds and blew the flood away. Wardell says that Genesis shows that "life was on this planet before the six day creation period. The darkness and flooding mentioned in the first verses had apparently killed much of the former growth but sufficient life remained to be a basis for the vigorous growth evidenced on the third day." Other forms were created during the six days. The earlier creations and geological ages helped prepare the earth for humans. Pre-Adamic human races existed as well. Fossil humans and paleo-Indians are not descendants of Adam, but preceded him. Refutes Flood Geology; argues for a tranquil Noahic Flood which left many human artifacts intact. Existence of death in world before Adam may be due to Satan's prior Fall. "Evolutionary teachings that man evolved from apes and thus is 'animal' in nature can lead to a low evaluation of life resulting in such concepts as favoring abortion." Drawings, diagrams; bibliog.

1707 Warington, George. *The Week of Creation, or The Cosmogony of Genesis Considered in Its Relation to Modern Science.* 1870. London: Macmillan. Much of text prev. pub. in *J. of the Victoria Inst.* (a creationist journal). Listed in Schadewald files.

1708 Warren, Erasmus. *Geologia; or, A Discourse Concerning the Earth Before the Deluge.* 1690. London: R. Chiswell. Listed in Schadewald files.

1709 Warren, William F. *Paradise Found: The Cradle of the Human Race at the North Pole.* 1885. Boston: Houghton, Mifflin. *Warren:* president of Boston Coll.; Methodist minister. Scholarly book of over 500 pages. Eden was located at the Pole. M. Gardner says that Warren "draws on the sciences of geology, climatology, botany, zoology, anthropology, and mythology to prove that the climate at the North Pole was once exceedingly warm and pleasant. Here Adam and Eve were created. Later, the Deluge of Noah submerged the land on which Eden was situated, and changed the climate to its present frigidity."

1710 Warriner, D.A., Jr. *What Is Life?* (pamphlet). 1962. Priv. pub. (Troy, Mich.). "This fine booklet presents consideration of some theological and biological evidences and facts; conclusion points out that scientific facts do not support logically only the theory of evolution or refute logically the concept of creationism" (**J.N. Moore,** in 1965 *CRSQ*).

1711 Warring, Charles Bartlett. *Genesis I and Its Critics.* 1851. New York.

1712 ________. *The Mosaic Account of Creation, the Miracle of Today; or, New Witnesses to the Oneness of Genesis and Science.* 1875. New York: J.W. Schermerhorn. "To which are added an inquiry as to the course and epoch of the present inclination of the earth's axis, and an essay upon cosmology." Suggests a pole shift for the Flood. Also wrote *Studies Upon the Inclination of the Earth's Axis* (1876; Poughkeepsie, N.Y.).

1713 ________. *The Relation of the Mosaic Cosmogony to Science* (pamphlet). [1878]. Orig. pub. 1878 in *Penn Monthly.*

1714 ________. *Professor Huxley versus Genesis I* (pamphlet). 1892. Orig. pub. in *Bibliotheca Sacra.*

1715 Warshofsky, Fred. *Doomsday: The Science of Catastrophe.* 1977. New York: Pocket Books (Simon and Schuster). Pub. by arrangement with Reader's Digest Press. (*Reader's Digest* pub. a favorable summary of Velikovsky in 1950). Endorses and presents good summary of **Velikovsky;** similarly interprets myths and legends as factual reports. Genesis and other creation stories are memories of cosmic cataclysms. Appeals to R. Thom's mathematical catastrophe theory, Vsekhsvyatskii's theory that comets are recent volcanic ejections from planets, Cesare Emiliani's evidence of sudden rise in ocean level 11,600 years ago, and Santillana and von Dechend's thesis in *Hamlet's Mill* that Flood myths refer to astronomical changes. Agrees with Velikovsky that radiation showers resulting from cosmic cataclysms produced simultaneous and radical mutations of many characters—the only way evolution could have proceeded. Warns of dangers of genetic manipulation: "The irony is that by ruling out catastrophe as a natural, creative force, we may unleash a biocatastrophe far worse than any natural disaster would achieve." Movement of continents not always slow drift; they show evidence of cataclysmic shifts of position. Presents favorably **Patten**'s theory of catastrophic Ice Age in 2800 B.C. caused by near collision of astral body with earth which dumped huge amounts of ice; mentions that Patten is a preacher but not that he is a creationist. Well-written. Bibliog., index.

1716 Watchtower Bible & Tract Society. *Did Man Get Here by Evolution or by Creation?* c1967. Brooklyn, N.Y.: Watchtower Bible and Tract Society. Jehovah's Witnesses. One of the most influential anti-evolution books; 18,000,000 copies in thirteen languages. Packed with quotes from both scientific and popular sources; also biblical references, especially in final chapters. 248 references cited. Includes most of the popular anti-evolution and anti-Darwinian quotes (many are quoted out of context). Plain, earnest style; appealingly insistent and naive. Presents standard creation-science arguments. Evolution is not a fact; life comes only from design; complexity of cells and organs; sudden appearance of life and gaps in the fossil record; mutations are destructive, not creative; uniqueness of man; proposed 'ape-men' all either true humans or apes; 'primitive' humans have degenerated from originally created perfection. Old-earth creationism: man was created 6000 years ago, but is predated by animal fossils. Permits Gap Theory view: Genesis "allows for thousands of millions of years that the material of the earth could have been in existence before being inhabited by living things." Additionally, insists on Day-Age interpretation: creation 'days' were not literal days, but long periods. Endorses Pre-Flood Water Canopy, which shielded earth from cosmic rays (resulting in misleadingly 'old' C-14 readings). Collapse of Canopy caused the Deluge and the sudden Ice Age. Decries false religious teachings of an eternal Hell and of young-earth creationism as barriers to acceptance of Bible. The world's wickedness is due mostly to belief in evolution, which has led to atheism, communism, German militarism, and the breakdown of morality. Satan, who began the lie of evolution, has ruled the world since Adam's Fall, but the Last Days began in 1914, and this generation will see the end of world, followed by a glorious perfect new world. Describes the earthly paradise to come: no death or disease, disappearance (annihilation) of the wicked. Drawings.

1717 _______. *Accidents of Evolution? Or Acts of Creation?* (booklet). 1981. Brooklyn, N.Y.: Watchtower Bible and Tract Society. Special issue (9/22/81) of widely-distributed Jehovah's Witnesses magazine *Awake!* Standard Creation-Science arguments. Includes: "Could Chance Create Bacteria?" "The Incredible Cell," "The Fossil Record—Their Best Proof" (quotes scientists on its lack of proof of evolution), "God Did It First," "Evolution's Revolution" (new theory of punctuated equilibria proves desperation of evolutionists), "Design Requires a Designer," "When It Touches on Science the Bible Is Scientific," etc. Length of creation days: Day-Age theory. Discusses demand for equal time for creation-science in class. Says that Jehovah's Witnesses shun politics, preferring to teach creationism at home—but are available to lecture on it.

1718 _______. *Truth: Beyond the Reach of Science* (booklet). 1982. Brooklyn, N.Y.: Watchtower Bible and Tract Society. Special issue (11/8/82) of *Awake!* Stresses the provisional, non-absolute nature of scientific knowledge, the non-predictable nature of scientific applications, and the confirmations of scientific discovery in the Bible. Counsels wait-and-see attitude towards apparent contradictions between Bible and science: presumably science will come around eventually.

1719 _______. *Life—How Did It Get Here? By Evolution or by Creation?* c1985. Brooklyn, N.Y.: Watchtower Bible and Tract Society. Sequel to and updated version of the classic 1967 book. Two million copies first edition. Same style and format, but bigger. Lots of drawings; many photos and illustrations in color. Adds some recent anti-evolution quotes to the old standards: **Hitching,** Tom Bethell, **Jastrow, Hoyle,** etc. Adds study questions to each page. Reiterates opposition to young-earth creationists: Genesis states earth could have existed for billions of years before first Genesis 'day'; creation 'days' were probably millennia. Describes harmony between creation days and scientific findings. Expanded sections on examples of design in nature, animal instinct, uniqueness of planet earth, modern scientific inventions found in nature, miracle of human brain, fulfilled biblical prophecy. Attractive volume, nicely presented. Relentless barrage of anti-evolution quotes and arguments, but remains easy to read.

1720 Watlington, Francis. *The Log Book of Noah's Ark: A Scientific Investigation and Analysis of the Voyage by Noah's Ark and Incidents of the Flood, by a Practical Navigator.* 1906. Boston, Mass.: Mayhew.

1721 Watson, David C.C. *The Great Brain Robbery.* 1976 [c1975]. Chicago: Moody Press. Orig. pub. c1975 (Worthing, Sussex, England: Henry E. Walter). Acknowledges help of **Enoch, Gower, M. White.** Presents this book as a nontechnical summary of creation-science arguments, but presents many theological arguments. Says **Whitcomb** and **Morris**'s *Genesis Flood* is one of the greatest books of all time; quotes it extensively. The "monstrous error" of evolution is the biggest delusion—the "greatest brain robbery" ever. This error goes undetected because adaptation of species is confused with evolution. Defends strict young-earth creationism, worldwide Flood, pre-Flood Water Canopy. Says we must face the fact that scientific theories flatly contradict Genesis. "Darwinism," in particular, "contradicts nearly everything written in the first eleven chapters of Bible history: it denies instant creation, the fixity of species, the special creation of man and woman, the Fall, the curse, the universal Flood, the miraculous confusion of tongues, and the young age of the earth." There is no acceptable harmonization between evolution and Genesis. Non-literal interpretations weaken faith. Science is "knowledge based upon observation and the observer's testimony." There will never be any scientific explanation of man's origin, which was an instantaneous miracle. Summarizes Flood Geology arguments. Recommends CRS; presents **Whitelaw**'s C-14 argument. Mass/energy equivalence shows that light was visible before sun was activated on fourth day. Darwinism is simply Epicureanism revived; modern gnosticism. The Bible must be the only basis for science education. Concludes by quoting Churchill: "We believe that the most scientific view, the most up-to-date and rationalistic conception, will find its fullest satisfaction in taking the Bible story literally ... [etc.]" Bibliog.

1722 ________. *Myths and Miracles?* 1976. Worthing, Sussex, England: Henry Walker.

1723 Watts, David C., and David J. Tyler (eds.). *Focus on Creation* (booklet). 1978. Normanton, Derby, England. Rainbow Press. Cited in **R. Smith** 1979. Papers given at 1978 Creation and Origins Conference, "dealing in a concise and somewhat technical manner with basic aspects of creation and origins."

1724 Watts, Newman. *Why Be an Ape? Observations on Evolution, by a London Journalist.* 1936. [London]: Marshall, Morgan & Scott. *Watts:* journalist; also wrote religious books. Quoted in **Field** 1941.

1725 ________. *Why I Believe in Creation, Not Evolution* (pamphlet). Undated [England:] Evolution Protest Movement. EPM pamphlet #11.

1726 [Webb, Samuel]. *The Creation and the Deluge, According to a New Theory; Confirming the Bible Account, Removing Most of the Difficulties Heretofore Suggested by Sceptical Philosophers, and Indicating Future Cosmological Changes Down to the Final Consummation and End of the Earth.* 1854. Priv. pub. (Philadelphia). Pub. anonymously by Webb, who says he wrote portions of it 40 years previously. "The Creation was the result of natural law—there was a physical and moral necessity for the Deluge; both of which phenomena must have taken place about the time and in the way and manner described by Moses. Neither the Creation nor the Deluge was 'a miracle,' or a deviation from the known laws of Nature..." Defends Genesis against unbelievers by proposing this quite novel theory, which provides scientific answers to objections directed against the Bible by skeptics. If Genesis is false, then so are Moses' other books and the rest of the Bible. Webb denies that his theory leads to materialism, pantheism, or atheism. Insists that the six days of creation were literal 24-hour days, and occurred a few thousand years ago, in strict accordance with biblical chronology. Earth was initially molten, surrounded by vapor clouds, which froze into a great luminous Saturn-like ring.

The antediluvian world was flat and marshy; the rings produced a uniformly tropical climate by a greenhouse effect. Rain alone could not have caused the Deluge—it was caused by the collapse of rings. Saturn, in fact, will undergo a similar Deluge when its rings collapse. The habitable portion of the antediluvian world (primarily the Mediterranean seabed) is now under water. The catastrophic ring collapse completely altered earth's topography: higher land sank, and mountains were raised. The antediluvian population, which perished in the Flood, consisted of only 10–15,000 people. Collapse of the frozen ring accounts for the Siberian mammoth burials. From **Newton**'s laws of universal gravitation and motion, Wells predicts that the earth will shortly undergo other catastrophic changes, and will be replaced by "new heavens and a new earth." "The type of animal life is progressing; *a new race of animals,* as much superior to man as man is to monkey, will hereafter appear..." The sun will shrink until it appears no larger than a first magnitude star. At Creation, it was immensely larger—as big as the orbit of Venus. Suggests there may be life on other planets, even the moon. Drawings.

1727 Wells, Carveth. *Kapoot: The Narrative of a Journey from Leningrad to Mount Ararat in Search of Noah's Ark.* 1933. New York: Robert M. McBride. *Wells:* member of Explorer's Club; author of travel and adventure books. "I had read in a book that somewhere in Armenia there is a part of Noah's Ark and as I always like to have some definite object in my travels, I decided to try to find this ancient piece of wood." Describes the deplorable and impoverished conditions in Soviet Russia (everything is "kapoot"). Decries communist efforts to "kick God out of Russia." Traditions of Noah and the Flood very much alive in Armenian Caucasus. The Ark relic is housed in the monastery at Echmiazin, where Noah settled after the Flood, at the base of Ararat, just inside the Soviet border. The monastery is the home of the Armenian Patriarch; its monks have supposedly "spent their lives from time immemorial trying to climb Mount Ararat in search of Noah's Ark." Archbishop Mesrop allowed Wells to see the gorgeously decorated relic: a petrified piece of wood from Noah's Ark—the rudder. Also saw Jesus's crucifixion coat, pieces of the Cross, the crucifixion spear, etc. Photos.

1728 Werner, Abraham Gottlobb. *Short Classification and Description of the Various Rocks.* 1971 [1787]. New York: Hafner. Orig. 1787 (Dresden); in German. Werner taught at a German mining school and developed a classification of minerals. Recognizing the succession of rock strata, he proposed the "Neptunist" (water) theory of geology, arguing that most strata were sedimentary deposits. (The Neptunist theory was opposed by the "Plutonic" theory which attributed most strata to volcanic action.) Earth was originally covered by vast ocean; the "primary" rocks were deposited out of solution as seas retreated. Later, the first land was exposed; "transitional" rocks were formed from erosion of sediments into the sea plus crystallization from the ocean. "Secondary" (Floetz) rocks formed after further exposure of land and erosion. These, and the latest, "alluvial" deposits, were exposed with further retreat of seas. This explains rock strata and types. Werner's geological theories were non-biblical and assumed an old earth, but part of its great appeal was its compatibility with the Flood account. Many of his followers used it to support the Genesis story of Creation.

1729 Wesley, James Paul (ed.). *Progress in Space-Time Physics, 1987.* Blumberg, W. Germany: Benjamin Wesley. Collection of 38 original papers (37 in English) by European physicists who oppose Einstein and advocate absolute space. Authors include **Marinov,** Wesley, **T. Barnes,** E.W. Silvertooth, **Bouw,** others. Cited and recommended by Bouw in *Bull. Tychonian Soc.* and by Barnes in *CRSQ. Wesley:* "The corrupt journal system of communication with its anonymous censors generally permits only sterile 'establishment' ideas to be heard... Only such a free exchange of ideas [as expressed in this vol.] can lead to the necessary progress in physics that all of us researchers in space-time physics so strongly desire." (Barnes points out that creationists suffer from the same censorship.) Book lists 55 space-time scientists "who do not subscribe to Einstein's special theory of relativity." Wesley castigates special relativity as a "sick and deformed" absurdity.

1730 West, Bob. *Evolution Vs Science and the Bible* (booklet). 1974. Orlando, Fla.: Bob West Publications. Set of thirteen four-page pamphlets, plus "Teacher Guide" booklet and set of homework sheets and tests for each unit. Designed as 13 week course of study for church schools, but also useful for private or family study. West says the student materials can also be widely distributed as tracts, and suggests that the course be taught from fifth grade all the way to adult level. Intended as counter-attack against the evolutionist propaganda children are subjected to even before starting school. Strongly creationist. Denounces evolution as unscientific and because it contradicts the Bible. "The Bible record is the only record that harmonizes with scientific fact." Advocates Gap Theory creationism followed by literal six-day re-creation. Quotes white-supremacist literature as evidence that evolution is racist; equates it with Nazism and atheism. Claims evolutionists argue that dog-to-horse blood transfusion is fatal, proving no relationship, but that ape-to-man transfusion gives no reaction, proving blood relationship; however, transfusions from other animals to man show only weak reactions. (This is taken unacknowledged from **P. Johnson** 1938.) Cites many other creation-science works. Presents standard creation-science arguments, Flood Geology. "If organic evolution is true," then: (1) "There is no God. (2) The Bible account of creation is a myth or fantasy. (3) The scriptures are not from God. (4) Jesus is not our saviour. (5) Man is only an animal. (6) There is no such things [sic] as sin, or morality. (7) There is no God-given law against anything, including murder and adultery." Drawings.

1731 Westberg, V. L[uther]. *Mystery of the Buried Redwood* (tract). Undated [1963]. Priv. pub. (Napa, Calif.). *Westberg:* engineer; head of Westberg Manufacturing Co. (Sonoma). Redwood tree found by well driller in Napa. Westberg explains it as evidence of the Flood. He tells how he was strongly influenced by reading **T. Schwarze**'s 1957 Canopy Theory book. Summarizes canopy theory and explains how its collapse produced the Flood. "The direct cause of this first cataclysmic event on earth was SIN."

1732 _______. *The Master Architect* (booklet). Undated [1972]. Priv. pub. (Napa, Calif.). Canopy Theory. Earth created 6000 years ago; Flood 1600 years later. When God lifted the waters above the firmament, it was in the form of ice crystals. These "eventually interlocked" to form a spherical shell of ice spears containing 450 feet of water, suspended by centrifugal force 350 miles above the earth's surface. (Elsewhere he says it could have been 3000 miles high.) Canopy ice spears transmitted light to earth by fiber optics. God caused its collapse, to produce the Flood, by sending out a great solar flare which melted a section. It "woofed down at the poles"; its remnants are the Antarctic ice cap.

1733 _______. *Created Perfection Lost!* (tract). Undated. Priv. pub. (Napa, Calif.). Ice shell canopy was created on the second day—the only day not pronounced "good" by God. He knew it would start deteriorating, and eventually collapse to produce the Flood. After the Flood, the world, created perfect, was devastated and much harsher. Moon was originally full every night. "Archeological finds confirm the Biblical accounts and disprove the Satanical promotion of the evolution theory with its Long ages, Ice ages, Stone ages, Cavemen, Ape ancestry concepts." Evolution is the foundation of communism.

1734 _______. *Miracle Clouds of Today and the Cloud Miracles of Yesterday* (tract). Undated. Priv. pub. (Napa, Calif.). NASA photos of earth's cloud cover reminds us that there were no clouds (plural) and no rain until the Flood. Pre-Flood Earth was covered by an immense Ice Canopy.

1735 _______. *Mystery of the UFOs* (tract). Undated. Priv. pub. (Napa, Calif.). UFOs have proliferated since 1945. Suggests they are inhabitants from other dimensions coming to investigate atomic explosions. Born-again Christians will likewise be able to translate across the phase gap after death. The biblical

Firmament or Heaven may be one of these phased earths.

1736 Westermeyer, Anton. *The Old Testament Vindicated from Modern Infidel Objections.* Quoted and discussed in A. White 1896. Dr. Westermeyer lived in Munich. Opposes geology; elaborates on theology of Gap Theory creationism. Generations of creatures from the original creation succumbed to Satan's corruption and became demons. During the six-day re-creation, God destroyed these demons or drove them from their original habitat. "By the fructifying brooding of the Divine Spirit on the waters of the deep, creative forces began to stir; the devils who inhabited the primeval darkness and considered it their own abode saw that they were to be driven from their possessions, or at least their place of habitation was to be contracted, and they therefore tried to frustrate God's plan of creation and exert all that remained to them of might and power to hinder or at least to mar the new creation." They produced "horrible and destructive monsters, these caricatures and distortions of creation" of which fossils are the remains.

1737 Wharton, Edward C. *Genesis: Historical ... Or Mythological?* (pamphlet). Undated. West Monroe, La.: Howard Publishers. *Wharton:* Bible teacher at Sunset School of Preaching (Lubbock, Texas). Mostly theology. Only a literal interpretation of Genesis, with its account of the destruction of the world by Flood, will alert us to the coming destruction of the world by fire.

1738 Wheeler, Gerald W. *The Two-Taled Dinosaur.* 1975. Nashville, Tenn.: Southern Publishing Association. *Wheeler:* M.A.—Univ. Michigan. Strict creationism: presents many standard creation-science arguments (not just about dinosaurs). Includes chapter on California science textbook controversy. Includes section on the Flood Paradigm.

1739 _______. *Deluge* (booklet). c1978. Nashville, Tenn.: Southern Publishing Association.

1740 Wheeler, Ruth. *Dinosaurs.* 1978. Mountain View, Calif.: Pacific Press. Seventh-day Adventist. Strict creationism. For children.

1741 Whewell, William. *Astronomy and General Physics, Considered with Reference to Natural Theology.* 1833. London: Pickering. Also pub. 1836 (Philadelphia). One of the Bridgewater Treatises. *Whewell:* Cambridge Univ. scholar; Anglican priest; logician, trained in math and geology; respected for contributions to philosophy of science and as author of *History of the Inductive Sciences;* coined the terms 'scientist,' 'physicist,' and the names of several of the geological epochs. Whewell abandoned belief in Flood Geology but reaffirmed catastrophism and the fixity of species.

1742 _______. *Indications of the Creator.* 1845. Philadelphia. A refutation of Chambers' *Vestiges of the Natural History of Creation.* Millhauser says it is "little more than a collection of extracts from Whewell's previously published work, tending to prove that science does not concern itself with final causes—that 'our Morphology cannot prejudice our Teleology'; and it has the further decided peculiarity of answering *Vestiges* without mentioning it by name."

1743 Whiston, William. *A New Theory of the Earth, from Its Original, to the Consummation of All Things, Wherein the Creation of the World in Six Days, the Universal Deluge, and the General Conflagration, As Laid Down in the Holy Scriptures, Are Shewn to Be Perfectly Agreeable to Reason and Philosophy.* 1697. London: B. Tooke. Six editions pub. up to 1755. Intended to replace **Burnet**'s 1681 theory, and was likewise an attempt to reconcile Genesis with science. Whiston attempted to wed the Bible with the new Newtonian conception of the universe, insisting on literal interpretation of the Bible, but also explanation by natural law whenever possible. Whiston's mechanism for change in an otherwise stable universe was the catastrophic influence of comets. Earth began as a comet which became a planet when its elliptical orbit smoothed out and its unformed atmosphere (the "chaos" of Gen. 1:2) settled down. Earth didn't rotate at first, so the Creation 'days' lasted a year. Sun and moon became visible on the fourth day when the atmosphere cleared. Cometary atmosphere settled onto surface in strata according to density. Earth began to rotate at the Fall, when it was struck by a comet, which also tilted its axis. The worldwide Flood of Noah was caused by close approach of a comet in 2349 B.C.; previously earth had 360-day year. Water from comet's tail fell on earth, and gravitational stress distorted and cracked earth's crust, allowing waters of the deep to spill out. Destruction of world by fire, ushering in the Millennium, will also be caused by comet impact. Whiston's book so impressed **Newton,** whose disciple he was, that he inherited Newton's mathematics chair at Cambridge. Newton, however, later decided Whiston's intepretation was not literal enough, and that his naturalistic explanations were too radical and mechanistic. Whiston was later dismissed from Cambridge and tried for heresy (Arianism).

1744 Whitcomb, John C., Jr. *The Origin of the Solar System: Biblical Inerrancy and the Double-Revelation Theory* (booklet). 1975. Philadelphia: Presbyterian and Reformed. Expanded version of 1962 address at Moody Bible Institute; published 1963 in *Grace Journal* Further rev. for publication as booklet (Phillipsburg: Presbyterian and Reformed; undated [1964]). *Whitcomb:* B.A.—Princeton Univ.; Th.D.—Grace Theological Seminary (dissertation on the worldwide Flood); Old Testament prof. at Grace Theological Seminary (Winona Lake, Ind.); Grace Brethren; respected theologian and author, best known as co-author with **H. Morris** of *The Genesis Flood.* An essay on the nature of biblical inspiration, using astronomy for illustration. Whitcomb rejects the "double revelation" theory: the popular theological notion that God's truth is revealed equally in His two books—the Bible and Nature. Double-Revelation advocates argue that each is authoritative in its own realm; the theologian, that is, must defer to the scientist in the interpretation of nature. Whitcomb, however, insists that absolute primacy be given to scripture, even when scientific theories contradict the Bible. God does reveal Himself in nature, but many truths remain outside scientific investigation—especially theological truths and one-time supernatural acts of creation. Most of booklet consists of criticism of scientific theories of origin of solar system, with many scientific references. Photo, drawing. Whitcomb's criticisms seem to be directed in part at Morris and his followers, who now maintain that creation-science can be purely scientific and need not refer to the Bible.

1745 ________. *Creation According to God's Word* (booklet). Undated [1966?]. Grand Rapids, Mich.: Reformed Fellowship. First section on "The Nature of Biblical Creation": Creation was supernatural, literal, and *ex nihilo.* It can be understood only through the "special revelation" of the Bible (not via the inferior "general revelation" of nature). Creation also includes "superficial appearance of history or age" (discusses **Gosse**). Other sections on "Creation of Plants and Animals" and "Creation of Mankind": standard creation-science arguments; many scientific references.

1746 ________. *The Early Earth.* c1986 [1972]. Grand Rapids, Mich.: Baker Book House. Orig. 1972. Foreword by **H. Morris.** Expanded version of 1966 booklet; includes same topics (nature of biblical creation, creation of plants and animals, and of mankind), plus "Creation of the Universe" and a section refuting Gap Theory creationism ("Was the Earth Once a Chaos?"). Insists on recent, sudden creation *ex nihilo* in six literal days. Earth was created before sun and stars to emphasize uniqueness of earth and God's role as Creator. Standard creation-science arguments. Many theological and scientific references. Discusses **Chalmers, Scofield,** and **Custance** in Gap Theory chapter. The "Son of God created the earth as a dynamic, functioning, fully equipped home for man in a fantastically short period of time." Scientific discoveries in all fields prove naturalistic theories of origins to be increasingly untenable. In this fallen world, it is hard to imagine the perfect world of creation, with no decay or degeneration, no bloodshed, no death; "a foretaste of the New Earth which God will some day create." But God has given us "fascinating and adequate hints of these primal realities in nature itself," in addition to the "clear and self-authenticating account" in the Bible. Photos; bibliog., index.

1747 ________. *The World That Perished.* 1973. Grand Rapids, Mich.: Baker

Book House. Intended as a sequel (also an updated and a popular restatement) to Whitcomb and **Morris**'s *Genesis Flood.* "Biblical and Scientific Evidence for the Genesis Flood as a Global Catastrophe." Foreword by H. Morris. Back cover has ad for Films for Christ movie of same title, a "documentary" on the Genesis Flood, which is loosely based on this book. "The world that then was, being overflowed with water, perished" (II Peter). The sinful world was destroyed by God; the rocks and fossils thus give witness to the power of the Creator God in these Last Days before the coming destruction of the world by fire. First section: "God Destroyed the World—Supernaturally." Refutes various compromise theories and attempts to reconcile Genesis with science by natural law explanations (discusses **Ramm, Filby, Custance, Patten**, etc.); insists that "Genesis is *consistently supernatural* in its presentation." Second section: "The Flood Destroyed the Entire World."—refutes local-flood interpretations (especially **J.W. Montgomery**). Third section: "The Effects of the Flood Are Visible Today"—Flood Geology summary. Uniformitarianism and evolution represent deliberate rejection of the biblical record of catastrophism and creation, as prophesied in II Peter. Fourth section: "The Basic Issue: Is the Bible Truly God's Word?"—includes rebuttals to negative reviews of *Genesis Flood,* especially in *JASA*. Photos; bibliog., index.

1748 ________. *The Bible and Astronomy* (booklet). c1984. Winona Lake, Ind.: BMH Books. Separate publication of Whitcomb's chapter in **Mulfinger** (ed.) 1983. "The ultimate purpose and significance of the astronomic universe is *theological*. . ." Vastness and diversity of celestial objects proves God is Creator of universe. Joshua's Long Day and Hezekiah's Sign were local miracles: God did not stop earth's rotation, but supernaturally altered the sunlight in Palestine. Bible says intelligent physical life exists only on earth: Christ's incarnation on earth was not an "insignificant event in the career of the son of God, stopping briefly on earth, as it were, on His way to other planets and galaxies to carry on a cosmic ministry of revelation and redemption." Theological and scientific references. Photos.

1749 Whitcomb, John C., and Donald B. DeYoung. *The Moon: Its Creation, Form and Significance.* 1979. Winona Lake, Ind.: BMH Books. Foreword by Larry Redekopp (Ph.D.; aerospace engin. prof. at USC). Acknowledges 64 people who read manuscript drafts. Wants to correct relative neglect of astronomy by modern creation-science. Space program has disappointed many by futile attempts to prove extraterrestrial evolution. "By the very nature of the case, a concept of ultimate origins involves basically unprovable presuppositions that are religious in character. . . . The issues are, at bottom, moral and ethical, rather than merely academic and intellectual." Criticizes double-revelation theory; science is not an independent and equally authoritative source of truth. Refutes various theories of moon's origin. Evidence for young age of moon. Cratering may have occurred during the first three days of Creation to prepare crust, or from bombardment when earth's Water Canopy collapsed. Discusses reports of various transient lunar phenomena and anomalies; interprets these as evidence of young, geologically active moon. God deliberately designed moon to appear same size as sun. Eclipse records help to date historical accounts. God waited until fourth day to create moon as sign of His superiority; this despite the prevalence of moon-worship. Photos (many NASA color photos), charts, diagrams; bibliog., index.

1750 Whitcomb, John C., and Henry M. Morris. *The Genesis Flood: The Biblical Record and Its Scientific Implications.* 1961. Phillipsburg, N.J.: Presbyterian and Reformed. Also pub. 1961 by Baker (Grand Rapids, Mich.). Foreword by John McCampbell (geology prof. at Univ. S.W. Louisiana). This book served as catalyst for the modern creation science movement and led to the formation of the Creation Research Society. It was largely responsible for the popular revival of creationism. Moody Press wanted to publish Whitcomb's 1957 Th.D. thesis on the Flood; Morris reviewed the scientific chapters, then Whitcomb and Morris decided to collaborate.

Moody backed off, but **Rushdoony** got P&R to publish it. Chapters 1–4 and appendices are by Whitcomb. Morris wrote Introduction and chapters 5–7. 518 pp., bristling with scientific references and footnotes, it appears extremely scientific, though its Flood Geology arguments are mostly updated versions of **G.M. Price**'s. Whitcomb's chapters present biblical arguments for worldwide Flood, and refute local Flood theories. Chapter on "Uniformitarianism and the Flood: A Study of Attempted Harmonizations" discusses many of the older creationist and Flood theories. Blames **Cuvier**'s acceptance of multiple catastrophes for eventual success of uniformitarianism, and criticizes many other compromise theories. Includes photos of **Burdick**'s carved Paluxy manprints (proclaims them genuine). Notes that Bible speaks of four other epochs besides the Flood which produced some geological strata; initial act of Creation (basement and some Precambrian rocks), third day of Creation (land/water separation), some effects from pre-Flood Water Canopy, and some post-Flood effects. But insists that major geological formations and virtually all fossil-bearing strata were formed by the Flood. Earth's core and mantle are essentially unchanged since Creation. Uniformitarian science cannot aid in understanding creation. "The geologic record may provide much valuable information concerning earth history *subsequent* to the finished Creation. . . , but it can give no information as to the processes or sequences employed by God *during* the Creation, since God has plainly said that those processes no longer operate—a fact which is thoroughly verified by the two universal laws of thermodynamics!" Evolution is simply assumed, though no evidence supports it—"one of the most astounding paradoxes" of scientism. "A real understanding of origins requires, as we have repeatedly emphasized, divine revelation." Thus only the Bible, and not science, can tell us about creation, and any historical science must be based on the primary facts of the Creation, Fall, and Flood. Creation necessarily includes initial appearance of age; to deny this is to affirm atheism. Discusses climatic and other effects of Water Canopy. Bulk of Flood waters came from internal sources —the rupturing of the "fountains of the deep." Presents standard Flood Geology arguments: hydrodynamic sorting of organisms and sediments to produce various strata; differential mobility of organisms fleeing Flood which results in sorting according to type; some layered strata due to currents. Oil and coal formed by Flood. Discusses fossil graveyards; Ice Age as consequence of Flood. Says there are gaps in biblical genealogies, but Flood cannot have occurred more than 5,000 years before Abraham. Photos, charts; index.

1751 Whitcombe, J.C., and C.E.A. Turner. *God's Double Revelation: Scripture and Science* (pamphlet). Undated. [England:] Evolution Protest Movement. EPM pamphlet #112.

1752 White, A.J. Monty. *What About Origins.* 1978. Kingsteignton, England: Dunestone Printers (priv. pub.?). *White:* gas kinetics Ph.D.—Univ. Coll. (Wales); chemist; academic registrar at Univ. Wales Inst. of Science and Technol. (Cardiff). White had a religious conversion in 1964. Later, he married a creationist, and realized he interpreted Genesis—creation —differently than he did the rest of the Bible, so he became a creationist because the rest of the Bible depends on creation. (Cited in **R. Smith** 1979.)

1753 ________. *How Old Is the Earth?* 1985. Welwyn, Herts., England: Evangelical Press. Biblical and scientific arguments for young-earth creationism. If earth is young, then evolution is false. If evolution were true, then Bible must be false. Discusses Genesis genealogies; allows gaps, but insists creation cannot have been more than 20,000 years ago. Stresses creation with "superficial appearance of age." Refutes Gap Theory, Revelatory Theory and Day-Age Theory creationism. Criticizes radiometric dating; presents many dating methods pointing to recent creationism. Includes many standard creation-science arguments; quotes **Whitcomb** and **Morris, Barnes, M. Cook, Setterfield,** and other creationists. Bibliog.

1754 White, Arthur K., and Ray B. White. *A Toppling Idol: Evolution.* 1933. Zarephath, N.J.: Pillar of Fire. *A.*

White: minister. Illustrations; bibliog.

1755 White, Ellen G. *Patriarchs and Prophets.* 1958 [c1890]. Washington, D.C.: Review and Herald. Orig.(?) c1890 (Battle Creek, Mich.). "The Conflict of the Ages Illustrated in the Lives of Holy Men of Old." White is the Seventh-day Adventist prophetess; her writings are considered divinely inspired. This book is a narrative presentation of Old Testament history. Includes strong emphasis on recent, literal six-day creation and effects of worldwide Flood. The biblical account of creation "is so clearly stated that there is no occasion for erroneous conclusions. God created man in His own image. Here is no mystery. There is no ground for the supposition that man was evolved by slow degrees of development from the lower forms of animal or vegetable life. Such teaching lowers the great work of the Creator to the level of man's narrow, earthly conceptions. Men are so intent upon excluding God from the sovereignty of the universe that they degrade man and defraud him of the dignity of his origin." The Creation Week consisted of seven literal days, which we are to commemorate in our ordinary week and observance of the Sabbath. To assume these were long ages is to deny the Fourth Commandment. The fossils found by geologists which appear to deny the Mosaic chronology are of immense and fantastic antediluvian creatures, all buried by the Flood. Geology cannot tell us their age, only the Bible can. Earth was created exceedingly beautiful and bounteous, but it began to deteriorate as a result of Adam's Fall. It was completely devastated by the Flood, and much remains desolate. "The entire surface of the earth was changed at the flood." Violent winds and currents buried the remains of the pre-Flood inhabitants; mountains were heaped up; useful minerals were hidden underground. The Flood buried immense forests which formed coal and oil deposits. Stresses that the Flood is warning of coming destruction of the world. Illustrations. White's insistence on recent, literal creation and the effects of the worldwide Flood inspired **G.M. Price**'s Flood Geology and many other Adventist anti-evolutionists.

1756 ________. *Principles of True Science: or Creation in the Light of Revelation.* 1986 [1920s]. Payson, Ariz.: Leaves-of-Autumn Books. Compiled from the writings of Ellen G. White by Marion E. Cady. Orig. 1900. Revised and combined with *Science in the Bible;* pub. in 1920s. Reprinted 1986 (Leaves-of-Autumn). Alphabetical arrangement of 1600 terms and topics, which are explained by excerpts from various writings of White. Includes many excerpts advocating recent creation and denouncing evolution. "The sophistry in regard to the world's being created in an indefinite period is one of Satan's falsehoods." "Evolution and its kindred errors are taught in schools of every grade... Thus the study of science, which should impart a knowledge of God, is so mingled with the speculations and theories of men that it tends to infidelity." "We need to guard continually against those books which contain sophistry in regard to geology and other branches of science... [T]hey need to be carefully sifted from every trace of infidel suggestions. ... It is a mistake to put into the hands of the youth books that perplex and confuse them." Also includes much on diet and health.

1757 White, Joseph L. *The Creation of God: The Struggle of Life for Perfection in a Spiritual World of Science.* 1975. Madison, Tenn.: White Pub.

1758 Whitehead, John W. *The Separation Illusion.* 1977. Milford, Mich.: Mott Media. "A Lawyer Examines the First Amendment." Foreword by **R. Rushdoony.** *Whitehead:* J.D.—Univ. Arkansas; founder and president of Rutherford Institute (Manassas, Va.), which deals with Christian legal issues, abortion, school prayer, and creation-science. Acknowledges assistance of Rushdoony. America is in decline because it has largely forsaken the Christianity of its Founding Fathers. Denies that Founding Fathers were mainly deists. The Federal gov't. has usurped the rights of the states. The "wall of separation" between church and state violates the intent of the Constitution. This separation "illusion" is based on Satan's false promise that man is autonomous and can become as God. The Constitution did not institute a democracy, which levels everything to the lowest common denominator and leads to mob tyranny. Genesis advocates monarchy. Prior to the Civil War, the South had the next best thing—a Christian aristocracy. The abolitionists—heretical revolutionaries—called the Constitution a "covenant with death" and "agreement with Hell." "The abolitionists, while utilizing the slave issue as a base, had a more fundamental motive than slavery for attacking the South." Whitehead quotes Rushdoony that the Civil War was largely the result of a "Unitarian statist drive for an assault on its Calvinist enemy, the south... [T]he anti-Christian, Jacobin attack on slavery had to be fought, and slavery defended, because the revolutionary reordering of society would be far worse than anything it ought to supplant." Slavery wasn't that bad anyway, and the "caste system" that replaced it after the war was worse. The Unitarian-abolitionist movement was predicated on the assumption that humans could build their own utopia. Whitehead calls for a total Christian reformation of society. Stresses presuppositional nature of attitudes and worldviews, and emphasizes the constant and inevitable warfare between the godly and ungodly. Americans have abandoned the Bible and therefore knowledge; they trust in evolution—that they were made in the image of the beast—rather than in God the Creator. Analyzes court decisions. Bibliog., index. Whitehead was co-author, with John Conlan (former Ariz. congressman), of influential 1978 *Texas Tech Law Review* article "The Establishment of the Religion of Secular Humanism and Its First Amendment Implications," which denounced the teaching of evolution.

1759 ________. *The Freedom of Religious Expression in the Public High School.* 1983. Westchester, Ill.: Crossway Books. Rutherford Inst. Reprint: Vol 1.

1760 ________. *The Stealing of America.* 1983. Westchester, Ill.: Crossway Books (Good News). Acknowledges help from **W. Bird** and **F. Schaeffer.** Atheistic, humanistic Big Brother is stealing our nation and eroding our freedoms. American society is now perilously similar to Germany just before the Nazi takeover. We are "On the Road to Auschwitz" and totalitarianism. Evolution is leading the way: it results in abortion, genocide, and dictatorship. **Hitler** and Stalin based their views on evolution. "The historical foundation for freedom was the belief in the Creator who established moral absolutes and gave to man absolute rights..." "Modern materialistic science cannot understand man because it does not understand the Creator." Much of book is devoted to the "ominous parallels" between our humanistic society and Nazi Germany. Bibliog., index.

1761 ________. *The End of Man.* 1986. Westchester, Ill.: Crossway Books. Recommended as creationist book in 1986 *Acts & Facts.*

1762 Whitelaw, Robert L. *Does Christian Faith Depend on Scientific Fact?* (pamphlet). N.d. [1971?]. Bible-Science Assoc. *Whitelaw:* prof. of mech. and nuclear engineering at Va. Polytechnic Inst. Rebuttal to 1971 article in *The Banner* which said Christian faith can't be destroyed by science since it isn't dependant on science. Whitelaw asserts that Christianity is built on facts—it is scientific. Evolution is speculative, and is less probable than the likelihood that the universe doesn't really exist. "In short, Christian faith is rooted in actions that are reported in a Biblical record that satisfies all the canons of scientific evidence." The Bible must be rejected if any fact or event in it is false. Denial of Adam and Eve, the Flood, etc., is therefore denial of the Resurrection. All new findings or theories of science must be scrutinized in light of the Bible. If they do not conform to scripture they must be rejected.

1763 ________. *The Geological Age-Names* (tract). 1971. Vancouver, B.C.: Creation Science Assoc. of Canada. Orig. in *Bible-Science Newsletter.* Geological column and age-system "represent nothing more than the discredited onion-coat theory masquerading in biological form..." They are pure make-believe, "invented to prop up an equally fanciful notion of evolution, without adducing a shadow of scientific evidence." Worse, they contradict the clear revelation of God. Urges that creationists expose the absurdities of geological age-system to the public, and presents arguments which refute it. Geol. column is constructed upon assumption of evolution—index fossils. "A creation date of about 7,000 years ago, and a flood date almost exactly 5,000 years ago can be confidently asserted, supported by direct Biblical testimony, and by radiocarbon dating. These dates cannot be controverted either by secular history or by geological evidence."

1764 ________. *Evolution and the Bible in the Light of 15,000 Radiocarbon Dates* (booklet). 1981. Huntsville, Ontario, Canada: M.B.C. Publications. 6th ed. Orig. 1969 radio broadcast; then 1970 *CRSQ* article. M.B.C.: Muskoka Baptist Conference. If evolution is true, there is no purpose, no destiny, no truth or error, no right or wrong, no hope. Analyzes distribution of dates published in journal *Radiocarbon*. There are hardly any dates around 5,000 years ago; they reappear prior to that time, and disappear about 7,000 years ago. This corresponds exactly to the Flood (4940 B.P.) and the original Creation.

1765 Whitney, Dudley J. *The Case for Creation*. c1946. Malverne, N.Y.: Christian Evidence League. Published as series of five booklets. *Whitney:* agricultural chemistry B.S.—Univ. Calif. (Berkeley); editor and contributor to several agricultural journals; wrote for *Proc. Victoria Inst.* and *Bible Champion;* published newsletter *The Creationist;* member of Religion and Science Assoc. with **Higley, Price** and **B. Nelson;** deacon of Community Rural Church (non-denom., charismatic).

1766 ________. *The Face of the Deep: A Defense of Divine Creation*. c1955. New York: Vantage Press.

1767 ________. *Genesis versus Evolution: The Problem of Creation and Atheistic Science*. 1961. New York: Exposition Press. Foreword by **H. Slusher.** Strong argument for literal, young-earth creation. Science affirms the Genesis account of recent six-day creation. Naturalistic descent of plants and animals from simple one-celled form millions of years ago is an "absurdity." Stresses unbridgeable dichotomy between naturalistic evolution operating over geologic ages and belief in Creator God as revealed in Genesis. Endorses Flood Geology: a single event restructured earth's surface; evidence from catastrophic burial of fossils. If the Flood occurred, then recent divine Creation must be true. Cites examples of human footprints in Paleozoic strata. (Also cites Glen Rose [Paluxy] dinosaur fossils, but for different purpose.) Advocates creation date about 4224 B.C.; Flood 1565 years later. Noah's sons: Japheth is progenitor of Aryans; Chinese are probably Semites (descendants of Shem); Nimrod, Phoenicians, and Africans are Hamites. Evidence refuting evolution includes similarity of form of organisms separated by oceans—continents couldn't have moved. Suggests they were placed there separately by a second creation following the Flood—"a very reasonable supposition provided the concept of creation can be allowed." The Flood resulted in uplift of continents. Fresh water runoff gradually accumulated and froze in Arctic, finally producing sudden temperature drop about 2500 years ago (the Ice Age). Argues that vulcanism is key to earth history. Great diversity of North American fossil forms explainable only by the Flood. Atmospheric helium and rates of mineral deposits in oceans point to young earth. "Reason positively demands a decision in favor of divine creation, which is only another way of saying that commonsense science positively proves the fact of God."

1768 ________. *The Noachian Deluge*. Undated. CHR Evidence. Listed in SOR CREVO/IMS.

1769 Wiester, John L. *The Genesis Connection*. 1983. Nashville, Tenn.: Thomas Nelson. *Wiester:* geology B.A., business admin. M.A.—Stanford Univ.; taught historical geology at Stanford; former president of Astro Industries (high temp. equipment for nuclear and aerospace industry); now owns cattle ranch in Buellton, Calif.; co-author of **Amer. Sci. Affil.** 1986 (which incorporates much material from this book). Acknowledges assistance of paleontologist Preston Cloud (who has recently written his own, naturalistic account of the same topic: *Oasis in Space: Earth History from the Beginning*). Wiester was recently converted to Christianity by his wife. Real issue is choice between purely naturalistic interpretation which presupposes closed universe and Christian view of supernatural acts by personal God. "We owe our existence either to the creative acts of God or to *random chance.*" Insists that science harmonizes perfectly with Bible. Accepts entire evolutionary chronology of universe and earth but argues that certain events are explainable only as creative acts by God (God-of-the-gaps view). "Each creation command in Genesis correlates with a scientific puzzle or gap." Presents Day-Age interpretation. 1st day/era: from Big Bang up to formation of earth. 2nd day/era: outgassing of atmosphere and water vapor from earth; condensation of water to form shallow seas. 3rd day/era: uplift of crust and formation of continents; age of blue-green algae. 4th day/era: transformation of sun's energy from ultraviolet to radiation beneficial to animals as a result of photosynthesis, and consequent accumulation of oxygen and ozone shield. 5th day/era: age of marine life (Cambrian through Devonian). 6th day/era: age of land animals. Scientific record and Genesis sequence are the same, but Genesis also treats creation of vegetation and of birds "topically" as well as chronologically. Science does not support "chance," "mechanistic" origin of life. Darwinian evolution is refuted by lack of transition to multi-cellular forms. Punctuated equilibrium theory brings science "remarkably closer to the biblical view." Public is being "duped" by a "hominid hoax": the claim that humans are descended from fossil hominids. Science is moving ever closer to the "unchanging biblical pattern." Genesis contains a "step-by-step account of changes that God made in the geologic and biologic forms on the Earth" to fulfill God's plan, but doesn't describe how these changes occurred. Other than the creative acts already noted, these may have occurred via evolution. Explains dating methods proving ancient earth. Attractively presented; well-written. Photos, maps, drawings, charts; bibliog., index.

1770 Wigand, A. *Der Darwinismus und die Naturforschung Newtons und Cuviers.* 1874–1877. Brunswick, Germany. ("Darwinism and the Natural Science of **Newton** and **Cuvier.**") 3 vols. *Wigand:* prof. of botany at Univ. Marburg, Germany; did important research in several areas (plant physiology, morphology, systematics, microbiology). This book is cited as creationist in **Lever** 1958, although Wigand apparently believed in some form of evolution. According to the *Dict. Sci. Biog.,* Wigand was strongly influenced by his religious beliefs. He "was one of Darwin's most ardent opponents in Germany, although he always tried to oppose Darwinian theory exclusively on scientific grounds." According to **Graebner** 1943, Wigand contends that "Darwin's doctrine is based on false premises and that its results do not agree with actual observation; that it is not even a scientific hypothesis but philosophical speculation; that it grossly offends against the principle of Causality and organic development." *Wigand:* "Parading in the guise of natural science it is really a perversion which bears within it a menace to true science."

1771 ________. *Der Alternative: Teleologie oder Zufall?* 1877. Kassel, Germany. ("Teleology or Chance?")

1772 ________. *Der Darwinismus: Ein Zeichen der Zeit.* 1878. Heilbronn, Germany. ("Darwinism: A Sign of the Times.")

1773 Wight, George. *Geology and Genesis, a Reconciliation of the Two Records.* 1857. London: John Snow. Modified Gap Theory: elaborates on regional destruction-and-re-creation scheme of **Pye Smith.**

1774 Wilder-Smith, A.E. *Man's Origin, Man's Destiny: A Critical Survey of the Principles of Evolution and Christianity.* 1975 [c1966]. Minneapolis: Bethany House. Orig. c1966; in German (*Herkunft und Zukunft des Menschen;* Brunnen: Giessen). First English ed. 1968 (Wheaton, Ill.: Harold Shaw). *Wilder-Smith:* organic chem. Ph.D.—Univ. Reading (England); two other doctorates in pharmacology (Dr. es Sc. and D.Sc.) from E.T.H. Zurich and Univ. Geneva (Switz.); cancer research at Univ. London; former pharmacology prof. at Univ. Ill. Med. Cntr.; research director of Swiss pharmaceutical co.; taught in Turkey, also lectured U.S. servicemen as NATO advisor on drug abuse; scientist host of six-part creationist films series *Origins: How the World Came to Be* (1982; Films for Christ); British-born, now lives in Switzerland; frequent contributor to

Swiss creation-science journal *Factum* (in German). Strict young-earth, six-day creationism, but allows for earth considerably older than 6,000 years. Scholarly, scientific style, many references; also openly biblical and Christian. Chief argument is that Intelligence (Logos) is necessary for creation of life; life is impossible without prior Plan; random systems cannot by themselves produce Design. Insists that the Argument from Design has "never been adequately refuted." Since God created supernaturally from outside our physical world, we cannot observe His creative plan directly (we can only see the "natural" results, except in the case of manifest miracles). "Only an examination of the 'end product' (man, or any of God's creatures or creations) will give us an indirect and faint idea of the overall grand concept." Includes most standard creation-science arguments. Strongly endorses Paluxy manprints as evidence man co-existed with dinosaurs. Stresses that Darwinism has been used as the "main weapon against the biblical doctrine of origins." Emphasizes use of evolution by communists and Nazis; quotes **Hitler** at length, but notes he used evolution as a convenient means to pursue his own ends. (Beate Wilder-Smith, a German, has written a book *The Day Nazi Germany Died.*) Races of man derive from Noah's sons. Man is biologically different from animals. Endorses pre-Flood Water Canopy. Index fossils: assumption of evolution is used to prove evolution. The "illusion of age ... lies in the very nature of creation *ex nihilo.*" Humans could be selectively bred back to original Adamic state. Evolutionary convergence is impossible: "How should evolution, if it is a purely materialistic concept, *know* what is *needed* by a habitat?" Declares flatly that evolution violates laws of thermodynamics, presenting detailed analogies to prove this. Quotes **Jeans, C.S. Lewis,** H. Blum, **Velikovsky, Kerkut.** Refutes **Teilhard** and several German theistic evolutionists. Long discussions of nature of God and Christ, and afterlife. Photos; bibliog., index.

1775 _______. *The Creation of Life: A Cybernetic Approach to Evolution.* c1970. San Diego: Master Books. Orig. c1970 (Wheaton, Ill.: Harold Shaw). DNA code is more than mere pattern: it is writing—an actual script. Such an information code proves supernatural Creation. "Energy and entropy relationships *must be accounted for* when dealing with programs and codes. ... It is our thesis that ... hindrances to accepting the postulate of an exogenous intelligence to account for nature's coding have been finally and completely overcome by quite recent advances in cybernetic science." Computers have refuted theory of random origin of life. Knowledge of cybernetics allows us to dispose of anthropomorphic concept of God and view Him as Intelligence. Matter is the product of Thought, not thought of matter. This helps explain reality of ESP. Suggests that drugs might enable us to see into the future. Telepathy is proof of primacy of consciousness. Lack of physical evidence of initial Intelligence is "surely a proof of its transcendental nature." Denounces materialistic science and laments bias against supernaturalists. "The materialistic *view* of life brings with it a superficial and ... brutalizing, lawless *way* of life. Why have law and order deteriorated so rapidly in the United States? Simply because for years it has been commonly taught that life is a random, accidental phenomenon with no meaning except the purely materialistic one. The older supernatural views taught that life was a plan and code, which needed for its government a plan of supernaturally given codes or laws." Darwinism "amounts to a denial of the laws of thermodynamics and indeed of all laws—for randomness is not subject to laws." Materialistic origin-of-life theories are used as "propaganda for an atheistic *Weltanschauung.*" Creation of life in the lab is possible, but would require intelligent planning. This proves the original creation was supernatural. According to the Bible, "refusal to accept something which is self-evident (such as the relationship between design and designer)" causes brain damage. Despite claims to the contrary, scientific materialists do not respect the laws of nature; supernaturalists do. Cites **Paley,** P. Mora, **Moorhead** and Kaplan. Bibliog., index.

1776 _______. *A Basis for a New*

Biology. c1976 [1975]. St. Louis, Mo.: Telos International. Orig. 1975 (Neuhausen-Stuttgart, Germany: Hänssler-Verlag [Telos Series]); in German. Discusses missing links; long life of biblical patriarchs (still possible today); "cloning" of Eve from Adam; and the Creation.

1777 ________. *Die Demission des Wissenschaftlichen Materialismus*. 1976. Neuhausen-Stuttgart, Germany: Telos-Verlag (Hänssler). Argues that biology cannot progress if it continues to attribute origin of biological structures to chance rather than to "know-how."

1778 ________. *The Natural Sciences Know Nothing of Evolution*. c1981 [1978]. San Diego: Master Books. Orig. 1978; in German (*Die Naturwissenschaften Kennen Keinen Evolution;* Basel: Schwabe). "Evolution is thus basically an attempt to explain the origin of life from matter and energy without the aid of know-how, concept, teleonomy, or exogenous (extramaterial) information. It represents an attempt to explain the formation of the genetic code from the chemical components of DNA without the aid of a genetic concept (information) originating outside the molecules of the chromosomes. This is comparable to the assumption that the text of a book originates from the paper molecules on which the sentences appear, and not from any *external* source of information..." Chance cannot program information. "Thus according to the laws of physics it is impossible for matter to have organized itself without the aid of energy and of teleonomic machines!" Paleontology gives no evidence that evolution occurred; information theory proves positively that it cannot occur. Because scientists always add "know-how" and energy in their experiments, they have been "successful in their attempts to create artificial life." Lengthy discussion of Paluxy manprints, which Wilder-Smith endorses strongly. Cites other anomalies fatal to evolution; e.g. **Burdick**'s trilobite/human fossil. Refutes biochemists M. Eigen and J. Monod at length. Cites W. Penfield on mind/body duality; Chomsky on mystery of origin of "concept and information"; E. Fromm's criticisms of K. Lorenz's deification of evolution. Discusses radiometric dating; origin-of-life theories and 'right-' and 'left-handed' amino acids. Diagrams; index.

1779 ________. *He Who Thinks Has to Believe*. c1981. San Diego: Master Books/ Minneapolis: Bethany House (co-pub.). "A Thought-Provoking Allegory on the Origin of Life." First half of book is a story about Neanderthals in New Guinea who witness a plane crash. They examine the wreck; later, when rescue party arrives, the Neanderthals discuss the Creator and origin of life with them. The Neanderthals realize that life could not arise without the input of Intelligence. They expound Wilder-Smith's sardine can analogy (proposed in *Man's Origin*): every sealed can refutes evolution—they are thermodynamically open systems, but never produce life. Life requires input of information: *Logos*—genetic code. Only one formula for life: "matter plus energy plus ideas = life." Neanderthals explain that faith in Creator is fully rational; modern evolutionary belief is purely emotional, removed from reality—schizophrenic and mentally damaging. Second half of book consists of philosophical arguments and affirmation of theism and Christianity. Einstein and other great scientists realized the existence of the Creator.

1780 ________. *The Reliability of the Bible*. c1983. San Diego: Master Books. Jesus affirmed the truth of the Old Testament. Bible prophecy proven true; archeology and science prove historical accuracy of Bible. Project Mohole confirmed worldwide Flood 11,600 years ago. Chapter on cloning of Eve from Adam: "The entire report reads exactly like a historical description of surgery under normal physiological conditions for surgery." Y chromosome was deleted from Adam's rib cell and X chromosome doubled to produce female. If creation in six 24-hour days is not true, then the Ten Commandments (which affirm it) are frivolous. Students are brainwashed into rejecting literal interpretation of Creation, Adam and Eve, and the Flood because of relentless propaganda of materialistic science, and consequently do not take the Bible seriously.

1781 Williams, Arthur F. *Is the Day-Age Progressive Theory of Creation Heretical?* (pamphlet). Undated.

Crawfordsville, Ind.: Pleasant View Baptist Church. Listed in Ehlert files.

1782 Williams, Emmett, and George Mulfinger. *Physical Science for Christian Schools.* c1974. Greenville, S.C.: Bob Jones University Press. *Williams:* M.S.—Virginia Polytechnic Inst.; Ph.D.—Clemson Univ.; former prof. metallurgical engin. at VPI; then head of physics dept. at Bob Jones Univ.; now in private industry; former V.P. of Creation Res. Soc. and editor of *CRSQ;* led to Christ after hearing **H. Morris** speak at VPI. Large comprehensive textbook for 9th grade; covers standard physical science material. Even though it deals primarily with physics and chemistry and not life sciences, it is emphatically creationist, taking frequent (and gratuitous) swipes at evolution. It is officially *dedicated* to creationism. "Evolutionism spawns a disrespect for authority, for moral values, and for God Himself. Evolutionism basically destroys man by convincing him he is a mere accident of nature, a clever animal at best." Endorses strict young-earth creation, Flood, Water Canopy, creation with appearance of age, Paleozoic human fossils, CRS, BSA; recommends many creation-science books. Evolutionists are "thoroughly uncientific and shamefully dishonest." Attribute origin of metalworking and chemistry to Tubal-cain in Genesis. Emphasizes that scientific theories, unlike the Bible, change (and are therefore unrealiable), and that science must be based on direct observation. States that pollution threat is exaggerated: God will send far worse afflictions in the coming Judgment. Recommends nuclear power; destructive power of atomic weapons much less than God's wrath. Evolution violates laws of thermodynamics, which are based in scripture. Includes biog. sketches of Bible-believing scientists, intended to dispel notion that scientists are atheists. Tells parables illustrating fundamentalist virtues. Includes study questions and activities. Photos, drawings, diagrams, charts, tables.

1783 _______ (ed.). *Thermodynamics and the Development of Order.* 1981. Norcross, Ga.: Creation Research Society Books. Creation Res. Soc. Monograph Series: No. 1. Authors: **Mulfinger** on history of thermodynamics; Harold Armstrong (*CRSQ* editor) on arrangement of order; David Boylan (chem. engin. Ph.D.; head of engineering dept. at Iowa State Univ.) on development of order; Williams on fluctuations, resistance of living systems to degeneration, and creation model for natural processes (three articles orig. in *CRSQ*); **Gish** on origin of biological order; Ralph E. Ancil on creationist interpretation of genetics; **H. Morris** on thermodynamics and biblical theology. First Law of thermodynamics says creation is finished (no new energy), but evolution says creation is still going on. Second Law says direction of all natural processes is towards disorder, but evolution requires increasing order and complexity. "Scientific laws overrule the process of evolution. ... Therefore it is concluded that evolution could not have occurred, since the first and second laws of thermodynamics would prevent any process that consistently produces greater order and complexity in the physical universe" (Williams). Morris discusses many biblical miracles—supernatural suspensions of natural laws. God declared Second Law of Thermodynamics as a consequence of Adam's Fall, and will repeal it after the coming Judgment and destruction of the world by fire. Some chapters fairly technical; equations. Diagrams, charts, tables.

1784 Williams, F.J. *Evolution ... The Hydrogen Bomb and You.* c1963. Hollywood, Calif.: Cloister Press. *Williams:* surgeon. "Evolution attempts to eliminate the First Adam so that the 'Second Adam' (that is Jesus Christ) and his claim upon your life can be eliminated!" People are going to Hell because of evolution. "Evolution doesn't care what happens to you!" Society doesn't really believe in evolution, or else it wouldn't take care of its weak. "Does nature really need a creature smart enough to discover the hydrogen bomb?"—this passage is the only reference to the Bomb in the whole book. Strident tone.

1785 Williams, John. *The Natural History of the Mineral Kingdom.* 1789. **B. Nelson** 1931 devotes an entire chapter to Williams' Deluge theory of coal formation, quoting him at length. Williams surveyed and described British coal strata. He argued that coal formed from antediluvian timber. Most of the earth was covered with a luxuriant growth of trees before the Flood: enough to account for all coal deposits. Timber floated on the turbulent chaos of the Flood, becoming largely mushy. Then it was deposited on the ocean bottom, often in finely laminated strata alternating with other deposits. The many fine strata result from the great tidal currents of the Flood, "several miles in perpendicular depth," sweeping back and forth, exposing dry land in between the currents. Emphasizes that the various strata are "promiscuously" arranged with respect to gravitational sorting, which must be result of successive streams of water.

1786 Williams, Jon Gary. *The Other Side of Evolution.* c1970. LaVergne, Tenn.: Williams Bros. Publishers. 5th ed. Orig. ed. had Foreword by Russell Artist. This ed. has Foreword by **B. Thompson;** book is distr. by his Apologetics Press. *Williams:* Church of Christ minister in LaVergne, Tenn.; protested BSCS textbooks in Kentucky. "Students have the right to hear all the facts; they have a right to be exposed to all available evidence. It shall be the purpose of this work to present a serious analysis of the evolution theory giving long overdue emphasis to *the other side.*" Standard creation-science. Consists largely of quotes by scientists; includes many pre-1970 anti-evolution references and quotes. Examples of design in nature; vestigial organs; etc. Mentions some less familiar hominid fossils. Last two chapters on harmful effects of belief in evolution, and "Why Christians Cannot Believe Evolution." Index.

1787 Williams, Lindsey. *The Energy Non-Crisis.* 1980. Wheatridge, Colo.: Worth Publishing. 2nd ed.; enlarged. Orig. 1980. Distr. by CLP (ICR). "As told to" **Clifford Wilson.** Foreword by ex-senator Hugh Chance (Colo.). *Williams:* chaplain to Trans-Alaska Oil Pipeline 1974–1976; Baptist. Alaskan oil, especially the vast Gull Island find, makes the U.S. energy independent. Government "Big Brother" has conspired to cover this up, though, which is a "scandal bigger than Watergate." "There has never been an energy crisis . . . except as it has been produced by the Federal government for the purpose of controlling the American people." The government is trying to force us into socialism by making us energy-dependent: by hiding oil finds and by forcing prices up. Williams urges support of Christian schools and Bible training. Also advocates "infiltration" of government by Christians.

1788 Williams, William A. *The Evolution of Man Scientifically Disproved.* 1925. Camden, N.J.: Rev. William A. Williams. Also pub. 1928 (Waxahachie, Texas: J.K. Williams). *Williams:* D.D.; ex-president of Franklin Coll. (Ohio). Intended as an antidote to textbooks promoting evolution, infidelity and atheism. Presents fifty arguments refuting evolution, mostly by the "acid test" of mathematics. Declares that when evolution is subjected to rigorous scientific examination of facts, it fails utterly. This book acknowledged by **H. Morris** as the first presentation of anti-evolution probability arguments. First argument based on rate of human population increase. If earth were as old as evolutionists claim, its population would now be 2^{1240}: "Q.E.D." Full of big numbers. Asserts that statistics on design of human body, etc., disprove chance origin. Darwin used phrases indicating uncertainty 800 times; multiplied together, this means only 6 chances out of a quintillion that evolution is true. Vehemently argues that evolution is atheistic, therefore evil and untrue. "No one has a moral right to believe what is false, much less to teach it, under the specious plea of freedom of thought." Describes many gaps between animal types impossible for evolution to bridge. Genesis creation account allows for non-literal or non-consecutive 'days'; also, ages may have elapsed before the first 'day.' Quotes many anti-evolution statements by scientists. Part Two is refutation of various arguments for evolution, especially fossil hominids. Includes the serology (blood test) argument repeated in **P. Johnson** 1938 and recent *Laymen's Home Miss.* tract, which Williams hopelessly misunderstands. Last section: evidence of soul, immortality, conscience, sin, etc. disprove evolution. Praises soul-winning of A.S. McPherson and her opposition to evolution. Agrees that if not divine, the Bible is the greatest fraud ever perpetrated and Moses is the greatest liar in history. "If evolution wins, Christianity loses... We hope that scientists will consign to innocuous desuetude their camouflaged sesquipedalian vocabularies, and tell us what they mean in short words..." Alludes to Scopes Trial; quotes **Bryan.** Includes mathematical review problems, and hymn by Williams.

1789 Williamson, George Hunt. *Other Tongues, Other Flesh.* UFOs. Extraterrestrials visit earth. **Steiger** says that Williamson gives the interaction of "gods from the skys" with the primitive earthlings a biblical interpretation: Sons of God mating with the Daughters of Men.

1790 Willis, J.C. *The Course of Evolution.* 1940. Cambridge, England: Cambridge Univ. Press. *Willis:* botanist; director of Botanic Gardens in Rio de Janiero; F.R.S.; hon. Ph.D.—Harvard. **Hayward** 1985 says that "throughout his life he poured out a stream of books and papers attacking Darwinism"; this book constituting a "major attack." Argues that natural selection cannot affect plants very much, and is not responsible for evolution. F. Jenkin's objection to Darwin is still valid: crossing, hence dilution of traits, is unavoidable. For many traits no intermediates exist. Biogeography of plants and distribution of varieties is not explainable by natural selection. Evolution by chance is impossible. Willis suggests evolution by huge mutations: "Divergent Mutation." Evolution doesn't create new species out of varieties. New phyla or families appear suddenly; these then differentiate, working down the hierarchy. Willis does not explain how these major groups spring into existence: they appear as if created. Lack of transitional fossils between major forms. Fossil record doesn't show species which are "real intermediates between existing and fossil species. One finds, rather, examples of species that have some of the characters of one, some of another." Also suggests some guiding principle or thought behind evolution. (Also discussed in **G.R. Taylor, Rusch,** and **Rendle-Short** 1942.)

1791 Willson, Frederick A. *A Method of Teaching Creation/Evolution in the Secular School System* (report). Undated(?). Priv. pub. (Redondo Beach, Calif.). *Willson:* public school teacher; president of South Bay Creation Science Assoc. (Torrance, Calif.). This report is from Willson's own teaching experience. Recently, Willson joined **Bliss** of ICR in presenting Science Curriculum Workshops for Christian teachers.

1792 _______. *A Teaching Method and Materials for a Study of the Question: Man and the Apes—Different in Degree or Different in Kind?* Undated(?). Priv. pub.

1793 Wilson, Clifford A. *In the Beginning God...: Answers to Questions on Genesis.* 1975. Grand Rapids, Mich.: Baker Book House. Rev. ed. *Wilson:* archeology M.A.—Sydney Univ. (Australia); B.D.—Melbourne Coll. of Div.; psycholinguistics Ph.D.—Univ. S. Carolina; director of Australian Inst. Archeology; consulting ed. of *Bible and Spade;* taught at Monash Univ. (Melbourne); president of Pacific Coll. Grad. Studies (Australia). Argues for relatively recent creation, but not 4004 B.C. Biblical genealogies contain gaps.

1794 _______. *Rocks, Relics and Biblical Reliability.* 1977. Grand Rapids, Mich.: Zondervan/Richardson, Texas: Probe Ministries (co-pub.). Christian Free Univ. Curriculum Series. Includes fairly non-critical "Response" by R.K. Anderson (O.T. prof. at Wycliffe Coll. of Univ. Toronto). While with Australian Inst. of Archeol., Wilson was area supervisor for 1969 excavations at Gezer, Israel. Biblical archeology. The Bible gives accurate account of historical events. "So it is with the records of Creation and Flood: many scholars agree it is reasonable to assume that the narrative of Genesis 1-11 is far more factual than their counterparts would have conceded in the previous generation." Suggests earth formerly had Vapor Canopy; cites A.E. Ringwood. Attractive format. Photos.

1795 _______. *Monkeys Will Never Talk—Or Will They?* 1978. San Diego: Master Books. Unbridgeable gap between true human speech and animal communication. Uniqueness of human speech and language; difference of kind as well as degree. Creation explains these differences, but evolution cannot. Uniqueness of man: describes differences between man and animals.

1796 ———. *Ebla Tablets: Secrets of a Forgotten City.* c1979. San Diego: Master Books. 3rd ed.; rev. and updated. Thousands of clay tablets from Tell Mardikh in Syria—ancient Ebla—dated about 2300 B.C. Wilson asserts translations prove accuracy of Bible: the names match. Creation account proves Genesis is not derived from Babylonian sources. Tablets disprove Documentary Hypothesis of Bible's origin. Tablets use biblical personal names—endings change from '-il' to '-ya' (as in "Yahweh"). The language resembles Hebrew [also Arabic]. May have been written right after Babel. Tablets include Flood story; Supreme God. Relies heavily on Freedman's report of informal talk with Pettinato, and Freedman's conclusions.

1797 ———. *Creation or Evolution: Fact or Fairytales.* c1985. Dandenong, Victoria, Australia: Pacific College. "Answering Young People's Questions." Aimed at high school students. Short book; standard creation-science arguments. Includes summaries of arguments from several of Wilson's previous books; also Paluxy dig with **Baugh,** population increase argument, probability argument, capacity and design of Ark, Flood Geology, fire-breathing dinosaurs, Nebraska Man, etc.

1798 ———, **and John Weldon.** *Close Encounters: A Better Explanation.* 1978. San Diego: Master Books. *Weldon:* B.A.—Cal State Univ.; teacher at San Diego Sch. of Evangelism; research ed. for *Wake Up America;* co-author with **Z. Levitt** of *Psychic Healing.* Foreword by **Rushdoony,** who says: "The authors conclude that the UFO phenomena are real; they are demonic, and they are totally anti-Christian." "The UFO mythology is a radical denial of God and of Scripture." Connection of UFOs with occult; manifestations of Satan to counterfeit the Second Coming. Cites **F. Salisbury**'s arguments against chance origin of life. Discusses **Downing**'s criticisms of Wilson's view of UFOs and the Bible. Bibliog. Wilson also wrote *Crash Go the Chariots* refuting **von Däniken** (1976 [1973]; San Diego: Master Books), by far the largest selling critique; *Crash Goes the Exorcist* (1974; Melbourne and New York: Word of Truth); and many other books debunking occultic, paranormal interpretations while insisting on biblical supernaturalism.

1799 Wilt, Paul. *Physics for Christian Schools.* 1987. Greenville, S.C.: Bob Jones Univ. Press. *Wilt:* physics and computer science teacher at Bob Jones Univ.

1800 Winchell, Alexander. *Creation the Work of One Intelligence.* 1858. *Winchell:* LL.D.; geology, zoology and botany prof. at Univ. Michigan; director of State Geol. Survey; later chancellor and geol. prof. at Syracuse Univ. In 1878 Winchell was ousted by fundamentalist Southern Methodists from Vanderbilt Univ. "for holding questionable views on Genesis"; he returned to Univ. Mich. Winchell helped popularized geology in America; helped organize it as a scientific discipline, and made important paleontological observations. Winchell always attempted to harmonize science with the Bible. This book mentioned by Millhauser as critical of Chambers' *Vestiges.*

1801 ———. *Sketches of Creation: A Popular View of Some of the Grand Conclusions of the Sciences in Reference to the History of Matter and of Life.* 1870. New York: Harper & Bros. "Together with a Statement of the Intimations of Science Respecting the Primordial Condition and the Ultimate Destiny of the Earth and the Solar System." "Science interpreted is theology. Science prosecuted to its conclusions leads to God." "We have witnessed the progressive development of the physical world—its successive adaptations to its successive populations, and its completion and special preparation for the occupancy of man, and have learned that the whole creation is the product of one eternal, intelligent *master purpose*—the coherent result of ONE MIND." Shows that the Flood could not have deposited all sedimentary strata; argues that earth has changed considerably since Creation, shaped by fire and water over long ages, and that many forms have emerged and become extinct. Evolution of solar system and of earth. Takes reader on tour of the geological ages. Describes four basic animal plans: radial, mollusc, articulated, and vertebrate; and succession of various life-forms as manifestations of "dominant ideas" of each age. Man's embryological development recapitulates these "ideas"—but this proves (contra Chambers' *Vestiges*) overseeing Intelligence. Chapter on "Primeval Man": man first appeared at end of Ice Age. "Primeval man ... was a barbarian, but he was by no means the stepping-stone between the apes and modern man." Earth was gradually prepared for man—the final product of evolution. Muses on future of universe, as stars die out. Many engravings.

1802 _______. *Reconciliation of Science and Religion.* 1877. New York: Harper & Bros. Concordist, Day-Age interpretation of Genesis. Accepts some evolution; endorses Cope's views of development. Argues for local, not worldwide, Flood, but presents Flood tales from many countries. Index.

1803 _______. *Preadamites, or a Demonstration of the Existence of Man Before Adam.* 1890. Argues for existence of humans before Adam. Presents "an anthropological account of the evolution of the human family without, in his view, contravening the Scriptures" *(Dict. Sci. Biog.).* (Quoted frequently in **Carroll** 1900.) Winchell also wrote *The Doctrine of Evolution* (1874; New York), and *Sparks from a Geologist's Hammer* (1881; Chicago).

1804 Wingren, Gustaf. *The Flight from Creation.* 1971. Minneapolis: Augsburg. Short book. *Wingren:* theologian; apparently not a literalist.

1805 Winrod, Gerald Burton. *Science, Christ and the Bible.* c1929. Chicago: Fleming H. Revell. *Winrod:* founder of Defenders of the Christian Faith, a group organized to combat evolution and modernism; editor of *Defender* magazine (published articles by **Riley**); described by Marsden as an "especially vocal" anti-communist, anti-Semitic and pro-Nazi fundamentalist. This book based on Winrod's sermons and lectures. "Christianity is the science of things spiritual": it is "demonstrable truth" which rests upon immutable spiritual laws. "Philosophy has become a dangerous thing in our schools since the coming of Darwinian evolution." It reduces everything to natural law; man becomes merely a cog in a heartless, soulless machine. "All true science, and Christianity, the one true religion, starts at the same place with the first four words of the Bible—'In the beginning God.'" "Between the proved facts of science and the truth of Christianity there is perfect harmony, but between the guesses of scientists and the dogma of religionists there is discord." "God wrote two books—the Book of books and the book of nature. There is no discord between the two books." "Science is exact knowledge, gained and verified. True science never asserts; it always proves." God revealed the modern law of heredity when He said the sins of the parent are visited upon the children until the third and fourth generation. Presents other examples of modern science in the Bible. Explains how Lot's wife was encrusted with salt. Asserts angels are always masculine. Noah prophesied that Ham's descendants—the black race—would be a servant people. Nimrod was black. Mentions favorably the Anglo-Israel doctrine that America and England descended from the Israelite tribes of Mannasseh and Ephraim. Satan ruled over earth long before the creation of Adam (Gap Theory creationism). After Satan defied God, "The earth was mashed and wrapped in darkness for, no doubt, many milleniums." The five geological strata, and the fossils, are often found out of order. Cites human fossil found in Triassic layer. "When the theory of evolution hits the rocks of geology, it goes to pieces." Transmutation of species is impossible. No organism can cross over the boundary—the "charmed circle"—of its species. "If the transmutation of species were true, we would see about us all manner of hideous monstrosities. There would be creatures with heads like men and necks like giraffes and bodies like horses." Goes on to discuss the geology and biology of the Church, and the coming Tribulation and Translation of the Church. Winrod also wrote many other books on topics such as Bible prophecy, Satan, communism, the Protocols of the Elders of Zion.

1806 Wintrebert, P. *Le Vivant: Createur de son Evolution.* 1962. Paris: Masson. *Wintrebert:* hon. prof. at Sorbonne. Discusses **Cuenot.** Neo-Lamarckian.

1807 Wiseman, P.J. *Creation Revealed in Six Days.* 1949 [1948]. London: Marshall, Morgan & Scott. 2nd ed. Orig. 1948. Also pub. 1958. "The evidence of Scripture confirmed by Archaeology." *Wiseman:* air commodore; C.B.E. Perhaps the best-known exposition of the Revelatory Theory of creationism. The six days of creation refer not to the actual process of creation, but to God's revelation of them to man. The creation narrative was probably written on six tablets, containing the words revealed by God and presented one at a time on six literal, consecutive days. Wiseman points out that Babylonian creation accounts are generally inscribed on six tablets. He argues that study of the literary structure —the parallel arrangement of two three-day sequences, and comparison of the Genesis closing formula with the colophons of Mesopotamian tablet series—supports this interpretation. Criticizes Day-Age and Gap theories on scientific and biblical grounds. Also criticizes Revelatory theory version involving visions rather than written account. God revealed Creation over a six day period, then declared a day of rest, in order to establish this weekly pattern for man. Comparison with Babylonian creation account indicates Genesis is not a variant, but the original source. Book of Enoch alludes to revelation of creation. The date of Creation is unknown; the Revelatory theory avoids, however, the scientific difficulties of strict young-earth creationism and the unbiblical nature of the evolutionary scheme. Quotes **J. Jeans, A. Fleming,** other scientists. "It seems obvious that only by accepting the Bible account can we account for the immortality of man." Includes detailed commentary on translation of Genesis creation story.

1808 _______. *Clues to Creation in Genesis.* 1977. London: Marshall, Morgan and Scott. Rev. ed. More on the Revelatory Theory.

1809 Wlodyga, Ronald R. *Health Secrets from the Bible.* 1979. Altadena, Calif.: Triumph. Hippocrates and Aesculapios were instructed by Satan, who appeared as Apollo. The serpent staff (medical symbol), and the knowledge of healing herbs are Satanic. Wlodyga endorses the diet, health, and agricultural practices of Hunza Valley in the Himalayas (follows the accounts of Renee Taylor and Robert McCarrison), which he equates with biblical teachings and practices. Includes sections on reflexology (psychic massage) and iridology (diagnosis of all disease from examination of the eye), of which his wife is a practitioner.

1810 _______. *The Ultimate Source of All Super Natural Phenomena.* c1981. Altadena, Calif.: Triumph. Wlodyga, a follower of **H.W. Armstrong,** here presents the views and doctrines of his Worldwide Church of God. Also thanks publisher **Dankenbring.** The supernatural is real, and cannot be understood by science. "Knowledge of the supernatural can only come through SUPERNATURAL REVELATION, since it is above and beyond natural law." Occult and paranormal phenomena are real; they are Satanic deceptions intended to lure us into worship of a false messiah. Discusses UFOs, ESP, divination, astrology, other psychic and occult manifestations. God revealed His plan for the Redeemer in the stars before His written revelation. Chapter "In the Beginning" presents Gap Theory creationism: Lucifer's rebellion; the pre-Adamic reign of Satan on earth. Traces spread of Satan's false religion from origin by Nimrod in Babylon and the apostasy of the Roman Church. Explains descent of Armstrong's Worldwide Church of God from true believers persecuted as heretics: Waldenses, Cathars, Puritans, etc. Rejects as heresy Sunday worship, concept of the Trinity, immortality of soul, heaven and Hell. Explains Bible prophecy, Armageddon and the Millennium. Discusses demonic possession of **Hitler** and Uri Geller. Equates 'Santa' with 'Satan.' Rejects Shroud of Turin as Satanic deception. Evolution is a ridiculous fairy tale aimed at tricking us into rejecting the Bible. Bibliog.

1811 Wolf, Fred Alan. *Star Wave: Mind, Consciousness, and Quantum Physics.* 1984. New York: Collier Books (Macmillan). God via quantum mechanics. Quantum mechanics is the basis of evolution and creativity. "It is my attempt to explain as best as I can the real reason we human beings are here in the first place, as conscious, spiritual creatures." Psychic phenomena are manifestations of parallel worlds. Endorses **von Ditfurth**'s view of evolution by creation; argues that molecules evolve. Cites his argument that cytochrome C demonstrates evolutionary history but not formation by chance.

1812 Wolfe, Samuel T. *A Key to Dooyeweerd.* 1978. Nutley, N.J.: Presbyterian and Reformed. *Wolfe:* B.D.–Faith Theol. Sem.; has contr. to *CRSQ;* lives in Santa Barbara, Calif. A presentation of **Dooyeweerd**'s "Cosmonomic" philosophy, especially from his *New Critique of Theoretical Thought.* Praises Dutch Calvinists; cites former Dutch Prime Minister Kuyper, **J. Smuts, H. Taylor, van Til, De Wit.** Universe is composed of hierarchy of progressive complexities, or "spheres": time, space, motion and energy, biotic, the human sensorium, understanding and logic, the will and the cultural mandate, semantics, social intercourse, economics, esthetics, law, ethics, and faith (the last not explicitly dealt with by Dooyeweerd). Each sphere is based upon God's laws. True freedom comes from recognition and acceptance of limits set by God. Man's mind is not autonomous. Dooyeweerd's most important work was in the biotic sphere; he revolutionized biology, demonstrating that mechanism and vitalism were both wrong. Appeals to Burr and Northrop's electrodynamic theory of life, **Driesch,** and Einstein as demonstrating inadequacy of materialistic view. Information code of outer cell layer–cytoplasm–is more basic and important than DNA. Says Dordt Coll. in Iowa is center of Cosmonomic movement in U.S.; *Vanguard* magazine promotes it in Toronto. Urges that Cosmonomic and creation-science movements join forces; praises **Rushdoony** for effectively combining them. Discusses his own attempts, aided by **G. Howe,** to acquaint Americans with Cosmonomic movement. Index.

1813 Wollaston, Thomas Vernon. *On the Variation of Species, with Especial Reference to the Insecta; Followed by an Inquiry into the Nature of Genera.* 1856. London: Van Voorst. *Wollaston:* English entomologist who "rejected transmutation as contrary to both known facts and sound reasoning" (Gillespie). Tried to keep science and theology separate, but chose special creation over evolution, which he considered unprovable. Species vary, but not past certain limits. Also believed in Atlantis. This book pre-dates Darwin's *Origin,* but Wollaston remained a critic of Darwin even afterwards. (Quoted and discussed in Hull.)

1814 Wolters, Albert M. *Creation Regained: Biblical Basics for a Reformational Worldview.* c1985. Grand Rapids, Mich.: William B. Eerdmans. Not concerned primarily with science. Deals with *all* of "creation"–society, aesthetics, culture, politics, etc.–not just physical matter. Biblical view considers law as a condition of freedom, contrary to humanism, which treats law as an impediment to freedom. Christ and Satan both make totalitarian claims; nothing is neutral. Wolters calls for a "restoration" rather than a "repristinization"–it is not possible to return to the original Edenic conditions. Emphasizes that Creation is "good," and warns against the Gnostic heresy of shunning the physical world as evil. Wolters denies *ex nihilo* creation, and argues that God created out of pre-existing unformed earth.

1815 Wolthuis, Enno. *Science, God, and You.* 1963. Grand Rapids, Mich.: Baker Book House. *Wolthuis:* chemistry prof. at Calvin Coll. (Grand Rapids). "This book sets forth our view of the way in which one can profess the Christian faith and at the same tie welcome and promote scientific progress." Reviews history of science, stressing contribution of Calvin and other Reformers. Criticizes temptation fostered by modern science to rely solely on naturalistic explanation. Emphasizes that "Science Has Its Limitations." Discusses various positions regarding relation of science and faith. Rejects mechanist view, "dualist" view which assumes they are unrelated, and "ascetic" view which denies the physical world. Stresses sovereignty of God over His creation. "All we can know about God's work of creation must be supplied by God Himself in His Word. That is our only source of information regarding the *fact* of creation. The doctrine of creation is a major one with respect to the relation between the Christian faith and science. That is to say, once a person accepts this doctrine on faith, there are many logical consequences for his scientific activity." Says no consensus exists among Bible scholars on what the Bible teaches about the "scientific significance of the Genesis account," but asserts that "any theory of origins is false which fails to anchor this universe in the will of God." Points out that Genesis is "not a detailed account of creation in scientific language." Discusses meaning of God's command to have "dominion over the earth"—says science is a "Christian duty."

1816 Wonderly, Dan E. *God's Time-Records in Ancient Sediments.* c1977. Flint, Mich.: Crystal Press. Foreword by **R.C. Newman.** Book is expansion of 1975 *JASA* article also included in Newman's 1977 book. *Wonderly:* grad. degrees in biology and theology; science teacher in Christian colleges (inc. Grace Coll.); lives in Maryland. Wonderly, an old-earth creationist, here refutes young-earth creationism, and gives convincing explanations of non-radiometric proofs for an ancient earth. Most of book is good scientific discussion of sedimentation layers, erosion features, deep drilling, reefs and coral formation as proof of ancient earth. Also presents philosophical and theological arguments for rejecting recent creation, which he considers extra-biblical. There is abundant scientific evidence for ancient earth, and God is not deceptive. Includes well-documented chapters on early Christian views regarding age of the earth. Praises and quotes many Christian scientists who attempted to harmonize findings of geology with the Bible, such as **Hitchcock, H. Miller, A. McCaul, T. Birks, Jamieson, W. Dawson,** and others. Speaks favorably of both Day-Age and Gap Theory creationism, but prefers the former. Includes Appendix on length of creation 'days' (at least some were longer than 24-hour days). Urges that distinction be made between evolution—which he rejects—and age of earth. "The separating of these two issues can be of untold value, both in promoting mutual understanding between Christians and in helping to present the Biblical account of creation to the public. . . . Whenever we attempt to 'throw out' *both* evolutionary theory and the established facts concerning the age of the earth, we will find unrelenting resistance. Public school teachers and pupils should be, and can be, alerted to the transitory nature of evolutionary theory if we will not at the same time deny the geologic evidences for age." Photos, maps, diagrams; bibliog., index.

1817 Woodman, J.M. *God in Nature and Revelation.* 1875. Chico, Calif.: John G. Hodge. Also pub. 1888 (subtitled *The Grand March of Time Complete*) as part of a single-volume set which includes other works by Woodman: *The Neptunian or Water Theory of Creation* (1888; San Francisco: Bacon), and *The Song of Cosmology: or The Voice of God in the Science of Nature* (1880). *Rev. Woodman:* teacher of natural, mental and moral philosophy, and of natural and revealed religion at Chico Academy (Calif.). Intended to provide opportunity to study the Bible in connection with modern scientific discoveries. Woodman describes the 'days' of creation as long ages (Day-Age creationism), but also refers to the "visions" received by Moses from God revealing aspects of the creation days: i.e. the Revelatory theory of creationism. Earth developed from unformed matter during first creation 'day,' which lasted perhaps 18 million years—the duration of revolution around the galactic center. Book arranged as a hundred "lessons," followed by study questions. Promotes the old "Neptunist" theory (earth's geological deposits all formed by action of water—the Flood) and refutes rival Plutonic theory (geology explained as the result of vulcanism). Volcanoes are caused by burning coal and oil. Does not mention Darwin or any contemporary theories, but denies that life could have originated without creation by God. "The development theory contradicts universal observation. . . . The origin of all things is revealed in the Bible." Higher orders of animals did not descend from the lower. Much space devoted to philosophical and theological affirmations of the Bible, moral and civic lessons. Includes arguments from design, physiological adaptations, Bible prophecy, Satan. The Flood was caused by tilt in earth's axis, and consisted of great tidal currents. Asia, Europe, and North Africa were submerged by trapped waters, but animal life in other regions survived. Many engravings.

1818 Woodrow, Ralph Edward. *Noah's Flood, Joshua's Long Day, and Lucifer's Fall.* 1984. Riverside, Calif.: Ralph Woodrow Evangelistic Assoc. Advocates a regional Flood; refutes universal Flood with scientific and practical arguments. States that worldwide Flood violates literal intepretation of Bible by presuming much earlier date. Skeptical discussion of many Ark-eology tales. Reprints, and refutes, tract version of **Roskovitsky** account. Argues that Joshua's Long Day is based on mistranslation. Joshua did not command sun to stand still; rather, he sought respite from midday heat (hence the hailstorm). Reprints, and refutes, tract version of **H. Hill**'s NASA/IBM tale. Suggests God caused only shadow to move on Hezekiah's sundial. Argues against Gap Theory creationism. Maps, and many drawings, old and new. Woodrow also wrote *Babylon Mystery Religion* (1966) and other fundamentalist books.

1819 Woods, Andrew J. *The Center of the Earth.* c1973. San Diego: Inst. for Creation Res. ICR Tech. Mono. No. 3. *Woods:* M.S.; physicist with Gulf Energy and Environmental Sciences (San Diego). Much of brief text is "Discussion" by **H. Morris,** who summarizes argument: "since God intended for man to 'fill the earth' after the Flood. . . , and since the ark 'rested upon the mountains of Ararat' the very day that God restrained the Flood from further destruction. . . , wouldn't it be reasonable to think that God had arranged for the 'port of disembarkation' to be located somewhere near the geographical center of the land which man was commanded to fill?" Woods' computer study confirms that Ararat is indeed near the center of earth's land area, as determined by average distances to all other unit sections of the earth's land. Many equations, tables, diagrams, map.

1820 Woodward, John. *An Essay Towards a Natural History of the Earth . . . with an Account of the Univeral Deluge: and of the Effects It Had Upon the Earth.* 1695. London: R. Wilkin. 3 eds. up to 1723. *Woodward:* Cambridge Univ. M.D.; physic (medicine) prof. at Gresham Coll., London; Fellow of Royal Soc. Woodward "asserted unequivocally" that fossils were natural remains of organic creatures, and explained how fossil deposits were formed by Noah's Flood. Contrary to **Burnet,** he maintained that the Flood did not ruin a once-perfect world, but that it was required to render the earth suitable for fallen man. He also rejected the theory that land and sea had exchanged places in the past. The Flood submerged the whole earth, and dissolved or held in suspension all the mineral matter of the earth's crust. Organic remains were mixed in with this slurry. These materials subsided into the various stratigraphic layers observed now, according to their specific gravity. "By this conception of the Deluge, Woodward managed to account for fossil deposits without conceding perpetual change on the earth's surface. Moreover, by making the Deluge the explanation of these puzzling phenomena he strengthened belief in the historicity of that event and in the full inspiration of Scripture" (J.C. Greene). Woodward suggested that extinct fossil creatures might still be found alive in ocean depths. Woodward's classifications of minerals and fossils were important scientific contributions. He also established the first geology chair, at Cambridge.

1821 Woolsey, John Martin. *The Discovery of Noah's Ark, Final and Decisive* (booklet). 1910. New York: Cochrane. Listed in Schadewald files.

1822 World's Fundamental Conference. *Scriptural Inspiration versus Scientific Imagination.* 1922. Los Angeles: Biola Book Room. Fourth Annual Great Christian Fundamentals Conference, held in Los Angeles, 1922. Includes anti-evolution chapters by **Riley, Keyser,** and **A.C. Dixon.** Riley argues for Day-Age creationism; Dixon for the Gap Theory.

1823 Wren, Edward Conway. *Evolution-Fact or Fiction?* 1936. London: Thynne. *Wren:* British major.

1824 ________. *The Case for Creation.* 1949. [Malverne, N.Y.:] Christian Evidence League. Flood Geology and recent creationism. Wren also wrote *Four Worlds* (1946; Northam).

1825 Wright, George Frederick. *Scientific Aspects of Christian Evidences.*

1898. New York: D. Appleton. *Wright:* B.A. – Oberlin Coll.; M.A. – Oberlin Theol. Sem.; Congregational pastor; later became professional geologist; taught both theology and geology at Oberlin, where he was prof. of the "harmony of science and revelation"; a founder of Geol. Soc. of Amer. "Receptive to Darwin's views, Wright was the foremost early champion of a Christian Darwinist theology" *(Dict. Sci. Biog.)*. But Wright, who consistently advocated a mediating view, came to be considered a conservative by the turn of the century. He was editor of *Bibliotheca Sacra,* which aligned itself with the new fundamentalist movement, and wrote about evolution for the famous series *The Fundamentals* (**Meyer**, ed.). He allowed for limited evolution and a Day-Age interpretation, but "argued strongly against Darwinian claims that evolution could explain the origins of life or the uniqueness of humans" (Marsden). Though he died some years before the Scopes Trial, Wright was the only scientist **Bryan** could name during the trial besides **Price** who disputed evolution. This book, like his others, is a reconciliation of science and Christianity. Natural selection cannot operate on chance; there must be a "Contriver" for Darwinian evolution to work – must be purpose behind evolution. Wright rejects atheistic evolution, but allows theistic evolution. Suggests "paroxysmal" evolution. Includes biblical references. Index.

1826 ________. *Scientific Confirmations of Old Testament History.* 1906. Oberlin, Ohio: Bibliotheca Sacra. Wright advocated a relatively recent end of the Ice Age (which he considered a single period of advancing and receding glaciers). (He wrote an important work *The Ice Age in North America and Its Bearings Upon the Antiquity of Man* in 1889.) In this book he says the end of the Ice Age resulted in a Flood catastrophe which temporarily submerged Europe, northern and central Asia, and much of North America in up to 3,000 feet of water.

1827 ________. *Origin and Antiquity of Man.* 1912. Oberlin, Ohio: Bibliotheca Sacra. Archeology shows that life is probably 24 million years old, but not more than 50 million years old. Wright disputes uniformitarianism, and argues that the glacial epoch is connected with great extinctions. Man's mental and moral capacity comes from God, not from evolution. Admits a genetic connection – a descent relationship – between man and animals, but denies that man evolved from existing primates. Argues that 'primitive' man was highly developed. *Pithecanthropus erectus* was human, not a connecting link between apes and man. Discusses fossils and artifacts demonstrating advanced status of early man, including Nampa figurine and Calaveras skull (notes doubts as to its authenticity, but thinks skeptics are biased). Stresses differences between man and apes. "While the antiquity of man in the world cannot be less than 10,000 years, it need not be more than 15,000. Eight thousand years of prehistoric time is ample to account for all the known facts relating to his development." Developmental changes of organisms resulted from "paroxysms of nature" – not from gradual evolution. Divine intervention was necessary for these changes. Insists on Creator God, and on "foresight" of creation, citing **A.R. Wallace.** Spontaneous generation is impossible. Also discusses Bible chronology; argues that Genesis contains condensed genealogies. Drawings; index.

1828 Wright, W.E. *Searching Science and Scripture* (booklet). Undated. Jos, Nigeria: SIMLIT Publications. *Wright:* missionary in Nigeria; chairman of Niger Challenge Publications. SIMLIT: Sudan Interior Missions. "Based upon writings of **H.M. Morris**" – in particular, *Science, Scripture, and Salvation.* Morris's creation-science is here "blended with" Wright's writings. "When scientific theories conflict with the Scriptures, the scientist must rethink his conclusions and bring them into line with the revealed Word of God. Proved scientific fact and the Word of God cannot and do not disagree." "It is an obvious fact that man is morally wrong and in rebellion against God, just as the Bible says.... In the evolutionary hypothesis there is, as we said earlier, an attempt to by-pass God and escape our accountability to Him." Mentions standard creation-science arguments. Alleged ape-men are degraded people, not ascending apes. Creation with "apparent age" solves apparent contradictions between biblical account of creation and supposed great age of earth.

1829 Wysong, Randy L. *The Creation-Evolution Controversy.* 1976. Midland, Mich.: Inquiry Press. *Wysong:* D.V.M. —Mich. State Univ.; veterinary surgery and medicine at Hagadorn Vet. Clinic (S. Lansing, Mich.). Book purports to be fair and objective presentation of both sides, presented mostly as an imaginary debate. Urges lay "jury"—book's readers—to rationally weigh expert scientific evidence and arguments he presents for both positions. Wysong never admits he is a creationist (though he does concede he may have some unstated bias). Elsewhere he has denied being a creationist, though other creationists routinely refer to him as one. His concluding summary of evidence for evolution is barely one page; summary for creation is ten pages. Evolutionist arguments are often misrepresented or misleadingly presented. Emphasizes that opinion on origins affects rest of one's worldview. This book is a comprehensive collection of all standard creation-science arguments; includes virtually all common anti-evolution quotes. Packed with scientific references; cites many critics of Darwinian evolution (or those who seem to be) such as Mora, G. Wald, Blum, **Kerkut;** also quotes several dozen creationists, but does not identify them as such. Discusses origin-of-life theories and experiments, chance formation or modification of life and probability arguments, DNA and information science, reductionism, age of the earth, assumptions of radiometric dating, many other dating methods which give young age, thermodynamics, lack of transitional fossils, harmfulness of mutations, design of the eye, impossibility for natural selection to allow survival of incompletely-evolved organs, "predictions" of neo-Darwinism refuted by science, circularity of geological reasoning (assuming evolution, fossils are used to date rocks, then rocks are used to date fossils), geological evidences for catastrophism and Flood, out-of-order fossils (includes Paluxy and most common examples), pre-Flood Water Canopy. Includes listing of anti-evolution resources and literature. Photos, drawings, diagrams, charts, tables; bibliog., index. This book is widely praised by creationists as a careful, non-religious work which demonstrates the scientific weakness of evolution.

1829a Yockey, Francis Parker. *Imperium: The Philosophy of History and Politics.* 1962 [1948]. Torrance, Calif.: Noontide Press. Orig. 1948; priv. pub. ("Westropa Press"); written under pseudonym "Ulick Varange." *Yockey:* linguist, lawyer; worked for War Crimes Tribunal after WWII but quit, then wrote this book in Ireland; U.S. passport revoked; arrest in Calif. in 1960; died in jail. Philosophy of book based on Hegel; historical conception on Spengler; also influenced by Nietzsche, Goethe, Carlyle. Essentialist, historicist view of history as cultural/racial Destiny. Rejects materialist thought as obsolete in science and all other spheres; advocates "organic" interpretation of culture—"cultural vitalism" and spirituality. History is cyclic, not linear; true explanation is "Destiny-thinking" rather than rationalist, scientific, causal. Each great Culture is a superorganic being which arises independently, with its unique spiritual Type or Genius—its "World-Idea" or "Culture-soul." The previous great Cultures succumbed by attempting to maintain empires embracing alien racial/cultural entities. Calls for new united European Empire (Yockey's pseud. "Ulick Varange" is Irish-Viking-Russian comb.) which will expel unassimilable racial/cultural elements: this new Empire of the West will triumph in its death-struggle against Russia, Asia, and other alien cultures. The unified genius of the West flowered under Gothic Christianity, only to dissolve in petty nationalism. Napoleon initiated a new Imperial stage of European Culture. In this century, Germany attempted to fulfill the Destiny of the West (Yockey usually refers to Germany simply as "Europe"), but "Culture-Distorters" and "parasites" in America (mainly Jews, Negroes) and alien Russian Communists and Asians forced Europeans into a suicidal war. Appeals for triumph of Religion over mechanistic science and rationalism; Authority, Order and Hierarchy over democracy and equality. Describes decline of America due to racial/cultural amalgamation; praises KKK for resisting aliens; asserts Negroes were better off as slaves; denounces liberalism. Hitler valiantly tried to unify Europe and mobilize it against racial/cultural aliens; culture-distorters in the U.S. and Jewish-Asiatic Russians responded with vile propaganda barrage. Germans did not mistreat Jews, but the Allies waged war cruelly and unfairly against innocent Germans, persecuting them savagely after the war. Yockey denounces the War Crimes Trials as illegal acts of vengeance. Exhorts the coming Empire of the West to conquer Russia in order to preserve its "biological existence." Includes chapter on "Darwinism" denouncing naturalistic evolution. Yockey concedes that evolution itself ("organic" rather than materialistic) has merit, but refutes its popular conception—"journalistic Darwinism"—as the materialistic "animalization of Culture-man." Compares natural selection with Calvinist predestination; asserts Darwinism is teleological—a religious projection of capitalism onto the natural world. Denies struggle for existence in nature. Cites criticisms of Darwinism by vitalists. Species do not evolve by adaptation; they arise suddenly and persist as stable forms. No transitional forms, only distinct species. Primitive forms still exist, contrary to Darwinist assumptions. Half-evolved organs contradict natural selection. Darwinism is an obsolete biologico-religious doctrine which served to justify the mechanical view of life as formed by accidental, external causes. The new view recognizes the primacy of inner forces: Will, Soul, Destiny. Entropy and atomic physics also refute materialism. Curiously, Yockey also argued that race is "fluid"—it is not entirely genetic, but is also partly influenced by the "soil" (the geographic and spiritual environment). The editor of the 1962 ed. tries to explain away this notion as well as Yockey's anti-Darwinism. Noontide Press is the pub. arm of the pro-Nazi Inst. for Historical Review. Index.

1830 Young, Arthur M. *The Reflexive Universe: Evolution of Consciousness.* c1976. Delacorte Press (Merloyd Lawrence). *Young:* inventor of Bell helicopter; now with Inst. for Study of Consciousness. Acknowledges assistance from Pelletier, Ivan Sanderson, Hapgood; blurbs by paranormalist Puharich, Head Lama Ringpoche, nuclear physicists and others. Presents a theory of the universe which incorporates truth of cosmogonic myths. "The universe is a *process* put in motion by purpose." Every aspect of this process occurs in seven stages, each adding a new power. Seven kingdoms: light (potential), subatomic particles (substance), atoms (form, identity), molecules (combination), plants (organization), animals (mobility), and man (dominion). All process involves a descent to the fourth stage: a loss of freedom—a "Fall" into determinism; then an ascent to the seventh. Quantum uncertainty principle proves freedom and purposive nature of first stages; followed by descent into inert and determinate matter; finally, human conciousness. Genesis account describes these seven stages: Creation 'days' are really "generations." Revealed Scripture is necessary in addition to reason. Life requires an "act of will," contrary to Darwinian explanation. Evolution theory is unscientific, and largely political propaganda. Accepts reality of paranormal phenomena: ESP, psychic healing, mediumistic ectoplasm, Carlos Castaneda, reincarnation. Animals have a "group soul": this explains inheritance of instinct, which DNA does not. Man, however, possesses immortal individual soul. Explains genius as inheritance of skills learned in past lives. Advocates teleology and inheritance of acquired characteristics. Evolution by chance mutation is impossible; intermediate forms are impossible. Tautology of natural selection; survival of primitive forms refutes Darwinian evolution. Darwinian evolution is reversed in man: the less fit multiply. Accepts fabulous ages of pre-Flood patriarchs. Miracles are evidence of Christ's superhuman powers. Discusses seven stages of all myths, including (and especially) Genesis. Admires **Besant**'s Theosophy. Drawings, diagrams; index.

1831 Young, Davis A. *Creation and the Flood: An Alternative to Flood Geology and Theistic Evolution.* 1977. Grand Rapids, Mich.: Baker Book House. *Young:* attended Princeton, Penn State,

Brown; geology prof. at Calvin Coll. (Grand Rapids); former prof. at NYU, Univ. N. Carolina. Argues against Flood Geology and recent creation; advocates old-earth (Day-Age) creationism instead. Presents biblical evidence in this book; his later book focuses on scientific evidence. Young, an orthodox evangelical Christian, insists that faith does not require a recent creation or Flood Geology.

1832 _______. *Christianity and the Age of the Earth.* c1982. Grand Rapids, Mich.: Zondervan. Contemporary Evangelical Perspectives series. Young, who has had extensive field experience in geology, refutes Flood Geology and recent creation. This book is primarily a response to **Whitcomb** and **Morris**'s *Genesis Flood,* but also refutes **Whitelaw, Barnes, Nevins (Austin),** and **M. Cook** at length. First chapter is well-documented discussion of attitudes regarding age of the earth by early church authorities. Other chapters deal with early geological investigations and theories, scientific realization of earth's antiquity, theological accommodation and harmonization theories. Comprehensive discussion; many references and quotes, both scientific and theological. Also discusses rise of modern Flood Geology and young-earth creationism. Second part of book is good presentation of scientific evidences of earth's age: stratigraphic methods, sedimentation, geochemical processes, and radiometric dating methods. Refutes all standard creation-science arguments for young earth. Third part concerns philosophical and theological arguments. Points out that geology does not rely on "substantive" uniformitarianism (assumption of unvarying rates), and accuses strict creationists of relying on uniformitarian assumptions themselves. "The faith of many Christian people could be hindered when they ultimately realize that the teachings of the creationists are simply not in accord with the facts. . . . Imagine the trauma and shock of finally realizing that Flood geology, which has been endorsed so enthusiastically by well-meaning Christian leaders, is nothing more than a fantasy. . . . Furthermore, creationism and Flood geology have put a serious roadblock in the way of unbelieving scientists." "May I plead with my brethren in Christ who are involved in the young-Earth movement to abandon the misleading writing they provide the Christian public. . . . We Christians need to stop expending our energies in defending a false creationism and in refuting a false creationism. Let us spend our energies on interpreting the Bible and the world that God in His mercy and grace has given us. A vigorous Christian science will be of far more service in meaningful evangelism and apologetics than the fantasies of young-Earth creationism." Charts; bibliog., index. Morris's *Science, Scripture and the Young Earth* (1983) is a rebuttal of this book. In 1987, Young, **Van Till,** and a third ASA member and colleague at Calvin Coll. came under investigation by Christian Reformed Church officials on charges of heterodoxy for suggesting that earth has an evolutionary history. The three have written a "book-length manuscript decrying distortion of science by both 'young-earth creationism' and scientific naturalism," though the charges (apparently instigated by **D. Gish**) were based on their previous writings.

1833 Young, Edward J. *Studies in Genesis One.* 1964. Philadelphia: Presbyterian and Reformed. Bible exegesis. Argues for *ex nihilo* creation. Says Creation days were not ordinary days, but does not commit himself to any particular theory.

1834 Young, George. *Scriptural Geology.* 1838. Also 2-vol. ed. in 1840. *Rev. Young:* D.D. This work "devoted principally to an attack agaisnt the rising uniformitarian and evolutionary theories of geology." Vigorously attacks Lyell for relying on unwarranted assumptions. Also opposes the new schemes of reconciliation. Allows extinctions but not development of new forms: says that some fossils in the lowest strata are identical to living forms. Fossils prove there was a universal warm climate, before the Flood. High mountains were located in equatorial regions; oceans and scattered archipelagos near the poles; and land of intermediate altitude in between, resulting in an even climate worldwide. Examination of fossils indicates formation by catastrophe and rapid deposition. Young argues that fossil remains were carried far from their orignal habitats. (Described and quoted in **B. Nelson** 1931.)

1835 Young, Norman. *Creator, Creation and Faith.* 1976. London: Collins. Attempted reconciliation of creation and evolution. Presumably old-earth.

1836 Youngblood, Ronald. *How It All Began: A Bible Commentary for Laymen/Genesis 1–11.* c1980. Ventura, Calif.: G/L Regal Books (Gospel Light). *Youngblood:* grad of Fuller Theol. Sem., Valparaiso Univ. and Dropsie Univ.; Semitic languges specialist; dean of Wheaton Coll. Grad. Sch.; now at Bethel Sem. West (San Diego). Mostly theological arguments. The Bible "categorically rules out evolution on the grand scale overwhelmingly claimed for it by its supporters." Creation may involve "secondary means": there may be variation within Genesis 'kinds,' but only within strict limits. The date of Creation is not known, but it is very old. Suggests that the creation 'days' are "indefinite and timeless." Stresses literary form and analysis of Genesis. The Genesis account is historical but not fully chronological. The creation days are arranged in verse pattern: three days of forming, then three days of filling (populating); this literary sequence may not be the actual chronological order. Opening verse of Genesis contains the four fundamental concepts of physics: "In the beginning [time] God created [causality] the heavens [space] and the earth [substance]." Pre-Adamic hominids may be as old as scientists claim, but at some relatively late point God intervened to produce "biblical" man. Accepts description of Eden and the Fall as literal. Cain may have feared the pre-Adamic hominids. Interprets long ages of antediluvian patriarchs as literary devices. Argues for regional Flood; discounts claims that the Ark can be found on Ararat. Teacher's Manual and Student Discovery Guide for Bible Study groups using this book also available. Bibliog.

1837 _______ (ed.). *The Genesis Debate: Persistent Questions About Creation and the Flood.* 1986. Nashville: Thomas Nelson. "Persistent Questions About Creation and the Flood." Pro and con arguments on eleven topics. Terence Freitheim (O.T. prof. at Luther Northwestern Theol. Sem., St. Paul) defends literal creation days; **McCone** argues against, preferring a literary interpretation. **R.C. Newman** defends Genesis as a chronological account; Mark Throntveit (Luther N.W. Theol. Sem.) opposes, stressing its literary form instead. Steven R. Schrader (Hebrew and O.T. prof. at Falwell's Liberty Univ., Vir.) defends recent creation, quoting many creation-scientists; **D. Young** opposes, saying that "Genesis 1 should be regarded as a highly structured theological cosmology that extensively employs a royal-political metaphor. . ." **J.N. Moore** denies that evolution was involved in creation. George Kufeldt (O.T. prof. at Sch. of Theol., Anderson Coll., Ind.) denies the existence of pre-Adamic people. James Borland (Liberty Baptist Theol. Sem., Vir.) affirms that the antediluvians lived hundreds of years. F.B. Huey (Southwestern Baptist Theol. Sem., Texas) argues that the "Sons of God" in Gen. 6 were angels. **S. Austin** defends a worldwide Flood; Donald Boardman (emer. geology prof. at Wheaton Coll.) argues for a regional Flood. **C. Henry** argues that Genesis supports capital punishment.

1838 Zamora, Margarita T. *The Nature of Origins: Man and the Earth Within the Universe.* c1973. New York: Vantage Press. *Zamora:* M.A.–Univ. Havana; Cuban; lives in Fla. Endorses parapsychology. "The history of mankind is very short and very simple"; so is this book. Mishmash of pseudoscience and the occult; lots of diagrams and drawings. Metaphysical evolution: Zamora describes its stages, past and future. Man's first stage (bi-sexual) lasted 700 million years. First Africa, then Asia, then Europe will vanish in future stages, like Lemuria and Atlantis did in the past. Cycles, levels of occult wisdom, hybrid energy matrix of the universe, fusion of the poles, lots of vibrations.

1839 Zarr, Benjamin. *Preamble to "God vs. Evolution."* c1983. Priv. pub. (Worcester, Mass.). This *Preamble,* and *God vs. Evolution,* were both issued separately. Zarr's 1984 book combines them.

1840 _______. *Evolution.* 1984. Priv. pub. (Worcester, Mass.). Bizarre, crackpot biology. The first section is the same as Zarr's 1983 *Preamble.* Title page: "No one can break down my theory!" "There isn't a single book in the world that does justice to the subjects of God and evolution!" God miraculously created a living protoplasmic yolk out of crystals; these became bacteria. Adult bacteria swallowed infants to produce sperm. These sperm gave birth to earthworm kingdom. Other plants and animals resulted from cohabitation of sperm with "ferm." "Man consists of two, separate, four-segmented earthworms..." Totally rejects mitosis, chromosome and gene theory. Right testicles form "spermo"; left form "sperma." These then form organs on the right and left sides of their descendants. The outer cell wall figures prominently in Zarr's notion of "embryolization." Life goes through 50,000-year cycles, then "self-destructs," then starts anew. Zarr says his revelations come directly from God. "CREATIONISTS ARE RIGHT! ... Creationists WILL win with my theory! They will be able to present a complete method of creationism that can be taught in every school. My theory of creationism and evolution, in the true sense of the word, will be acclaimed as the best presentation of organic life to the present." Rejects standard Adam and Eve story, but proclaims that "a newly adapted Bible can be accepted as the LEGAL BASIC THEORY OF EVOLUTION!" Many drawings. Zarr has sent his book to many university libraries; copies are stamped "Please accept this book with my compliments hopeful that it will make a biological improvement in your library."

1841 Zepp-LaRouche, Helga (ed.). *The Hitler Book.* 1984. New York: New Benjamin Franklin House. *Zepp-LaRouche:* German wife of Lyndon LaRouche, the "Democratic" presidential candidate; studied philosophy, history and political science at various German universities; founder of the European Labor Party, the Club of Life, and the Schiller Inst. (a LaRouche think-tank). This book exposes the sinister, international neo-Nazi oligarchy which operates through the Int'l Monetary Fund, Club of Rome, the peace movement (German Green party), etc. This conspiracy is allied with Moscow, western oligarchs and a widespread surviving Nazi network. It is based on worship of nature and the Earth Mother, eternalist cycles of death, purification and rebirth, the "myth" of over-population, and it is anti-technology. (LaRouche advocates nuclear power, SDI, and maximum technological development.) The conspiracy's fullest manifestation is the Darwinist-racialist-Gnostic Thule Society of the Nazis. **Hitler** and **Rosenberg** only hinted at this Thule mysticism publicly, but in 1942 Hitler said to Heim: "We would do well to assume that what is reported by the mythology of the deities is in fact a memory of a previous *reality.* At the same time, within every body of legends we always find the story of the collapse of the heavens... My only explanation for this is that an immense natural catastrophe destroyed a race of human beings who possessed the *highest* level of culture. *What we find on earth today must be a small remainder which, living for that memory, will gradually find its way back to culture.*" Opposed to this conspiracy are LaRouche's neo-Platonic republican ideals, "negentropic evolution," continuous growth and continuous creation as embodied by Christ, and realization of the uniqueness of man. "It follows that the adherents of the Conservative Revolution consider Charles Darwin's theories to be the most important 'scientific' idea of the nineteenth century. For if the eternal cycle is to remain unbroken, then according to Darwin the strong must eliminate the weak and destroy those who 'do not deserve to live,' just as the race theoreticians Jahn and Hans Grimm, and finally the Nazis themselves, imagined themselves to be members of a master race, possessing the right to exterminate inferior races." Zepp-Larouche, focusing on Germany, calls for a return to classical German republican ideals. Photos. L. LaRouche, a former Marxist who makes sensational claims about international dope-peddling by Western leaders, has been indicted on conspiracy charges related to fraudulent campaigning and fund-raising, and obstruction of justice. In his paper *New Solidarity,* LaRouche advocates biblical "dominion" over the earth rather than "stewardship." Pagan earth-mother worship, the "back to nature movement" and other anti-Christian cults, he says, degrade man into a creature of nature. Man, with his creative will, which is "coincident with the Creator," is "absolutely" above the beasts; he is made in the image of God. The present oligarchical conspiracy derives from the ancient Babylonian cult. "For the worshippers of Satan (Great Mother), there must be no religious freedom."

1842 Zerbe, Alvin Sylvester. *Christianity and False Evolutionism.* c1925. *Zerbe:* at Central Theol. Sem. (Dayton, Ohio). Listed in Ehlert files. **Graebner** 1943 quotes an unnamed work by Zerbe: "Science starts with the pantheistic postulate that the universe had no beginning in time but has always existed, its existence being due, not to a Supreme Being, or God, but to some resident force or energy. To support this claim all sorts of theories, hypotheses, conjectures and make-shifts are advanced. Naturally under this view the old Bible doctrine of the existence of a Supreme Being, the Creator of the universe and of man, must be given up, together with practically everything that is distinctive of the Bible and the Christian religion."

1843 ________. *Evolution in a Nutshell: The Pros and Cons Briefly, Clearly, and Fully Presented.* c1926. Chicago: Laird and Lee.

1844 Zimmerman, Dean R. *Evolution: A Golden Calf.* 1976. Salt Lake City: Hawkes. *Zimmerman:* B.A., M.A.—Brigham Young Univ.; grad. study at Cal State Long Beach and USC; presumably Mormon. Back cover: "Either equal time must be granted—or the theologians of atheism should be gagged." Standard creation-science. Many scientific references. Consists mostly of quotes by scientists critical of evolutionary theory, anti-Darwinists, and creationists. Chapters on difference between man and apes, morphology, embryology, paleontology, scientific method, evolution as a faith, natural selection, spontaneous generation, life as more than DNA, mutations, the Cambrian boundary, the myth of the geologic column, radiometric dating, evidence for catastrophism, human history as refutation of evolution, the evolution of evolutionary theory, and philosophical and theological arguments. "If there had been no Fall to mortality, as the evolutionists would maintain, then there would be no need for Jesus Christ." Evolution leads to atheism and thus immorality. If evolution is a "law," it is "indisputably the least documented, least substantiated law in all science." It is "a faith, a dogma, a philosophy, but certainly not a science." Appeals to the Bible, but does not state his own creationist interpretation. Tables; index.

1845 Zimmerman, Paul A. *We Are the Offspring of God* (pamphlet). Undated [1963?]. Address presented at S. Calif. District Conv. of Lutheran Church—Missouri Synod, 1963. *Rev. Zimmerman:* Ph.D.; president of Concordia Lutheran Coll. (Ann Arbor, Mich.); former president of Concordia Teachers Coll. (Neb.); chemist and theologian. Insists on strict, recent creation; literal interpretation of Adam and Eve. Largely theological arguments, but includes many scientific references. Quotes many anti-Darwinists. "In the final analysis it is the subject of man that represents the crux of the entire question. Man is the crown of creation." "It is indeed in the basic philosophy of evolution that we find the greatest objection to evolution." Evolution claims that man has "risen from the swamp" and needs no Savior. This is not science, and it contradicts the Bible.

1846 ________. *The Christian and Science* (booklet). Undated [1964?]. Caldwell, Idaho: Bible-Science Newsletter. Address presented to N.W. Dist. Conv. of Lutheran Church—Missouri Synod, 1964. Insists on verbal inspiration and inerrancy of Bible. Rejects non-literal creationist interpretations as well as evolution and radiometric dating. Scientific references.

1847 ________. *A Brief Catechism on the Theory of Evolution* (pamphlet). 1975. [St. Louis, Mo.?:] Concordia Tract Mission. Listed in SOR CREVO/IMS.

1848 ________ **(ed.).** *Darwin, Evolution, and Creation.* 1959. St. Louis, Mo.: Concordia Publishing House. Authors: Zimmerman, **Klotz, Rusch,** Raymond F. Surburg (theol. prof. at Concordia Tchrs. Coll., Neb.). God reveals Himself both in nature and the Bible. "Neither form of revelation can possibly contradict the other." Scientific truth is relative and changing; the Bible's truth is absolute, but may be misinterpreted. The authors acknowledge some limited evolution, not fixity of species. "But there is much in evolutionary theory, in its accompanying philosophy, and in its denial of creation that we must reject and oppose. We hold that Christians must not confuse scientific fact, theory, and just plain scientific speculation." Rusch reviews history of evolutionist ideas and Darwinism. Surburg discusses various interpretations of Genesis creation account, quoting many theologians, and contrasts these with evolutionist ideas; also the negative influence of Darwinism on theology, philosophy, psychology, education, and sociology. "From a Christian point of view all forms of naturalism must be rejected..." Zimmerman presents evidence for creation (design, and criticism of evolution), and arguments for a young earth (allowing for slightly longer chronology than **Ussher**'s). Klotz criticizes evolutionist arguments. Bibliog., index.

1849 ________. *Creation, Evolution, and God's Word.* c1972 [1966]. St. Louis, Mo.: Concordia Publishing House. Orig. presentations at various Missouri Synod Lutheran conferences, pub. 1966 in limited ed. titled *Essays from the Creationist Viewpoint.* Authors: **Klotz, Rusch, Zimmerman, Lammerts,** Richard G. Korthals (M.S.; physics prof. at Concordia Jr. Coll., Ann Arbor; USAF Lt. Col. ret.; former astronautics prof. at Air Force Acad.). Klotz, Rusch, and Lammerts criticize evidence for evolution. They accept limited change ("microevolution"), but reject macroevolution. Includes most standard creation-science arguments. Zimmerman discusses various positions regarding the status of biblical inspiration and inerrancy, quoting many theological sources, plus opinions regarding science and the Bible. Korthals concedes that Darwinism is correct if we assume naturalism, but he rejects this assumption. If we assume the truth of the Bible, evolution must be false. Christians are "people with religious beliefs—people who not only use reason but also have faith in God's revelation to us. As such you are told in the revealed Word that the earth was created by God and that He created us in His image..." Contends that creation with appearance of age makes it impossible to disprove special creation.

1850 Zuch, Roy. *Creation Evidence from Science and Scripture* (booklet). c1976. Wheaton, Ill.: Scripture Press. "Provides teaching aids such as transparencies, duplicating masters, and instructions for teaching five sessions on Creationism" (BSA catalog).

1851 Zuelow, Walter. *Creation or Evolution* (pamphlet). Undated. Priv. pub. Naive little polemic. If evolution is true, then why aren't apes still evolving? How can spiders build webs if they have such tiny brains? Assumes that evolution requires inheritance of acquired traits.

1852 Zwemer, Samuel M. *The Origin of Religion.* 1945. New York: Loizeaux Bros. 3rd ed., rev. Based on the 1935 Smyth Lectures at Columbia Theol. Sem. (Decatur, Ga.); also article by Zwemer "The Origin of Religion: By Evolution or by Revelation" (1935; *Trans. Victoria Inst.*). *Zwemer:* Presbyterian; history of religion and missions prof. at Princeton Theol. Sem.; missionary to Moslems. **H. Morris** calls this book "a creationist masterpiece which should have devastated all evolutionary concepts of religion..." Zwemer argues that all primitive cultures acknowledge or worship a Supreme Deity, and that this Supreme Deity has everywhere the same attributes. This proves that monotheism is not an evolutionary development, but that knowledge of the true God was the original religion. Most cultures have Creation myths concerning a Golden Age, and allude to the Fall of man. Universality of prayer, and of the institution of marriage, worship of fire, and belief in the immortality of the soul likewise point to a common primitive tradition and revelation. Presents ethnographic data and discusses various anthropological theories regarding primitive religion and the evolution of religion. Praises Andrew Lang's theory of primitive monotheism and Wilhelm Schmidt's *Origin of the Idea of God;* criticizes the evolutionist schemes of Tylor, Max Müller, Frazer, Lubbock, Spencer, Durkheim, and others. Argues that evolutionary theories of religion stem from uncritical attempts to apply Darwinian theory. "Our conclusion, then, is that we need no longer cross a 'Rainbow Bridge' to find a cave-man who by evolutionary processes become a *homo sapiens;* but that on the threshold of human history and in the earliest cultures he greets us made in the image of God, conscious of his Creator, aware of moral impulses... One cannot read the mass of evidence in recent books on ethnology without finding again and again corroboration of the truth of Revelation: 'God created man in his own image...'" Bibliog., index.

Name Index

A **boldface** *number indicates that the person cited is author of that work; other numbers indicate that the person's name is mentioned in the entry.*

A

B

C

D

E

F

G

H

I

J

K

L

N

O

P

R

S

T

U

V

W

Y

Z

Title Index

A

B

C

D

E

F

G

H

I

J

K

L

M

N

O

P

Q

R

S

T

U

V

W

Y

Subject Index

A **boldface** *number indicates that the institution cited is listed as author of that work.*

A

B

C

D

E

F

G

H

I

J

K

L

M

N

O

P

Q

R

S

T

U